T0335844

# Mid-Ocean Ridges

The world's mid-ocean ridges form a single, connected global ridge system that is part of every ocean, and is the longest mountain range in the world. Geologically active, mid-ocean ridges are key sites of tectonic movement, intimately involved in sea floor spreading. This coursebook presents a multi-disciplinary approach to the science of mid-ocean ridges – essential for a complete understanding of global tectonics and geodynamics. Designed for graduate and advanced undergraduate students, it will also provide a valuable reference for professionals in relevant fields. Background chapters provide a historical introduction and an overview of research techniques, and following chapters cover the structure of the lithosphere and crust, and volcanic, tectonic and hydrothermal processes. A summary and synthesis chapter recaps essential points to consolidate new learning. Accessible to students and professionals working in marine geology, plate tectonics, geophysics, geodynamics, volcanism and oceanography, this is the ideal introduction to a key global phenomenon.

- Supports students and professionals new to technical aspects or geographic areas with a full glossary and extensive directory of feature names.
- Avoids jargon and fully introduces and defines technical concepts and terms.
- Richly illustrated, including colour figures and comprehensive data tables.
- Extensive references provide detailed starting points for further study, and a valuable resource for professional researchers from many different fields.

**Roger Searle** is Emeritus Professor of Geophysics at Durham University. He has spent 40 years studying mid-ocean ridges, and was a pioneer in the use of side-scan sonar to study their geodynamic, tectonic and volcanic processes. In his research he also uses topographic analysis and gravity and magnetic modelling to understand ridge structures. He was awarded the Royal Astronomical Society's Price Medal in 2011 and elected a Fellow of the American Geophysical Union in 2012. Professor Searle has worked in many of the world's major oceanographic institutions, participated in 37 research cruises and led 18. He was first full chairman of the international research organisation InterRidge, and has served on national and international committees, including chairing the International Ocean Drilling Program's Site Survey Panel.

'This volume provides a comprehensive, up-to-date and authoritative account, extensively illustrated and referenced, of the geology, the morphology, the tectonics and the chemistry of the ridges, relating these to the underlying mantle movements. It also describes in detail the techniques used in these studies. Professor Searle has been at the forefront of research on the mid-ocean ridges throughout his career, and has produced an ideal textbook both for students and those currently researching the geology of the ocean floor.'

– Sir Anthony Laughton, FRS, *formerly Director of the Institute of Oceanographic Sciences, UK*

'Professor Searle has done a superb job of summarizing and analyzing the history of, and the latest insights into, mid-ocean ridges, ranging from ultra-slow to fast spreading rates and including the tectonics, geophysics, geochemistry, volcanism and hydrothermal activity of this "longest mountain range in the world." This is an essential volume for any student or researcher studying mid-ocean ridges, both those in the Earth sciences and those with backgrounds in marine biology, chemistry oceanography, physical oceanography and other related fields.'

– Ken C. Macdonald, *Emeritus Professor of Marine Geophysics, University of California at Santa Barbara*

# Mid-Ocean Ridges

ROGER SEARLE
Emeritus Professor, Department of Earth Sciences,
Durham University

Shaftesbury Road, Cambridge CB2 8EA, United Kingdom

One Liberty Plaza, 20th Floor, New York, NY 10006, USA

477 Williamstown Road, Port Melbourne, VIC 3207, Australia

314–321, 3rd Floor, Plot 3, Splendor Forum, Jasola District Centre, New Delhi – 110025, India

103 Penang Road, #05–06/07, Visioncrest Commercial, Singapore 238467

Cambridge University Press is part of Cambridge University Press & Assessment,
a department of the University of Cambridge.

We share the University's mission to contribute to society through the pursuit of
education, learning and research at the highest international levels of excellence.

www.cambridge.org
Information on this title: www.cambridge.org/9781107017528



First published 2013

*A catalogue record for this publication is available from the British Library*

*Library of Congress Cataloging-in-Publication data*
Searle, Roger, 1944–
Mid-ocean ridges / Roger Searle, Emeritus Professor, Department of Earth Sciences, Durham University.
pages cm
Includes bibliographical references and index.
ISBN 978-1-107-01752-8 (hardback)
1. Mid-ocean ridges. 2. Plate tectonics. 3. Sea floor spreading. I. Title.
QE511.7.S45 2013
551.1′36 – dc23 2013017281

ISBN 978-1-107-01752-8 Hardback

**To my family.**

'Could the waters of the Atlantic be drawn off so as to expose to view this great sea-gash which separates continents, and extends from the Arctic to the Antarctic, it would present a scene the most rugged, grand and imposing. The very ribs of the solid earth, with the foundations of the sea, would be brought to light . . . '

Matthew Fontaine Maury (1860)

# Contents

# Preface

Mid-ocean ridges are where the oceanic crust, which covers over 60% of the Earth's surface and is renewed every 200 million years or so, is generated. They are thus features of first-order importance in the Earth system. Mid-ocean ridges were discovered some 150 years ago, and have been studied with increasing intensity and detail since then. We are now beginning to have an outline level of understanding of their structures and processes. Ridges are primarily studied by geophysicists and geologists. But chemists are interested because ridge crest hydrothermal systems exchange chemical elements between the rock of the oceanic crust and the overlying ocean waters; physical oceanographers are concerned with how ridge topography and geothermal heat influence ocean waters and currents, and biologists study the unique ecosystems that inhabit hydrothermal vents, which may hold clues to the origins of life and the nature of the 'deep biosphere' of microbes that live deep in crustal rocks.

This book attempts to set out an overview of the current understanding of mid-ocean ridges across most of the scientific disciplines involved. I have tried to make it reasonably comprehensive, while admitting that an encyclopaedic coverage is certainly beyond my ability. I intend the book to be suitable for a wide audience, in terms of both their level of prior knowledge and the nature of their disciplines. Thus I hope it can be used as a general introduction and reference by senior undergraduates and starting postgraduate students taking courses in, for example, geodynamics, Earth systems or oceanography, by doctoral students as a starting point for their researches, and by both academic and other professionals who may need an introduction or reference to areas outside their immediate specialties.

My aim has been to highlight at least some of the milestone papers that have influenced our understanding of ridges, and to use illustrations from them. The bibliography is by no means comprehensive, but I hope it contains enough key references to serve as a useful starting point for further research. I have provided a brief historical background to ridge studies, and have included two appendices to aid the reader new to this field. Appendix A is a glossary of technical terms used, and Appendix B is a directory of feature names, briefly giving the nature of each feature referred to in the book and its geographical location.

The book includes brief mathematics, including some critical equations where appropriate, but is largely non-mathematical. However, I have tried to make clear the physical principles involved in the various processes described. There is no detailed discussion of petrology or biology, although I have tried to give an outline of key petrological and biological issues where required.

The book starts with an introduction followed by a brief historical review of techniques. It then follows a logical path through the lithosphere, ridges as plate boundaries, crustal structure, volcanism, tectonism and hydrothermal systems. Each chapter has a brief summary of the main topics covered. Each can be read more-or-less independently, especially with the help of the appendices, and ample cross-references are provided. The final chapter summarises the descriptions given in the earlier chapters, and attempts to set them in a unified conceptual model that synthesises current thinking. The reader seeking a quick introduction to mid-ocean ridges might start at this final chapter, before referring to earlier chapters for details.

SI units are used throughout, and temperatures are given in degrees Celsius (°C). Years are indicated by 'a', with thousand years and million years denoted 'ka' and 'Ma'. Figures generally have scale bars or, if not, latitude scales. A useful guide is that one degree of latitude is approximately 111 km, and one minute of latitude (1/60th of a degree) is about 1 mile or 1.8 km. North is to the top unless otherwise indicated.

I am greatly indebted to the many colleagues and students who over the years have informed, encouraged, argued and generally contributed to my nevertheless sadly limited understanding of mid-ocean ridges. There are too many individuals to name, and it would be invidious to list just a few, but thanks for your friendship on this exciting journey. I must, however, particularly thank Suzanne Carbotte, Colin Devey, Gretchen Früh-Green, Rachel Haymon, Marvin Lilly and Ken Macdonald for reading and providing invaluable comments on draft chapters, thereby saving me from a number of howlers. Any remaining errors and omissions are, of course, my responsibility alone. I am also indebted to John Gould, Mark Holmes, Dave Sandwell, Martin Sinha and Adam Soule, who supplied original versions of illustrations. The writing was supported in part by a Leverhulme Trust Emeritus Fellowship.

I am grateful to all rightsholders who kindly gave permissions for re-use of figures in this book. In every case, reasonable effort was made to establish the correct rightsholder, and credit to the source is given for all figures, but in the case of any unfortunate omission, the rightsholder should contact the publisher to arrange correction.

Finally, I must thank my family for their forbearance during the writing of this book; until I began it I did not realise how appropriate this traditional acknowledgement is! They have suffered many weeks of my self-imposed isolation with computer, books and reprints, not to mention the four years of my life spent on research expeditions at sea. Thank you to them for their enduring support and encouragement.

# 1 Introduction

## 1.1 The global mid-ocean ridge system

Mid-ocean ridges (MOR)s are a product of the separation and spreading of tectonic plates, and are a major component of the Earth system. Their role as plate boundaries is discussed in detail in Chapter 4. Back-arc spreading centres are functionally similar, and for many purposes can be grouped together with ridges, although they will not be considered in detail in this book. This chapter introduces MORs, outlines their discovery and the theory of plate tectonics (since ridges form an important class of plate boundaries), and briefly introduces some of the institutions and organisations that have been critical for understanding ridges.

The MOR system spans the world, extending to some 65 000 km in length (Figure 1.1). Plate separation rates range from only a few millimetres per year to some 160 mm $a^{-1}$, and even faster at some past times (Müller *et al.*, 2008; Teagle *et al.*, 2012). MORs are characterised by shallow ocean floor, narrow bands of shallow seismicity and high heat flow. Their crests generally lie some 2.6 km below sea level; their flanks may be thousands of kilometres wide and reach depths in excess of 6 km. They contain arguably the largest array of active volcanoes in the world, and host extensive hydrothermal vent fields where unique ecosystems based on chemosynthesis flourish. They are the places where new oceanic lithosphere is continuously generated, and where this newly created lithosphere is flexed, faulted and chemically altered. This book aims to present an overview of all of these processes.

Many aspects of ridges are at least in part related to spreading rate. Strictly speaking, 'spreading rate' is the rate at which each tectonic plate grows, or spreads away from the ridge axis. However, it is sometimes also used to mean the rate at which the two plates diverge (which for symmetric spreading is twice the spreading rate). To avoid confusion, the two rates should be identified by different names. Although not entirely accurate, the custom has developed of using 'half spreading rate' for the rate of plate growth and 'full spreading rate' for the rate of plate separation. Alternative uses would be 'plate accretion rate' and 'plate separation rate'.

Because ridge characteristics tend to depend on spreading rate, ridges are often grouped into broad classes as a function of this rate (Table 1.1). Nevertheless, spreading rate varies continuously, and the given class boundaries are approximate and may vary between authors. Other factors, particularly mantle temperature and fertility, may also control ridge morphology, structure and processes.

**Table 1.1** Spreading rate categories of mid-ocean ridges

| Class | Total opening rate, km $Ma^{-1}$ | Reference | Example |
|---|---|---|---|
| Superfast | >130 –150 | Sinton *et al.* (1991) | East Pacific Rise 20° S |
| Fast | 90–130 | Lonsdale (1977) | East Pacific Rise 13° N |
| Intermediate | 50–90 | Lonsdale (1977) | Juan de Fuca Ridge |
| Slow | 20–50 | Lonsdale (1977) | Northern Mid-Atlantic Ridge |
| Ultra-slow | <20 | Grindlay *et al.* (1998) | Southwest Indian Ridge, Gakkel Ridge |

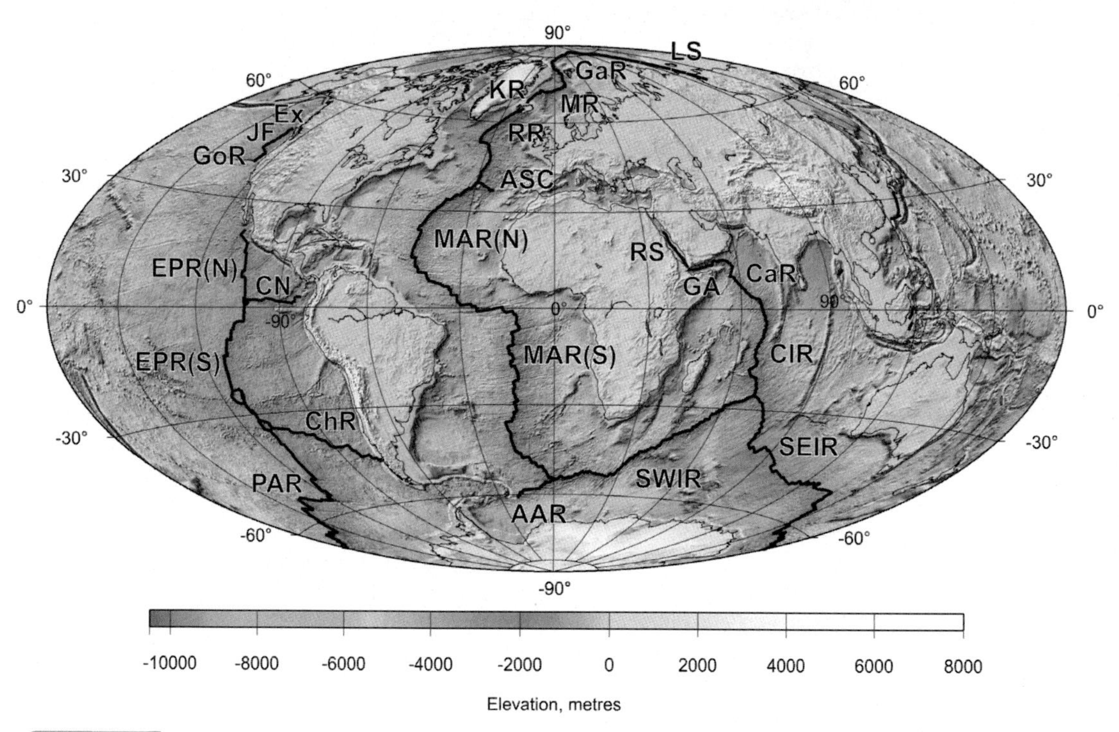


**Figure 1.1** World topography with mid-ocean ridges superimposed (heavy black lines). Back-arc spreading centres have been omitted for clarity. AAR, American–Antarctic Ridge; ASC, Azores Spreading Centre; CaR, Carlsberg Ridge; ChR, Chile Rise; CIR, Central Indian Ridge; CN, Cocos–Nazca Spreading Centre; EPR, East Pacific Rise (north and south); Ex, Explorer Ridge; GA, Gulf of Aden; GoR, Gorda Ridge; GaR, Gakkel Ridge; JF, Juan de Fuca Ridge; KR, Kolbeinsey Ridge; LS, Laptev Sea Rift; MAR, Mid-Atlantic Ridge (north and south); MR, Mohns Ridge; PAR, Pacific–Antarctic Rise; RR, Reykjanes Ridge; RS, Red Sea; SEIR, Southeast Indian Ridge; SWIR, Southwest Indian Ridge. For colour version, see plates section.

Each ocean contains an MOR (Figure 1.1). Most are named for the ocean they are in (e.g., Mid-Atlantic Ridge – MAR), but some have specific regional names (e.g., Gakkel Ridge in the Arctic Ocean), and some are named for the plates they separate (e.g., Cocos–Nazca Spreading Centre). The fast-spreading East Pacific Rise (EPR) and Pacific–Antarctic Rise are called 'rise' because their morphology is gentler than that of the slower-spreading

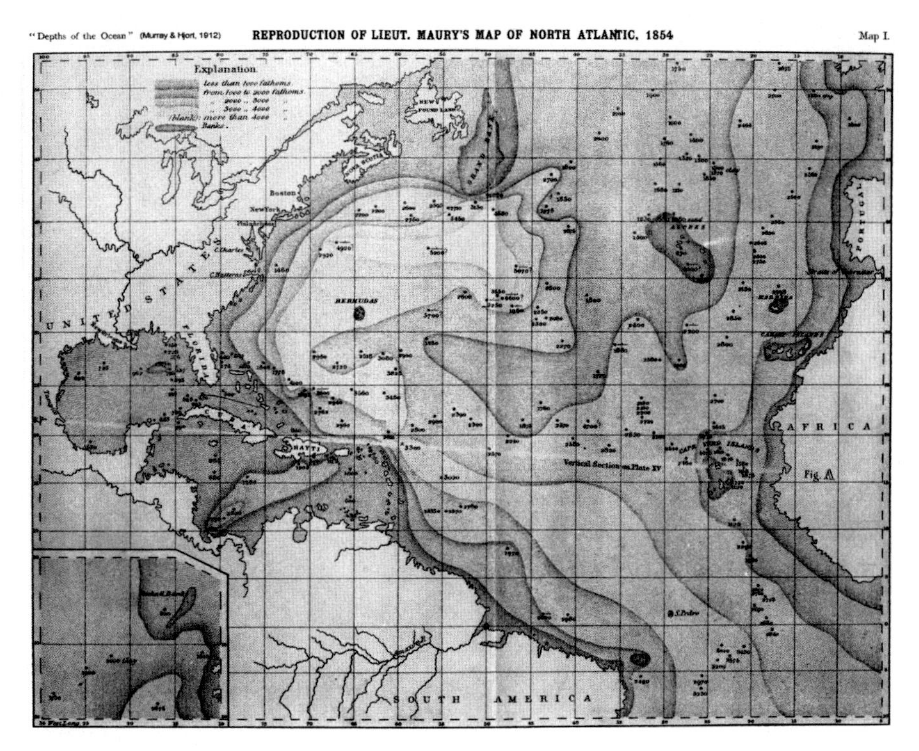


Figure 1.2 The first map of the North Atlantic, showing the position of the MAR after Murray and Hjort (1912), reproduced from Maury (1860).

'ridges'. Some ridges are not centred in their ocean, having come about by rifting of a pre-existing oceanic basin rather than of the bounding continent, so are not 'mid-ocean' in the strict sense; however, their structures and processes are common to the MORs *sensu stricto*.

## 1.2 The discovery of MORs

Mid-ocean ridges were discovered in the nineteenth century. The American hydrographer Matthew Fontaine Maury developed improved methods of deep-sea sounding, and produced the first map showing the varying depth of the North Atlantic Ocean (Maury, 1860). His Plate VII shows a broad, shallow region in the position of the MAR between 20° N and 52° N (Figure 1.2). The northernmost part of this was called Telegraphic Plateau, and was discovered by surveys for the first trans-Atlantic telephone cables. The rest of this broad rise, south of 45° N, was known as 'Middle Ground'.

The first systematic oceanographic expedition was the circum-global cruise of HMS *Challenger* from 1872 to 1876 (Thomson, 1877). The *Challenger* expedition's studies

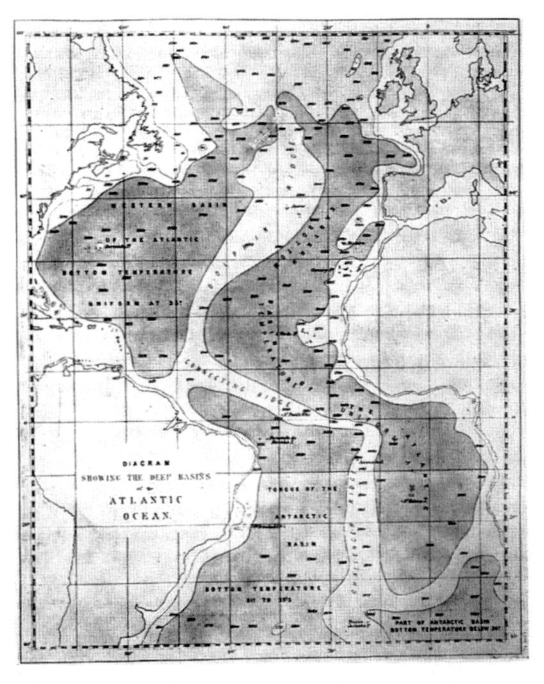


Figure 1.3 Map of the MAR, after Tizard (1876).

included both depth soundings and measurements of water temperature. On the basis of a difference between the deep-water temperatures between the western and eastern Atlantic, Tizard (1876) inferred the existence of a topographic ridge separating the two basins. His Plate 6 shows the existence of this ridge, named 'Dolphin Ridge' in the North Atlantic and 'Challenger Ridge' in the south (Figure 1.3). They are joined by 'Connecting Ridge' which comprises what are now known as the equatorial Atlantic fracture zones.

Early in the twentieth century, Murray and Hjort (1912, p. 135) produced a map showing many of the world's MORs, although only one, the MAR, was yet named as such. In the Pacific, the US ship *Albatross* had discovered the Albatross Plateau (now known to be part of the EPR) between 1880 and 1905. Thousands of soundings had been made world-wide by 1923, when electronic echosounding began to be introduced (Shepard, 1959). The German ship *Meteor* ran 14 echosounding lines across the south Atlantic in 1925–1927 (Marmer, 1933), clearly revealing the southern MAR in detail (Shepard, 1959, pp. 162–163). These were the only detailed profiles available prior to World War II (Heezen and Menard, 1963). At about the same time, the Danish vessel *Dana* (1928–1930) discovered the Carlsberg Ridge in the Indian Ocean (Tharp, 1982). Following World War II, continuously recording precision echosounders came into widespread use (Section 2.2.2), and by the middle of the twentieth century the general outline of the global MOR system was well established (Shepard, 1948; Figure 1.4).

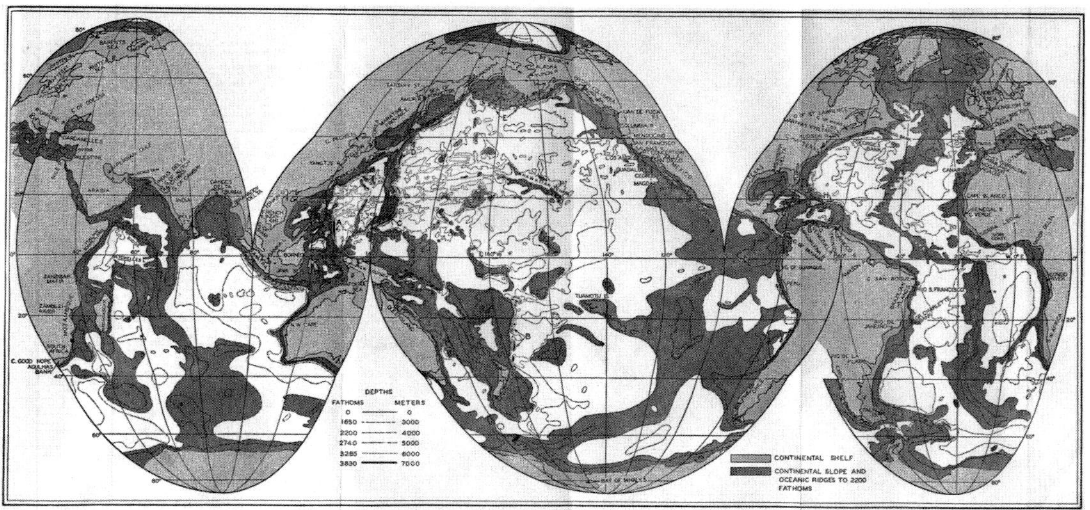


Figure 1.4 Bathymetric chart of the world's oceans, after Shepard (1948).

## 1.3 Sea floor spreading and plate tectonics

The theory of sea floor spreading views tectonic plates as being created at MORs and then 'spreading' away from them on either side. The theory developed from the ideas on continental drift proposed by Wegener (1912, 1966), based on the matching coastlines and geological features, and the divergence of evolutionary trends, on either side of the Atlantic. Dietz (1961) proposed that oceanic seismic layer 3 (the lower crust, see Chapter 5) and the uppermost mantle are chemically the same, and introduced the term 'lithosphere' to describe the outermost, rigid part of the Earth (Chapter 3). Importantly, he suggested that the sea floor represents the tops of convection cells; the MORs mark the up-welling sites and the trenches are associated with down-welling. This contains the essence of sea floor spreading and plate tectonics as now understood. Hess (1962) also accepted mantle convection, and suggested that the MAR is spreading at about 1 cm $a^{-1}$ half-rate. He proposed that ridges' elevations reflect their thermal expansion, and that their low seismic velocities are partly due to raised temperatures. Crucially, he suggested that continents ride passively on the convecting mantle rather than having to plough through oceanic crust, providing a more acceptable mechanism for continental drift.

Critical evidence for sea floor spreading came from detailed marine magnetic surveys. Detailed surveys by Scripps Institution of Oceanography in conjunction with the United States Coast and Geodetic Survey off the west coast of the USA and Canada (Mason, 1958; Mason and Raff, 1961; Raff and Mason, 1961) revealed extensive, parallel, linear magnetic anomalies (Figure 1.5) in the region of the Gorda, Juan de Fuca and Explorer Ridges (though these ridges were not recognised at that time). Subsequent studies elsewhere

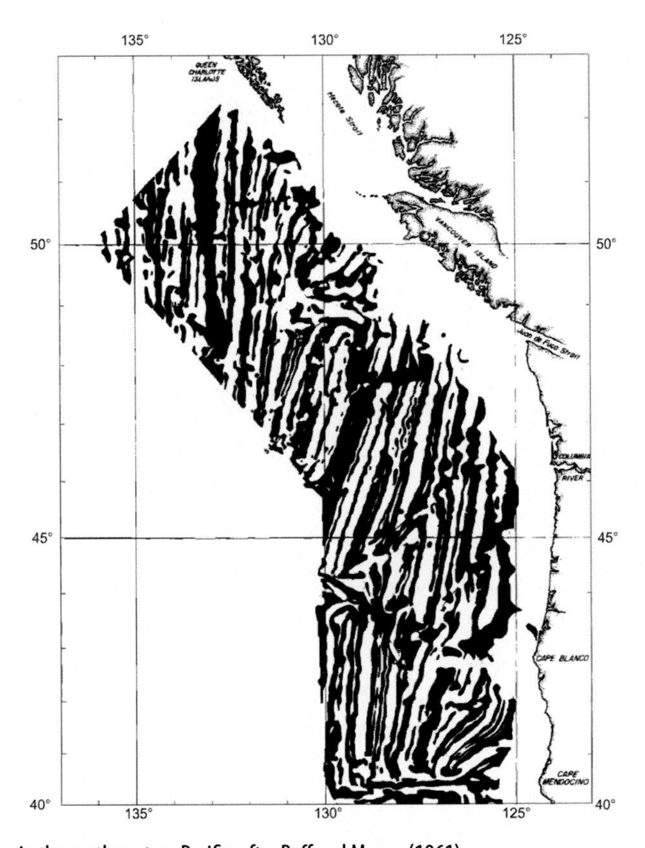


**Figure 1.5** Magnetic lineations in the northwestern Pacific, after Raff and Mason (1961).

showed the remarkable symmetry of magnetic lineations about ridge axes (e.g., Heirtzler *et al.*, 1966; Figure 1.6).

Vine and Matthews (1963) of Cambridge University conducted a small but very detailed survey over part of the Carlsberg Ridge. They demonstrated that the observed magnetic lineations could be explained if approximately 50% of the sea floor was underlain by normally magnetised material and 50% by reversely magnetised material. By combining this observation with the recognition of periodic reversals of the Earth's magnetic field (Cox *et al.*, 1963) they were able to explain the magnetic lineations in terms of sea floor spreading. A much greater body of supporting evidence was later published by Vine (1966). Essentially the same interpretation had been put forward more or less simultaneously by Lawrence Morley, who was unable to get his ideas published immediately (Glen, 1982) but did so later (Morley and Larochelle, 1964).

In summary, the theory of sea floor spreading states that tectonic plates diverge from MOR axes, at separation rates ranging from $<10$ km Ma$^{-1}$ (Dick *et al.*, 2003) to $>160$ km Ma$^{-1}$ (Naar and Hey, 1989). This causes the underlying ductile mantle to rise and, at higher spreading rates, to partially melt producing magma that solidifies to form a volcanic crust (Figure 1.7); at slower spreading rates, the mantle is extruded directly onto the sea floor (Chapter 7).

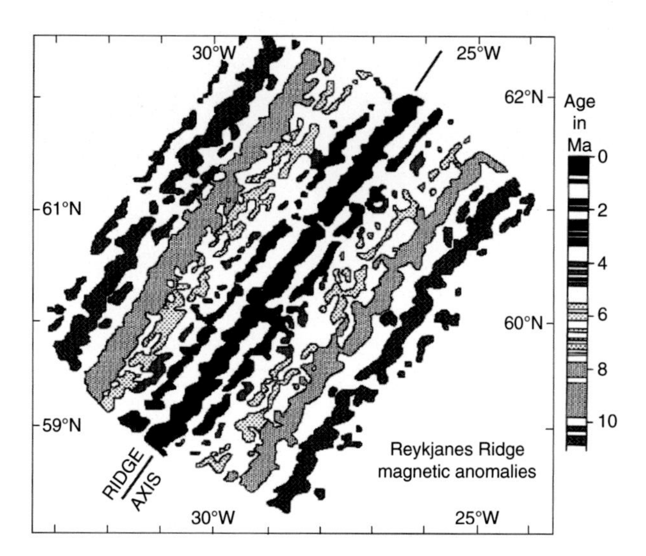


**Figure 1.6** Magnetic anomalies over the Reykjanes Ridge, reprinted from Heirtzler *et al.* (1966), © 1966, with permission from Elsevier.

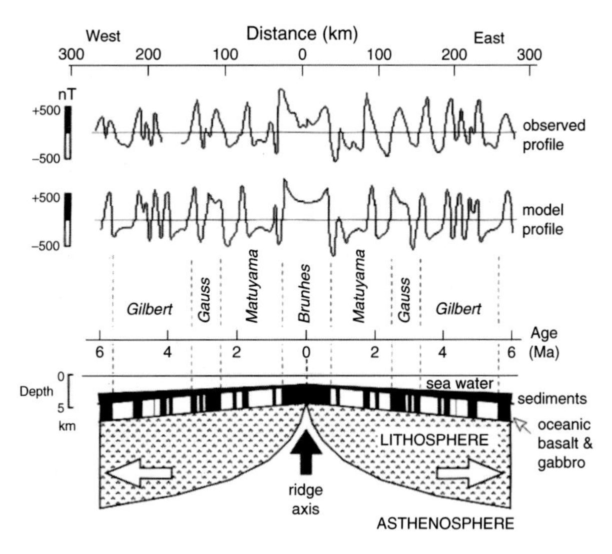


**Figure 1.7** Diagram of sea floor spreading, after Lowrie (1997).

The recognition of the origin of lineated magnetic anomalies, coupled with radiometric dating of the field reversals (Cox *et al.*, 1963), provided a means of both dating older sea floor and determining spreading rates. A detailed magnetic reversal timescale is now available back to the late Cretaceous (118 Ma; Cande and Kent, 1995). The most prominent positive magnetic anomalies have conventionally been numbered from 1 (axial anomaly, zero age) to 34 (83 Ma, end of Cretaceous magnetic normal epoch), with a separate numbering scheme for older anomalies. Periods of constant field direction (normal or reversed) are called ‘chrons’, and have been numbered to fit in with the pre-existing

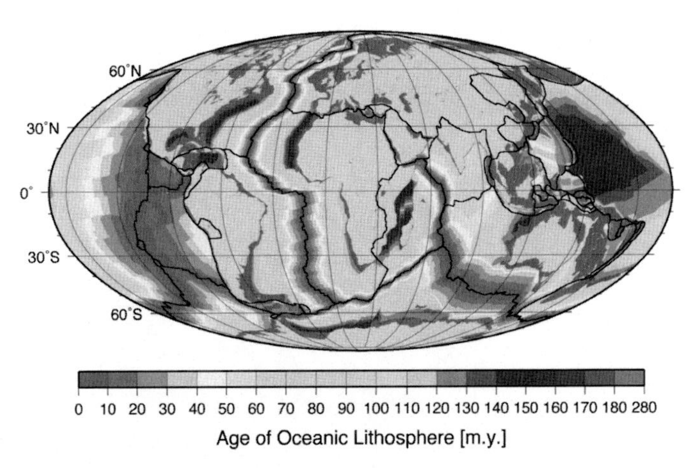


Figure 1.8 Age of the ocean floor, after Müller *et al.* (2008). For colour version, see plates section.

anomaly numbering scheme (e.g., Cande and Kent, 1995). Global measurements of marine magnetic anomalies, coupled with plate kinematic modelling, has allowed the age of the ocean basins to determined (Müller *et al.*, 2008; Figure 1.8).

## 1.4 Oceanographic institutions

Marine research is unusual, and possibly unique, in being extremely multi-disciplinary, involving geologists, geophysicists, geochemists, physicists, biologists, engineers, mathematicians and others. This may partly reflect the hostile medium in which they all work, the need to share expensive vehicles and platforms, and perhaps the youthfulness of these sciences where the inter-relations of sea floor rock, water motion, chemistry and biology are still being worked out. As a result, specialised multi-disciplinary oceanographic research institutions have played an important role in developing MOR studies.

Early on, several nations developed oceanographic institutions, and these gave impetus to the new oceanographic sciences by concentrating funding and effort. In the USA, Scripps Institution of Oceanography, Woods Hole Oceanographic Institution and the Lamont–Geological Observatory (now Lamont–Doherty Earth Observatory) were all founded in the first half of the twentieth century. They have provided important technological support, including developing the first deep-towed geophysical instrument (SIO), developing manned and unmanned submersibles (WHOI) and maintaining the world's only academic research ship with a fully 3D-seismic capability (LDEO), as well as making major advances in MOR and other marine science.

Britain's National Institute of Oceanography (now renamed the National Oceanography Centre, Southampton) was founded in 1950, has carried on important investigations of MORs since its founding, and has played a particularly important role in the development and use of side-scan sonars. Many other nations have developed regional or national oceanographic research centres, for example IFREMER in France, GEOMAR and the

Institut für Meereskunde in Germany, and the Japanese Marine Science and Technology institute (JAMSTEC). There are numerous university and a few independent research centres around the world, including the Monterey Bay Aquarium Research Institute (MBARI), founded by David Packard in 1987. It focusses on the development of instruments and research systems, and has made important contributions to, for example, understanding its neighbouring Gorda, Juan de Fuca and Explorer ridges.

## 1.5 Dedicated MOR research programmes

The study of MORs increasingly requires the use of pooled resources to provide expensive ships and instruments, and focussed approaches to make rapid progress and engage funding bodies. It is more of a global enterprise than many science disciplines, and this has encouraged high degrees of international collaboration.

Recognising a need for focussed research, the US Ridge initiative began in 1987, with the aim of understanding the 'geophysical, geochemical and geological causes and consequences of energy transfer' along the global spreading centre network (Delaney, 1989). It was extended into the RIDGE 2000 programme, and funded focussed research on a small number of selected sites, supported by the National Science Foundation. Several other national programmes, including the British BRIDGE (Cann *et al.*, 1999), French Dorsales, and German D-Ridge were developed along a similar model (Anonymous, 1993).

It was recognised that there was also a need for an international organisation to coordinate MOR research around the world. As a result, InterRidge (www.interridge.org) was founded in 1990. It aims to promote interdisciplinary, international studies of oceanic spreading centres by creating a global research community, planning and coordinating new science programmes that no single nation can achieve alone, exchanging scientific information, and sharing new technologies and facilities. It operates by means of a number of working groups, by convening workshops, acting as an information exchange, publishing a newsletter, and maintaining a web page. Notable achievements have included focussing research interest onto the remote and poorly studied ultra-slow spreading South West Indian Ridge and Arctic Gakkel Ridge. InterRidge is dedicated to reaching out to the public, scientists and governments, and to providing a unified voice for ocean ridge researchers worldwide. For example, it has recently sponsored the 'InterRidge statement of commitment to responsible research practices at deep-sea hydrothermal vents' (http://www.interridge.org/IRStatement).

Another important international organisation involved (in part) in MOR research is the Integrated Ocean Drilling Program and its predecessors, the Ocean Drilling Program and Deep Sea Drilling Project (http://www.iodp.org/). IODP provides a vital facility for drilling deep into the oceanic crust (Section 2.13.4).

## 1.6 Outline of this book

In order to allow readers from different disciplines to use the book easily, I have tried to make each chapter as self contained as possible, with cross-referencing where appropriate.

In addition, the appendices include a glossary of terms and a directory of place names used. Because the history of ridge science is intimately bound up with the development of technology, Chapter 2 provides a brief historical review of the relevant major developments in instruments and techniques. Chapter 3 describes the deep foundations of ridges in the oceanic lithosphere. Chapter 4 looks at ridges as tectonic plate boundaries. In Chapter 5, I focus on the structure of the oceanic crust. This leads to the description of the magmatic construction of the crust in Chapter 6. Chapter 7 describes the deformation of the newly formed crust by tectonic processes such as faulting. Chapter 8 covers hydrothermal vents, and finally Chapter 9 provides a summary and synthesis, and attempts a unified model of sea floor spreading.

# 2 Techniques of MOR study: a brief historical review

## 2.1 Introduction

Like most sciences, the history of MOR studies is closely linked to advances in technology, ranging from the lead-weighted sounding line in the nineteenth century to modern, autonomous robotic instruments. A detailed account of most current geophysical surveying techniques is given by Jones (1999).

## 2.2 Depth measurement

A fundamental measurement is the ocean depth, which yields vital information on the structure of the sea floor and the processes that shape it.

### 2.2.1 Line and weight sounding

Initial attempts to measure full ocean depth using the 'lead lines' employed for navigating shallow waters proved inaccurate, until Maury (1860) made a systematic study to improve such soundings, for example by measuring the speed of fall of the weight to judge when it reached the bottom. These methods were rapidly adopted by other navies. William Thomson (Lord Kelvin) improved the sounding line method by replacing hemp line by steel piano wire in 1870 (Shepard, 1959). This was trialled on the *Challenger* expedition (1872–1876) and was used on the contemporaneous USS *Tuscarora* expedition (Murray and Hjort, 1912). Kelvin introduced his 'Sounding Machine' in 1878, using a hand-operated, and later mechanised, winch (Figure 2.1).

### 2.2.2 Electronic echosounding

Electronic echosounders work by timing the passage of sound from an emitter on the ship via its sea floor reflection and back to a ship-based receiver.[1] Echosounding began to be introduced following the development of anti-submarine sonar during World War I

[1] 'Sounding' (verb 'to sound') here means measuring depth, and is derived from the same old English root as 'sound' meaning 'a channel'. It is distinct from 'sound' in an acoustic sense, which is derived from Latin 'sonus'. Sounding does not necessarily imply depth measurement by acoustic means.

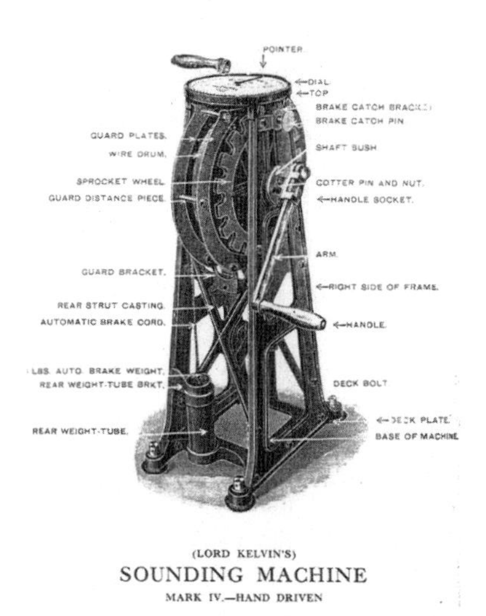


Figure 2.1 Kelvin Mk IV sounding machine, from the Kelvin and James White catalogue of 1907.

Figure 2.2 The UK National Institute of Oceanography's Precision Echosounder in use in the 1960s (after Tucker and McCartney, 2010).

(Shepard, 1959). Although there were problems measuring the time with sufficient accuracy, the ease and speed of the method, without the need to slow or stop the ship, led to its rapid adoption.

Continuously recording deep-sea echosounders were developed during World War II. Constant-frequency drivers provided accurate time (depth) resolution, and graphic recorders presented two-way travel time, or depth, on one axis with time (along track) on the other (Figure 2.2). Usually a 10 or 12 kHz sound source was used, providing efficient transmission through kilometres of sea water. These instruments routinely measured depth to accuracies of better than 1 in 3000 (Heezen and Menard, 1963; Tucker and McCartney, 2010). However,

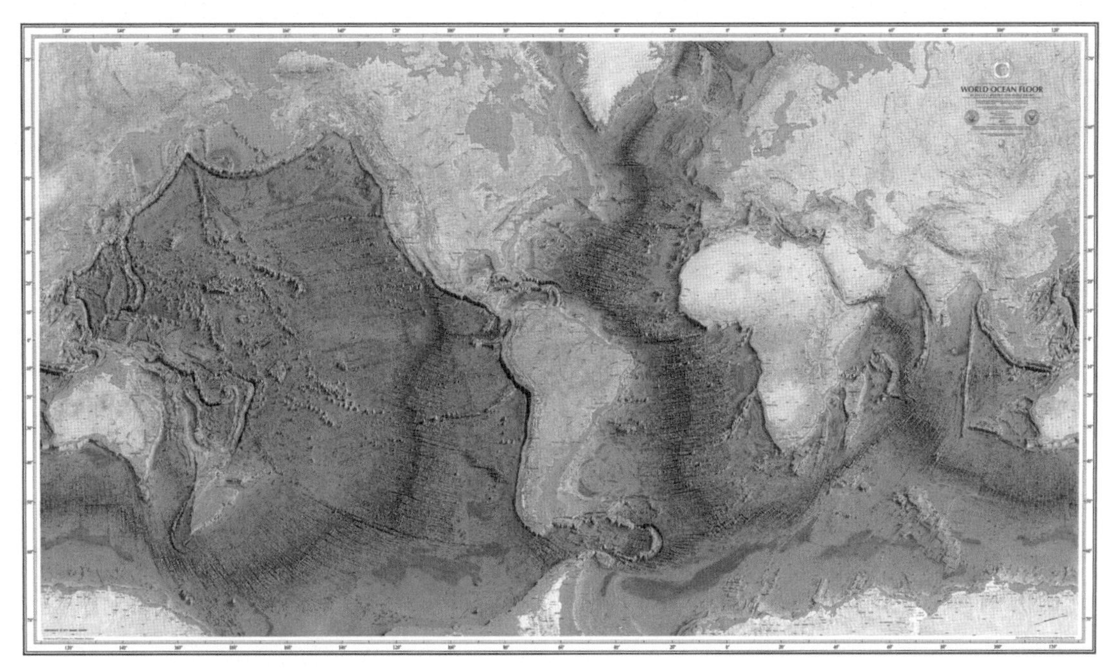


Figure 2.3 World Ocean Floor Panorama by Bruce C. Heezen and Marie Tharp (1977). Copyright by Marie Tharp 1977/2003. Reproduced by permission of Marie Tharp Maps, LLC 8 Edward Street, Sparkill, New York 10976. For colour version, see plates section.

these echosounders were limited by the need to read and digitise data manually, and by their wide (~30°) beams, limiting their deep-water horizontal resolution to many hundreds or even thousands of metres.

Transferring the detail available on the profile to the wide spaces between required careful interpretation. In particular, Bruce Heezen and Marie Tharp developed the technique of making 'physiographic diagrams' (Heezen and Tharp, 1954; Heezen and Menard, 1963; Tharp, 1982), producing remarkable physiographic maps of all the world's oceans (Heezen and Tharp, 1957; 1961; 1964; 1971; 1977); Figure 2.3.

### 2.2.3 Multi-beam echosounding

A major advance was the multi-beam (or 'swath') echosounder, developed by the US Navy in the 1960s (Glenn, 1970). The first commercial version, SeaBeam, became available in 1977 (Renard and Allenou, 1979). These instruments use electronic beam-forming to create a fan-shaped array of very narrow (~1°) beams in the thwart-ships or cross-track direction (Figure 2.4). A wide range of models is now available. Most modern oceanographic ships are equipped with a deep-sea version, typically using ~120 beams in a 120° fan, covering a swathe several kilometres wide with lateral resolution of <100 m and operational at full cruising speed. Data are automatically digitised and corrected for variations in sound speed and refraction. They can then be combined into digital terrain models and visualised in a variety of ways (Figure 2.5).

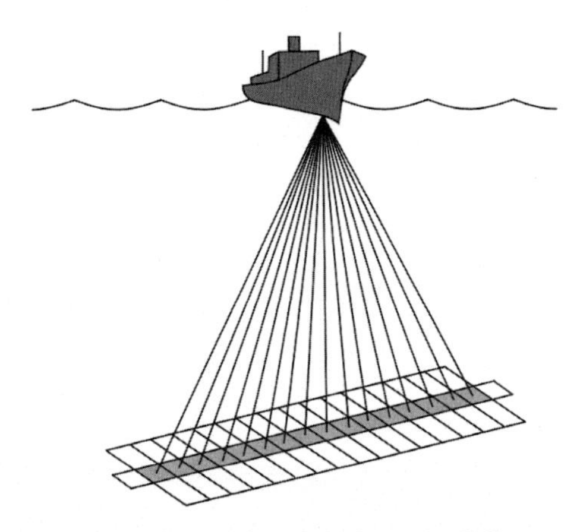

Figure 2.4 The beam pattern of the first SeaBeam multi-beam echosounder (Renard and Allenou, 1979).

With their high spatial resolution and broad coverage, multi-beam echosounders enabled the precise geometry of MOR plate boundaries to be defined (Macdonald *et al.*, 1984), facilitating understanding of transform faults (Gallo *et al.*, 1986; Madsen *et al.*, 1986; Pockalny *et al.*, 1988) and the neovolcanic zone (Smith and Cann, 1990), and leading to the discovery of overlapping spreading centres (Macdonald and Fox, 1983), propagating rifts (Hey *et al.*, 1985), and Oceanic Core Complexes (OCCs; Cann *et al.*, 1997). Short-range, high-resolution multi-beam echosounders using frequencies ~200–300 kHz can resolve to ~1 m and allow detailed imaging of volcanic morphology (e.g., Soule *et al.*, 2005).

## 2.3 Magnetic field

Measurements of the magnetic field over the oceans played a pivotal role in the development of sea floor spreading and plate tectonics theory (Chapter 1).

### 2.3.1 Instrumentation

Early measurements at sea were confined to determining the declination – the horizontal angle between geographic or true north and the direction of the north magnetic pole (Bullard and Mason, 1963). Following the introduction of steel ships in the nineteenth century, shipboard measurement of the magnetic field became restricted to special non-magnetic ships such as *Galilee* and *Carnegie* operated by the Carnegie Institute, Washington. Such ships traversed the main oceans several times (Fleming, 1937), and some measurements were made by aircraft (Bullard and Mason, 1963).

The big breakthrough was the development of magnetometers that could be towed behind ships and thus escape the effects of the ships' fields (Bullard and Mason, 1961). Initially fluxgate (Miller and Ewing, 1956), but later proton precession magnetometers (Hill, 1959), were used. The latter utilise the fact that protons (which have a magnetic moment) precess

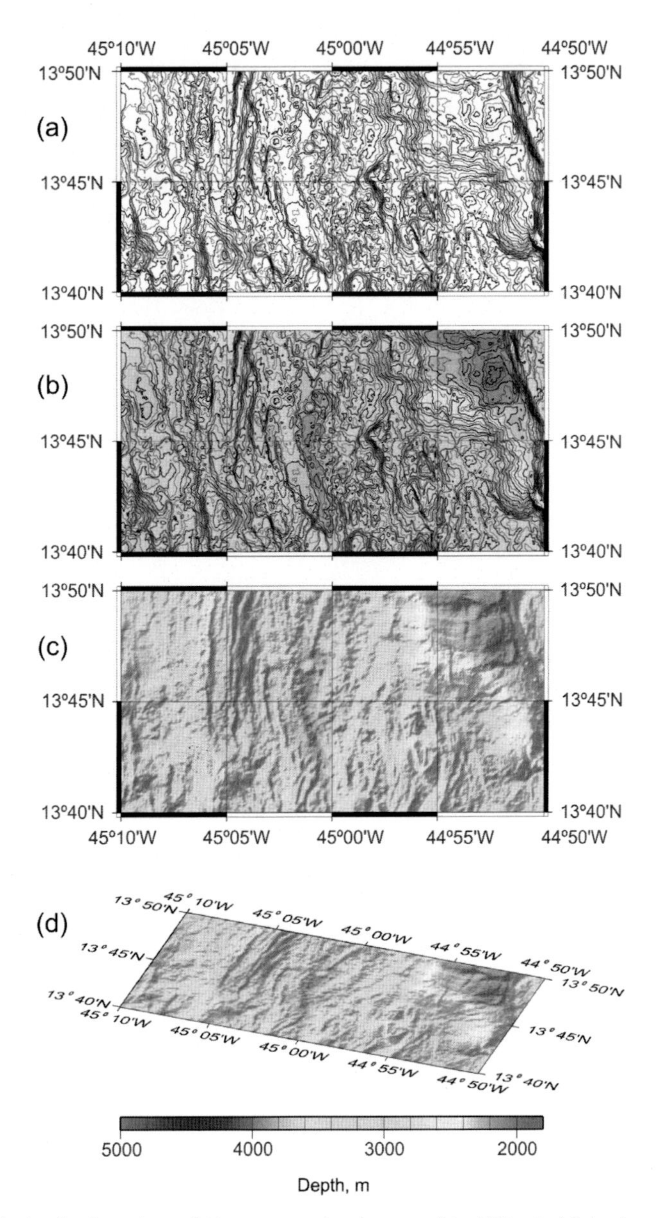


**Figure 2.5** Examples of depth visualisation using multi-beam sonar, showing part of the MAR axis: (a) simple contour plot, with 50 m contour interval; (b) the same contours with colour-coded depth; (c) colour-coded shaded relief image with illumination from the NW; (d) same data as (c) displayed in an oblique view. For colour version, see plates section.

about the local magnetic field direction with a frequency proportional to the field strength. A hydrogen-rich fluid such as paraffin provides the protons, and an electric current passed through a coil aligns them. When this current is switched off, the protons precess, inducing a current whose frequency is measured. The measurement cycle takes only a few seconds, so measurements are almost continuous, and routinely achieve an accuracy of about 1 nT

Figure 2.6 The towed sensor of a modern marine magnetometer, courtesy Mark Holmes.

or 2 parts in $10^5$ of the Earth's total field strength. The detailed surveys that first revealed magnetic lineations related to sea floor spreading were carried out with a fluxgate magnetometer, regularly calibrated against a proton precession instrument (Mason, 1958; Mason and Raff, 1961; Raff and Mason, 1961; Figure 1.5).

Recently, Overhauser effect magnetometers (which also measure the precession frequency of protons) have been introduced (Figure 2.6). They have greater sensitivity, faster cycling, less directional dependence and lower power consumption. Fluxgate magnetometers, often measuring all three field components, have been installed in submersibles, deep-towed vehicles and autonomous underwater vehicles (e.g., Flewellen *et al.*, 1993; Tivey, 1996; Searle *et al.*, 2010). Finally, Isezaki (1986) developed shipboard three-component magnetometers, made possible by continuously recording GPS to correct for the motion, and hence the magnetisation, of the ship.

## 2.3.2 Interpretation

Magnetic field observations are normally reduced to 'magnetic anomalies' by subtraction of the regional value of the Earth's field, usually derived from the International Geomagnetic Reference Field (IAGA Working Group V-MOD, 2010). The IGRF, updated every 5 years, consists of a spherical harmonic series that best fits observations of the Earth's field around the world, including data from ships, satellites and land observatories.

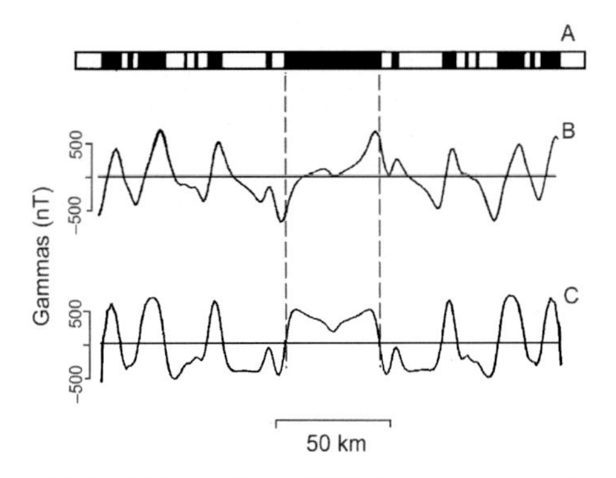


Figure 2.7 The effect of 'reduction to the pole', after Blakely and Cox (1972). A: standard model of crustal magnetic polarisation (black, normal) for the past 3.5 Ma, with top and base of model at 3.5 and 5.0 km below sea level; B: total field magnetic anomaly that model A would produce at sea level if the magnetised blocks were on a 45° trending ridge at 27° N); C: the effect of transformation to the pole, which deskews the anomalies and aligns positive anomalies over normally magnetised blocks.

Modelling the magnetic field is complicated by the need to take into account both the strength and direction of the field, but is straightforward using digital computers. Details are given, for example, by Telford *et al.* (1990). Initially interpretation was restricted to forward modelling, i.e. calculating the magnetic field that would be produced by a body with specified properties. Two-dimensional models were common, in which the magnetised body is represented by an infinitely long prism of prescribed cross-section (e.g., Talwani and Heirtzler, 1964), while 3D models could be dealt with by computing the effects of individual slices through the body and integrating them numerically (e.g., Talwani, 1965).

An important development was the introduction of Fourier transform techniques, which allow very rapid calculations, the use of continuously varying magnetisations, and digital filtering (Parker, 1972; Schouten and McCamy, 1972). This facilitated, for example, determination of the finite widths of magnetic polarity transitions (Macdonald, 1977), detailed modelling of volcanic crustal construction (Schouten and Denham, 1979; Figure 5.15), and representation of the rapid decrease in magnetisation in young oceanic crust (Allerton *et al.*, 2000). It also allows correction for the inherent asymmetry of magnetic anomalies when a linear ridge axis is oblique to the Earth's field direction (Schouten and McCamy, 1972) and a technique called 'reduction to the pole', which transforms the anomalies to the shape they would have if the same source body were transferred to the magnetic pole where the field is vertical (Blakely and Cox, 1972; Figure 2.7). A further important development was 'inverse modelling': inverting observed magnetic measurements directly to the best-fit model with an estimate of its uncertainty (Parker and Huestis, 1974; Macdonald *et al.*, 1980b). However, as with all potential field methods, there always remain ambiguities in interpretation (e.g., Telford *et al.*, 1990). These Fourier techniques assume the magnetic field is observed on a horizontal plane. Data from variable depths can be transformed to the equivalent field on a horizontal surface by 'upward continuation' (Parker and Klitgord, 1972; Guspi, 1987; Hussenoeder *et al.*, 1995).

Further analysis techniques enable the direct determination of the positions and directions of magnetic boundaries using three-component measurements (Seama *et al.*, 1993; Kitahara *et al.*, 1994; Yamazaki *et al.*, 2003).

### 2.3.3 Rock magnetism and palaeomagnetism

Rocks can become magnetised by a variety of mechanisms. The most important for MOR studies is thermo-remanent magnetisation (TRM), in which minerals (mainly magnetite and titanomagnetite) acquire a magnetisation when cooled below their Curie and blocking temperatures (Tauxe, 1998). The acquired field is proportional and parallel to the Earth's field at the time of cooling. The strength of the TRM is closely related to the amplitudes of marine magnetic anomalies, enabling the thickness of the magnetic source layer to be estimated. The Curie temperature is about 580 °C for magnetite, and ranges between −150 °C and almost 600 °C for titanomagnetite, depending on the exact composition. Blocking temperatures are slightly lower. Investigating these effects enables the magnetic mineralogy of rocks to be determined. Magnetite is also produced when the mineral olivine is hydrated to form serpentine (Chapter 8), an example of chemically acquired remanent magnetisation (CRM).

In palaeomagnetism, the direction of the remanent magnetisation is used to estimate the past position and orientation of the rock sample relative to the geomagnetic field, and requires the use of fully oriented samples. Palaeomagnetism provides important confirmations of the predictions of plate kinematics, and has also been used to determine structural rotations of crustal blocks at MORs, for example in the evolution of OCCs (e.g., Morris *et al.*, 2009; see also Chapter 7).

## 2.4 Gravity

The strength of the gravitational attraction reflects subsurface mass and density distributions; marine gravity has been important in elucidating crustal structure and crustal thickness variations. The first gravity profile over a MOR was made in 1958 (Worzel, 1959; Talwani *et al.*, 1961; Figure 2.8).

### 2.4.1 Measurement of gravity at sea

The main problem in measuring gravity at sea is in correcting for the accelerations of the ship; in heavy seas, these can be a significant fraction of the Earth's gravitational acceleration. Early measurements were made by timing the oscillations of a pendulum in submerged submarines to avoid the worst water motion (Vening Meinesz, 1929). The late 1950s saw the development of gravity meters capable of operating on surface ships in reasonable sea states (Worzel and Harrison, 1963). These were essentially very sensitive spring balances, where extension of the spring reflects varying gravity. They operated on gyro-stabilised platforms, with ship accelerations recorded independently to correct

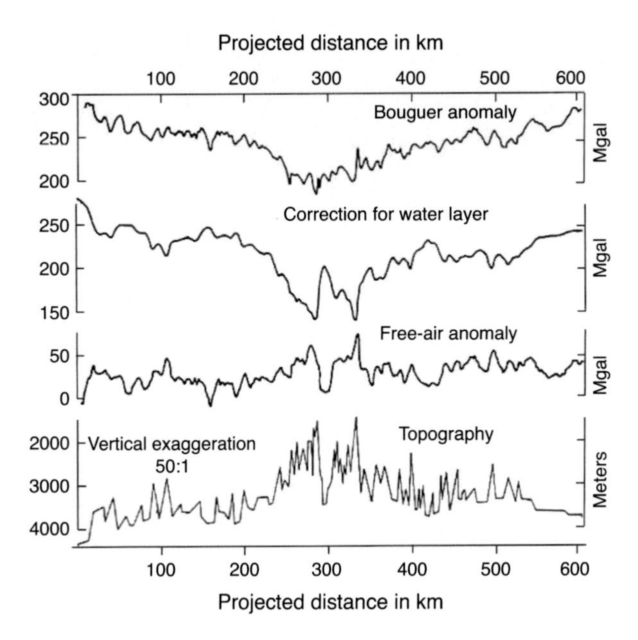


Figure 2.8 Free-air and Bouguer anomaly profiles across the MAR at 30° N, after Worzel and Harrison (1963). This material is reproduced with permission of John Wiley & Sons, Inc.

the meter readings. Other sensors such as a vibrating string have also been used (Wing, 1969).

Gravity measurements have been made in manned submersibles by Spiess *et al.* (1980) and others, typically using a portable land meter, although the instability of the submersible makes such measurements difficult and imprecise. Continuously recording instruments, allowing averaging over several minutes, have improved the resolution (Ballu *et al.*, 1998).

A very important development has been satellite altimetry, which enables the free-air anomaly to be derived from global datasets of sea-surface elevation (Haxby and Weissel, 1986; Sandwell, 1991). Sea-surface elevation is measured by a satellite-mounted radar altimeter, and corrected for effects such as wave motion and tides. The resultant global view (Figure 2.9) enables ridges to be investigated where no ship data exist, for example at high and low latitudes.

### 2.4.2 Gravity interpretation

Gravity measurements are routinely corrected for the effects of the Earth's spin and oblate spheroidal shape. Together these cause gravity at the poles to be greater than at the equator by about 0.5%. The 'theoretical gravity', $g_0$, which accounts for these effects but otherwise assumes a homogeneous Earth, is given by the International Gravity Formula and is based on the shape of the best-fitting sea-level ellipsoid. The current value (using the World Geodetic System 1984 ellipsoid) is

$$g_0 = 9.780\,326\,7714\frac{1 + 0.001\,931\,851\,386\,39\sin^2\lambda}{\sqrt{1 - 0.006\,694\,379\,990\,13\sin^2\lambda}}, \tag{2.1}$$

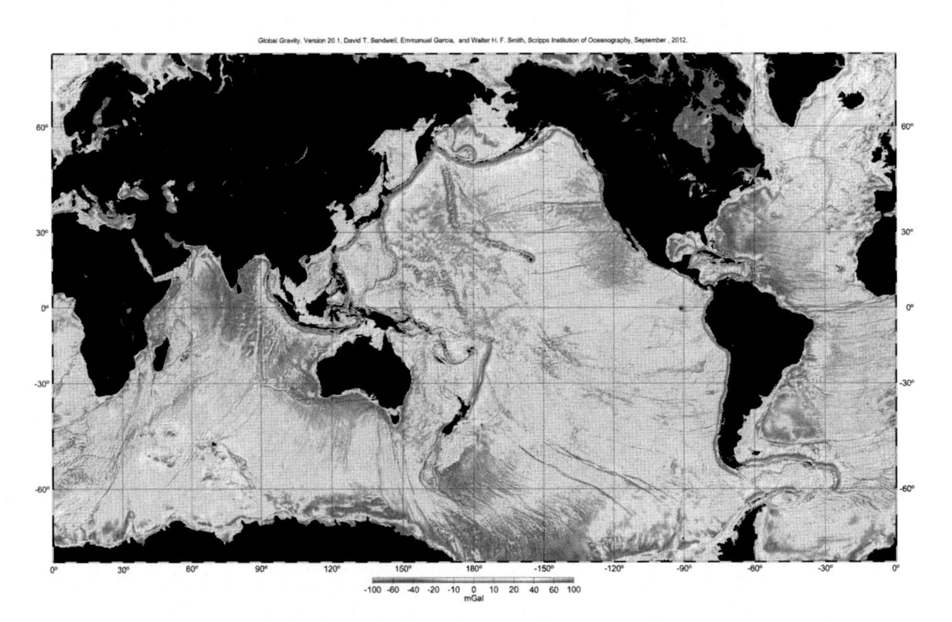


**Figure 2.9** Global free-air gravity field obtained from satellite altimetry, after Sandwell and Smith (2009). Image courtesy of David Sandwell. For colour version, see plates section.

where $\lambda$ is the latitude (Moritz, 1980). In the past, slightly different formulae, such as the International Gravity Formula 1967, have been used, so care must be taken when comparing different surveys.

When marine gravity is measured in ships there is no need to correct for height above sea level, as is done in land surveys, so the observed gravity minus $g_0$ gives the 'free-air anomaly'. The varying water depth of the oceans has a significant effect on observed gravity, so this effect is often calculated and removed from the free-air anomaly, producing the 'Bouguer anomaly', which should then reflect variations in the density distribution below the sea floor (Figure 2.8).

Early interpretation employed forward modelling programs similar to those used in magnetic interpretation (Talwani and Heirtzler, 1964), and could suggest simple density distributions to match observed anomalies (Figure 3.11). Again, an important breakthrough was the introduction of Fourier methods (Parker, 1972), allowing the gravitational attraction of a three-dimensional surface to be quickly calculated. Kuo and Forsyth (1988) used these methods to calculate the so-called mantle Bouguer anomaly (MBA), which corrects for the effects of the water/crust and crust/mantle interfaces, assuming these interfaces are parallel (i.e. assuming a constant thickness crust). Removing the effect of hotter, less dense mantle under the ridge axis (Figure 3.12) produces the residual mantle Bouguer anomaly (RMBA; e.g., Figure 5.2). Density variations in the mantle are relatively small, so variations in the MBA mainly arise from variations in crustal thickness and can be used to rapidly map them (Lin *et al.*, 1990). Similar methods allow the rapid computation of quite complex models of crustal structure at a scale of just a few kilometres (e.g., Blackman *et al.*, 2008). As with magnetics, gravity interpretation involves inherent ambiguities, requiring additional information (such as seismic data) to resolve.

The admittance technique (McKenzie and Bowin, 1976; Watts, 1978) compares the spectra of gravity and bathymetry data, using the techniques of time series analysis. An observed gravity profile $g(x)$ can be predicted by applying a filter, $f(x)$, to the corresponding bathymetry profile $b(x)$, where $x$ is distance along the profile. Mathematically,

$$g(x) = f(x)^* \, b(x), \tag{2.2}$$

where * indicates convolution. Profiles against distance, $x$, are converted into spectra against wave number, $k_n$ (where $k_n = 2\pi/\lambda_n$ and $\lambda_n$ is the $n$th component of wavelength), by taking the Fourier transform. Then the 'admittance function' between gravity and bathymetry, $Z(k)$, is the Fourier transform of $f(x)$, defined as

$$Z(k_n) = [G\,(k_n)]/[B(k_n)], \tag{2.3}$$

where $G(k_n)$ is the Fourier transform of $g(x)$ and $B(k_n)$ is the Fourier transform of $b(x)$. This technique has been very important in investigating the degree of isostasy at MORs (Section 3.6).

## 2.5 Heat flow

Measurements of conductive heat flow through the oceanic crust were an important contribution to early ideas on sea floor spreading, and provided strong evidence for the plate model of oceanic lithosphere (Section 3.2). The first oceanic heat flow measurements were described by Bullard (1952, 1963).

Heat flow, $Q$, is measured by simultaneously determining the temperature gradient $dT/dt$ through a layer of sediment together with the thermal conductivity, $k$, of the sediments:

$$Q = k\frac{dT}{dt}. \tag{2.4}$$

Temperature gradient is typically measured using a 'bow-string' probe, in which an array of thermistors in a narrow tube a few metres long is supported by a corer (Gerard *et al.*, 1962) or a rigid pole that is lowered into the sediments (Figure 2.10). Thermal conductivity is determined by briefly heating a wire in the tube and modelling the temperature decay as the heat dissipates through the sediment. Similar but smaller probes have been deployed from manned submersibles (e.g., Johnson *et al.*, 1993).

An important additional component of oceanic heat flow is carried in advecting hydrothermal fluids (see Chapters 3 and 8). Measuring this component has proved challenging, but a number of instruments have been developed (e.g., Schultz *et al.*, 1992; Pruis and Johnson, 2004). They utilise various methods to trap or funnel fluids as they exit the sea floor, and then to measure their fluxes and temperatures.

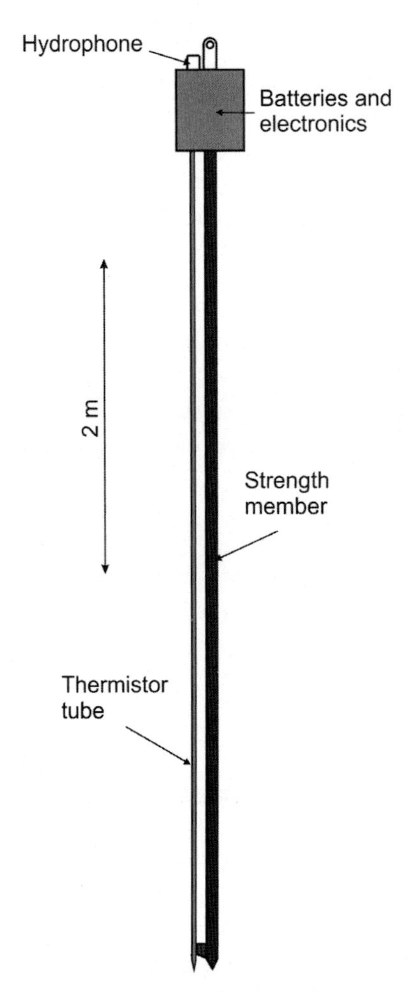


Figure 2.10 Diagram of a heat flow probe. Note thin sensor tube alongside thicker strength member.

## 2.6 Earthquake seismology

The distribution of earthquakes delineates active plate boundaries, while the analysis of seismic records can reveal the nature of the plate motions and faulting that produce them. Observations by detailed networks of recorders can be analysed to show subsurface distributions of seismic activity reflecting areas of active faulting, brittle deformation, magmatic and hydrothermal activity.

### 2.6.1 Global seismograph networks

The general distribution of earthquakes around the Earth was known by the middle of the twentieth century (Gutenberg and Richter, 1954). In 1961, the United States Coast

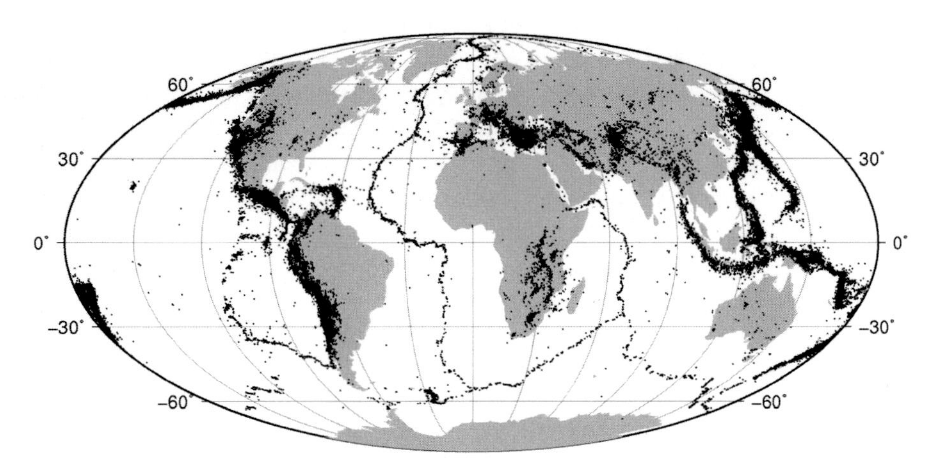


Figure 2.11 Global earthquake epicentres recorded from 1964 to 1973 by the WWSSN. Note narrow bands of earthquakes corresponding to the MOR axes. Data from the National Oceanic and Atmospheric Administration, International Seismological Commission database.

and Geodetic Survey established the World Wide Standardized Seismograph Network (WWSSN). This consisted of a global network of some 125 stations of three-component, standardised instruments. It rapidly built up a catalogue of earthquake locations with positional accuracies of ~20 km (Figure 2.11). This clearly associated earthquakes with the spreading plate boundaries (Isacks *et al.*, 1968). The network has been continuously upgraded, with the addition of new stations (including some located in boreholes in the ocean basins) and modern broad-band recorders, to form the Global Seismographic Network managed by the Incorporated Research Institutions for Seismology. Currently, the positional accuracy is ~10 km and the detection limit for MOR events is body-wave magnitude $m_b \approx 3.5$.

Focal mechanisms of earthquakes, i.e. the directions of movement of the sliding blocks causing the quake, provide important confirmation of the relative motions and extents of transform faults predicted by plate tectonics (Sykes, 1967; Isacks *et al.*, 1968; Section 4.2). Early focal mechanisms fitted only the direction of the first ground motions recorded (either toward or away from the hypocentre), but later the development of moment-tensor solutions, in which the amplitude as well as the directions of motion are fitted, provided more robust results (Langston, 1981).

## 2.6.2 Ocean bottom seismometers

A further important technique is the use of recording ocean bottom seismometers (OBS) capable of deployment in sea floor arrays to observe small magnitude local earthquakes at high resolution (e.g., Lilwall *et al.*, 1977; 1981; Kirk *et al.*, 1982; Toomey *et al.*, 1985; Wilcock *et al.*, 1992; Wolfe *et al.*, 1995; Sohn *et al.*, 1998; Barclay *et al.*, 2001; deMartin *et al.*, 2007). Locations may be determined with a precision ~1 km. Frequency against magnitude plots can help distinguish 'volcanic' from 'tectonic' events because the two yield different slopes (e.g., Lilwall *et al.*, 1977; Kong *et al.*, 1992).

Recently hydroacoustic arrays have been used to study MOR seismicity (Fox *et al.*, 2001; Smith *et al.*, 2002). Hydrophones are deployed on deep-water moorings and detect seismic T-waves (water compressional waves converted from crustal body-waves at the sea floor). Analysis of the arrival times locates the origins of the T-waves, which are close to (but not necessarily at) the seismic epicentres. Precision is ~10 km or better (Smith *et al.*, 2002; Smith *et al.*, 2006), with detection thresholds several magnitude levels below those of land-based seismograph networks (Fox *et al.*, 2001), allowing finer details of ridge tectonic processes to be analysed (Smith *et al.*, 2006).

## 2.7 Seismic refraction

Seismic refraction is arguably the most powerful technique for exploring crustal structure at ridges. Its development began in the 1930s (Ewing *et al.*, 1937), followed by several groups in Britain, Russia and the USA in the early 1950s (Hill, 1952; Raitt, 1956; Galperin and Kosminskaya, 1958; Officer *et al.*, 1959).

### 2.7.1 Data acquisition

The energy from controlled explosions is transmitted to the sea bed and refracted through higher velocity rocks before returning to a recording seismometer. Initially, chemical explosives were used, but increasing expense and restrictive shipping regulations have led to most modern work being carried out using 'air guns' as sources, in which highly compressed air is rapidly vented into the sea. The resultant energy has a dominant frequency dependent on the volume of air released, so 'tuned' arrays of different sized guns can be built to control the frequency content and shape of the pulse. Larger guns produce more energy at lower frequency, which gives better penetration, but small guns produce higher frequencies at lower power, so yield improved resolution but less penetration. Typical frequencies are in the range 20–200 Hz, giving resolutions of several hundred metres. Some experiments have been carried out using on-bottom shots (Christeson *et al.*, 1994; Blackman *et al.*, 2009) detonated with the NOBEL device (Koelsch *et al.*, 1986), but the technique has not been used extensively.

To image lower crustal features, recorders must be several tens of kilometres from the shots. This was initially achieved using either floating sonobuoys that radioed data back to the shooting ship (Hill, 1963) or via hydrophones deployed from a second ship. It is now normally done using ocean bottom seismometers placed on the sea floor (Whitmarsh, 1970), with the OBS being precisely located by sound ranging from the shooting ship. At the end of an experiment an acoustic signal commands the instruments to release their anchors and return to the surface for recovery.

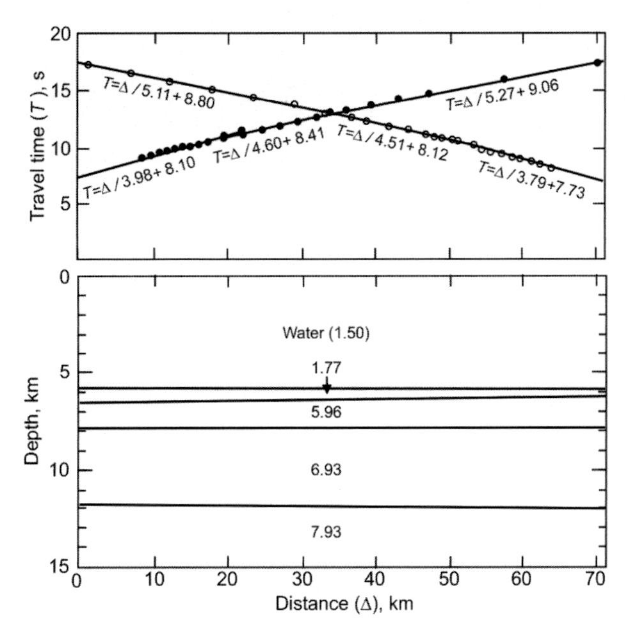


Figure 2.12 Early, three-layer model of oceanic crust in the western North Atlantic, after Raitt (1963). This material is reproduced with permission of John Wiley & Sons, Inc.

## 2.7.2 Interpretation

Early interpretation techniques modelled stacks of planar layers, either horizontal or slightly dipping. By 'reversing' the profiles (placing shots at each end of the recording array) the layer dips could be determined. The data appeared to fit a simple layered structure (Raitt, 1956; Figure 2.12).

Later interpretation techniques modelled amplitudes as well as travel times, allowing more precision (Fuchs, 1968; Helmberger and Moms, 1969; Fuchs and Müller, 1971). Further theoretical development allowed for the modelling of lateral variations. Inversion techniques were developed (Detrick *et al.*, 1982), following the introduction of the so-called tau-p transformation. This converts a plot of offset (shot–receiver distance) vs. travel time to one of intercept time vs. ray parameter (see Appendix A for definitions). This innovation freed interpretation from the restriction of discrete layers bounded by velocity discontinuities, and showed for example that 'layer 2' (Section 5.3.2) is a region in which velocity increases rapidly with depth, while 'layer 3' has a small vertical velocity gradient (Spudich and Orcutt, 1980). Ray tracing algorithms (Chapman and Drummond, 1982; Zelt and Smith, 1992) enabled inversion of fully two-dimensional models (e.g., Figure 2.13), and allowed confidence limits to be placed on the models.

Finally, the development of seismic tomography (Thurber, 1983), coupled with the deployment of two-dimensional arrays of OBS, has led to the production of fully three-dimensional velocity models (e.g., Toomey *et al.*, 1994; Wolfe *et al.*, 1995; Sohn *et al.*, 1997; Dunn *et al.*, 2005; Figure 2.14).

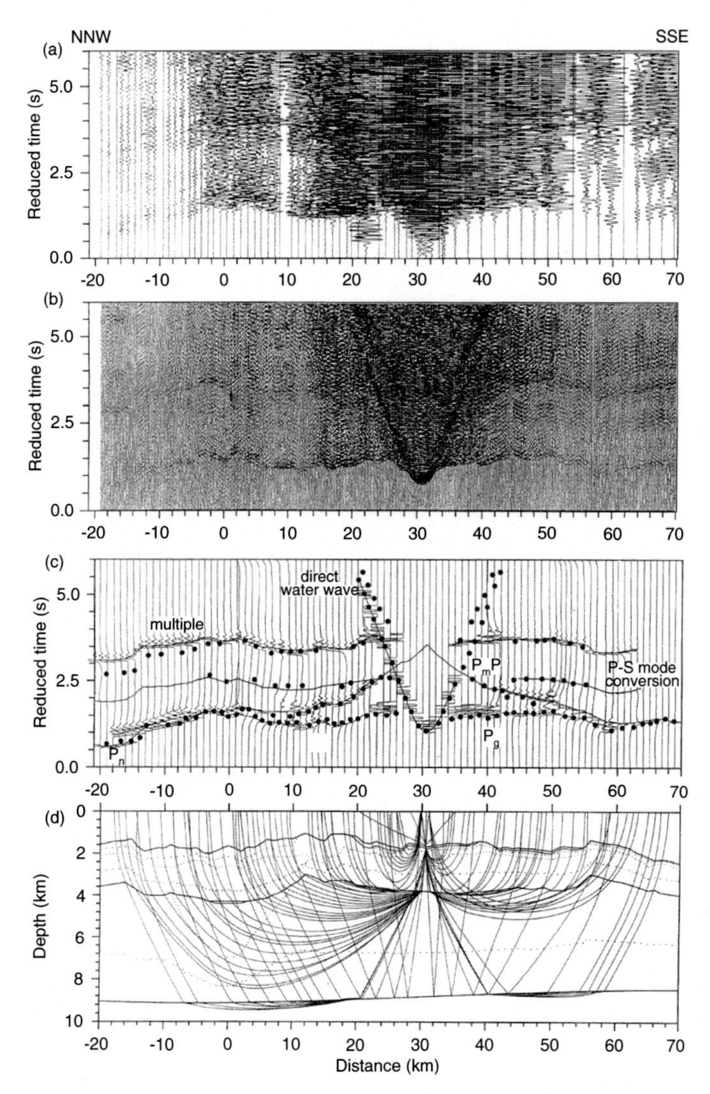


Figure 2.13 Two-dimensional wide-angle seismic model across the Reykjanes Ridge, after Navin *et al.* (1998, Figure 4). Observations of vertical motion from explosive (a) and airgun shots (b); (c) computed synthetic seismograms with observed (dots) and modelled (continuous line) arrivals; (d) vertical section through model showing contours of P-wave velocity and computed ray paths.

## 2.8 Seismic reflection

The seismic reflection method is an extension of echosounding, but using lower-frequency sound to penetrate thousands of metres into the sea bed. It provides high-resolution images of inter-crustal interfaces, such as sediment horizons, magma chambers and faults.

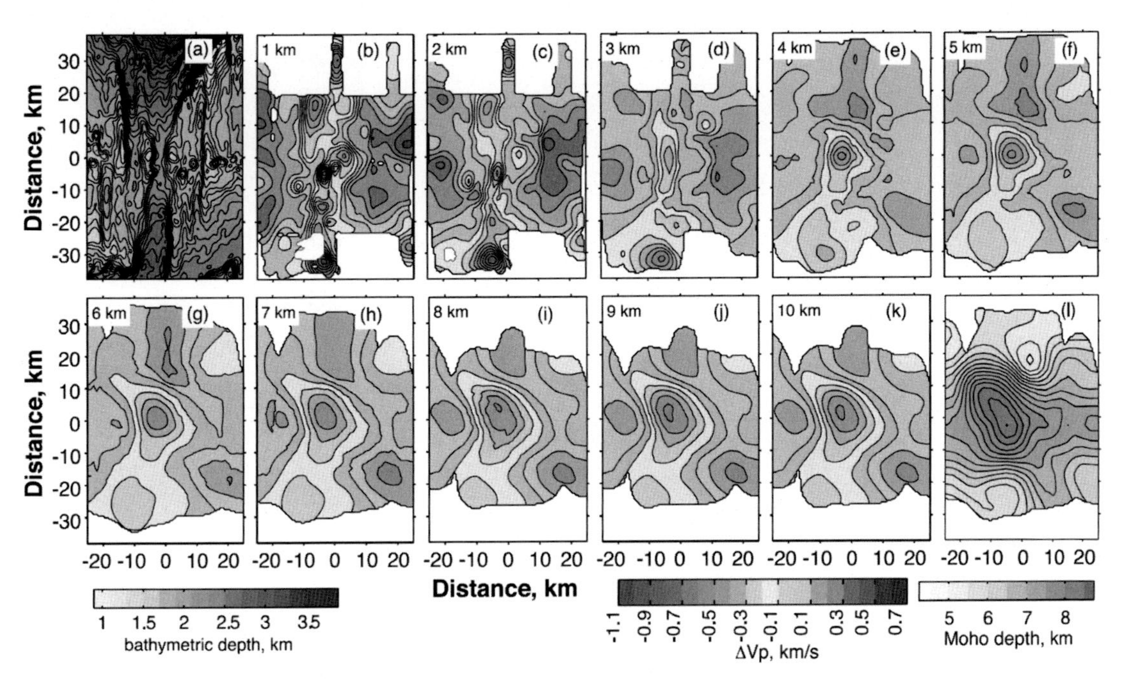


**Figure 2.14** Three-dimensional (3D) seismic tomographic model of a segment of the MAR near 35° N, after Dunn *et al.* (2005). (a) Bathymetry of the segment. (b) to (k) Two-dimensional (2D) horizontal sections through the model at 1 km depth increments, showing computed departure of P-wave velocity from a simple one-dimensional (1D) starting model. (l) Crustal thickness inferred from the model. For colour version, see plates section.

Shallow-water reflection experiments were carried out in the 1930s to 1950s using echosounders and a variety of sound sources (Hersey, 1963). The Swedish *Albatross* expedition conducted some of the first deep-water reflection measurements in 1947 using depth charges, and showed that sediment thickness increases with distance from the MOR (Reinke-Kunze, 1994). Other early sound sources included 'pingers' (where an electrical pulse is converted into sound), 'sparkers' in which a rapid discharge of electrical current vaporises water which then implodes, and gas exploders and 'airguns' that produce underwater explosions. Early airguns were described by Ewing and Zaunere (1964) and Jones (1967). Airguns and the related 'water gun' have been extensively developed and used by both academia and the offshore hydrocarbon industry.

Seismic reflection surveys aim to produce 2D (or, recently, 3D) images of crustal structure. Both source and receiver are towed, usually from the same ship (Figure 2.15). Early work used one or a few airguns with receivers comprising relatively few hydrophones encased in an oil-filled, neutrally buoyant, towed tube or 'streamer'. Analogue recording was used, with little or no signal processing (Ewing and Tirey, 1961). Nevertheless, such 'single channel' reflection profiles provided powerful views into the sea bed, often revealing the presence of a strongly but diffusely reflecting 'acoustic basement' interpreted as the top of the volcanic crust, beneath hundreds of metres of sediment (Figure 2.16).

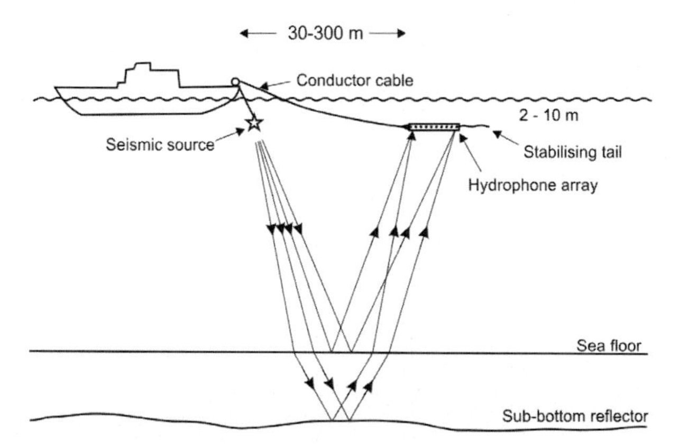


**Figure 2.15** The basic seismic reflection system, after Jones (1999, with permission).

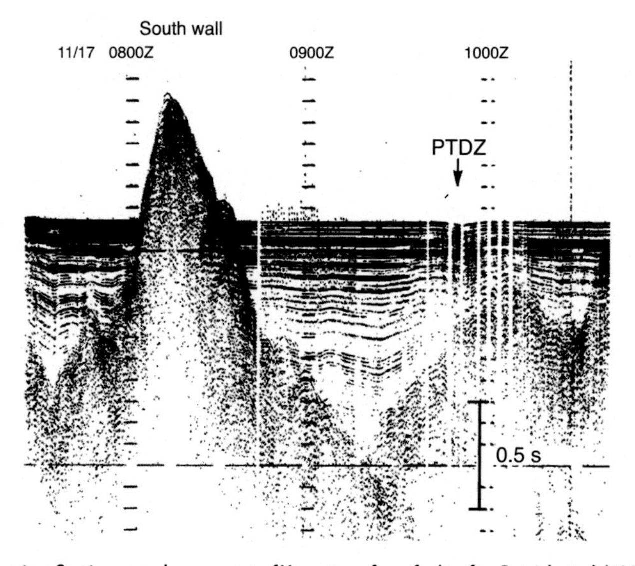


**Figure 2.16** Single-channel seismic reflection record across part of Vema transform fault, after Detrick *et al.* (1982). PTDZ: Principal transform displacement zone (see Section 4.4.1). Note rough acoustic basement (top of igneous crust) marked by broad, fuzzy reflection at base of section, and flat-lying turbidite sediments infilling the transform valley. Section length ~30 km. Scale bar of 0.5 s two-way travel time would represent 500 m for a velocity of 2 km $s^{-1}$.

Nowadays, tuned arrays of airguns and similar devices are the usual sound source. Receivers are 'multi-channel', comprising arrays of hydrophones up to several kilometres long towed behind the shooting ship. The great advantage of multiple receivers is that different shot–receiver pairs have different length travel paths and, as the ship progresses, the same subsurface point is imaged by different travel paths (Figure 2.15). This allows (a) individual records to be added to improve the signal-to-noise ratio, and (b) an array of seismic processing techniques to be used to correct and enhance the data (Kearey and

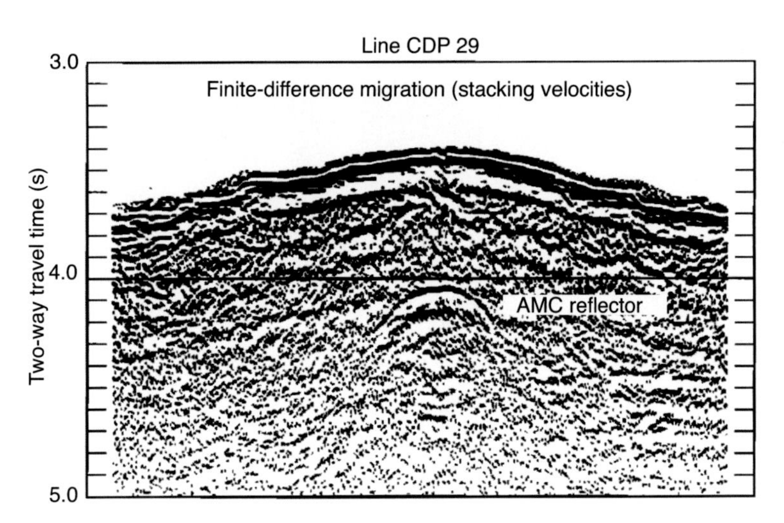


Figure 2.17 Seismic reflection image of the sub-axial magma chamber under the EPR axis near 9.5° N, after Kent *et al.* (1993).

Brooks, 1991). Such techniques have enabled, for example, the imaging of crustal melt lenses (e.g., Morton and Sleep, 1985; Kent *et al.*, 1993; Figure 2.17).

Some experiments have been carried out with deep-towed seismic reflection (Bowen and White, 1986), but the technique has not been used extensively. Fully 3D seismic acquisition is now available to the academic community, and has been used to image the 3D structure of the crust and the melt sills within it (Singh *et al.*, 2006a, b; Figure 2.18).

## 2.9 Compliance

A relatively new method of investigating the ocean floor is the compliance technique (Crawford *et al.*, 1998). Long-period ocean waves create a fluctuating pressure field on the sea floor, which causes the sea bed to be deformed. The relationship between pressure and deformation is called the compliance, and depends on the distribution of density and elastic properties in the underlying crust and lithosphere. The method has been used to investigate the structure under several MOR locations (Crawford *et al.*, 1991; Crawford *et al.*, 1999; Crawford and Webb, 2002; Crawford *et al.*, 2010).

## 2.10 Side-scan sonar

A different approach to imaging the sea floor is to use side-scan sonar. This was developed, mainly in Britain, from World War II anti-submarine sonars (Stride, 2010), and was first applied to geological mapping in the early 1950s (Chesterman *et al.*, 1958; Donovan

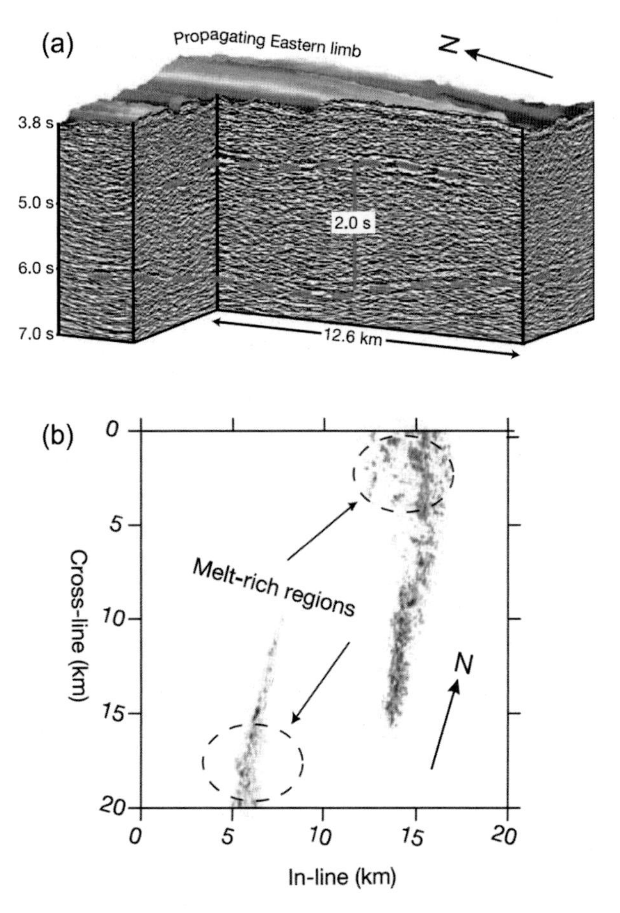


**Figure 2.18** Three-dimensional (3D) seismic image of the EPR overlapping spreading centre near the 9° N, after Singh *et al.* (2006b). (a) Section through the data volume. Coloured upper surface is sea floor topography. Dashed orange lines indicate reflections from axial magma chamber (upper line) and base of crust (lower line). (b) Amplitude map of magma chamber reflections, emphasising positions of melt-rich regions. For colour version, see plates section. Reprinted by permission from Macmillan Publishers Ltd: *Nature* © 2006.

and Stride, 1961). Unlike echosounding, the main sonar beam is directed sideways, not vertically, and the beam pattern is arranged to be narrow in the direction of travel but broad in the cross-track direction (Figure 2.19). When the transmitted sound pulse interacts with features on the sea floor, some is 'backscattered' towards the source (like illuminating a road with car headlights), and this scattered energy is received either by the transmitting transducer or by a separate receiver. The intensity of backscattered sound is recorded and plotted against travel time or range, building an image from a succession of closely-spaced pulses.

Early side-scan sonars had limited range. In the late 1960s, the British National Institute of Oceanography developed a long-range side-scan called GLORIA (Geological Long

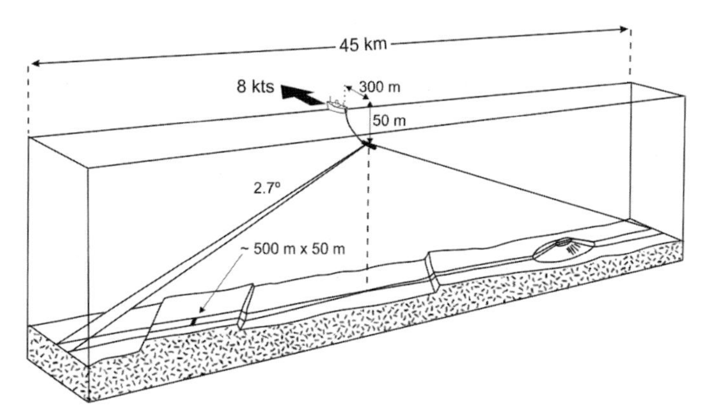


Figure 2.19 The principle of side-scan sonar, exemplified by the GLORIA system, after Searle *et al.* (1990) with kind permission from Springer Science and Business Media.

Range Inclined Asdic[2]; Rusby *et al.*, 1969). This used relatively low-frequency sound (6–7 kHz) to give ranges up to 30 km. GLORIA was used extensively for imaging many details of MORs (e.g. Laughton, 1981). An updated GLORIA II was launched in 1977. A similar system, developed somewhat later in the USA, was SeaMARC II, which operated at 10–12 kHz with a range of 5 km but the important addition of a swath bathymetry capability (Blackinton *et al.*, 1983).

Higher-frequency side-scans provide greater resolution; they have limited range, but can be towed near the sea floor to provide high-resolution images (Spiess and Mudie, 1970). An acoustic frequency of around 30 kHz has proved particularly effective for mapping MORs, and has been implemented in the US SeaMARC I system (Chayes, 1983) and the British TOBI (Towed Ocean Bottom Instrument; Flewellen *et al.*, 1993; Figure 2.27).

## 2.11 Electrical methods

Electrical techniques can elucidate the electrical resistivity (or conductivity) structure, and are important because conductivity is more sensitive to the presence of hydrothermal or magmatic fluids than are seismic techniques. To measure resistivity, electric current is injected into the sea bed using a powerful source deployed near the sea floor. Some minerals display an electric polarisation ('self-potential'), which can be measured with a voltage meter.

Francis (1985) made electrical resistivity and self-potential measurements from a submersible over a sulphide ore deposit on the EPR, and similar measurements have subsequently been made on the MAR with submersible Alvin (von Herzen *et al.*, 1996).

[2] GLORIA is a double acronym: Asdic comes from Anti-(or Allied) Submarine Detection Investigation Committee.

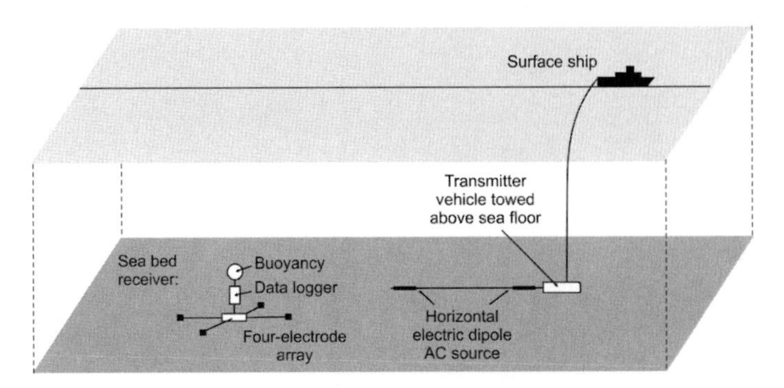


**Figure 2.20** Diagram of a controlled-source electromagnetic sounding system (Sinha *et al.*, 1990).

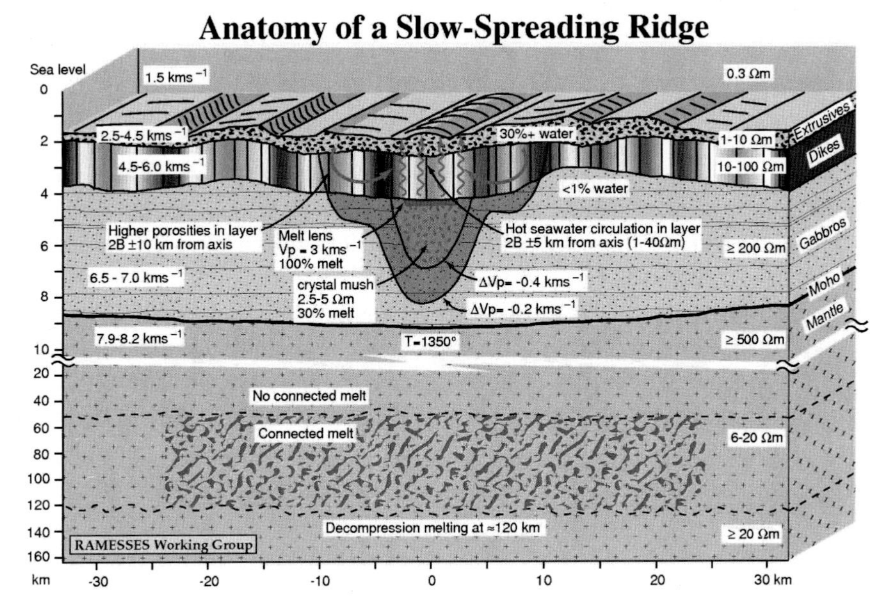


**Figure 2.21** Lithospheric structure and melt distribution beneath the Reykjanes Ridge from a combination of controlled-source electromagnetic sounding, seismic refraction and reflection and other geophysical data, after Sinha *et al.* (1998, Figure 5). Note change of depth scale at 10 km. For colour version, see plates section.

Self-potential has been used extensively to survey for metallic sulphide deposits (e.g., Beltenev *et al.*, 2007).

Electromagnetic sounding makes use of electromagnetic induction, using towed, controllable current sources with sea floor magnetometers and potential meters (Spiess *et al.*, 1980; Edwards *et al.*, 1981; Chave and Cox, 1982; Figure 2.20). Early studies of MORs were carried out by Spiess *et al.* (1980) and Evans *et al.* (1991). Unsworth (1994) reviewed early applications of the method to MOR studies. A later example is its combination with other geophysical techniques to determine the lithospheric structure and melt distribution beneath the axis of the Reykjanes Ridge (Figure 2.21).

Figure 2.22 The WHOI Acoustically-Navigated Geological Undersea Surveyor ANGUS. Photo © Woods Hole Oceanographic Institution.

## 2.12 Visual imaging

A fundamental method of investigation is visual imaging. Sea floor photography began in the USA in 1939 (Ewing *et al.*, 1967), and soon underwater cameras were developed elsewhere. Early cameras often had a device to trigger them at a set distance from the sea floor. Later, pingers and then more sophisticated echosounders were used to determine height. The camera, lights and any associated electronics are mounted on a rugged frame towed slowly over the sea floor. Camera developments are reviewed by Huggett (1990).

The main difficulty is that light is easily absorbed in seawater and tends to be strongly backscattered by it. Ninetyeight percent of red light is absorbed within 1 m, and 98% of blue and green light within 15 m (Huggett, 1990). In practice, visual observation requires placing the camera (or human observer) within a few tens of metres of the target. To reduce backscattering, the light source and camera can be separated horizontally, but a practical limit is about 2 m (Huggett, 1990). The US Naval Research Laboratory developed the LIBEC (LIght BEhind Camera) system, in which powerful down-looking lights are positioned ~10 m above the camera (reducing backscatter), which can then be up to ~25 m above the sea floor affording a wide field of view (Patterson, 1972). An influential application was a large areal coverage of the FAMOUS study of the MAR (Ballard and Moore, 1977).

The towed-camera system ANGUS (Acoustically-Navigated Geological Undersea Surveyor) was developed specifically for MOR studies (Phillips *et al.*, 1979; Figure 2.22). Acoustic navigation (Section 2.15.4) allowed images to be located with a precision of 5–10 m. ANGUS and similar systems have been widely used on MORs.

In addition to towed camera systems, still- and video-cameras, including high-definition cameras, are routinely deployed on manned submersibles (Kohnen, 2009), giving more control over positioning and selection of images (Figure 2.23). More recently, they have been attached to Remotely Operated Vehicles (Ballard, 1993; German *et al.*, 2003; Figure 2.24; Section 2.14.4).

**Figure 2.23** The French manned submersible *Nautile*. The manned sphere with three forward-facing portholes is seen centre, with cameras and lights above and to the sides. Two robotic sampling arms are folded beneath the sphere, and retracted below them is the sample basket.

**Figure 2.24** The British National Oceanography Centre's ROV *Isis*, showing cameras, lights and sampling arms.

Figure 2.25 A modern dredge.

## 2.13 Sampling

One of the most basic methods of studying the sea floor is to collect samples for further investigation. These can then be subjected to the full array of physical, chemical, petrological and other analyses available.

### 2.13.1 Dredging

Dredging – the dragging of a steel frame with a sampling bag over the sea floor – was used at least as early as the *Challenger* expedition (Section 1.2), and is still an important technique for recovering large amounts of rock from steep or rugged outcrops (Figure 2.25). It has the advantage of being simple and robust, and is usable in rough weather when other operations are impossible. The disadvantages are imprecision of sample location (a single dredge haul can cover several hundred metres or more, and may contain rocks that have rolled from further afield), lack of sample orientation, and restriction to loose, surface rocks.

### 2.13.2 Wire-line drills

The British Geological Survey developed a number of wire-line drills, which are lowered to the sea bed on a cable that also carries a power supply and returns digital data. One was used by MacLeod *et al.* (1998) to obtain hard-rock cores up to 3.8 m long, but has a depth limitation of 2000 m. A separate development produced a rotary corer designed specifically for MOR work, usable to 4000 m, and fitted with an orienting device (MacLeod *et al.*, 1998; Allerton and MacLeod, 1999). This obtains oriented cores up to 1 m long, which have been used for palaeomagnetic studies (MacLeod *et al.*, 2002). Another is the MeBo drill developed by the University of Bremen (Freudenthal and Wefer, 2007).

### 2.13.3 Other devices

Attempts have been made to develop rotary rock corers for deployment from submersibles and ROVs (Stakes *et al.*, 1997), but so far these have not been widely used.

Manned submersibles and ROVs use 'robot' sampling arms, capable of breaking off small rocks and operating small sediment corers (Figures 2.23 and 2.24). Samples can be oriented using the 'Geocompass' devised by J. A. Karson and R. Catanach (Cogné *et al.*, 1995).

A novel sampling device is a wax-tipped corer (Reynolds *et al.*, 1992). This can chip fragments off the surfaces of glassy lava flows; the chips then become embedded in the wax and are thus recovered.

### 2.13.4 Ship-based ocean drilling

A major sampling facility is provided by the deep-drilling capability of the international Integrated Ocean Drilling Program (www.iodp.org). Early deep-ocean scientific drilling was undertaken by Project Mohole (Horton, 1961) with the aim of drilling to the top of the mantle. Although this was not achieved, the project proved the feasibility of such drilling, and laid the grounds for the formation of the Deep Sea Drilling Project (1968–1983), which subsequently developed into the Ocean Drilling Program (1985–2004) and then IODP. IODP and its predecessors use technology largely derived from the hydrocarbon industry to drill hundreds to thousands of metres into the sea bed, recover samples of sediment, hard rock and fluids, and emplace instruments (Coffin *et al.*, 2001). The programme operates two permanent drilling ships: *JOIDES Resolution* using mainly rotary drilling, and *Chikyu* with a riser capability allowing deeper, more stable holes to be achieved. Another arm of the programme uses 'Mission Specific Platforms' that are typically chartered on an ad-hoc basis appropriate for the particular target area: for example, icebreakers for arctic work and jack-up drilling rigs for continental shelves.

## 2.14 Ships and other platforms

### 2.14.1 Ships

In marine science, like space science, instrument platforms are as important as the instruments themselves. From the earliest days, good ships with highly skilled crews have been essential for successful deployment of instruments.

Figure 2.26 The British research ship RRS *James Cook*, courtesy NERC Marine Facilities – Sea Systems, National Oceanography Centre, UK.

Many early oceanographic ships, like HMS *Challenger*, were adapted from their original purposes. Naval ships made important contributions. Prince Albert I of Monaco used his own ship, the *Princesse Alice* for research voyages (Carpine-Lancre, 2001). Lamont Geological Observatory's schooner R/V[3] *Vema* was converted from a pleasure yacht in 1953. Following the end of World War II, several US naval vessels were converted to civilian research ships. One of the first dedicated research ships was the UK's *Discovery II*, built in 1929 for whaling research. Woods Hole Oceanographic Institution's ketch R/V *Atlantis* was launched in 1930 and operated until 1966 (Woods Hole Oceanographic Institution, 2011).

Modern research ships (Figure 2.26) are designed for the specialist requirements of marine research. Most have diesel–electric propulsion for flexibility and control. The larger ones have 'dynamic positioning', which uses several computer-controlled thrusters to keep the ship in a fixed position or on a given track to a precision of a few metres, guided by GPS (Section 2.15.3). There are general-purpose and specialist laboratories, and usually deck space for containerised laboratories or instrument control rooms. Specialist winches (including some with electrically conducting cables) are provided for sampling or towing various instruments. There is usually a large working deck, with cranes and specialised gantries for deploying or towing equipment. The ship may provide services such as compressed air for seismic air guns.

### 2.14.2 Manned submersibles

China, France, Japan, Russia and the USA have all developed deep-ocean manned submersibles. One of the first to work on ridges was the French Navy's bathyscaphe, *Archimède* (Ballard *et al.*, 1975). However, early bathyscaphes had large buoyancy tanks and were bulky and unwieldy; current submersibles (Figure 2.23) are more manoeuvrable.

[3] Research Vessel.

The French *Cyana*, with a maximum depth rating of 3000 m, operated from 1969–2003, followed by *Nautile* with a 6000 m rating from 1984. The US Navy developed a series of submersibles called Deep Submergence Vehicles, including DSV-3 *Turtle* (Heezen and Rawson, 1977) and DSV-4 *Sea Cliff* (Rona *et al.*, 1997). Their initial depth ratings were similar to *Cyana*'s, but gradually increased. The vehicle most used in MOR research is DSV-2 *Alvin*, operated by WHOI, which was first used on an MOR during the FAMOUS study (Ballard *et al.*, 1975). Its current depth rating is 4500 m, planned to increase soon to 6500 m. The present generation of research submersibles also includes the Russian *Mir*, Japanese *Shinkai 6500*, and Chinese *Jiaolong*, all with depth capabilities of 6000–7000 m.

Research submersibles usually carry three persons, including either one or two scientific observers, in a 'shirt-sleeves' environment within an ~2 m diameter personnel sphere. Portholes are provided for direct observation, and typical instrumentation includes depth, altitude and speed measurements, still- and video-cameras, mechanical manipulators for collecting specimens and handling instruments and tools, and a 'sample basket'. Navigation is by a combination of acoustic navigation and dead-reckoning (Section 2.15). They may carry additional instruments including magnetometers, gravity meters, heat-flow probes, external thermometers, and devices for sampling soft sediment, water and biological specimens. A specialist 'mother ship' is required to transport, deploy and maintain the submersible. Each dive normally provides a few hours operation on the sea floor, enabling traverses of up to a few kilometres. Ground speed is limited to about 1 knot (0.5 m $s^{-1}$).

### 2.14.3 Towed platforms

To match the high-resolution observations achievable with manned submersibles, but without the heavy overhead of crew protection and life support, 'deep-towed' platforms were devised to carry instruments near the sea floor, towed by cable from a surface ship. The first was the Scripps Institution of Oceanography's Deep Tow, developed out of the search for the sunken US nuclear submarine *Thresher* (Spiess and Maxwell, 1964). Deep Tow had a high-resolution, 150-m-range side-scan sonar, near-bottom echosounder, proton magnetometer, cameras and other instrumentation, and revealed important details of MOR structure (e.g., Larson and Spiess, 1969). More recent developments include SeaMARC I and TOBI (Section 2.10; Figure 2.27) and a range of towed platforms developed by the Monterey Bay Aquarium and Research Institution. WHOI's ANGUS (Section 2.12) is another example.

### 2.14.4 Remotely Operated Vehicles and Autonomous Underwater Vehicles

Remotely Operated Vehicles such as *Jason* and *Isis* (Figure 2.24) were developed from the simpler deep-towed instruments that preceded them. They offer many of the benefits of manned submersibles but are smaller, lighter, cheaper to operate, and have up to several days' endurance. They are attached to a ship by an electrically conducting cable supplying power and operational commands and returning data, but have a degree of autonomous movement provided by on-board thrusters. Traverses of ~10 km are possible in one dive,

Figure 2.27 The British National Oceanography Centre's Towed Ocean Bottom Instrument (TOBI). The two long, horizontal tubes house the acoustic transducers; other instrumentation is within the frame.

with ground speeds up to about 0.5 m $s^{-1}$. Video-cameras provide both scientific data and guidance for the pilot. Instrumentation is similar to that carried on manned submersibles. Areal surveys can be made using high-resolution instruments such as magnetometers and multi-beam echosounders, yielding topographic resolution better than 1 m (e.g., Soule *et al.*, 2005).

Autonomous Underwater Vehicles need no physical connection to a ship (Figures 2.28 and 2.29). They travel and manoeuvre under their own power and guidance, over periods of days and distances of tens to hundreds of kilometres. They carry instruments such as high-resolution echosounders and magnetometers, and water-chemistry sensors such as CTDs and $E_h$ meters, but generally not rock sampling equipment. They can be programmed to fly various profiles, such as constant height or constant depth surveys. The first AUV in use for MOR research was the Automated Benthic Explorer (ABE; Yoerger *et al.*, 1998; Figure 2.28a). Other ROVs include WHOI's Sentry (White *et al.*, 2010 Figure 2.28b), the Canadian ROPOS (Embley *et al.*, 1995), British Autosub (Millard *et al.*, 1998; Figure 2.29), and a range of vehicles at MBARI (Paduan *et al.*, 2009).

## 2.15 Navigation

Mounting instruments on a moving platform inevitably brings problems of determining the precise location of the measurements, so the development of suitable navigation systems has been vital for the development of MOR science.

### 2.15.1 Celestial navigation

Until the middle of the twentieth century, almost all deep-water marine positioning employed celestial navigation. It involved measuring the positions (usually the altitude) of stars relative to the observer, yielding a location precision of about 2–3 km. Between

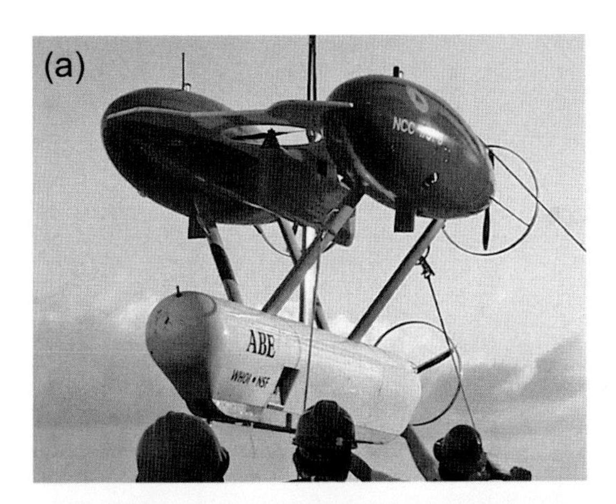

**Figure 2.28** (a) Wood's Hole Oceanographic Institution's Autonomous Benthic Explorer. Photo by Daniel Fornari © Woods Hole Oceanographic Institution; (b) ABE's replacement Sentry. Photo by Christophe Reddy © Woods Hole Oceanographic Institution.

**Figure 2.29** The British National Oceanography Centre's AUV Autosub in its launching gantry.

star sights one had to rely on dead-reckoning, and if no sight was achieved for several days (because of bad weather), positions could be in error by 10 km or more.

### 2.15.2 Electronic radio navigation

During World War II, a number of electronic positioning systems were developed, based on a comparison of the travel times between two or more radio pulses. From these developed LORAN (Powell, 1971), which was available until 2010. The last version, LORAN-C, employed a worldwide system of transmitters with ranges ~2000 km, providing almost global coverage and ~500 m accuracy. Russia operated a similar system called CHAYKA.

### 2.15.3 Satellite navigation

The first satellite navigation system, Transit, was introduced in 1964 by the US Navy and operated until 1991 (Danchik, 1984). Positions were determined by measuring the Doppler shift of radio signals transmitted from satellites whose orbits were precisely known. Accuracy was ~2 km, but fixes could only be obtained when suitable satellites were in view, typically every 1–2 h. This system revolutionised marine science, allowing measurements of all kinds to be plotted with ~2 km accuracy, and detailed surveys involving closely spaced ship tracks became possible.

Transit was followed by the US Air Force's Global Positioning System (GPS), which was released for civilian use in 1983 (Duven and Artis, 1985). GPS uses the same principle as LORAN, but with transmitters mounted on satellites in well-determined orbits. Initial civilian access had accuracy ~100 m, but currently GPS can provide continuous navigation with an accuracy ~1 m.

### 2.15.4 Acoustic navigation

Since electromagnetic radiation is rapidly absorbed in seawater, most underwater navigation relies on sound, although recently, high-precision inertial navigation systems have been introduced into AUVs.

Early aids to underwater navigation were 'pingers', whose height above the sea floor is given by the time difference between a direct and bottom-reflected sound pulse. More sophisticated is an acoustic 'transponder' which, on receiving a pulse, emits another, possibly at a different frequency or coded in some way. By measuring the intervals between these pulses, distances between pingers and transducers can be determined.

Long Base Line (LBL) acoustic navigation relies on an array of transponders set on the sea floor, between which a towed instrument fitted with a pinger or transponder is positioned. An initial survey determines the dimensions and position of the array relative to the surveying ship – and hence, via GPS, to the Earth. Positional accuracy is a fraction of the acoustic wavelength: $<1$ m at typical frequencies of ~10 kHz. Refraction in deep water causes sound to follow circular paths, concave upwards, so that more widely spaced transponders must be higher above the sea floor to avoid shadowing. This sets a practical limit of a few kilometres to the distance between transponders, and restricts LBL navigation to quite small surveys a few kilometres across.

An alternative is Short Base Line (SBL) navigation, in which the transponders are mounted on the survey ship. This avoids the need to deploy and survey a transponder net, and has no restriction on the limits of the survey area. It works well for vehicles close to a ship, but currently range is limited by ship noise and it cannot be used with instruments that may be several kilometres away.

Where LBL and SBL are impractical, towed vehicles are navigated by estimating their position relative to the towing ship. One way is to assume that the vehicle precisely follows the towing ship, with its position estimated from a combination of the length of the towing cable and the vehicle depth. A better estimate, with an accuracy of a few hundred metres, can be obtained by modelling the dynamics of the towing cable (Triantafyllou and Hoover, 1990; Hussenoeder *et al.*, 1995; Escartin *et al.*, 1999). Further precision can be obtained by matching features seen on deep-towed sonar images with the same features seen on ship-based bathymetry located with GPS, yielding absolute accuracies of ~100 m (Searle *et al.*, 2010).

## 2.16 Summary

Advances in the study of MORs have relied heavily on advances in technology. Many of the techniques in use today grew out of or were advanced by military developments during the two world wars.

Depth measurement progressed from the use of weighted sounding lines in the nineteenth century through electronic single-beam echosounding to the multi-beam echosounders in use today.

Measurements of the magnetic field are made easily and quasi-continuously using towed sensors such as fluxgate and proton precession magnetometers. Various digital techniques are available to reduce, interpret and model observations. Interpreting magnetic anomalies has been fundamental to developing sea floor spreading and plate kinematics theories. Palaeomagnetic methods can determine rotations of large crustal blocks.

Gravity is continuously measured using ship-mounted gravity meters that automatically correct for ship accelerations. It allows assessment of crustal structure; in particular, the use of the mantle Bouguer anomaly leads to a rapid assessment of crustal thickness variations. Admittance techniques can assess the isostatic state of the lithosphere.

Forward and inverse modelling are used to interpret both magnetic and gravity anomalies, with methods based on the fast Fourier transform being particularly important.

Conductive heat flow through the oceanic crust is determined by measuring the temperature gradient and thermal conductivity in soft sediments with an instrumented probe. Heat advected by circulating fluids can be measured by using devices that trap or funnel the fluids.

Earthquakes are located by using both global seismograph networks and local, temporary arrays of sea floor seismometers. They reveal the patterns of plate boundaries and other active faults and the thickness of the oceanic lithosphere. Modelling of ground motion can yield earthquake mechanisms, giving important confirmation to plate kinematic predictions.

Seismic refraction is a powerful method for determining crustal structure. Artificial sound sources are used to generate seismic waves that pass through the sea bed and are then detected in either towed arrays, floating sonobuoys or ocean bottom seismometers. Both 2D and 3D models can be derived showing the distribution of seismic velocities, and hence crustal structure. Further information on the properties of the lithosphere is obtained by measuring the compliance of the lithosphere to superimposed pressure loading from ocean waves.

The seismic reflection method uses low-frequency energy to penetrate the sea floor and image sub-bottom reflectors. Air guns are a common sound source, and buoyant streamers provide multi-channel receivers. Both 2D sections and 3D volumes can be imaged.

Side-scan sonar measures acoustic backscatter to produce images of sea floor texture. It can be ship-mounted or towed near the sea surface or sea floor. Higher frequencies give improved resolution but shorter range, and vice versa.

Electrical methods, particularly the controlled source electromagnetic method, are valuable for imaging conductive fluids, such as basaltic melt; DC resistivity and self-potential methods are also used.

Visual imaging comprises naked-eye observations from manned submersibles and photographic imaging from various platforms. Light attenuation and backscattering limit ranges to a few tens of metres.

Various sampling techniques are available, from dredging to highly sophisticated ship-based drilling. Wire-line drills, set on the sea bed but powered and controlled from ships, are beginning to be used. Robot manipulators are commonly used on manned submersibles and ROVs. Some devices allow orientation of samples for palaeomagnetic and other work.

Instrument platforms are a vital part of the available technology. They include specialist surface ships, towed platforms, semi-autonomous ROVs, independent manned submersibles and fully robotic AUVs.

Each platform needs a navigation system. Modern surface ships use GPS with an accuracy of $\sim$1 m. Other space- and land-based electronic systems have been used in the past. Underwater navigation requires acoustic techniques involving LBL systems with sea floor transponder arrays a few kilometres across or SBL systems utilising ship-mounted transponder arrays.

# 3 The oceanic lithosphere

## 3.1 Crust, mantle, lithosphere and asthenosphere

Traditionally, Earth scientists have divided the outer parts of the Earth into the crust and mantle. Broadly speaking these reflect a major change in chemical composition. Mantle rocks comprise a high proportion of minerals containing the elements iron and manganese, and have relatively high density ($\geq$3300 kg m$^{-3}$) and seismic P-wave velocity ($\geq$8 km s$^{-1}$); on the other hand crustal rocks, which are ultimately derived from melting (and associated fractionation) of the mantle, have lower proportions of iron and manganese, higher proportions of aluminium and silicon, and consequently lower densities and velocities. The oceanic crust is generally thinner ($\leq$8 km) and slightly denser ($\sim$2800 kg m$^{-3}$) than continental crust.

An alternative way of subdividing the Earth is based not on its composition but its mechanical properties. In this classification the outermost, relatively cool, layer is one where rocks behave in a strong, brittle or elastic manner, and is termed the lithosphere from the Greek *lithos*, 'rocky' (Dietz, 1961). Underlying this is a warmer, weak layer which, on geological timescales ($\sim$1 Ma), behaves as a plastic medium, and is termed the asthenosphere (from the Greek *asthenēs*, 'weak'). The viscosity of the asthenosphere is of the order of $10^{20}$ Pa s (Lowrie, 1997).

The crust–mantle boundary, being essentially compositional, does not coincide with the lithosphere–asthenosphere boundary, which is mechanical and usually deeper, except perhaps very near the ridge axis. The crust–mantle boundary is often referred to as the Mohorovičić discontinuity or simply 'Moho'. This is defined seismically as the depth at which the P-wave velocity first exceeds about 8 km s$^{-1}$, although a separate, 'petrological Moho' based on the inferred origin of the rocks is often also used (see Section 5.3.4).

Plate tectonics considers the lithosphere to consist of a small number of rigid, undeformable tectonic plates that slide over the surface of the Earth in relative motion to one another. Plastic mantle rises under MORs, turns over to diverge horizontally, and cools by conduction and hydrothermal advection, forming brittle tectonic plates. (In this process, some of the mantle may melt, rise, and solidify to form the crust – Chapter 6). MORs are one of the major types of lithospheric plate boundary, and many of their gross features are determined by the nature and evolution of the lithosphere.

Tectonic plates generally contain both the crust and the uppermost (lithospheric) mantle. Moreover they can, and usually do, comprise both continental and oceanic parts. For example, the North American plate includes both the continent of North America *and* the western part of the North Atlantic Ocean eastwards to the plate boundary at the MAR, including their crusts and underlying lithospheric mantle. The whole plate (crust and mantle, continental and oceanic) moves rigidly as one unit.

The base of the lithosphere is the depth at which mechanical behaviour changes from rigid to plastic. It depends mainly on the temperature, rock type and strain rate (how rapidly rocks are deformed), and has been thought to be gradational. It appears to roughly follow an isotherm; a common approximation in the sub-oceanic mantle is to take the brittle–plastic transition (and thus the base of the lithosphere) as the depth of the 750 °C isotherm (e.g., Searle and Escartín, 2004). However, recent work has suggested that in places the boundary may be quite sharp, indicating that it may be related to melting or hydration of the mantle, and may approximate an isotherm nearer 930 °C (Rychert and Shearer, 2011). The thickness of the oceanic lithosphere ranges from a few kilometres to over a hundred kilometres, depending on its age or distance from the ridge axis.

## 3.2 Oceanic heat flow and the thermal structure of the lithosphere

An early success of plate tectonics theory was to explain both the observed variations in conductive heat flow through the ocean floor and the characteristic shape of the MORs and flanking basins in terms of the conductive cooling and consequent thermal contraction of lithospheric plates.

### 3.2.1 Heat flow

Values of the conductive heat flow, derived from measurements of temperature gradient and thermal conductivity in sea floor sediments, range from several hundred mW $m^{-2}$ near the ridge axes to values of around 50 mW $m^{-2}$ in lithosphere ~100 Ma old (Bullard *et al.*, 1956; Lister, 1970; Stein and Stein, 1992; Figure 3.1a).

When new lithosphere is created at spreading plate boundaries, hot material at upper mantle temperatures (around 1300–1400 °C) is added to the trailing plate margin. As the plate drifts away from the boundary it ages and cools. Consequently its average density increases, and it gradually sinks into the asthenosphere to maintain isostatic equilibrium.

The simplest model of the cooling lithosphere can be obtained by assuming that the plate is an infinite half space, with one edge (the ridge axis) maintained at the deep mantle temperature $T_M$ by dyke emplacement, and with steady-state heat transport solely by conduction in the vertical direction. It can then be shown that the depth $z$ to a given isotherm $T$ is

$$z = 2\sqrt{\frac{kt}{\rho c_p}}\mathrm{erfc}^{-1}\left(\frac{T - T_M}{T_S - T_M}\right) \tag{3.1}$$

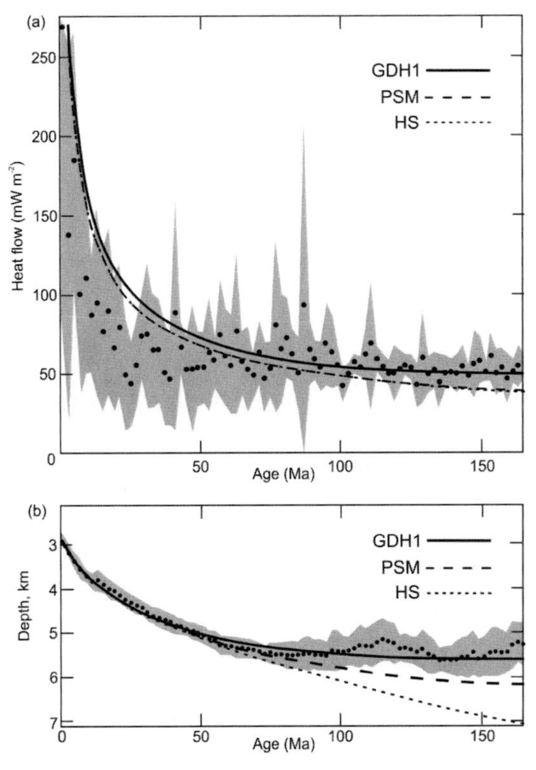


**Figure 3.1** Plots of (a) measured conductive heat flow and (b) ocean floor depth against lithospheric age, after Stein and Stein (1992). Dots and grey envelopes are averages and standard deviation ranges in 2 Ma bins. Curves give predictions from the half-space model (HS), the plate model (PSM, after Parsons and Sclater, 1977), and model GDH1. All models over-predict observed heat flow for lithosphere $< 50$ Ma, because these data do not include hydrothermal advection (Lister, 1972). Reprinted by permission from Macmillan Publishers Ltd: *Nature* © 1992.

(Parker and Oldenburg, 1973; Davis and Lister, 1974; Turcotte and Schubert, 1982), where erfc is the complimentary error function,[1] $k$ is the coefficient of thermal conductivity, $t$ is time since formation of the lithosphere, $\rho$ is lithospheric density, $c_p$ its specific heat at constant pressure, and $T_S$ the temperature at the surface of the plate (i.e. the ocean floor). Isotherms calculated from this model are shown in Figure 3.2a.

Rearranging Equation (3.1) and differentiating yields the vertical conductive heat flow

$$Q(t) = -k\frac{dT}{dz} = -T_M\sqrt{\frac{k\rho c_p}{\pi t}}. \tag{3.2}$$

Thus in this simple approximation the heat flow should be inversely proportional to the square root of lithospheric age. This model works reasonably well for sea floor ages from about 50 Ma to 100 Ma, but it seriously over-predicts observed heat flow at younger ages, and slightly under-predicts it at older ones (Stein and Stein, 1992; Figure 3.1a). The

[1] $\operatorname{erfc}(x) = \frac{2}{\sqrt{\pi}} \int_x^\infty e^{-\xi^2} d\xi$

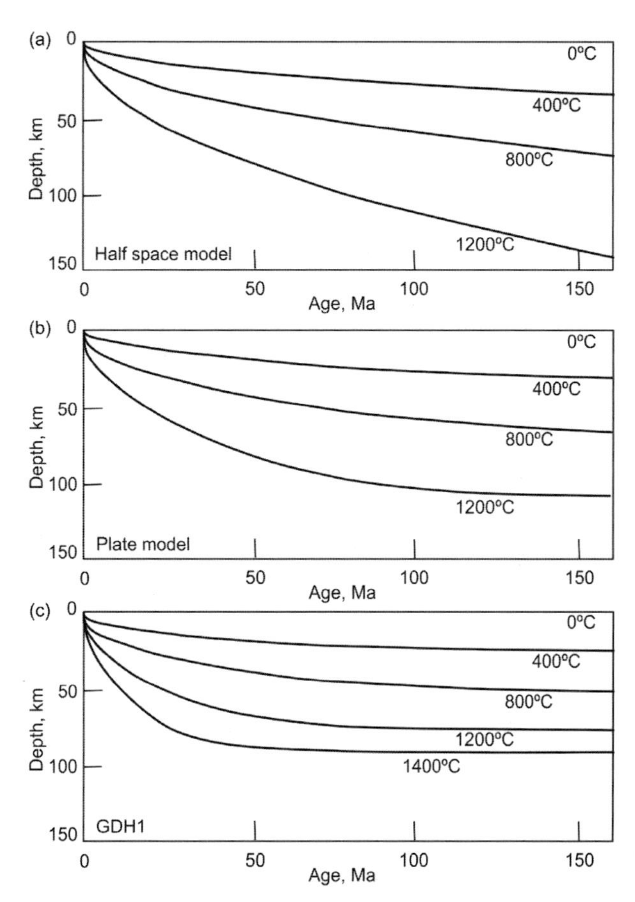


**Figure 3.2** Calculated lithospheric isotherms, from Fowler (2005) after Stein and Stein (1996). (a) Half-space cooling model (McKenzie, 1967); (b) simple plate model with no latent heat (Stein and Stein, 1992); (c) modified plate model GDH1 (Stein and Stein, 1992).

discrepancy at younger ages reflects the fact that the traditional method of measuring heat flow through the ocean floor measures only conductive heat flow, whereas a significant amount of cooling in younger lithosphere is accomplished through hydrothermal advection (Lister, 1972). This will be discussed in more detail in Chapter 8.

### 3.2.2 Depth of the ridge flanks

The flanks of MORs have a characteristic shape, becoming deeper but with a shallower slope as distance from the ridge axis increases (Figure 3.3). By rearranging Equation (3.1) to determine the average lithospheric temperature, and multiplying by the coefficient of thermal expansion $\alpha$ to yield its volume change, the average lithospheric density can be determined. Balancing columns of lithosphere with different ages (and therefore different

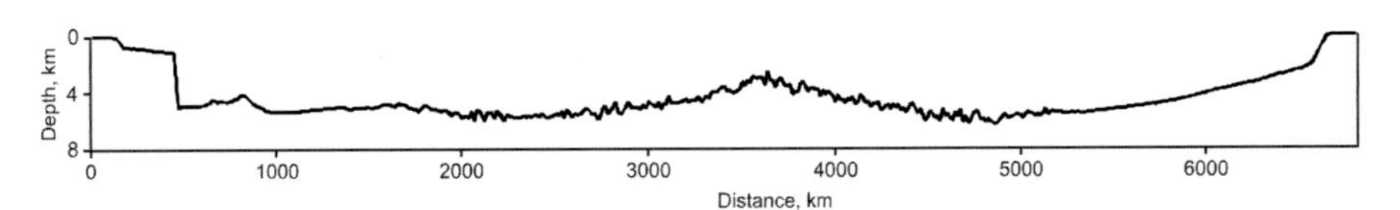


Figure 3.3 Cross section of the Atlantic from eastern USA to west Africa, crossing the MAR at 26 °41′ N. Data from GeoMapApp (http://www.geomapapp.org) (Ryan *et al.*, 2009).

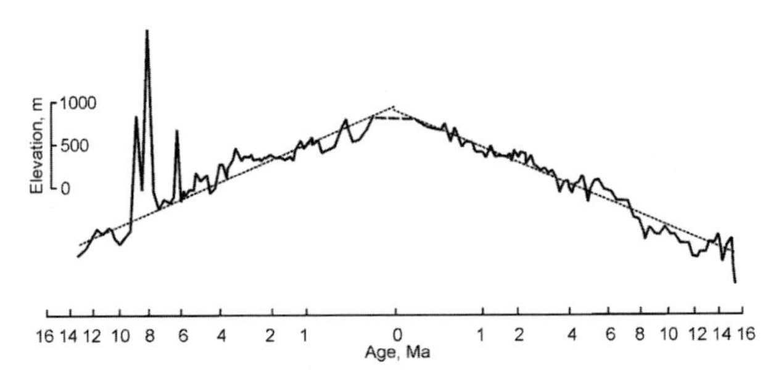


Figure 3.4 Depth of the ocean floor plotted against square root of age, reprinted from Tréhu (1975), © 1975, with permission from Elsevier. Profile crosses the southern EPR near 48° S. Dashed lines show least squares best fit to the data, ignoring major seamounts.

densities) and their overlying water columns to obtain isostatic equilibrium leads to an expression for the variation of water depth, $w$, with age:

$$w = \frac{2\rho_M}{\rho_M - \rho_W}(T_M - T_S)\left(\frac{kt}{\rho c_p \pi}\right)^{1/2}, \tag{3.3}$$

where $\rho_M$ and $\rho_W$ are the densities of mantle and water, respectively (Davis and Lister, 1974; Turcotte and Schubert, 1982).

Figure 3.1b shows sea floor depths plotted against lithospheric age, together with the predicted depth from Equation (3.3) as a short-dashed line. The half-space model fits well for ages less than ~50 Ma, but over-predicts the depth beyond that. (It might be thought from the discussion of heat flow that the half-space model would not give a good approximation to depth at young ages; however, it is the temperature distribution in the plate, rather than the details of how the heat escapes at the sea floor, which controls the plate thickness and depth, and this *is* well approximated by the half-space model at young ages).

For lithosphere <50 Ma, Equation (3.3) suggests that the depth of the sea floor should be proportional to the square root of age, and this is indeed the case (Figure 3.4). This was an important early test of sea floor spreading (Sclater *et al.*, 1971; Tréhu, 1975). Measured age–depth data from the Atlantic, Indian and Pacific oceans fitted slopes of 324–348 m $Ma^{-1/2}$ (Tréhu, 1975).

### 3.2.3 The plate model

Over older lithosphere, more heat flow is observed than is predicted by the half-space model, because at large ages the surface heat flow begins to 'sense' the finite thickness

**Table 3.1** Parameters of the GDH1 global depth and heat flow model of Stein and Stein (1992). Uncertainties where given are ± 1 standard deviation

| Parameter | Symbol | Value |
|---|---|---|
| Plate thickness | $L$ | 95 ± 15 km |
| Basal temperature | $T_M$ | 1450 ± 250 °C |
| Coefficient of thermal expansion | $\alpha$ | $(3.1 \pm 0.8) \times 10^{-5}$ |
| Specific heat | $c_p$ | 1.171 kJ kg$^{-1}$ K$^{-1}$ |
| Coefficient of thermal conductivity | $k$ | 3.138 W m$^{-1}$ K$^{-1}$ |
| Density of plate | $\rho_M$ | 3330 kg m$^{-3}$ |
| Density of water | $\rho_W$ | 1000 kg m$^{-3}$ |
| Depth of ridge axis | $z_0$ | 2600 m |

of the plate (i.e. the mantle temperature $T_M$ is reached at a finite, not infinite, depth). A better boundary condition is to assume a constant asthenospheric temperature $T_A$ at the base of a finite thickness plate (McKenzie, 1967; Parsons and Sclater, 1977). The isothermal boundary layer below the plate is thought to be maintained by small-scale convection in the asthenosphere. In this 'plate model' (again ignoring horizontal heat conduction, which is small compared to the vertical component), the temperature variation is of the form

$$T(t, z) = \frac{T_A}{H} z - \sum_{n=1}^{\infty} \frac{2T_A}{n\pi} \cdot e^{-\frac{t}{\tau_n}} \cdot \sin\left(\frac{n\pi z}{H}\right), \tag{3.4}$$

where $\tau_n = \rho_M c_p H^2/(\pi^2 n^2 k)$ is a thermal time constant, and other symbols are as defined above (McKenzie, 1967). Figure 3.2b shows lithospheric isotherms for the simple plate model, and its heat flow and depth predictions are given in Figure 3.1 by the long-dashed lines. Note how the deeper isotherms flatten with age. This model gives a significant improvement over the half-space model, but still under-predicts heat flow and over-predicts depth for older lithosphere.

### 3.2.4 Combined depth and heat flow

A thorough study of global data by Stein and Stein (1992) found that the combined depth and heat flow observations could be explained well by the plate model using a set of modified parameters (Table 3.1); they called this model GDH1 (for Global Depth and Heat flow). Isotherms from this model are shown in Figure 3.2c.

GDH1 predicts oceanic heat flow (in mW m$^{-2}$) against age (in Ma) as:

$$Q(t) = 510\, t^{-1/2} \text{ for } t \leq 55 \text{ Ma, and} \tag{3.5}$$

$$Q(t) = 48 + 96 \exp(-0.0278t) \text{ for } t > 55 \text{ Ma} \tag{3.6}$$

(Stein and Stein, 1992). These equations give an accurate approximation of heat flow around the world for $t > 50$ Ma (Figure 3.1, solid lines). Even the heat flow measured over hotspot swells (such as Hawaii) was found to be only slightly higher than the global average (von Herzen *et al.*, 1989; Stein and Stein, 1992).

For ocean depths, Stein and Stein (1992) suggest that a best fit overall reference line is

$$w = (2600 \pm 20) + (345 \pm 3)\, t^{1/2}, \tag{3.7}$$

while the preferred model GDH1 gives the following good approximations:

$$w = 2600 + 365\, t^{1/2}; \text{ for } t < 20\,\text{Ma, and} \tag{3.8}$$

$$w = 5651 - 2473 \exp(-0.0278\, t) \text{ for } t \geq 20\,\text{Ma}. \tag{3.9}$$

The age–depth predictions of GDH1 are also plotted in Figure 3.1, and fit the data well on a global scale.

Unlike heat flow, some regional ocean depths do differ significantly from the predictions of GDH1. Most notably, perhaps, the west flank of the EPR between 9° S and 22° S is subsiding at a rate of only 200–225 m $\text{Ma}^{-1/2}$ compared to the eastern flank, which is subsiding at a near-average rate of 350–400 m $\text{Ma}^{-1/2}$ (Cochran, 1986). Various explanations have been put forward, including one by Toomey *et al.* (2002), who suggest that this and similar asymmetries can be explained by asymmetric 'absolute' plate motion coupled with asthenospheric hotspot return flow under the west flank toward the migrating ridge axis.

### 3.2.5 Latent heat

Another important contribution to the thermal state of the lithosphere is the latent heat of crystallisation of magma at the ridge axis. For example, at a fast-spreading ridge, crystallising gabbro can supply 6 MW of latent heat per kilometre of ridge axis, compared with 19 MW $\text{km}^{-1}$ of specific heat (Cannat *et al.*, 2004). Parker and Oldenburg (1973) included latent heat of crystallisation in a simple way by assuming that melt solidifies all along the vertical boundary of the model at the ridge axis. This has a major influence on the distribution of temperature in the young lithosphere, although variations in the predicted heat flow are hidden by the larger effect of hydrothermal cooling. Nevertheless, the amount and distribution of this latent heat release is important, and depends critically on the detailed position, shape and size of MOR melt pockets and magma chambers. These are discussed in detail in Section 5.4.

## 3.3 Thickness of the oceanic lithosphere

At low temperatures and high strain rates, rocks below their yield strengths deform elastically; at higher temperatures (or lower strain rates), they undergo solid state creep, which allows them to deform plastically (e.g., Lowrie, 1997, pp. 312–319). If $T$ in Equation (3.1) is set to the temperature of the brittle–plastic transition (say 750 °C), then $z$ will be the thickness of the lithosphere. Since $\kappa$ is almost constant for most Earth materials, the lithospheric thickness should, like ocean heat flow and depth, be proportional to the square root of age:

$$z_L \propto \sqrt{t}. \tag{3.10}$$

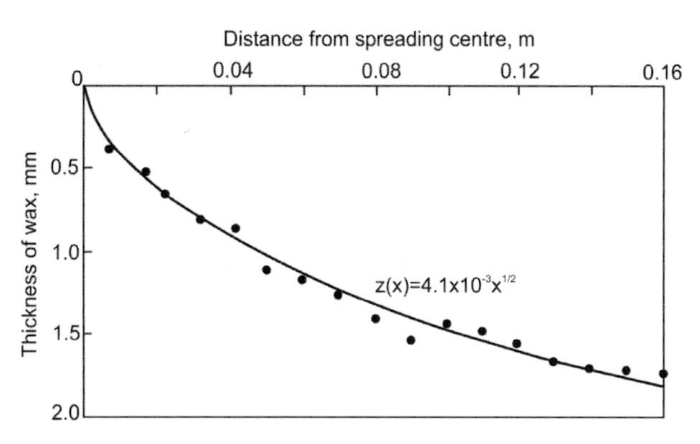


**Figure 3.5** Measured thickness of a wax model plate against distance from its origin (circles), compared to the best fitting square root of distance/age relation (continuous line), after Oldenburg (1975, Figure 7).

Oldenburg (1975), using the model of Parker and Oldenburg (1973), obtained an expression for the lithospheric thickness of

$$z_L = 35.4\alpha\sqrt{\frac{kt}{\rho c_p}} = 9.5\sqrt{t}\ \text{km}, \tag{3.11}$$

assuming $T_M = 1200$ °C, $c_p = 1.05 \times 10^3$ J kg$^{-1}$ °C$^{-1}$, latent heat $L = 4.2 \times 10^5$ J kg$^{-1}$, $\rho = 3.3 \times 10^3$ kg m$^{-3}$, and $k = 2.9$ W m$^{-1}$ °C$^{-1}$. This suggests that the plate thickness should increase from zero (in this approximation) to 9.5 km at 1 Ma, 19 km at 4 Ma and 95 km at 100 Ma.

Initially the base of the lithosphere was thought to reflect a gradual change of physical properties that would not produce a strong geophysical signal, unlike the sharp seismic refraction and reflection that occur at the base of the crust. An early attempt to demonstrate lithospheric thickening was made by Oldenburg (1975). He described an analogue experiment in which the growing lithosphere was modelled as an extending plate of paraffin wax that progressively froze as it moved away from a source of molten wax. By sectioning the frozen 'plate' at various distances (ages) he showed that its thickness did indeed increase as $t^{1/2}$ (Figure 3.5).

Subsequently, seismic studies revealed that the oceanic lithosphere is characterised by a high-velocity 'lid' overlying a 'low-velocity zone' (e.g., Gaherty *et al.*, 1999), which may contain small amounts of melt (Tan and Helmberger, 2007; Schmerr, 2012). Various seismological measurements have been used to estimate lithospheric thickness, including surface wave dispersion (Leeds *et al.*, 1974; Forsyth, 1975; Figure 3.6). Global seismic tomographic models show the progressive increase in shear wave velocity as the plate ages and cools (e.g., Ritsema *et al.*, 2011). Rychert and Shearer (2009, 2011) mapped a seismic reflection that they interpret as the base of the lithosphere. Its depth varies from 25 km to 130 km with distance from the ridge axis, with the best fitting isotherm being $930^{+90}_{-110}$ °C. Because the interface is quite sharp it may reflect changes in composition or degree of melting in addition to temperature.

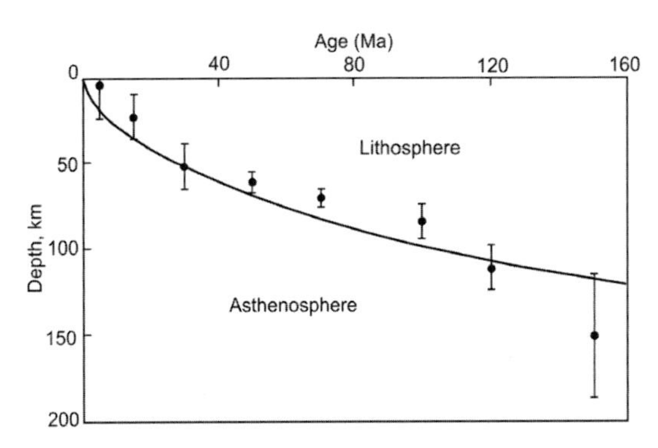


Figure 3.6 Estimates of lithospheric thickness versus age in the Pacific Ocean from seismic surface wave dispersion. Curved line is an approximate best fit $t^{1/2}$ law. From Leeds, A. R., Kausel, E. and Knopoff, L. (1974), Figure 1. Reprinted with permission from AAAS.

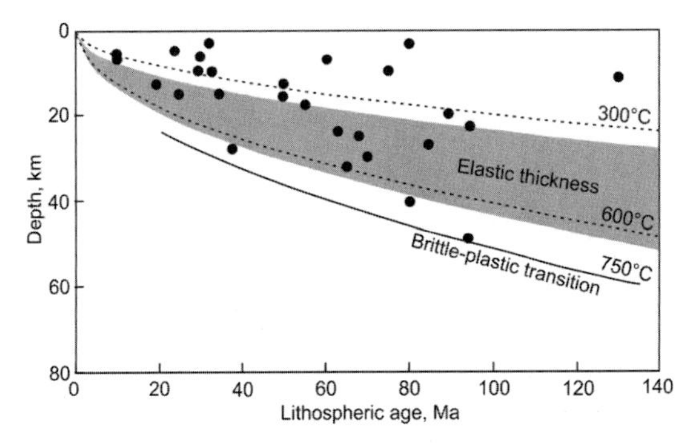


Figure 3.7 Oceanic intraplate earthquake depths as a function of lithospheric age, after Wiens and Stein (1983). Dashed lines correspond to 300 °C and 600 °C isotherms, calculated from the plate cooling model of Parsons and Sclater (1977). Solid line: predicted maximum depth at which dry olivine can sustain a 20 MPa stress for a strain rate of $10^{-18}$ $s^{-1}$. Grey: range of estimates of effective elastic thickness from plate flexure measurements of Watts *et al.* (1980).

A further means of estimating lithospheric thickness is to use the depth of the seismogenic zone, since earthquakes occur only where there is brittle deformation of the lithosphere. Wiens and Stein (1983) investigated the depths of oceanic intraplate earthquakes, and found they range from a few kilometres to about 50 km, but generally lie above the predicted 750 °C isotherm, corresponding to the brittle–plastic transition for dry olivine (Figure 3.7). Huang and Solomon (1988) show that centroid depths of teleseismically observed ridge-axis earthquakes deepen with decreasing spreading rate, from <2 km depth at fast-spreading rates (>20 km $Ma^{-1}$ half-rate) to ~6 km at slow rates (<5 km $Ma^{-1}$). As the centroid depth is an average of the focal volume, brittle deformation would extend twice as deep as these values. In MAR transform faults that cut older lithosphere, centroid depths indicate a maximum depth for brittle faulting of 20 km, corresponding to a lithospheric temperature of 900 ± 100 °C (Bergman and Solomon, 1988).

Global seismograph networks cannot easily resolve very shallow earthquake depths, but local arrays can. The thickness of the seismogenic zone near ridge axes has been estimated

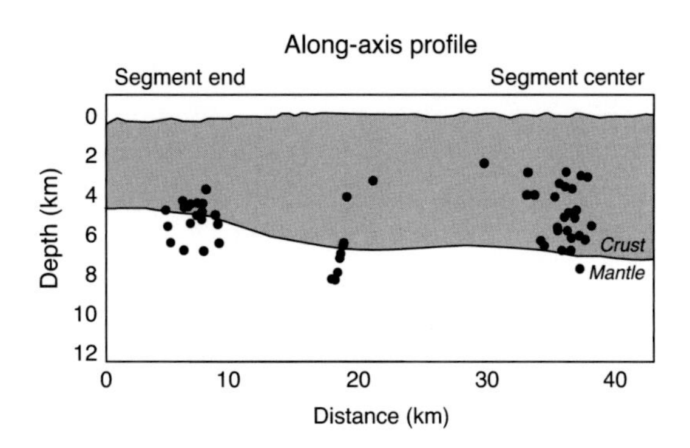


Figure 3.8 Distribution of microseismicity along the axis of a segment of the slow spreading MAR near 29° N, after Wolfe *et al.* (1995). Microseismicity clusters in three areas, at depths from 2 km to 9 km, mostly in the lower crust and upper mantle.

in a number of experiments using ocean bottom seismometer arrays. Brittle deformation at parts of the fast-spreading EPR may be confined to the uppermost 1 km of the crust (Sohn *et al.*, 1999), though elsewhere it may reach up to 5 km near transform faults (Lilwall *et al.*, 1981). Hypocentre depths are in the range 1.5–3.5 km at the intermediate-spreading Juan de Fuca Ridge (Wilcock *et al.*, 2002). At some parts of the slow-spreading MAR, the maximum hypocentral depths can be as shallow as 2 km (Wolfe *et al.*, 1995). Barclay *et al.* (2001) found microearthquakes between 3 km and 4 km below the sea floor in an anomalously shallow segment of the MAR at 35 °N, reflecting the higher crustal temperature and raised brittle–ductile transition there (Toomey *et al.*, 1993). Louden *et al.* (1986) found a similar depth range near Vema fracture zone. However, elsewhere on the MAR, earthquakes extend to 7–10 km depth (Lilwall *et al.*, 1978; Toomey *et al.*, 1985; Kong *et al.*, 1986; Louden *et al.*, 1986; Toomey *et al.*, 1988; Wilcock *et al.*, 1990; Kong *et al.*, 1992; Wolfe *et al.*, 1995; Figure 3.8). Similar hypocentral depths of 6–10 km occur on the very slow-spreading Southwest Indian Ridge (SWIR; Katsumata *et al.*, 2001). The deepest of these earthquakes are in the uppermost mantle (Figure 3.8).

## 3.4 Flexure and elastic thickness

A very important technique for investigating the oceanic lithosphere is to deduce estimates of its elastic thickness from measurements of the way it flexes under topographic loads (Watts *et al.*, 1980). The lithosphere is modelled as a thin elastic plate, supported isostatically by an inviscid substratum. Then the vertical deflection $d$ of the plate varies in the $x$ and $y$ horizontal directions according to

$$D\left\{\frac{\partial^4 d}{\partial x^4} + 2\frac{\partial^4 d}{\partial x^2 \partial y^2} + \frac{\partial^4 d}{\partial y^4}\right\} + (\rho_M - \rho_I)gw = L(x, y), \tag{3.12}$$

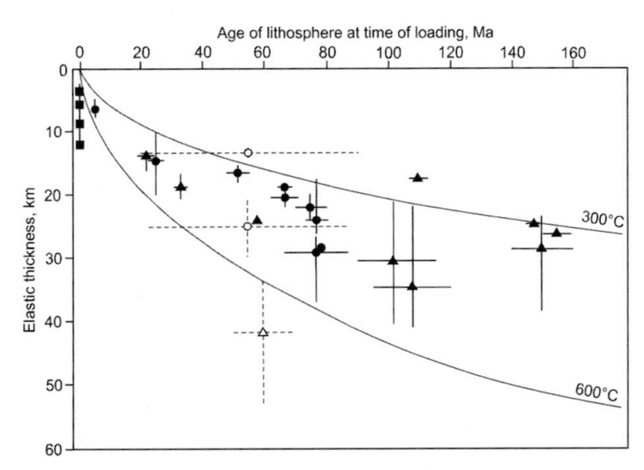


Figure 3.9 Elastic thickness, $T_e$, plotted against age of lithosphere at time of loading, after Watts *et al.* (1980). Squares: estimates based on MOR loadings; circles: seamount loadings; triangles: trenches. Crossed lines: ranges or uncertainties. Open symbols and dashed lines: larger ranges or uncertainties. Calculations of $T_e$ assumed Young's modulus = 100 GPa and Poisson's ratio = 0.25. Curves are 300 °C and 600 °C isotherms calculated from the plate cooling model of Parsons and Sclater (1977).

(Turcotte and Schubert, 1982), where

$$D = \frac{E}{12(1-\nu)^2} T_e^3 \tag{3.13}$$

is the flexural rigidity, $\rho_M$ is the density of the substratum, $\rho_I$ is the density of the medium infilling the depression caused by the load $L$, $g$ is the acceleration due to gravity, $E$ is Young's modulus, $\nu$ is Poisson's ratio and $T_e$ is the effective elastic thickness of the plate.

Watts *et al.* (1980) considered the flexure produced from a variety of loads such as ridges, seamounts, islands and subduction zone trenches, and showed that $T_e$ increases systematically with the age of the lithosphere at the time the load was applied (Figure 3.9). It ranges from just a few kilometres at the ridge axis to ~30 km at ages of 80 Ma or greater. These estimates are somewhat less than those obtained from surface wave dispersion (Figure 3.6). The discrepancy reflects the dependence of rheology on strain rate: the Earth deforms to seismic surface waves on a timescale of seconds to minutes, whereas the flexural estimates reflect relaxation over millions of years. Nevertheless, they display the same characteristic thickening with age, and lie between the 300 °C and 600 °C isotherms of the plate cooling model.

## 3.5 Gravity over MORs

The temperature and consequent density structure of the lithosphere has a profound influence on the gravity field over MORs. The free-air gravity anomaly on the sea surface is

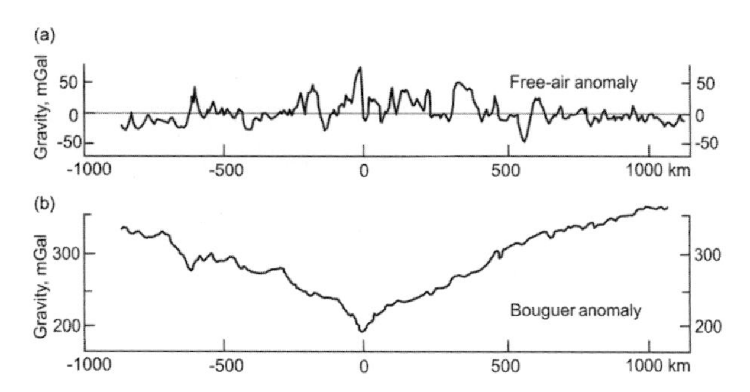


Figure 3.10 (a) Free-air gravity anomaly, and (b) Bouguer anomaly, from a profile crossing the MAR near 32° N, after Talwani *et al.* (1965).

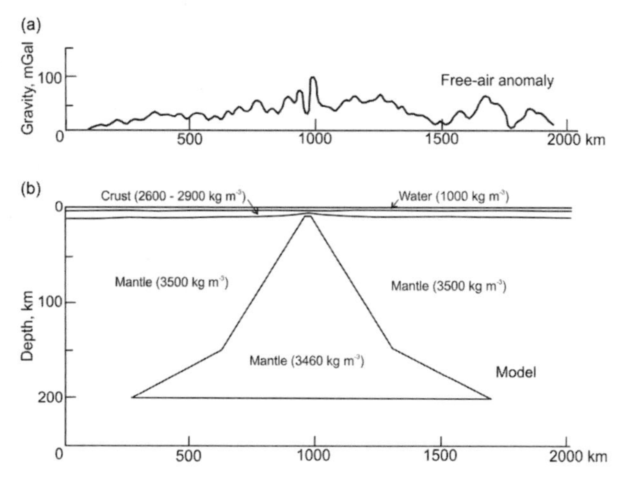


Figure 3.11 (a) Free-air gravity anomaly crossing the MAR near 46° N, and (b) a simple model that fits these observed data, after Keen and Tramontini (1970, Figure 8). The wedge-shaped low-density body would be interpreted as a region of hot asthenosphere under the ridge. In practice, temperature and density boundaries are gradational.

the observed gravity corrected for latitudinal variations resulting from the Earth's rotation. It effectively shows the cumulative effect of the density variations in and under the sea. Regionally the average free-air anomaly over MORs is close to zero (Figure 3.10a), indicating that ridges are broadly in isostatic equilibrium. In other words, the weight of the lithosphere is balanced by the buoyancy of the hot, low-density asthenospheric mantle, the excess mass of the ridge topography being compensated by the corresponding low-density body beneath it. This is seen more clearly in the Bouguer anomaly (Figure 3.10b), which additionally accounts for the variation in the attraction of the water layer, and so directly reflects sub-sea-floor variations in density. Figure 3.11 shows a simple model of the subsurface density that broadly fits these variations. Although gravity modelling cannot uniquely determine the shape of such a body, the observed gravity is certainly consistent with a broad, low-density region in the mantle. According to the plate model, this arises from the increased temperature and reduced density of the mantle below the ridge. The shape

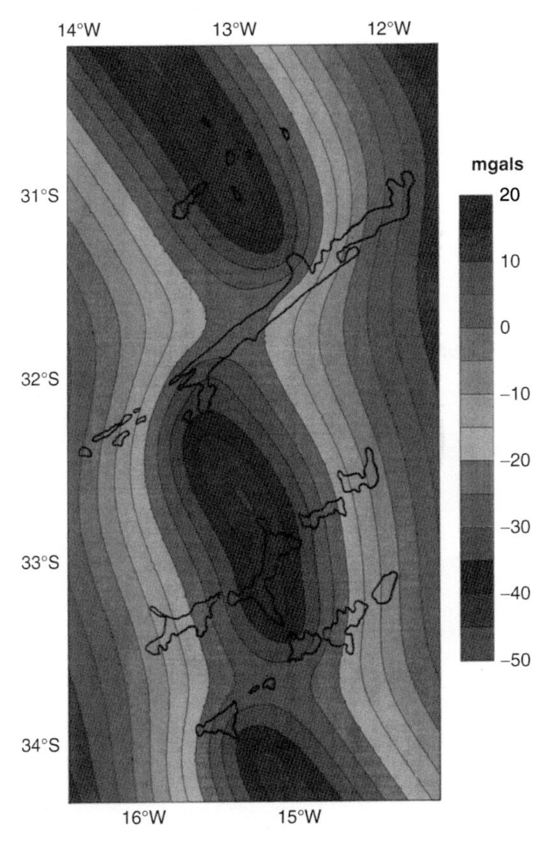


Figure 3.12 Predicted gravitational effect of the thermally induced density variations resulting from passive mantle upwelling under part of the southern MAR, after Kuo and Forsyth (1988, Figure 8) with kind permission from Springer Science and Business Media. Bold contours indicate 3600 m depth and partially outline the NW–SE spreading centres and their SW–NE transform offsets and associated fracture zones. For colour version, see plates section.

of the low-density body in Figure 3.11b reflects quite well the predicted distribution of temperature (and therefore density) shown in Figure 3.2.

A more detailed model of the gravitational effect of the temperature and density field below ridges is obtained by directly calculating the temperature variations induced by passive upwelling of hot mantle, driven by rigid-plate separation (Phipps Morgan and Forsyth, 1988; Figure 3.12). The gravitational effect is obtained by converting the modelled temperatures to density anomalies and then determining their gravitational attraction. As well as helping visualise the effects of lithospheric cooling, this provides an important correction to be incorporated in the so-called residual Mantle bouguer anomaly (Kuo and Forsyth, 1988; Prince and Forsyth, 1988), which is widely used in correcting observations of gravity over MORs (Section 2.4.2). Figure 3.12 shows how the predicted gravity increases away from very low values (reflecting hot, low-density rock) at the ridge axis, but that the pattern is modified by the step-like offset of the ridge axis by transform faults (see Chapter 4).

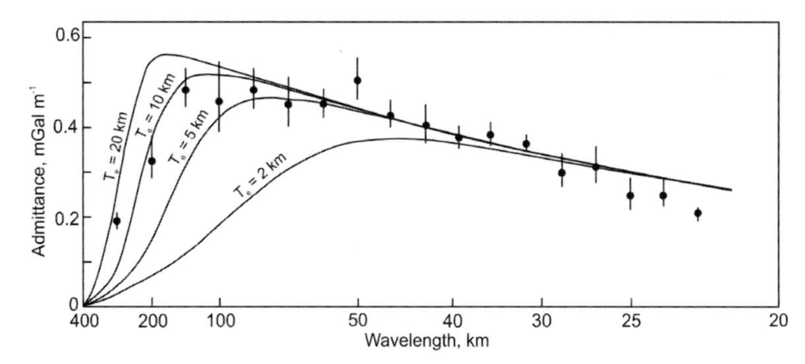


**Figure 3.13** Observed (dots) and modelled (curves) admittance functions obtained from the spectral ratio of gravity and bathymetry over the MAR axis, after Cochran (1979). Curves are for a plate model with effective elastic thickness ($T_e$) varying from 2 km to 20 km. Best fit is for $T_e = 9$ km.

## 3.6 Isostatic compensation

The observed gravity field can also yield information on the state of isostatic compensation, the depth at which the compensation is achieved, and the mechanism for achieving it, and so provides a valuable means of investigating the near-axis lithosphere. Many studies have used the admittance technique (Section 2.4.2; McKenzie and Bowin, 1976; Watts, 1978) to estimate the degree of coherence between gravity and topography and hence examine isostatic compensation by comparing with simple models.

McKenzie and Bowin (1976) used this technique over the slow-spreading MAR to show that topographic features begin to be isostatically compensated when their horizontal dimensions are greater than about 100 km, and deduced that the plate thickness at the time of compensation was just over 10 km (Figure 3.13). Larger-scale features are more completely compensated, but smaller-scale features are uncompensated, being supported by the strength of the rigid lithosphere (Figure 3.14). Cochran (1979) found that a flexed plate provided a better fit than simple Airy isostasy, and suggested that the best fitting elastic thickness for the MAR crest is in the range 7–13 km (Figure 3.13), with a somewhat lower value of 2–6 km for the fast-spreading EPR. More recent studies have found effective elastic plate thickness as low as 0.5–3 km at the MAR (Escartin and Lin, 1998; Smith *et al.*, 2008).

Fast-spreading ridges such as the EPR exhibit a small crestal high a few kilometres wide and a few hundred metres high, which appears to be supported by the buoyancy of sub-axial accumulations of magma. Madsen *et al.* (1984) modelled the EPR crest by a thin elastic plate that was broken at the axis and subject to a buoyant force there. The best fitting models had compensation depths of 6–7 km and a plate whose flexural rigidity ($D$) increased from $10^{18}$ N m at the ridge axis to $10^{21}$ N m 25 km away. These results suggested a buoyant body, assumed to be partially molten mantle, located at the base of the crust and in the uppermost mantle.

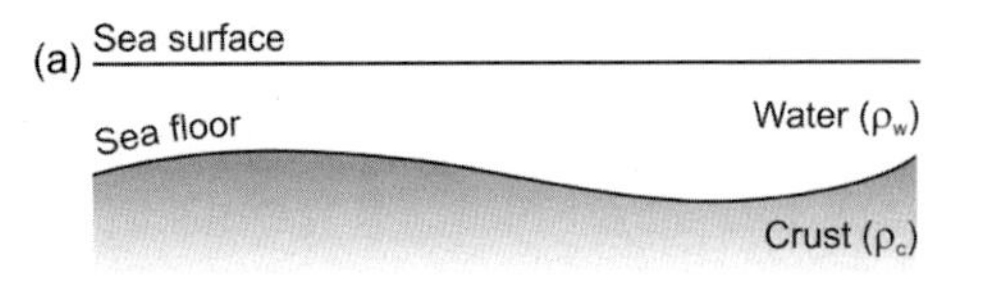


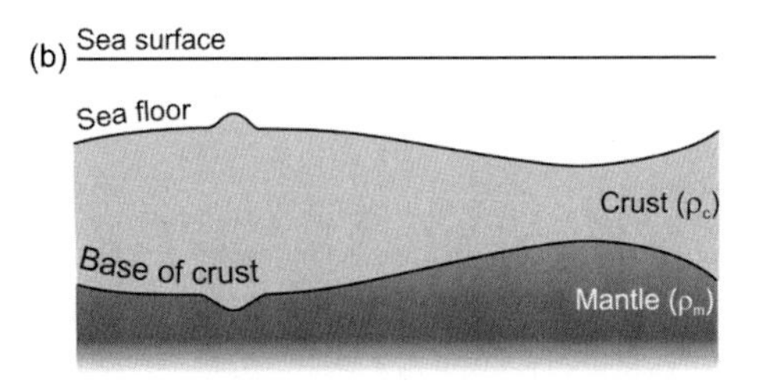


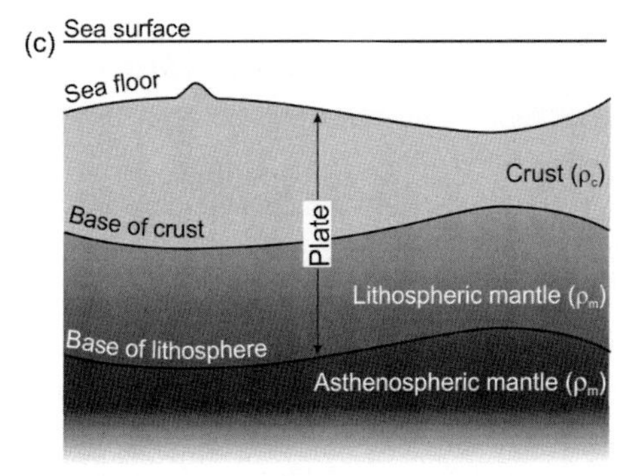


**Figure 3.14** Three models for isostatic compensation in the oceanic lithosphere, modified after McKenzie and Bowin (1976). (a) No compensation: observed gravity anomalies are produced by the density contrast at the sea floor. (b) Simple Airy isostatic compensation: both long- and short-wave topography is compensated by variations in the crustal thickness. (c) Compensation including an elastic plate: long-wavelength features are compensated because they bend the plate but short-wavelength features are not compensated. Model (c) best fits the observations.

Kuo *et al.* (1986) improved on this model by considering the effect of the plate thickening with age. Such a plate bends more easily than a uniform one, because the bending stress vanishes at its base. They found that this model fits the EPR data better than that of Madsen *et al.* (1984) if $T_e = 0.1$ km at the axis and the plate thickness $h$ is proportional to the square root of age, thickening at the rate of 5.5 km $\text{Ma}^{-1/2}$. Wang and Cochran (1993) also modelled the EPR crest with a thin, broken, buoyantly supported plate, and considered two different possibilities. In one, the plate has a constant flexural rigidity and yields a best fit thickness of 0.3–0.6 km. In the other case, the effective elastic thickness is given by

$$T_e(x) = Rx^{1/2} + T_e(0), \tag{3.14}$$

where $R$ is the rate of increase, $x$ the axial distance and $T_e(0)$ the thickness at zero age. Best-fitting values of $R$ were 0.2–0.3 $km^{1/2}$. The buoyant material supporting the crestal high was found to extend to 22–31 km below the sea floor for the constant rigidity plate and to 17–22 km for the growing plate.

Slow-spreading ridges do not exhibit a crestal high, but rather a broad median valley caused by tectonic thinning and necking of the axial lithosphere (Tapponnier and Francheteau, 1978; Chen and Morgan, 1990b; Poliakov and Buck, 1998; Buck *et al.*, 2005). Both gravity and seismic data indicate that the oceanic crust retains a roughly constant thickness across both the median valley and the ridge flanks, and therefore the negative topographic load represented by the median valley must be largely uncompensated and, therefore, dynamically supported (Freed *et al.*, 1995). Neumann and Forsyth (1993) examined this problem in detail, and concluded that a lithospheric stretching model is capable of generating the dynamic differential vertical stresses needed to create a median valley.

## 3.7 Summary

New oceanic lithosphere is created at MORs by the process of sea floor spreading. This is part of a mantle convection cycle in which hot, plastic mantle rises under ridges, cools to become rigid lithosphere, and drifts away from the ridge as a pair of spreading plates. The newly formed plates are hot, and as they spread away from the axis they cool, become denser, and sink isostatically, forming the characteristic topography of the ridge flanks. The shape of the flanks is predicted to be proportional to the square root of age, and fits this simple model well except for very young and very old lithosphere. A buoyantly supported axial high is superimposed at fast-spreading ridges, while a dynamically supported median valley occurs at slow-spreading ones. Very old lithosphere is consistently shallower than predicted, indicating continued heat input from the base of the plate. This plate model also explains variations in conductive heat flow observed at the sea floor, except near ridge axes where hydrothermal advection causes significant cooling.

Assuming that the brittle–plastic rheological transition occurs at a characteristic temperature, we expect the base of the plate to follow an isotherm, and this is largely supported by observations. The temperature of the transition depends on various factors including strain rate, and an approximate value of ~750 °C is often assumed. The maximum depth of earthquakes gives the thickness of the brittle seismogenic zone, which ranges from a few kilometres near the ridge axis to 40–50 km in old lithosphere. Generally the seismogenic zone is slightly shallower at faster-spreading ridges. Seismological methods respond to high strain rates and yield thicker estimates than do flexure measurements, which respond to long-term deformation. Measuring the flexure of plates under imposed loads gives lithospheric thicknesses ranging from ~1 km at zero age to ~30 km at ~100 Ma, corresponding to a brittle–plastic transition between about 300 °C and 600 °C.

Gravitational methods can be used to model the degree and manner of isostatic compensation. Ridge flanks are generally close to isostatic equilibrium, but their axial zones may depart from it. Fast-spreading ridges display small crestal highs that are best modelled by a thin elastic plate, broken at the axis and supported by a buoyant mass in the lower crust and upper mantle. Slow-spreading ridges, on the other hand, have large median valleys that are isostatically uncompensated and must be dynamically supported; such support appears to be provided by elastic stretching of the young lithosphere.

# 4 Ridges as plate boundaries

## 4.1 Ridges and plate kinematics

An important aspect of MORs is their function as divergent boundaries between tectonic plates (Figure 4.1). Plate tectonics considers the lithosphere to be broken into a number of thin, rigid caps, whose relative motions across the Earth are described by plate kinematics (McKenzie and Parker, 1967; Le Pichon, 1968; Morgan, 1968). Plate motions are described in terms of pure geometry, utilising Euler's rotation theorem that any motion on a sphere can be represented as a single rotation about an axis passing through the sphere's centre. This axis cuts the surface at two points called the Euler poles, or 'poles of rotation'. The relative motion between any two plates is completely described by the latitude and longitude of the relevant Euler pole and the rate of rotation. The most recent global description of plate motions is given by DeMets *et al.* (1990, 1994) .

The trace of an Euler rotation on the surface of the sphere is a 'small circle' centred on the pole (Figure 4.2). Such circles are analogous to lines of latitude, which are small circles about the geographic poles. Two plates sharing a common boundary aligned along such a small circle will move relative to one another by pure slip, so the azimuth of this small circle is the spreading direction. Geologically, such plate boundaries are transform faults (Wilson, 1965; Section 4.4). 'Great circles' (diameters of the sphere) that are normal to transform faults meet at the rotation pole (Figure 4.2). There are two other types of plate boundary: 'trenches' or subduction zones, where plates converge, and 'ridges', where they diverge. The ridges are spreading centres and are the main subject of this book. Since MOR axes represent divergent, or spreading, plate boundaries, they are often called 'spreading centres' or 'spreading axes'. The relative opening rate, or full spreading rate, $r$, is given by

$$r = \dot{\omega} \sin \theta, \tag{4.1}$$

where $\dot{\omega}$ is the angular opening rate and $\theta$ is the angular distance from the pole. Thus spreading rate increases from the pole to a maximum at 90° away. Measurements of transform fault azimuths and plate separation rates are important inputs to models of global plate motions (e.g., Minster *et al.*, 1974; DeMets *et al.*, 1990).

While transform faults must lie along small circles, ridges are not constrained to particular orientations. Nevertheless, the great majority of ridge segments are nearly orthogonal to the local spreading direction, so that ridge–transform plate boundaries often follow a 'stair step' pattern. This orthogonality has been elegantly reproduced in analogue experiments with

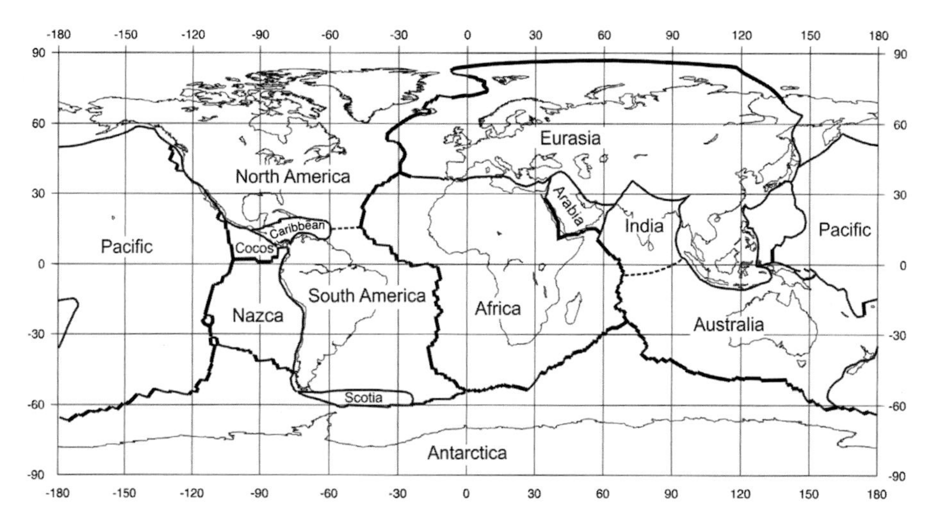


Figure 4.1 Map of major tectonic plates. Spreading (ridge) boundaries shown in bold; other plate boundaries by medium lines.

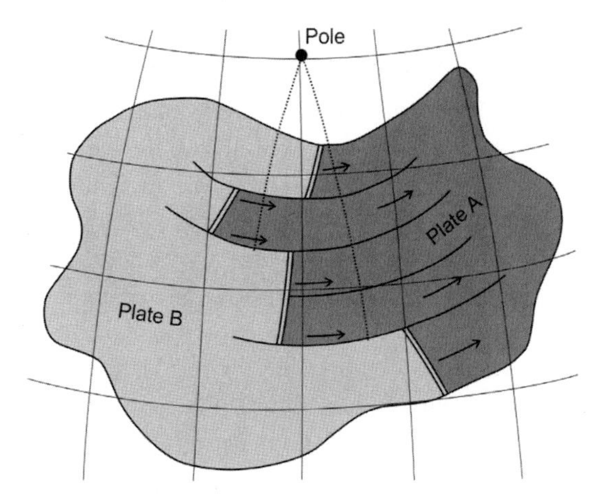


Figure 4.2 Two tectonic plates, A and B, rotating relative to each other (curved arrows) about their common pole of rotation (dot), after Morgan (1968). Double lines are spreading centres. Continuous lines are segments of small circles about the pole of rotation. Dotted lines are segments of great circles perpendicular to them, which meet at the pole. Fine lines represent the independent geographic coordinate system.

freezing wax (Oldenburg and Brune, 1972). It suggests that ridges normally re-organise their geometry to minimise the total ridge length at the expense of increasing the total transform length, implying that the resistance to spreading at ridges is greater than resistance to slip at transforms (Oldenburg and Brune, 1975). Departures from such orthogonality occur at a local scale (a few kilometres) near ridge offsets, but large-scale departures are relatively rare. One such is the Reykjanes Ridge south of Iceland, which spreads 36° obliquely over a distance of several hundred kilometres. Regions of significantly oblique spreading were discussed by Taylor *et al.* (1994).

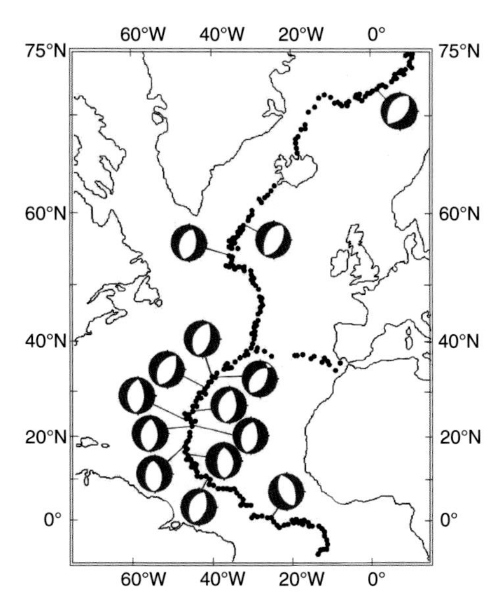


Figure 4.3 Earthquake epicentres and focal mechanisms along the northern MAR spreading centre, excluding mechanisms from transform faults, after Lowrie (1997). 'Beach ball' symbols show earthquake first motions (black compressional, white dilatational) on the lower focal hemisphere. Boundaries between the compressional and dilatational segments are called nodal planes, one of which is parallel to the active fault. Here, mechanisms are extensional (normal faulting) with nodal planes parallel to the ridge.

## 4.2 Seismicity and focal mechanisms

Isacks *et al.* (1968) documented the seismological evidence for plate kinematics. They showed that MOR earthquakes lie within very narrow bands marking the plate boundary (Figure 2.11 and Figure 4.3). At transform faults, earthquakes follow the line of the transform fault but are confined to the actively slipping plate boundary (Figure 4.4).

Earthquakes recorded by global seismic networks generally lie within a few tens of kilometres of spreading or transform plate boundaries, except in rare cases where the boundary is diffuse. Detailed studies of microseismicity recorded on ocean bottom seismometers (e.g., Wolfe *et al.*, 1995), show that microearthquakes generally cluster within a few kilometres of the ridge axis near segment centres,[1] but may extend 10–15 km away at segment ends. This may reflect a preponderance of volcanic and hydrothermal earthquakes at segment centres while more widely distributed tectonic events occur at segment ends (Wolfe *et al.*, 1995). Most MOR seismicity is focussed on transform faults and non-transform offsets. At slow spreading ridges, earthquakes tend to be clustered predominantly in areas characterised by asymmetric accretion associated with detachment faulting, where it may extend up to 20 km from the ridge axis (Smith *et al.*, 2003; Escartin *et al.*, 2008; Section 7.4.3).

[1] See Section 4.5 for discussion of spreading segments.

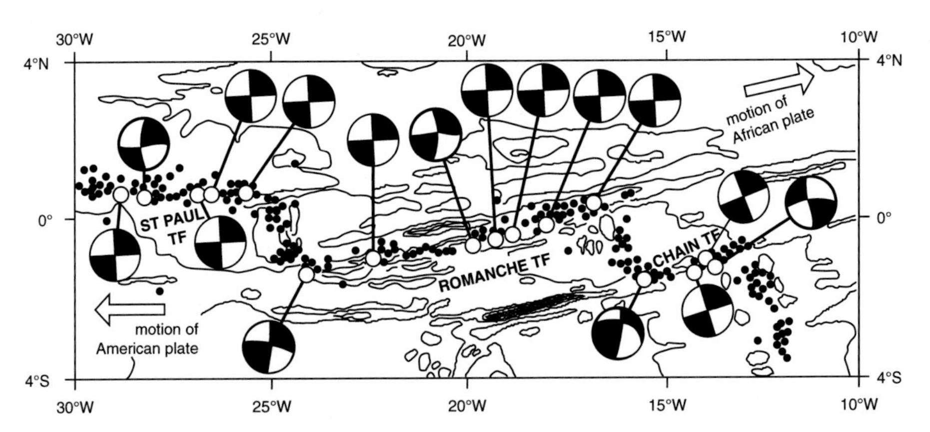


Figure 4.4 Earthquake epicentres and focal mechanisms along the equatorial Atlantic fracture zones, after Engeln *et al.* (1986). Outline bathymetry shown by contours. Symbols as in Figure 4.3. Here, mechanisms are dextral strike-slip with one nodal plane parallel to plate motion.

This and other evidence suggests that brittle deformation of the plate boundary is mostly confined to a narrow axial zone.

Earthquake focal mechanisms are entirely consistent with the relative motions and stresses predicted by plate kinematics (Sykes, 1967; Isacks *et al.*, 1968). Spreading centres are characterised by normal-faulting (extensional) mechanisms, and transforms by strike-slip mechanisms with one nodal plane along the transform direction (Figure 4.3 and Figure 4.4). Normal faults dip approximately 45° and strike parallel to the ridge axis (Huang and Solomon, 1988). Both normal and strike-slip faulting can occur at ridge–transform intersections. See Chapter 7 for a detailed discussion of faulting.

The almost complete dominance of extensional mechanisms at spreading centres shows that the brittlely deforming plate boundary is everywhere under tension, implying that the plates are being pulled apart by forces with a distant origin, such as 'slab-pull' and 'ridge-push' (Forsyth and Uyeda, 1975). The former arises from the weight of subducting lithosphere, while the latter has its origin in the outward gravitational sliding of the ridge flanks. Plates are not, as sometimes supposed, pushed apart by dyking at the axis; rather, dyking is a largely passive response to tensional fracturing.

MOR earthquakes have relatively low magnitudes, mostly less than body-wave magnitude 6.0. This reflects the relatively low stresses necessary to rupture the young, thin, weak lithosphere (Section 3.3) or to drive faults that are lubricated by weak minerals such as talc and serpentine (Section 7.3.3).

## 4.3 Spreading centres

Developments in the technique of bathymetry (that is, the measurement of ocean depth) have been fundamental to an understanding of the nature of ridges. As discussed in Chapter 3,

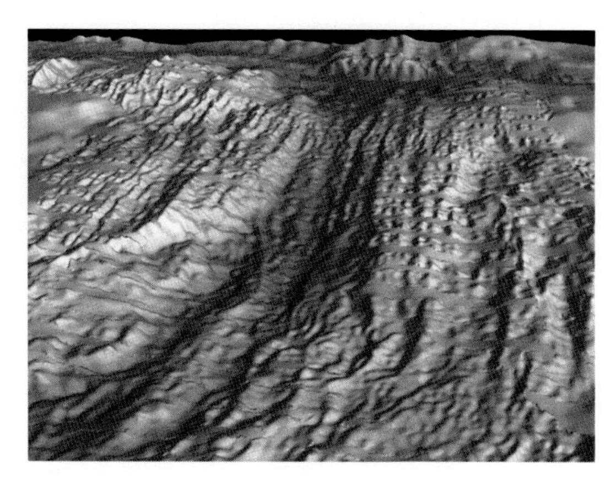

Figure 4.5 Shaded relief image of an ~100 km length of the MAR median valley at 24° N, looking south towards the Kane transform fault, seen in the distance. Vertical exaggeration 2×. Width of image ~20 km in foreground, 40 km at middle distance. Depth ranges from about 2600 m (white) to 5000 m (purple). Image produced using GeoMapApp (http://www.geomapapp.org; Ryan *et al.*, 2009). For colour version, see plates section.

the overall shape of MORs is a reflection of their thermal and density structure, themselves a result of steady cooling of the plates following their creation by igneous processes at the ridge axis.

A median valley (also called an 'axial valley' or 'rift valley') is present along the crest of most slow-spreading ridges. These valleys were first recognised (Heezen, 1960, 1969) from the long, single bathymetric profiles during the 1950s and 1960s. They are tens of kilometres wide and 2–3 km deep, with faulted sides clearly attesting to their rifted origin (Figure 4.5). The occurrence of rift valleys with high heat flow and seismicity at ridge axes led to the idea that these are the centres of sea floor spreading (Hess, 1962) and, ultimately, the places where oceanic plates are created.

The normal faults flanking the median valley are spaced a few kilometres apart and have throws (that is, vertical displacements) of several hundred metres (Section 7.3.2). The valley floor contains a neo-volcanic zone, the region where new oceanic crust is forming by volcanic activity, and often an elongated axial volcanic ridge (AVR) a few kilometres wide that marks the actual plate boundary (Section 6.7.3; Figure 4.5). The relatively shallow areas flanking the median valley are sometimes called the rift mountains.

While a median valley is common on the slow-spreading ridges of the Atlantic, Arctic and Indian oceans, it is notably absent from most of the fast-spreading EPR (Menard, 1960). Here, the flanks rise steadily towards the rise axis. On single-beam echosounder profiles the axis appears almost flat, though multi-beam echosounders have revealed a crestal high a few kilometres wide and a few hundred metres high that stands atop the main rise (Lonsdale, 1977; Larson and Spiess, 1969; Figure 4.6).

There is a first-order correlation of axial morphology with spreading rate (Figure 4.7). Small and Sandwell (1989) analysed free-air gravity anomalies over global ridge axes, and found that those spreading slower than 60 km $Ma^{-1}$ (full rate) are characterised by large amplitude negative anomalies, reflecting the existence of a dynamically supported

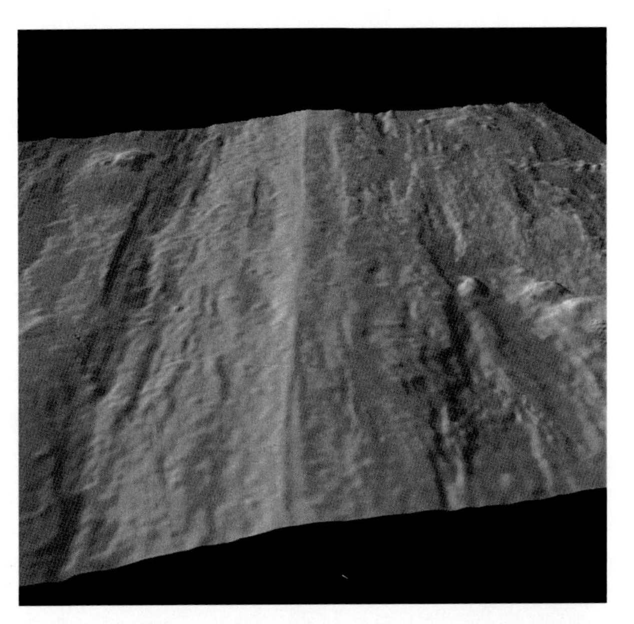

Figure 4.6 Shaded relief image of an ~60 km length of the EPR axial high at 17°30′ S, looking north. Vertical exaggeration 2×. Width of image ~35 km. Depth ranges from about 2600 m (pale grey) to 3100 m (purple). The crestal high can be seen as a narrow ridge in the centre of the pale grey strip. Image produced using GeoMapApp (http://www.geomapapp.org; Ryan *et al.*, 2009). For colour version, see plates section.

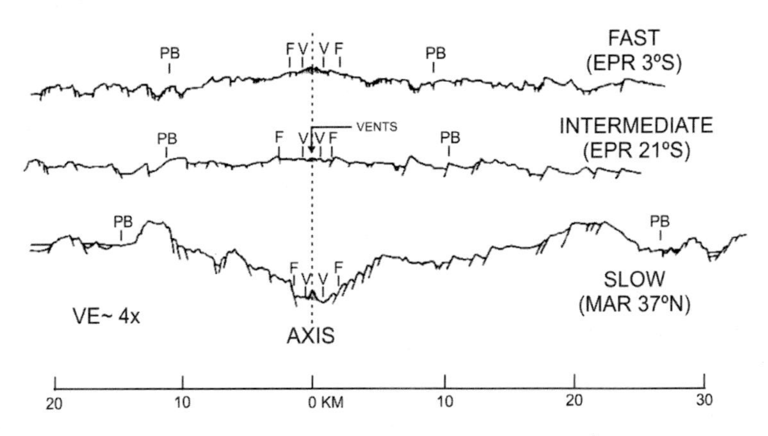


Figure 4.7 High-resolution deep-towed topographic profiles across the axes of fast (64 km $Ma^{-1}$ full rate), intermediate (42 km $Ma^{-1}$), and slow (23 km $Ma^{-1}$) spreading ridges, after Macdonald (1982). Reproduced with permission of *Annual Reviews*. V–V, neovolcanic zone; F–F, zone of active fissuring; PB, limits of 'plate boundary zone' (zone of active faulting). Vertical exaggeration ~4:1.

median valley, while those spreading faster than 70 km $Ma^{-1}$ have only small axial gravity anomalies reflecting isostatically compensated rises. There is an abrupt transition at 60–70 km $Ma^{-1}$, although there are anomalies: the slow-spreading Reykjanes Ridge (south of Iceland) has an axial rise similar to the fast-spreading EPR (Talwani *et al.*, 1971; Laughton

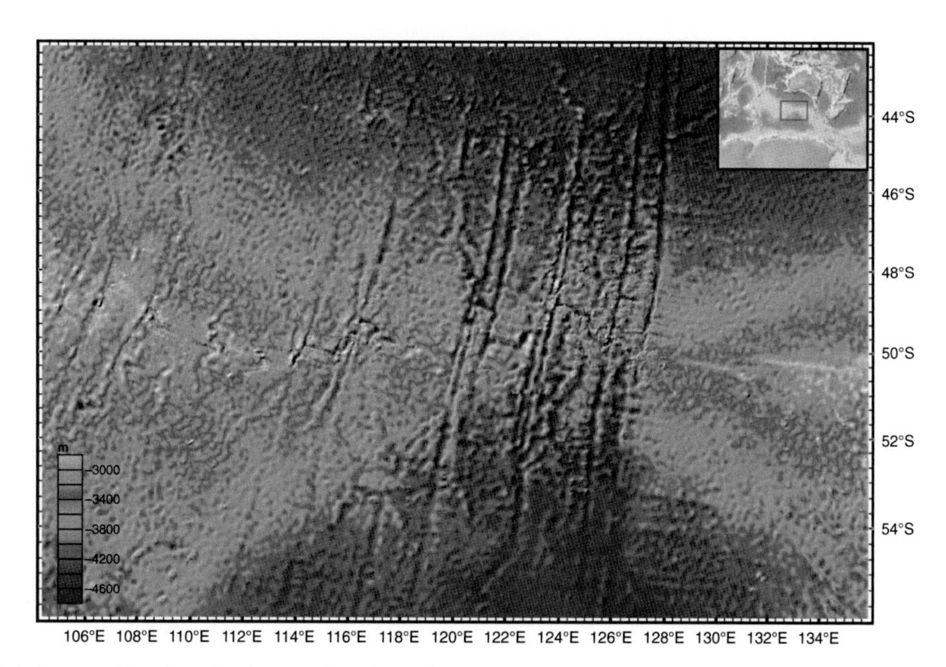


Figure 4.8 Bathymetry of the Australia–Antarctic Discordance, showing changes in along-axis character from axial highs to median valleys. Inset shows location of main figure. Image produced using GeoMapApp (http://www.geomapapp.org; Ryan *et al.*, 2009). For colour version, see plates section.

*et al.*, 1979; Searle *et al.*, 1998a), while the fast-spreading EPR develops a median valley as it approaches the Tamayo Transform near Mexico (Macdonald *et al.*, 1979; Lonsdale, 1995). Rapid variations of axial morphology, from axial high to median valley, also occur in the so-called 'Australia–Antarctic Discordance' south of Australia (Hayes, 1988; Sempéré *et al.*, 1997; Okino *et al.*, 2004; Figure 4.8). In all cases the transition takes place within a distance of only a few tens of kilometres along axis, with little or no change in spreading rate. It appears to reflect a change in the mechanism of isostatic compensation (Small and Sandwell, 1989).

The axial topographic and gravity highs over fast-spreading ridge crests have been successfully modelled as the result of an isostatic response to low-density partial melt (Madsen *et al.*, 1984; Wang and Cochran, 1993; Section 3.6). For slow-spreading ridges a series of models (e.g., Tapponnier and Francheteau, 1978) relate the formation of the median valley to 'necking' of the lithosphere under tension. Other models relate the axial topography to the balance between tectonic stretching and magmatic accretion (e.g., Ito and Behn, 2008). Chen and Morgan (1990a,b) explained the transition from rift valley to no rift valley as a function of the changing width of a weak, crustal 'decoupling region' overlying a ductile mantle. The size of the decoupling region is mainly a function of the temperature and crustal thickness, which itself depends on mantle temperature (McKenzie and Bickle, 1988; Bown and White, 1994). For slow-spreading ridges the decoupling region is small, allowing strong coupling between the brittle plate and underlying, diverging, ductile mantle that promotes accumulation of tensional strain and therefore necking; at fast-spreading

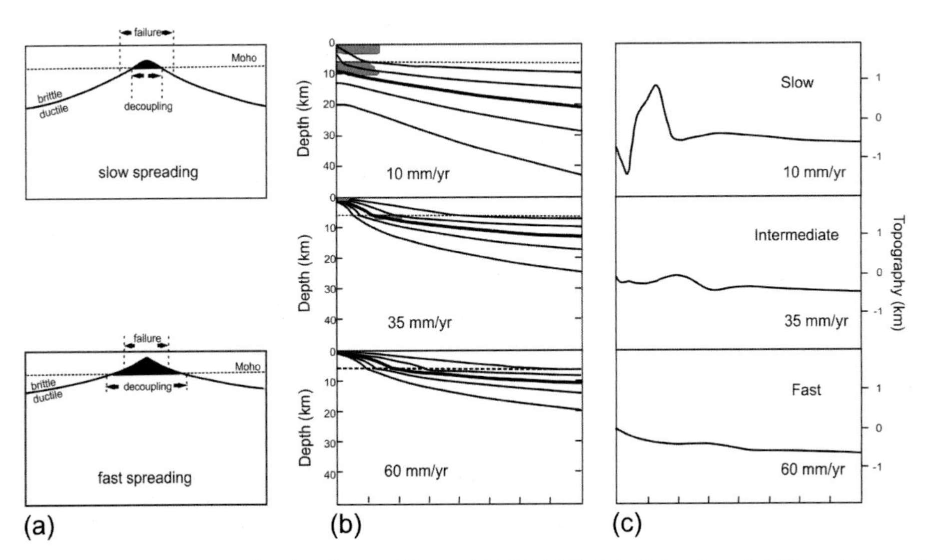


Figure 4.9 Formation of ridge axis topography, after Chen and Morgan (1990b). (a) Modelled lithosphere (bold line) with base of crust dashed, for slow- and fast-spreading ridges. Black: regions of weak lower crust ('decoupling zone'). (b) Isotherms at 250 °C intervals, with 750 °C isotherm (approximate brittle–plastic boundary) in bold, for three spreading rates. Shading: zones of upper crustal failure. (c) Computed sea floor topography for the three cases in (b). At slow spreading the lower crustal weak zone is too small to decouple the upper crust from the applied extensional forces, so a broad upper crustal zone fails under tension, producing necking and rifting. At fast spreading the decoupling zone is wider and decouples the overlying axial lithosphere from extensional forces, the region of crustal failure is negligible, and the lithosphere responds mainly to buoyancy. Although the assumption of dry olivine has been questioned, the principle of stretching-produced rifting remains valid.

ridges there is a large decoupling region, so the brittle plate is decoupled from the diverging mantle and its shape is determined largely by isostasy (Figure 4.9). Chen and Morgan (1990a,b) assumed a dry olivine rheology, which may be incorrect (Searle and Escartín, 2004), but a basic stretching model still seems to apply (Buck *et al.*, 2005). Recently, the gross morphology of slow-spreading ridges has been modelled numerically using elastic–plastic–viscous rheology to simulate the strain produced by tensional faulting (Poliakov and Buck, 1998; Figure 4.10).

These and other models make clear that the over-riding control on ridge morphology is the mantle temperature or degree of melt produced, rather than the spreading rate. Normally these are strongly correlated, but anomalous mantle composition affecting fertility or ease of melting may also be a cause. For example, the Reykjanes Ridge is near the Iceland hotspot so its mantle may be unusually hot (Searle *et al.*, 1998a; Smallwood and White, 1998), whereas there is thought to be a mantle cold spot under the Australia–Antarctic Discordance (Marks *et al.*, 1990; Okino *et al.*, 2004). At the Tamayo (and other transforms), old, cold lithosphere depresses the mantle temperature below the opposite spreading centre.

The morphology of the ridge axis is discussed in more detail in Chapters 5–7.

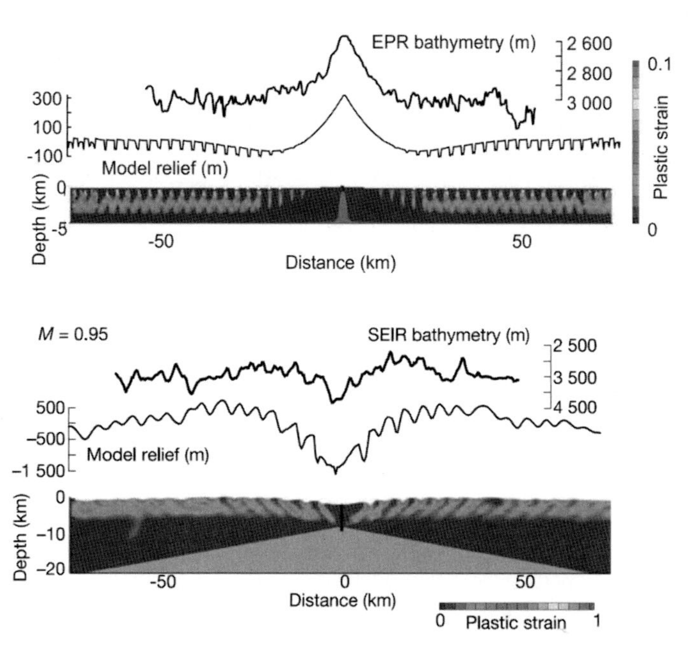


Figure 4.10 Numerical models of an elastic–plastic–viscous lithosphere overlying a weak asthenosphere for fast (top) and slow (bottom) MORs, after Buck *et al.* (2005). Coloured sections show computed strain in vertical, ridge-normal cross-sections. Thin black curves show sea floor topography computed from the models, compared with actual topographic profiles (bold curves). The fast-spreading case is dominated by axial buoyancy, while the slow-spreading case is dominated by stretching. Note the two sets of profiles have different vertical exaggerations. For colour version, see plates section. Reprinted by permission from Macmillan Publishers Ltd: *Nature* © 2005.

## 4.4 Transform faults and fracture zones

Spreading centres are offset at intervals by transform faults (Wilson, 1965), which are a key concept in plate tectonics (Morgan, 1968). The name arises because they allow plate boundaries of one type (ridge, transform or trench) or in one place to be 'transformed' into other boundaries elsewhere. Transform faults by definition follow small circles about plate rotation poles (Figure 4.2), and therefore exactly follow the azimuth of relative plate motion along them. Note that where a ridge–ridge transform offsets the spreading centre in a particular sense, the sense of relative plate motion is always in the opposite sense, i.e. left-lateral ridge offsets accompany right-lateral plate slip and vice versa (Figure 4.11).

The lengths of ridge–ridge transform faults range from ~30 km to ~1000 km at Romanche transform, which is currently the longest active ridge–ridge transform fault in the world. Ridge–ridge transforms normally leave traces, called fracture zones, that are approximately collinear with them; however, there is no relative motion along these traces (Figure 4.11). Although 'fracture zone' strictly applies to the inactive traces, it is often used for the whole structure including the active transform. 'Transform fault' is restricted to the actively slipping portion.

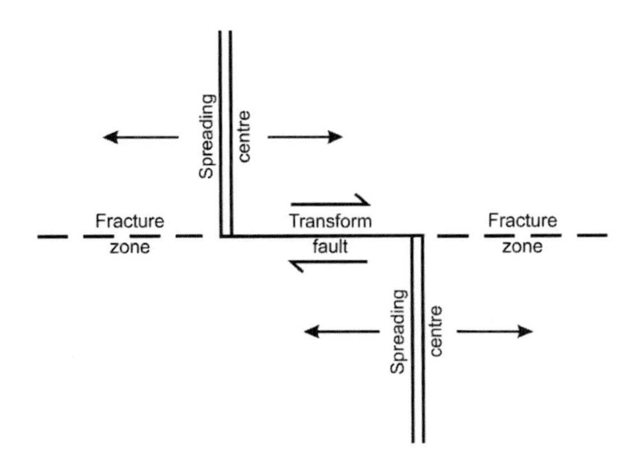


**Figure 4.11** A ridge–ridge transform fault. Double lines indicate spreading boundaries; single continuous line, transform fault; dashed lines, inactive fracture zone traces. A left-lateral offset of the plate boundary is accompanied by right-lateral slip on the transform, and vice versa.

Many transform faults reflect the original geometry of continental breakup. For example, the large equatorial Atlantic fracture zones such as Saint Paul (1° N), Romanche (0°) and Chain (1° S) can all be traced back to the continental margins, where associated structures can be seen in the continents (Le Pichon and Hayes, 1971). Because spreading centres tend to align perpendicular to the local spreading direction (Section 4.1), frequent offsets via small to medium offset transforms allow the ridge to maintain congruency with the original continent–ocean boundary (e.g., MAR from equator to Azores; see Figure 1.1). However, changes in spreading direction and major plate reorganisations can produce new transform faults or eliminate old ones, even in mid-ocean (Menard and Atwater, 1968).

There have been many detailed studies of ridge–ridge transform faults and some fracture zones (e.g., Fox and Gallo, 1984). Kane transform, with an offset of 150 km (10 Ma) at 23° N on the MAR is typical of many slow-slipping transforms (Pockalny *et al.*, 1988; Tucholke and Schouten, 1988/89; Figure 4.12), while Clipperton transform (offset 85 km or 1.5 Ma) at 10° N on the EPR (Gallo *et al.*, 1986; Figure 4.13) exemplifies many fast-slipping ones.

Ridge–ridge transform faults are characterised by linear 'transform valleys' tens of kilometres wide and kilometres deep. They generally increase in width and depth with the offset of the transform (Figure 4.14). The transform valley is partly an isostatic response to thinner crust at the transform (White *et al.*, 1984), and partly due to down-faulting resulting from thermal contraction and plate extension normal to the transform direction (Collette, 1974; Pockalny *et al.*, 1996). Often there is a flanking transverse ridge on one side (and rarely on both sides) of the transform, (for example, Vema transform, 11° N on the MAR; Heezen *et al.*, 1964a). Some of the largest transverse ridges expose significant quantities of serpentinised peridotite, whose buoyancy contributes to their elevation (Bonatti, 1976; 1978).

The transform valley deepens towards ridge–transform intersections, particularly at slow spreading, to produce specially deep basins, called 'nodal basins' (Figure 4.12). These have been interpreted as resulting from viscous head loss in rising asthenosphere (Sleep and Biehler, 1970), accentuated by the cooling effect of the opposing old, cold lithosphere, but they may also in part result from regional isostatic balancing of locally elevated lithosphere at the inside corners.

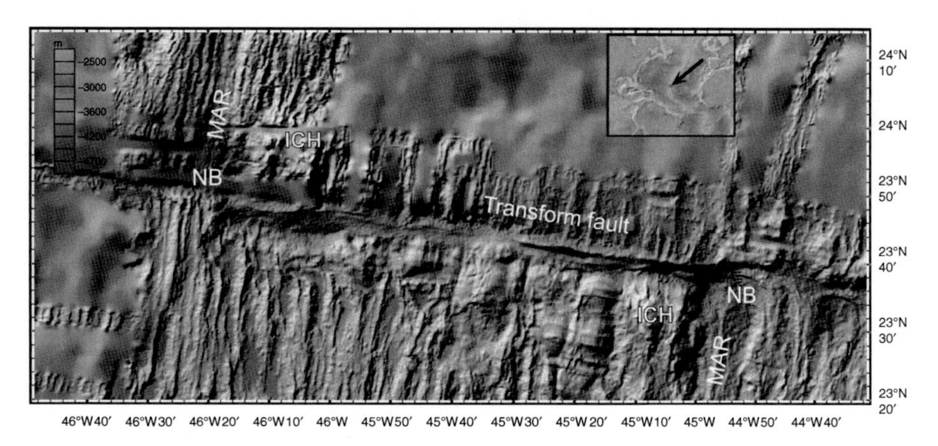


Figure 4.12 Shaded relief image of the slow-slipping, 150 km offset Kane transform fault, MAR. Inset shows location of main figure. ICH, Inside Corner High; MAR, MAR spreading axis; NB, Nodal Basin. Image produced using GeoMapApp (http://www.geomapapp.org; Ryan *et al.*, 2009). For colour version, see plates section.

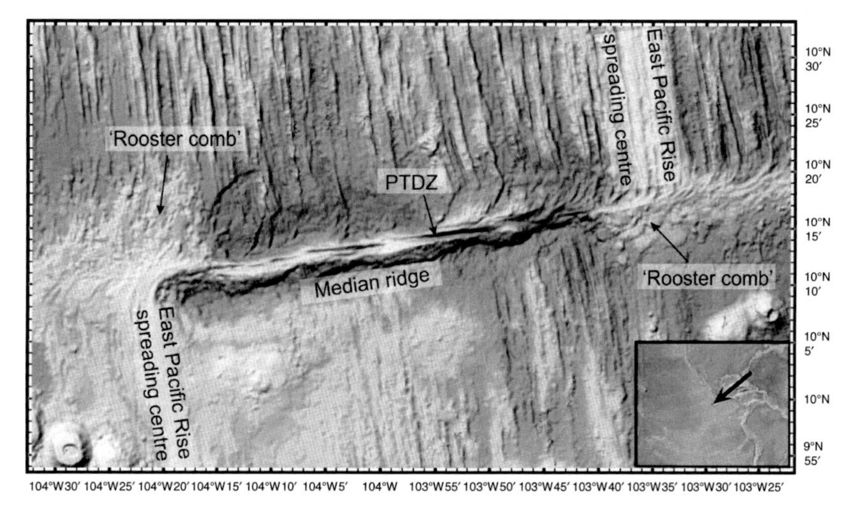


Figure 4.13 Shaded relief image of the fast slipping, 85 km offset Clipperton transform fault, EPR. Note prominent median ridge with superimposed scarps and troughs marking the principal transform displacement zone (PTDZ). Shallow 'rooster comb' areas opposite active spreading centres are thought to reflect reheating of older lithosphere. Inset shows location of main figure. Image produced with GeoMapApp (http://www.geomapapp.org; Ryan *et al.*, 2009). For colour version, see plates section.

The inside corners of slow-spreading ridge–transform intersections are anomalously shallow, and called 'inside corner highs', whereas the outside corners tend to be anomalously deep (Figure 4.12). Inside corner highs appear to be dynamically supported and have been explained by the effect of the transform decoupling the two opposing plates across the transform (Searle and Laughton, 1977; Severinghaus and Macdonald, 1988; Blackman and Forsyth, 1991). The necking that produces the median valley is accomplished via a force couple that simultaneously depresses the median valley floor and raises its flanks. At

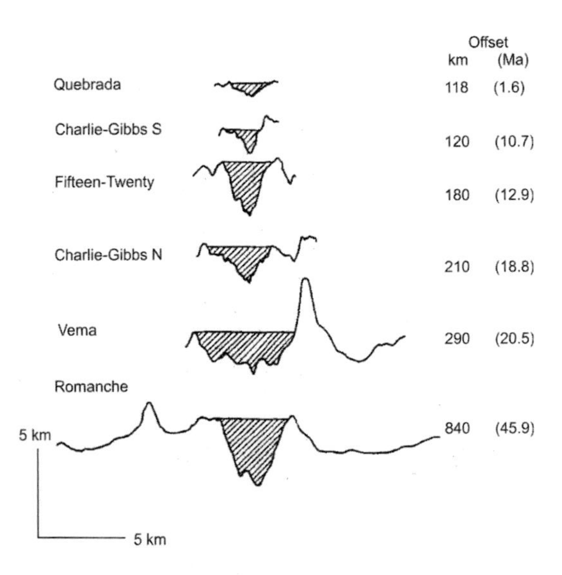


**Figure 4.14** Topographic profiles across transform valleys with different age and distance offsets. Transform valleys are shaded. Quebrada is a fast-slipping transform; the others are slow slipping.

a ridge–transform intersection, the lithosphere at the outside corner is continuous across the fracture zone, and is strongly coupled to the older lithosphere on the opposite side. This old, cold lithosphere will tend to be deeper because of its greater density, and so will tend to inhibit the formation of the rift mountains on the outside corner. In contrast, the inside corner is decoupled from the older lithosphere by the transform fault, allowing the couple that produces the rift valley to raise the inside corner. The effect may be enhanced by the superposition of a similar couple arising from normal faulting in response to extension across the transform as a result of ridge-parallel thermal contraction of the lithosphere (Turcotte, 1974). In addition, rift-flanking normal faults generally have larger throws near ridge offsets (Shaw and Lin, 1993), which will additionally enhance the topographic anomaly. Many inside corners are the sites of oceanic core complexes, whose formation will also contribute to the anomalous elevations (Section 7.4).

Fast-slipping transform faults are characterised by linear basins or troughs. They lack inside corner highs, but have anomalously elevated outside corners called 'intersection highs' or 'rooster combs' (Gallo *et al.*, 1986; Figure 4.13). These highs may be produced in part by the mechanism of Severinghaus and Macdonald (1988), since here the ridge topography is the inverse of the median valley of slow spreading ridges. However, Gallo *et al.* (1986) have suggested that intersection highs reflect enhanced thermal expansion and melt emplacement in the lithosphere opposite the offset spreading centre, and it seems likely that this is at least a contributing cause, if not the major one.

### 4.4.1 Elements of transform faults

Fox and Gallo (1984) divided transform faults morphologically into various elements. The 'transform domain' contains all transform-related tectonic elements such as the transform

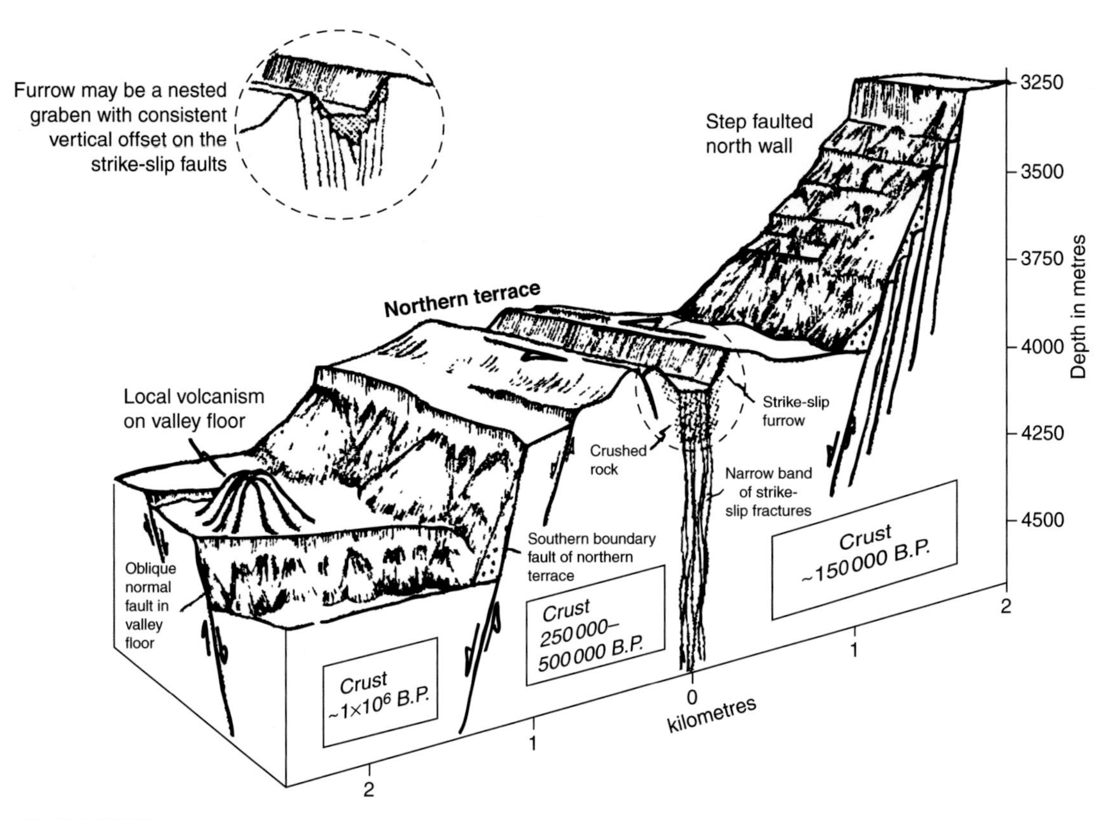


Figure 4.15 Block diagram showing the structure of part of the transform tectonised zone of the fast-slipping Quebrada transform fault, based on high-resolution near-sea-bed observations, after Lonsdale (1978).

valley, flanking ridge(s) and transform-related faults. The currently active strike-slip fault or narrow band of faults and other elements marking the instantaneous plate boundary is the principal transform displacement zone (PTDZ) and is usually a few hundred metres across. Over time the PTDZ appears to wander either side of its time-averaged position; this, together with other associated faults produces a band a few kilometres wide called the transform tectonised zone. The TTZ contains faults parallel to the transform axis that are interpreted as strike-slip faults and that take up, singly or jointly, the strike-slip motion of the transform. Other faults at high and low angles to the transform direction are interpreted as Riedel shears and tension gashes (Whitmarsh and Laughton, 1975; Macdonald *et al.*, 1986b; Section 7.6).

High-resolution studies using deep-towed side-scan and other instruments show that the PTDZ often contains a narrow median ridge a few hundred metres wide and high, typically with a small furrow along its crest that precisely follows the current plate boundary (Lonsdale, 1978; Kastens *et al.*, 1979; Macdonald *et al.*, 1979; Fox and Gallo, 1984; Gallo *et al.*, 1984; Macdonald *et al.*, 1986b; Figure 4.15). The ridge is composed of basaltic or serpentinite cataclastic rock formed by motion of the transform and forced up it by a small component of compression (Lonsdale, 1978; Macdonald *et al.*, 1979; Macdonald *et al.*, 1986b; Fornari *et al.*, 1989). Rarely, such as in Vema transform, the transform valley is

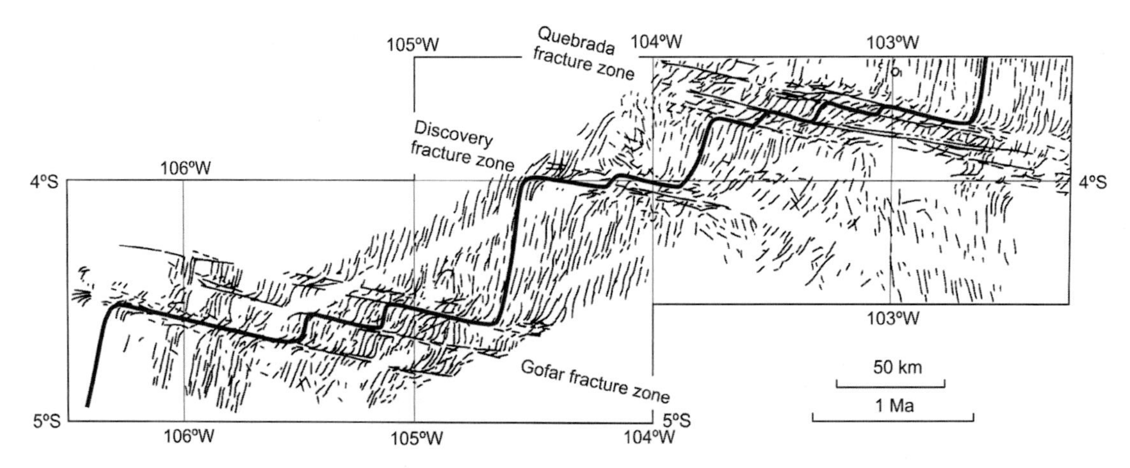


Figure 4.16 Tectonic features of the fast-slipping, multi-stranded Quebrada, Discovery and Gofar transform faults, after Searle (1983). Bold line marks the plate boundary; fine lines are fault scarps and other topographic lineations. Note tectonic fabric is rotated up to 45° oblique to the spreading direction near transform offsets.

flooded with sediments, allowing the PTDZ to be imaged subsurface by seismic reflection (Eittreim and Ewing, 1975; Bowen and White, 1986; Figure 2.16).

Slow-slipping transform faults mostly contain a single, well-defined PTDZ. However, fast-slipping transforms are often characterised by several transform strands only a few kilometres apart, separated by very short, often oblique, spreading centres or en echelon offsets (Searle, 1983; Gallo *et al.*, 1986; Figure 4.16).

## 4.4.2 Fracture zones

Fracture zones – the inactive traces of ridge–ridge transform faults – are also characterised by linear valleys, ridges and escarpments. They were easily detected by early single-beam bathymetry (Heezen, 1962; Heezen and Menard, 1963; Heezen *et al.*, 1964b; Menard, 1967; Menard and Atwater, 1969). Tucholke and Schouten (1988/89) mapped Kane Fracture Zone in detail across the whole width of the North Atlantic.

Most fracture-zone topography is similar to that described above for transform faults, with one important difference. Where there is a large age offset, the lithosphere on opposing sides of the fracture zone will have subsided thermally by different amounts (see Section 3.2.2), producing a bathymetric step with the older side deeper (Menard and Atwater, 1968; 1969). This is very obvious in the great Pacific fracture zones (Figure 4.17).

Like transform faults, fracture zones may also be flanked by a transverse ridge, usually on one side only. Such ridges do not appear to result from constructional volcanism, since normal sea floor tectonic fabric continues across them. Pockalny *et al.* (1996) suggested they are in part a flexural response of the lithosphere to transform-normal extension, while Abrams *et al.* (1988) concluded that they most likely arise from a combination

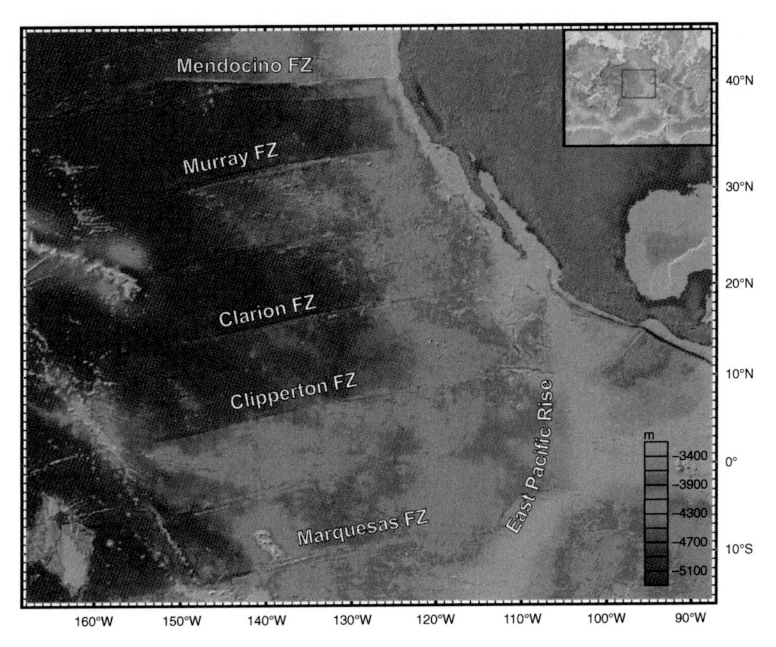


**Figure 4.17** Bathymetry of the north-eastern Pacific Ocean, showing the great Pacific fracture zones with their large topographic scarps. Inset shows location of main figure. Image produced using GeoMapApp (http://www.geomapapp.org; Ryan *et al.*, 2009). For colour version, see plates section.

of effects including thermal and viscodynamic forces operating near the ridge–transform intersection. It is not clear to what extent they are related to the rooster combs of fast-slipping transforms.

## 4.5 Ridge segmentation

The development and routine use of multi-beam echosounders in the 1970s provided a more comprehensive view of MORs. For example, the exact plate boundary was soon mapped along most of the EPR (Macdonald *et al.*, 1984; Macdonald *et al.*, 1986a; Figure 4.18).

The average spacing of major transform faults is usually several hundred kilometres, but the new multi-beam surveys showed many small, closer-spaced ridge offsets only a few tens of kilometres apart and with offsets of less than 30 km, producing a rich pattern of spreading centre segmentation. These small offsets often lack evidence of through-going strike-slip faults, so are technically not transforms; they are named non-transform offsets, or sometimes non-transform discontinuities (Lin *et al.*, 1990; Purdy *et al.*, 1990; Fox *et al.*, 1991; Grindlay *et al.*, 1992). One type, common on fast-spreading ridges, is the overlapping spreading centre (Macdonald and Fox, 1983; Figure 4.19).

Non-transform discontinuities divide the ridge axis into a number of 'segments', each a few tens of kilometres in length. Such segments, especially at slow-spreading ridges,

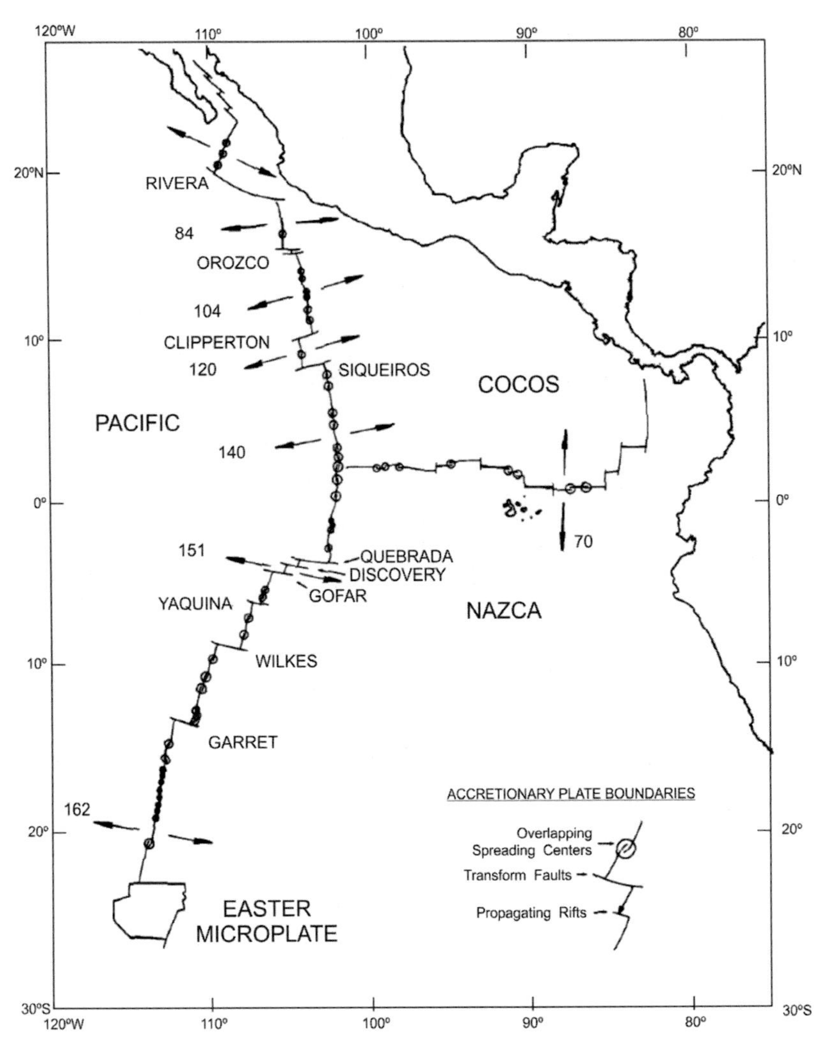


Figure 4.18 Plate boundaries in the eastern Pacific, showing spreading axes, transform faults, overlapping spreading centres and propagating rifts, after Macdonald *et al.* (1986a). Numbers are full spreading rates in km $Ma^{-1}$. Tectonic plates and transform faults are labelled.

may reflect small-scale convection cells arising from Rayleigh–Taylor density instabilities in the asthenosphere (Whitehead *et al.*, 1984; Schouten *et al.*, 1985; Lin *et al.*, 1990), although this effect is much reduced at fast-spreading ridges (Lin and Phipps Morgan, 1992). Nevertheless, Macdonald *et al.* (1991) suggest that different orders of segmentation (Section 4.6) reflect different degrees of mantle upwelling and melt delivery to the crust. As well as being bounded by transform and non-transform offsets, spreading segments, especially at slow-spreading ridges, are characterised by particular patterns of crustal thickness variation (Lin *et al.*, 1990) and faulting (Shaw, 1992), which will be discussed in detail in Chapters 5 to 7.

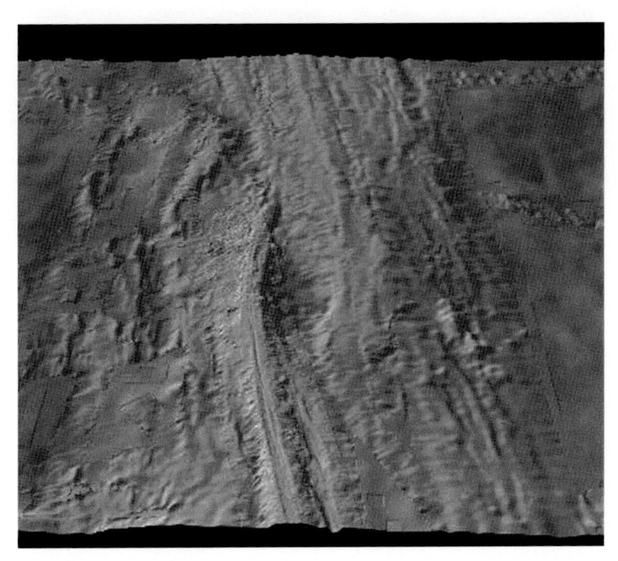

Figure 4.19 Oblique, shaded relief image of an overlapping spreading centre, EPR 11°45′ N, viewed towards the north. Depths range from approximately 2600 m (grey) to 3100 m (purple). Image is approximately 50 km wide at middle distance, and E–W gap between overlapping spreading centres is approximately 10 km at the widest point. Image produced using GeoMapApp (http://www.geomapapp.org; Ryan *et al.*, 2009). For colour version, see plates section.

## 4.6 The hierarchy of ridge axis discontinuities

Grindlay *et al.* (1991) and Macdonald *et al.* (1991) proposed a 'hierarchy' of ridge axis discontinuities and the segments they demarcate. Although there is continuous variation, they defined four 'orders' of structure, marking different stages in the continuum (Figure 4.20; Table 4.1). Segments are generally bounded by discontinuities of similar order. 'First-order' segments are hundreds of kilometres long and bounded by transforms; second-, third- and fourth-order segments are approximately 100, 50 and 15 km long, respectively, and mostly bounded by second- to fourth-order discontinuities.

### 4.6.1 First-order discontinuities

First-order discontinuities are transform faults (Section 4.4). They are regarded as 'rigid' since neither the offset geometry nor the position along the spreading centre varies with time. They contain a well-defined strike-slip fault parallel to the spreading direction, and have offsets of about 30 km or greater. This is similar to the effective thickness of young lithosphere, and is likely to be the minimum size of feature needed to develop a single, coherent and persistent through-going fault. Transforms are generally spaced several hundred kilometres apart, bounding first-order segments (sometimes called 'supersegments', such as that between the Kane and Atlantis transforms on the MAR (Figure 4.21). However, a few slow-spreading transforms such as the Charlie–Gibbs double transform in the North

**Table 4.1** The hierarchies of ridge-axis offsets, modified after Grindlay *et al*. (1991)

| Order of offset | Name | Offset length (km) | Defined by | Axial morphology | Off-axis trace | Stability |
|---|---|---|---|---|---|---|
| First | Transform fault | 30–1000 | Morphology | Linear valley parallel to spreading direction | Linear valley | Long lived, no along-axis migration |
| Second | Large overlapping spreading centre (fast spreading) | 3–~30 | Morphology | Overlapping spreading ridges flanking a basin | Zones of discordant topography, often oblique to spreading | Duration <~1 Ma; axial migration; 'self decapitation' |
| | Propagating rift (intermediate to fast spreading) | ~25 | Morphology | V-shaped neo-volcanic zone and pseudofault traces | V-shaped pseudofaults | Duration a few Ma; along-axis migration |
| | Oblique spreading centre (slow spreading) | ~5–30 | Morphology | 45° oblique median valley | Zones of discordant topography or non-linear valleys | Durations of several Ma, but can disappear or grow into transform faults; may migrate along axis |
| | En echelon jog (slow spreading) | <30 | Morphology | Offset neo-volcanic zones overlap slightly or separated by ridge-parallel basin | Zones of discordant topography, often oblique to spreading | Duration a few Ma; along-axis migration, may disappear or grow into transforms; may migrate along axis |
| Third | Small overlapping spreading centre (fast spreading) | 3–5 | Morphology | Overlapping spreading ridges flanking a basin | No off-axis trace | Short-lived |
| | Intervolcano gap (slow spreading) | <~5 | Morphology | Break in axial volcanic ridge with minor offset | No off-axis trace | Short-lived |
| Fourth | DEVAL[a] (fast spreading) | 0 | Morphology, geochemistry | Small deviation in axial trend | No off-axis trace | Short-lived? |
| | SNOO[b] (fast spreading) | <1 | Geochemistry | | No off-axis trace | Short-lived? |
| | Linked central volcano (slow spreading) | 0 | Geochemistry | | No off-axis trace | Short-lived? |

[a] DEViation from Axial Linearity.
[b] Small, Non-Overlapping Offset.

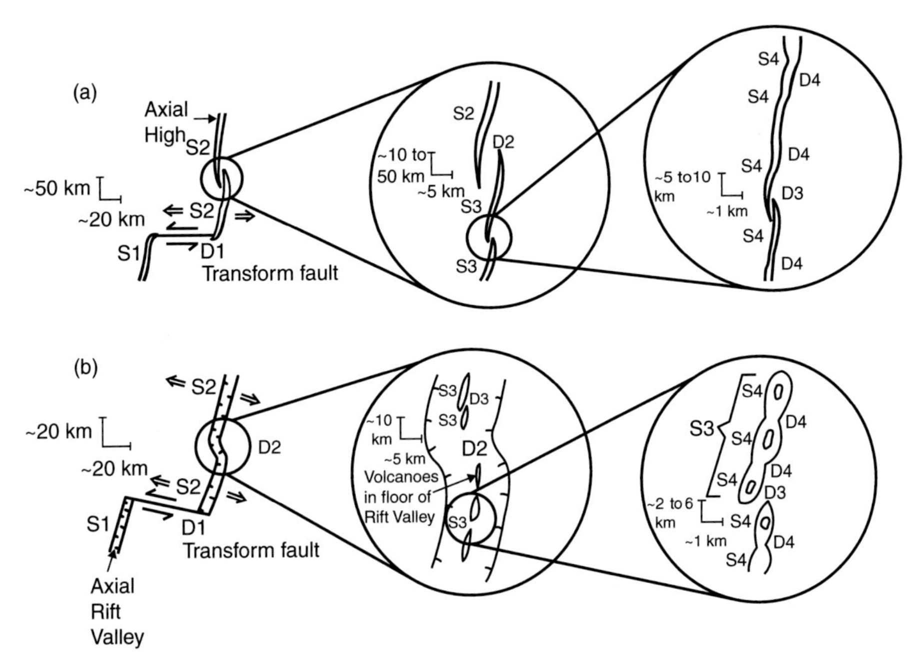


Figure 4.20 Hierarchy of spreading segments and offsets, after Macdonald *et al.* (1991). (a) Fast-spreading ridges; (b) slow-spreading ridges. D1 to D4 indicate ridge discontinuities of order 1 (transform fault) to 4 (DEVALS or inter-volcano gaps), while S1 to S4 indicate corresponding segments of order 1 to 4. From Macdonald *et al.* (1991). Reprinted with permission from AAAS.

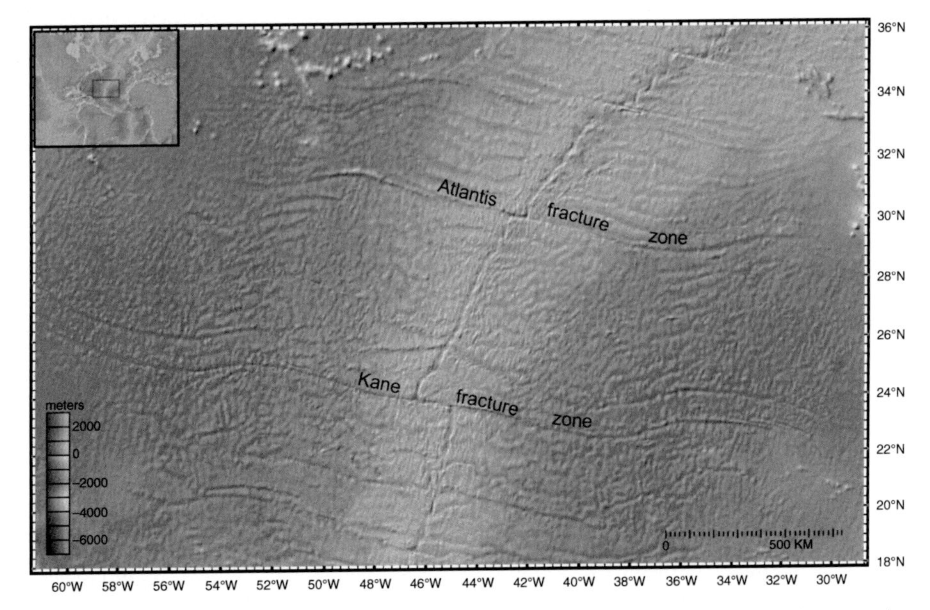


Figure 4.21 Shaded relief bathymetry of the central North Atlantic, showing off-axis traces of transform faults and non-transform offsets. Fracture zone traces show an early history of WNW–ESE spreading, followed by a period of WSW–ENE, then reverting to WNW–ESE. Traces of some but not all non-transform offsets approximately follow these trends, while others migrate along-axis. Image produced using GeoMapApp (http://www.geomapapp.org; Ryan *et al.*, 2009). For colour version, see plates section.

Atlantic (Searle, 1981) and the Oriente and Swan Islands transform pair in the Caribbean are no more than 100 km apart. Because transforms do not migrate, their off-axis traces – fracture zones – represent spreading flow lines, and readily show changes in plate motion (Figure 4.21).

### 4.6.2 Non-transform discontinuities

Second-order discontinuities have offsets from a few kilometres up to about 30 km. Their morphology differs markedly depending on spreading rate. At fast-spreading ridges they are represented by overlapping spreading centres (Macdonald and Fox, 1983; Macdonald *et al.*, 1984; Figure 4.19). Here, instead of being linked by a transform fault, the ends of the offset spreading segments overlap each other by about three times the offset length, curving in a characteristic way that is predicted by crack propagation theory (Sempéré and Macdonald, 1986). An 'overlap basin' forms between the spreading limbs. Clearly this overlapping must lead to deformation and/or rotation in the region of the overlap. These are 'non-rigid' offsets in the terminology of Grindlay *et al.* (1991), since they can readily change their geometry and migrate freely along the rise axis, leaving traces of disrupted topography that are V-shaped rather than flow-line-parallel. They can interact in complex ways, propagating through each other and cutting off their tips by 'self-decapitation' (Macdonald *et al.*, 1987; Cormier *et al.*, 1996; Cormier, 1997; Figure 4.22). Propagation of OSCs along a rise axis transfers material from one plate to the other and so leads to time-averaged asymmetric spreading; it can also provide a mechanism for ridge reorientation following a change in spreading direction (Cormier and Macdonald, 1994).

At slow-spreading ridges, second-order discontinuities take the form of oblique offsets in the spreading axis and median valley, such as occurs on the MAR at Kurchatov Fracture Zone (Searle and Laughton, 1977) and 33.5° S (Grindlay *et al.*, 1991; Figure 4.23). There is no transcurrent fault, rather a series of oblique-slip normal faults and tension gashes in the offset region. The formation of an oblique offset rather than an overlapping one presumably occurs because at slow spreading rates the axial lithosphere is thicker and stronger, resisting the propagation, rotation and deformation that overlapping spreading centres imply. The inactive traces of these offsets are valleys similar to fracture zone valleys, with two exceptions. First, these are also 'non-rigid' offsets, so can migrate along the ridge axis, forming V-shaped fracture zone traces or, in the case of smaller offsets, V-shaped series of basins (Figure 4.24). Secondly, the oblique faults that form in the offset region create a saw-tooth rather than a linear scarp along the edge of the fracture zone valley (Searle and Laughton, 1977).

Because non-transform offsets may migrate, spreading segments bounded by them can shrink or expand if the offsets migrate in opposite directions (Gente *et al.*, 1995; Tucholke *et al.*, 1997; Maia and Gente, 1998; Figure 4.24). Expanding segments are often shallower, with evidence of more robust volcanism and thicker magmatic crust, suggesting that increasing magmatism is a driving force for expansion.

The length of ridge offset can change as a result of asymmetric spreading in one or both of its flanking segments. This can lead to shrinking transforms turning into non-transform

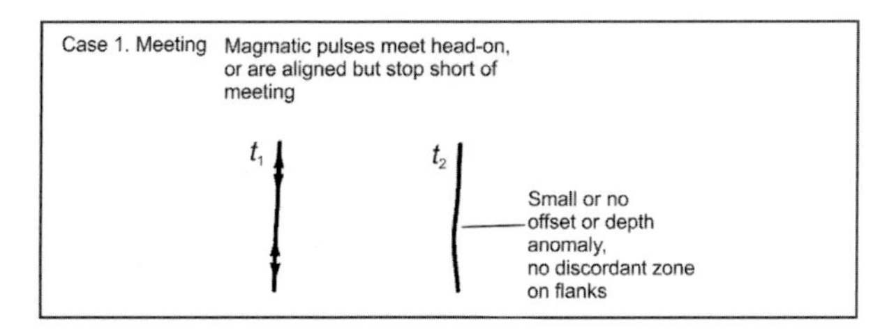


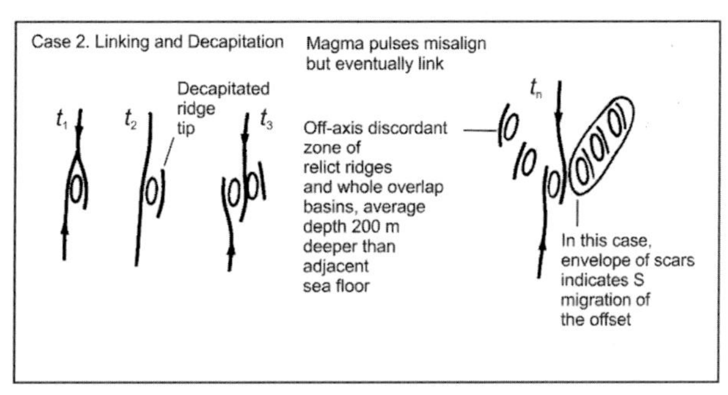


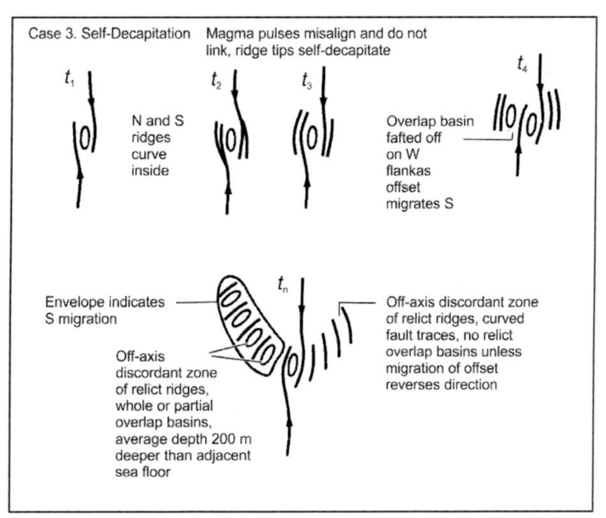


**Figure 4.22** Evolution of overlapping spreading centres, after Macdonald *et al.* (1988b). Reprinted by permission from Macmillan Publishers Ltd: *Nature* © 1988.

offsets and vice versa (Grindlay *et al.*, 1992), and even in the sense of offset reversing over time. Several examples can be seen in Figure 4.24. Some offsets shrink to and remain at zero offset for some time, before growing again, temporarily forming a 'zero-offset fracture zone' (Schouten and White, 1980). This persistence of segment boundaries in the absence of a ridge offset supports the idea that spreading segments reflect segmentation of upper mantle flow or melt delivery. Three-dimensional numerical modelling of melt production and migration has failed to predict mantle diapirs at segment centres, but shows that melt rising relatively uniformly beneath the axis is strongly focussed into segment centres by migration along the sloping lower boundary of the lithosphere (Magde and Sparks, 1997).

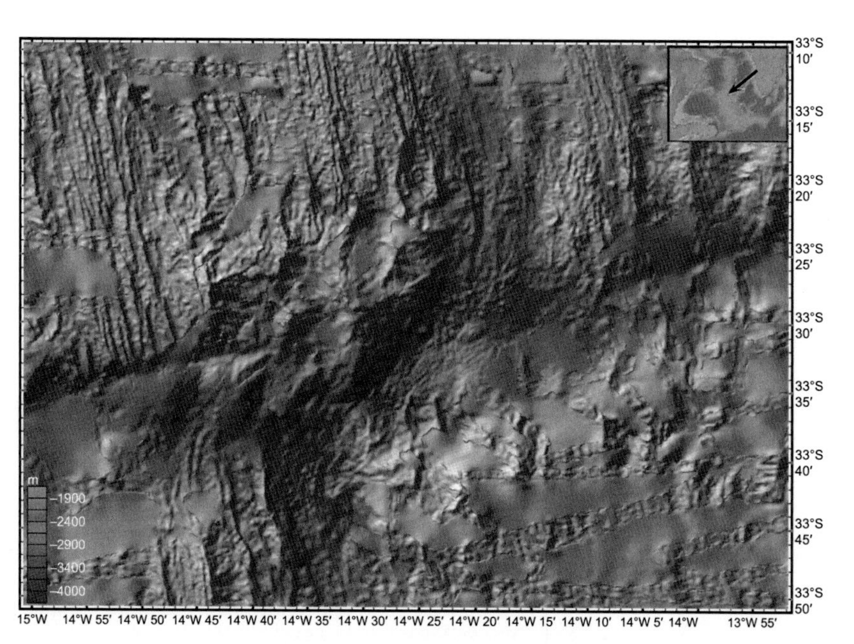


**Figure 4.23** Bathymetry of a non-transform offset: MAR at 33°30′ S. Inset shows location. Image is approximately 120 km wide. Spreading centre at 14°35′ W in the south is offset approximately 30 km to the NE at 14°20′ W. Image produced using GeoMapApp (http://www.geomapapp.org; Ryan *et al.*, 2009). For colour version, see plates section.

Grindlay *et al.* (1992), Macdonald *et al.* (1992) and others have described third- and fourth-order ridge discontinuities and associated segments. They are marked by offsets of axial volcanoes by less than 2 km or by discontinuities in the chemistry of erupted lavas (Table 4.1).

## 4.7 Triple junctions

Where any three tectonic plates meet, they do so at a geometrical point called a triple junction. McKenzie and Morgan (1969) analysed the geometry and stability of all possible configurations of triple junction, denoting the plate boundary types as ridge (R), transform fault (F) or trench (T). The stability of the configuration depends on both the geometry of the three plate boundaries and the relative velocities between them. McKenzie and Morgan found that quadruple junctions are always unstable and rapidly evolve into two triple junctions. Of sixteen possible triple junction configurations, RRR is the only one that is unconditionally stable, that is, it retains the RRR configuration as it evolves. Several other configurations are stable for certain special conditions.

Several actual triple junctions have RRR configuration, at least at the regional scale. These include the Bouvet (Southern MAR, SWIR and American–Antarctic Ridge), Galapagos (north and south EPR with Cocos–Nazca Spreading Centre), Chile (EPR, Pacific–Antarctic Rise and Chile Rise), Azores (MAR and Azores spreading centre) and Rodrigues (Central

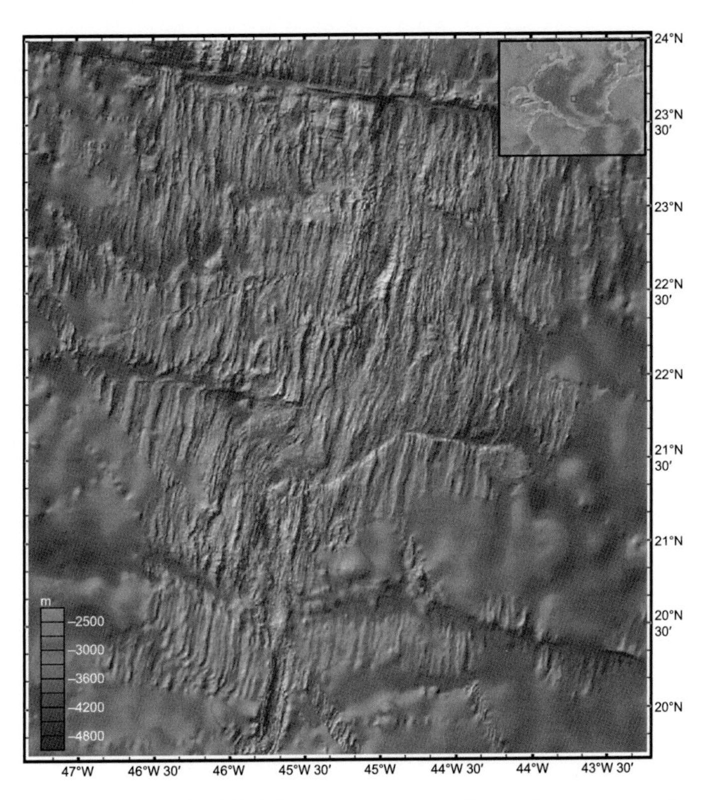


Figure 4.24 Shaded relief bathymetry of MAR 20° N–24° N, showing oblique traces of expanding segments and migrating non-transform offsets. Note that V-shaped wakes imply migration of offsets away from a centre near 22°30′ N, and that an erstwhile transform near 21°30′ N has recently converted to a southward-migrating non-transform offset. Image width approximately 400 km. Image produced using GeoMapApp (http://www.geomapapp.org; Ryan *et al.*, 2009). For colour version, see plates section.

Indian Ridge, SWIR and Southeast Indian Ridge). However, even these probably have different, and evolving, configurations at a local scale. The Galapagos (Lonsdale, 1988) and Chile (Bird *et al.*, 1998) triple junctions both have microplates at their cores (see Section 4.9 and Figure 4.30), so regionally three major plates and one microplate meet at three triple junctions, which may each evolve rapidly (Bird *et al.*, 1999). The Azores triple junction may have been FFR in the past (Searle, 1980), and the Rodrigues triple junction appears to cycle between RRR and RFR (Patriat and Courtillot, 1984; Figure 4.25). Indeed, Patriat and Courtillot suggest that most regional RRR junctions are also compatible with RFR geometry, and may alternate between the two depending on the relative dominance of magmatism or tectonism.

An interesting application of these ideas is the impingement of the old Pacific–Farallon RFR plate boundary onto the North American–Farallon trench at 30 Ma, which led to the creation of the San Andreas fault between a TFF and an RTF triple junction (Atwater, 1970; Figure 4.26). The triple junctions then migrated apart along the San Andreas fault, until the southern one met the eastern end of the Murray transform, whereupon it became a

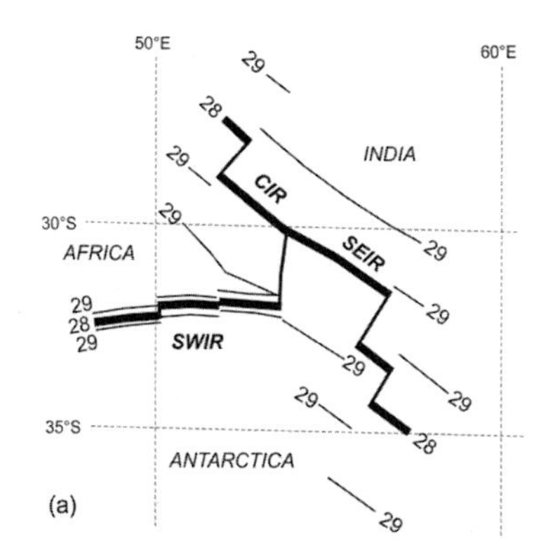


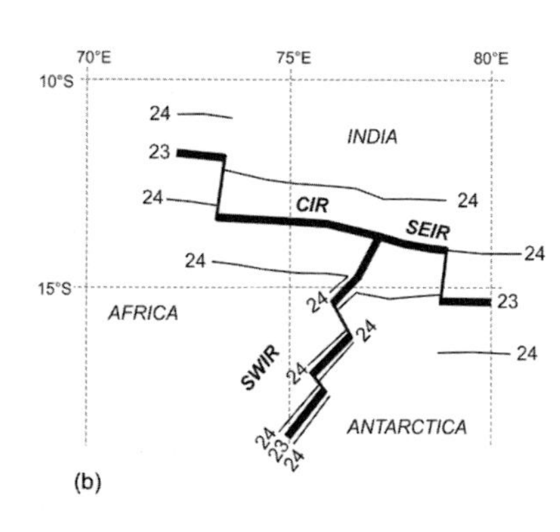


Figure 4.25 Evolution of the Rodrigues triple junction, central Indian Ocean, after Patriat and Courtillot (1984). Bold lines: spreading centres or 'ridges' (R); medium lines: transform faults (F); light lines with numbers: isochrons with magnetic anomaly number; dashed lines: palaeo-latitude and longitude in Indian reference frame. CIR, SEIR and SWIR are Central Indian Ridge, Southeast Indian Ridge and Southwest Indian Ridge, respectively. (a) RRF configuration at anomaly 28 time (58 Ma); (b) RRR configuration at anomaly 23 time (51 Ma).

second TFF and began moving north again (Figure 4.26c). Upon reaching the western end of Murray transform it changed back to RTF and began migrating south once more.

## 4.8 Propagating rifts

A particular form of non-transform offset is the propagating rift (Hey *et al.*, 1980). Grindlay *et al.* (1992) classify these as first-order discontinuities, but they lack a transcurrent fault and usually have offsets <30 km, making them more like second-order discontinuities. They are most common at medium-spreading ridges. Few have been clearly recognised at slow-spreading ridges, although migrating second-order offsets have some of their attributes. Large, propagating overlapping spreading centres on fast-spreading ridges closely resemble them.

At a propagating rift (Figure 4.27), one segment (the propagating rift) lengthens while the other (the 'doomed rift') shortens. The propagating spreading tip generates a clearly defined V-shaped wake composed of two 'pseudofaults', so-called because they offset magnetic isochrons even though there is never any strike-slip motion across them (Hey, 1977). (In fact, fracture zones can be considered a special case of pseudofaults in the limit when the propagation rate is zero.) The inner pseudofault lies on the inside corner or 'transform' side of the propagating rift tip, the outer pseudofault on the outside corner. As it propagates, the rift transfers lithosphere from one plate to the other. The dying tip of the doomed rift also leaves a topographic trace, usually a series of en echelon basins, collectively referred to as the 'failed rift' (Figure 4.28). The angle between the propagating rift axis and either pseudofault, and between the failed rift and the doomed rift, is

$$\alpha = \tan^{-1}(s/p), \tag{4.2}$$

where $s$ is the plate accretion rate and $p$ is the propagation rate (Figure 4.27).

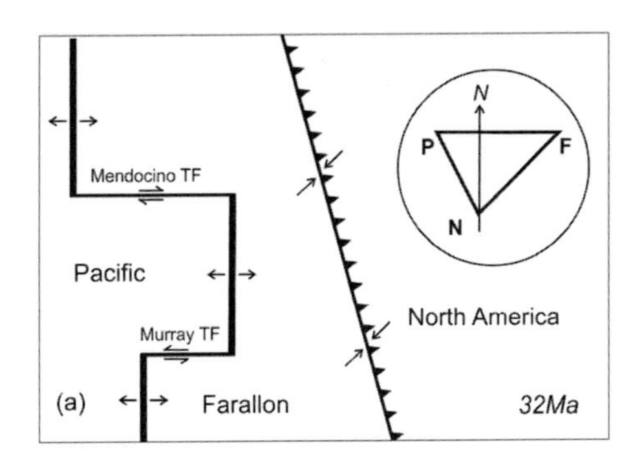


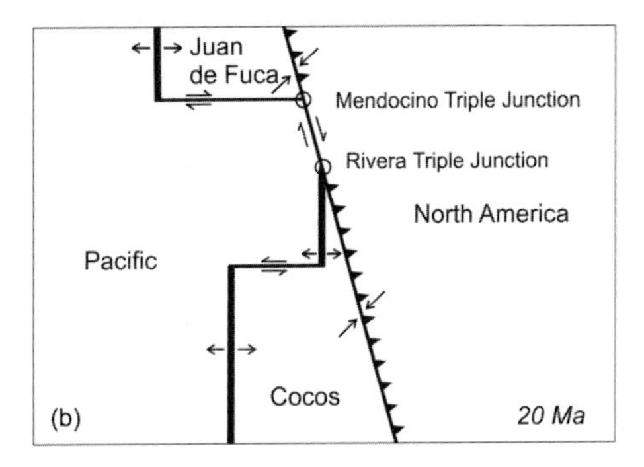


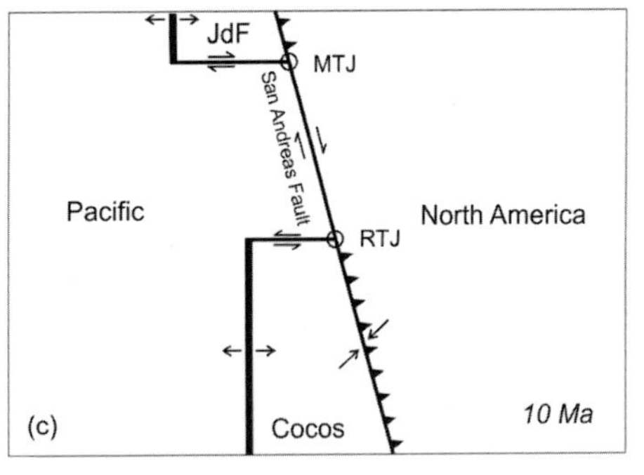


Figure 4.26 Evolution of San Andreas fault, after Cox and Hart (1986), with permission from John Wiley & Sons. Bold lines: spreading centres (ridges); plain fine lines: transform faults; toothed lines: subduction zones with teeth on over-riding plate; circles: triple junctions; double arrows: relative motion at the plate boundaries. Inset shows the relative velocity vector triangle between the three plates. (a) Approximately 32 Ma: the Pacific–Farallon Spreading Centre is approaching the Farallon–North America trench; (b) Approximately 20 Ma: part of the Pacific–Farallon Spreading Centre has been subducted, splitting the Farallon plate into the Juan de Fuca plate and the Cocos plate, and forming the Mendocino and Rivera triple junctions. Since the Pacific–North America relative velocity (P–N in the velocity diagram) is parallel to the new plate boundary, that boundary becomes a transform fault, the San Andreas. (c) The triple junctions move along the San Andreas fault as the new boundary evolves.

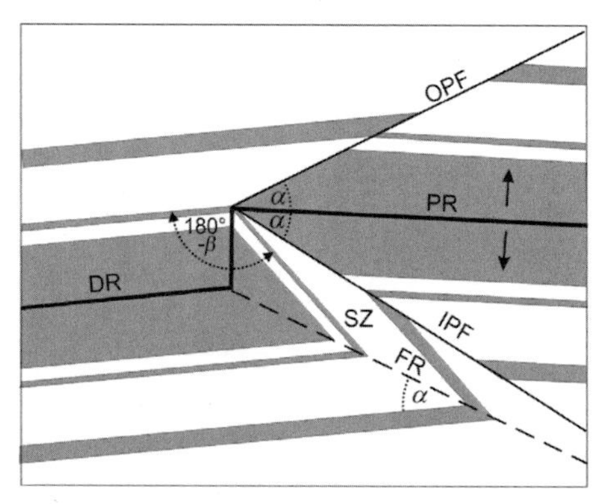


Figure 4.27 Geometry of a propagating rift, after Hey *et al.* (1986). Bold line: active plate boundary, with propagation to the left. Grey stripes: normally magnetised blocks; arrows: spreading direction; dashed line: failed rift; medium lines: pseudofaults; DR, doomed rift; FR, failed rift; IPF, inner pseudofault; OPF, outer pseudofault; PR, propagating rift; SZ, sheared zone. Angles $\alpha$ and $\beta$ are defined in text Equations (4.2) and (4.3).

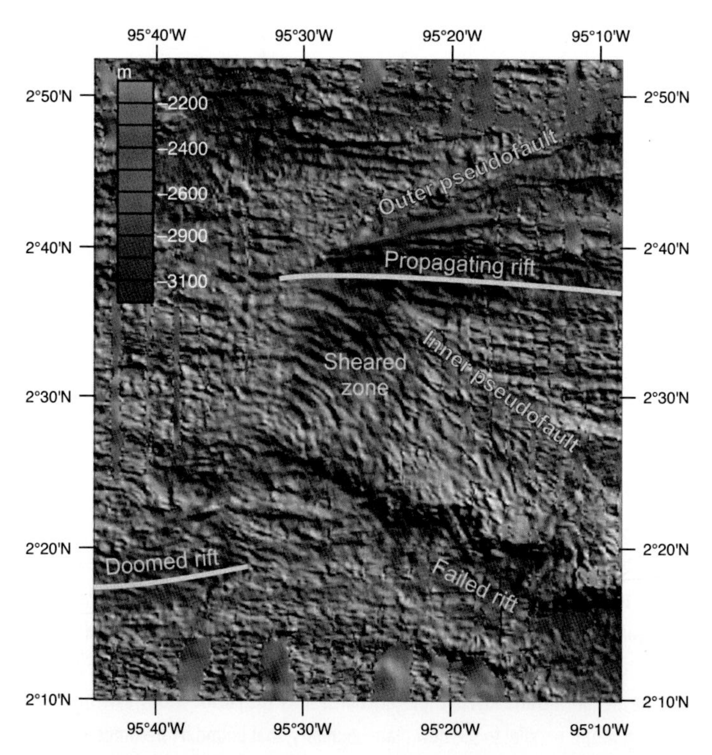


Figure 4.28 Shaded relief bathymetry of the propagating rift at 95° W on the Cocos–Nazca Spreading Centre, illuminated from NE. Yellow lines mark the propagating and doomed rift axes. Image produced using GeoMapApp (http://www.geomapapp.org; Ryan *et al.*, 2009). For colour version, see plates section.

In the original propagating rift model, the propagating rift tip and doomed rift tip are linked by an instantaneous transform fault that migrates along the ridge axis at the propagation rate (so-called 'bacon-slicer' model). However, detailed studies have found no evidence for the existence of even an instantaneous transform fault (Kleinrock and Hey, 1989a, b). Rather, the propagating rift deforms the region between the rift tips by bending pre-existing ridge-parallel fault blocks like driving a wedge into them, producing 'book-shelf faulting' (Searle and Hey, 1983). This results in a zone of obliquely lineated topography called the 'sheared zone' (Hey *et al.*, 1986; Figure 4.28). The angle between the rotated lamellae and the doomed rift is

$$\beta = \tan^{-1}(2s/p) \tag{4.3}$$

(see Figure 4.27). In practice, the propagating and doomed rifts often overlap to some extent, so that the spreading rate gradually diminishes towards the rift tips. There may be a continuum of structures between propagating rifts and migrating, large overlapping spreading centres. McKenzie (1986) discusses the detailed geometry of propagating rifts.

Rifts are observed to propagate away from regional topographic highs, and have been successfully modelled in terms of crack propagation driven by the excess topographic load (Phipps Morgan and Parmentier, 1985). Where there has been a change in regional stress or spreading direction, rifts may propagate normal to the new direction, sweeping away minor offsets inherited from the older regime to leave longer, offset-free ridge segments normal to the new direction (Searle and Hey, 1983). There are cases of adjacent rifts propagating back and forth (so called 'duelling propagators'), which can form complex tectonic and isochron patterns (Hey and Wilson, 1982; Macdonald *et al.*, 1988a; Cormier *et al.*, 1996).

Propagating rifts transfer small amounts of lithosphere from one tectonic plate to another. They are probably the major agents in the realigning of plate boundaries, since it may be easier to propagate an existing rift than to create an entirely new one by rifting older lithosphere. They have been closely involved in the evolution of new spreading centres by rifting oceanic lithosphere (Lonsdale, 2005), and probably in the original creation of oceanic spreading centres by continental rifting (Courtillot, 1982).

## 4.9 Oceanic microplates

An oceanic microplate can be thought of as an extreme case of an overlapping spreading centre. Here two spreading centres may overlap by several hundred kilometres, with the tips of the bounding ridges being joined to the major spreading plate boundaries by transform faults or more complex boundaries (Hey *et al.*, 1985; Anderson-Fontana *et al.*, 1986; Searle *et al.*, 1989; Larson *et al.*, 1992; Figure 4.29). Well-defined microplates are only seen at fast-spreading ridges, probably because their evolution requires thin, weak axial lithosphere to facilitate continuous readjustment of the microplate boundaries and the deformation of one or more plates as they evolve (Searle *et al.*, 1993).

As at overlapping spreading centres, the spreading rate along each overlapping arm must progressively slow towards its tip, so the total separation rate of the major plates remains

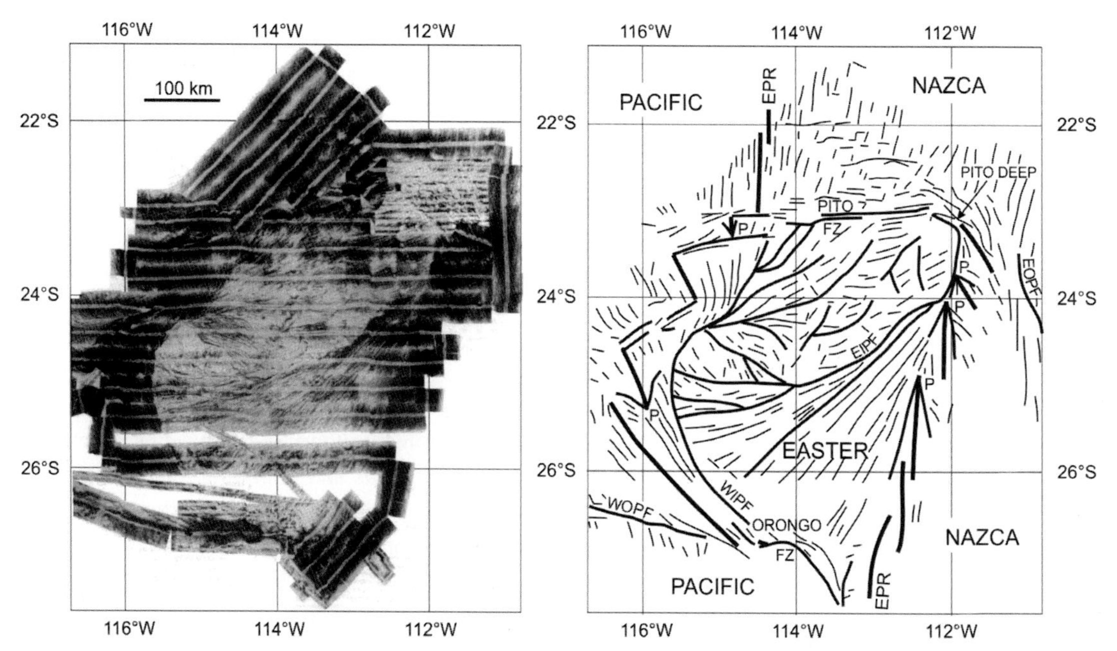


Figure 4.29 Easter microplate, after Searle *et al.* (1989). Left: GLORIA and SeaMARC II side-scan sonar mosaics, with darker tones for high backscatter. Right: tectonic interpretation. Heavy lines: spreading centres; medium lines: pseudofaults, transforms and fracture zones; light lines: representative tectonic fabric (parallel to the palaeo-ridge direction). EIPF and EOPF: eastern inner and outer pseudofaults; EPR: East Pacific Rise; FZ: fracture zone; P: propagating rift tip; WIPF and WOPF: western inner and outer pseudofaults; large capitals: plate names. Fanned lineaments in the southern half of the microplate reflect rapid rotation, and complex network of pseudofaults in the northern half reflect early microplate history.

constant. This causes the microplate to either shear or rotate. There is little evidence for interior shear, except early in microplate evolution (Bird *et al.*, 1998). However, long-range side-scan sonar imaging clearly shows fanned tectonic spreading fabric (fault scarps formed near and parallel to spreading centres; see Chapter 7) indicating progressive rotation (Searle *et al.*, 1989; Larson *et al.*, 1992; Rusby and Searle, 1995; Figure 4.29).

Microplate boundaries evolve rapidly. An evolving microplate must grow continuously unless spreading on its bounding arms is 100% asymmetrical. Both the eastern and western boundaries of Easter microplate contain several propagating rifts, each propagating in the direction of growth of the overlapping limbs, but also toward the interior of the microplate (Figure 4.29). This acts to limit the E–W growth of the microplate, enabling it to retain a roughly circular outline that facilitates continuing rotation. Nevertheless, the corners of the major plates immediately north of both Easter and Juan Fernandez microplates have suffered significant compression and have been deformed by folds and thrust faults (Rusby and Searle, 1993; Searle *et al.*, 2006).

Analysis of the tectonic fabric and magnetic anomalies at the Easter and Juan Fernandez microplates has yielded detailed tectonic histories of these microplates (Figure 4.30) that are very similar (Naar and Hey, 1991; Rusby and Searle, 1995; Bird *et al.*, 1998; Searle

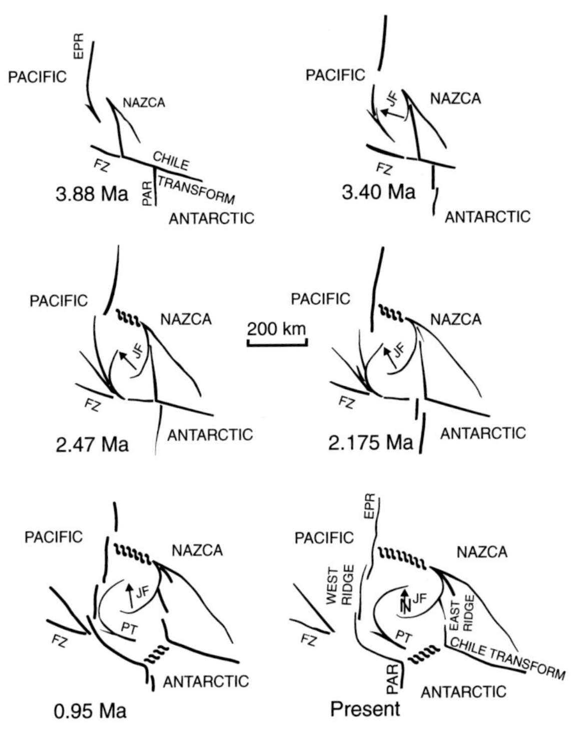


Figure 4.30 Tectonic evolution of Juan Fernandez microplate, after Larson *et al.* (1992). Plate boundaries and major pseudofaults are shown at six epochs. En echelon hatching: complex compression and other deformation at northern and southeastern microplate boundaries; arrows: palaeo-north direction in microplate.

*et al.*, 1993). This is remarkable since Easter lies on a single spreading centre (the EPR) and is bounded by two major plates, while Juan Fernandez lies at a triple junction between the Pacific, Nazca and Antarctic plates (Bird *et al.*, 1999). Each microplate grew from an initial, small overlapping spreading centre over the course of only a few million years, with the microplates rotating at rates $>10^\circ$ Ma$^{-1}$. These large rotations have been confirmed by palaeomagnetic measurements (Cogné *et al.*, 1995).

Schouten *et al.* (1993) produced an elegant kinematic model of microplate evolution, in which the microplate behaves like a ball-bearing rotating between two major plates (Figure 4.31). If there is no slippage along the microplate boundaries, the poles of rotation defining the motion of the microplate relative to each major plate must lie on the microplate boundary. As a consequence, the relative velocities and thus the nature of the boundary between the major plates and the microplate vary rapidly over short distances. This simple model accurately predicts the traces of the pseudofaults left by the propagating rifts as a microplate evolves (Figure 4.31c). Detailed analysis of Easter and Juan Fernandez microplates shows that the poles do indeed lie close to the microplate boundaries (Naar and Hey, 1991; Rusby and Searle, 1995). Neves *et al.* (2003) showed that the observed stress distribution is consistent with microplates being driven by shear tractions as in the

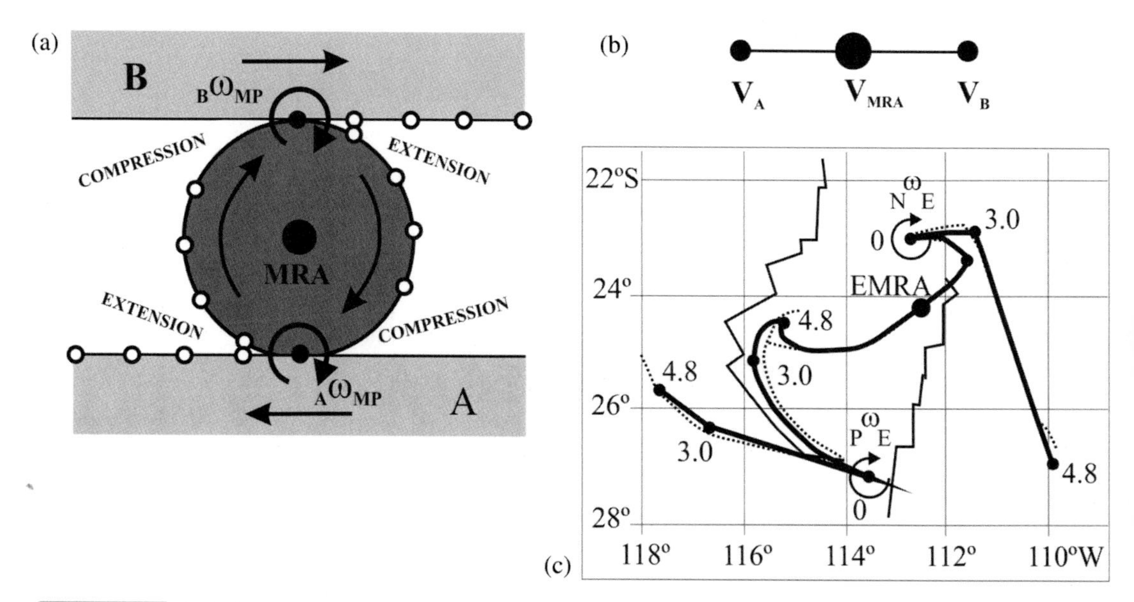


**Figure 4.31** Roller-bearing microplate model, after Schouten *et al.* (1993). (a) Shear between two plates A and B (straight arrows) causes a circular microplate between them (large filled circle) to rotate (long curved arrows) about the microplate rotation axis MRA. The relative motion between each major plate and the microplate is described by rotation vectors $_A\omega_{MP}$ and $_B\omega_{MP}$ about Euler poles located where the major plates touch the microplate (small filled circles and tightly curved arrows). Small open circles show earlier positions of these poles on the three plates, which define pseudofault traces. As the microplate approaches and leaves the instantaneous rotation poles it produces regions of compression and extension. (b) Velocity diagram for (a) showing relative velocities of the two major plates ($V_A$, $V_B$) and of the microplate rotation axis ($V_{MRA}$). (c) Application to Easter microplate. Dotted lines: observed pseudofaults; heavy lines: best fit roller bearing model; medium continuous lines: plate boundaries; numbered dots: prior positions of Euler poles on pseudofaults at 0, 3.0 and 4.8 Ma. Model involves two phases of evolution: from 4.8 to 3.0 Ma there is rapid northward propagation of the eastern rift tip, slower southward propagation of the western tip and progressive slowing of rotation; from 3.0 Ma to the present there is a constant rotation rate with propagation of both rift tips roughly along the Pacific–Nazca relative motion direction and rotation rate is constant.

roller-bearing model, but that the ridge resistance force that varies rapidly along the bounding ridges may also contribute to the driving force.

As microplates grow, their overlapping spreading arms initially propagate into progressively older lithosphere. However, curvature on these arms may allow one of them to reconnect to the opposite spreading centre, analogously to 'self-decapitation' of overlapping spreading centres. If so, the opposite arm fails, and the microplate is transferred entirely to one or other of the major plates and ceases to exist as a separate plate. Various defunct microplates in the eastern Pacific, such as Mathematician (Mammerickx *et al.*, 1988) and Bauer (Eakins and Lonsdale, 2003), developed by this process. Repeated episodes of microplate formation, evolution and extinction have been involved in the development and evolution of the current boundaries between the Pacific, Nazca and Antarctic plates (Lonsdale, 1994; Bird *et al.*, 1999; Lonsdale, 2005).

## 4.10 Summary

MOR spreading centres constitute one of the three types of tectonic plate boundary. Tectonic plates are created at ridges and diverge at rates ranging from a few to over 150 km $Ma^{-1}$. Ridge morphology is closely related to spreading rate: fast-spreading ridges have small crestal highs at their axes and deepen monotonically off-axis, while slow-spreading ridges are characterised by large median valleys several kilometres deep. The axial morphology changes abruptly, probably as a function of melt supply, at a full spreading rate of approximately 65 km $Ma^{-1}$, although the precise rate varies from place to place. Spreading centres are delineated by narrow bands of normal faulting and associated shallow earthquakes, whose focal mechanisms are consistent with extension along the spreading direction.

Spreading centres are offset by a hierarchical continuum of ridge-axis discontinuities, dividing the ridge into numerous spreading segments tens to hundreds of kilometres long. The largest offsets produce transform faults where plates slip past one another, and are characterised by linear valleys and flanking ridges parallel to the slip direction. They leave inactive scars called fracture zones on the adjacent plates that mark ancient spreading directions.

Offsets between ~30 km and a few kilometres are marked by 'second-order', 'non-transform' discontinuities. At fast spreading these are typically overlapping spreading centres, where slightly offset ridges propagate past each other producing an overlap some three times the offset distance; at slow spreading, the offset is typically characterised by a short, oblique spreading centre. Both types of offset can migrate along the ridge axis, and tend to leave oblique, V-shaped trails of disrupted topography. The differences between the two types probably reflect the thinner and weaker lithosphere at fast-spreading ridges, which can more easily respond to tectonic disruption.

At offsets of a few kilometres or less, higher-order discontinuities are marked by small offsets of axial volcanoes, small changes in the alignment of spreading centres, or geochemical variations in erupted lavas.

A particular type of second-order discontinuity is the propagating rift, where one spreading centre grows continuously at the expense of the other, in the process transferring lithosphere between tectonic plates. The propagating rift shears the transferred lithosphere, and its tip leaves a V-shaped wake whose arms are called 'pseudofaults'.

Three tectonic plates meet at triple junctions. Ridge–ridge–ridge triple junctions are unconditionally stable, but many combinations of different boundary types are not, and may evolve rapidly into more stable ones.

Oceanic microplates are similar to large overlapping spreading centres, but grow to a few hundred kilometres across while rotating rapidly like ball bearings between the major plates. Their boundaries may be complex, including many propagating rifts and areas of compressional deformation, and evolve rapidly.

Triple junctions, propagating rifts and microplates are important in facilitating the evolution of major plate boundaries.

# 5 Crustal structure and composition

## 5.1 Introduction

The oceanic lithosphere is a mechanical boundary layer, defined by its rigid, elastic properties and distinguished from the underlying asthenosphere, which has a plastic rheology (Chapter 3). On the other hand, the crust reflects the chemical differentiation of the Earth, and comprises rocks that are generally less dense and have lower seismic velocities than the underlying mantle. The oceanic crust is up to about 7 km thick. It consists mainly of products formed by partial melting of the mantle: basaltic lavas and the gabbros and dolerites that form their deep-seated counterparts, with similar chemistry but larger mineral grains resulting from slower cooling and crystallisation. Our knowledge of the structure of oceanic crust comes mainly from a combination of seismological, gravity and sampling studies (Cann, 1970, 1974; Carbotte and Scheirer, 2004).

## 5.2 Crustal thickness

The Mohorovičić discontinuity, or Moho for short, marks the base of the crust. It was originally defined seismically (on continents) as a velocity discontinuity at which the P-wave velocity first exceeds 8.0 km $s^{-1}$; sometimes this is called the 'seismological Moho'. In contrast, the 'petrological' Moho is defined as the base of the geological crust, irrespective of seismic velocity (Section 5.3.4). The earliest oceanic seismic refraction experiments detected the seismological Moho at depths of some 6–8 km below the base of the sediments – much less than the thickness of the continental crust (Raitt, 1956; Hill, 1957; Raitt, 1963; Figure 2.12).

White *et al.* (1992) found the mean thickness of oceanic crust from worldwide seismic refraction experiments was 7.1 ± 0.8 km, ignoring 'anomalous' regions such as fracture zones and hotspots. The observed thickness increases to 10.7 ± 1.6 km where ridges intersect anomalous mantle around plumes or hotspots, such as Iceland and the Galapagos archipelago, with a maximum of 20 ± 1 km on aseismic ridges (such as the Walvis Ridge) formed above the cores of mantle plumes. Crust is thin beneath many fracture zones (~2 to 3 km), as is crust formed at very low spreading rates such as the Southwest Indian and Gakkel Ridges (e.g., Minshull *et al.*, 2006), where it can be as thin as 1.5 km (Jackson *et al.*, 1982). More recent work on the EPR shows a mean thickness of 6.7–6.8 km, with

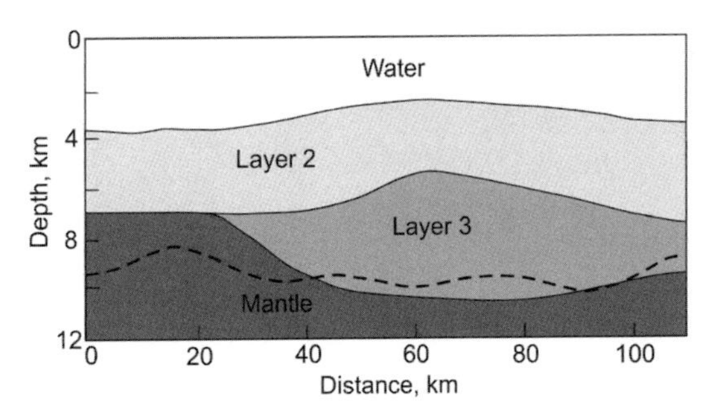


Figure 5.1 Two-dimensional (2D) crustal model along the axis of a single spreading segment on the MAR at 33° S. Solid lines: layer boundaries derived from seismic refraction measurements; dashed line: base of crust derived from gravity. Layer 2 is bounded by the sea floor and the 6.8 km $s^{-1}$ velocity contour. Top of the mantle layer determined by 2D seismic ray tracing. From Tolstoy *et al.* (1993). Reprinted with permission from AAAS.

somewhat less than 6 km near fracture zones (Canales *et al.*, 2003). Chen (1992) suggests that the thickness averaged over a spreading segment is almost independent of spreading rate, but that intra-segment variability is much greater at slower rates.

Detailed studies show that, especially at slow spreading ridges, the crustal thickness is greatest (~7–8 km) at segment centres, and thins to ~2–3 km at their ends. Most of the variation occurs in seismic layer 3 (Section 5.3.3; Detrick *et al.*, 1993a; Tolstoy *et al.*, 1993; Hooft *et al.*, 2000; Hosford *et al.*, *et al.*, 2001; Mutter and Mutter, 1993; Figure 5.1). At fast-spreading ridges the variation is only ~1 km (Lin and Phipps Morgan, 1992; Magde *et al.*, 1995; Van Avendonk *et al.*, 2001).

Gravity measurements offer a denser and more uniform coverage than is available with seismic refraction. Variations in the gravity anomaly over the MAR axis are well-correlated with the segmentation pattern (Section 4.6), with high gravity (interpreted as indicating thin crust) occurring over segment ends, and low gravity (thick crust) over segment centres (Kuo and Forsyth, 1988; Lin *et al.*, 1990; Detrick *et al.*, 1995; Figure 5.2). Lin and Phipps Morgan (1992) showed by numerical modelling that mantle density variations can have only an insignificant effect on the axial gravity signal except at large-offset transform faults, so the majority of the observed gravity variation is due to changes in crustal thickness. The 30–60 mGal variations observed at slow-spreading ridges require a reduction in crustal thickness of ~50% from segment centre to segment end, in agreement with seismic estimates.

At fast-spreading ridges the along-axis gravity variation is only 10–20 mGal, requiring less along-axis variation in crustal thickness. Lin and Phipps Morgan (1992) proposed that the difference is due to three-dimensional (3D), plume-like mantle upwelling below slow-spreading ridges, compared with a more two-dimensional (2D) or sheet-like upwelling under fast-spreading ridges (Figure 5.3a). This mantle structure would only have a small direct effect on the gravity, but would strongly influence the delivery of melt to the crust, with a 3D pattern of upwelling strongly enhancing melt delivery (and thus crustal thickness) at segment centres. Alternatively, there could still be 3D upwelling, with either a continuous

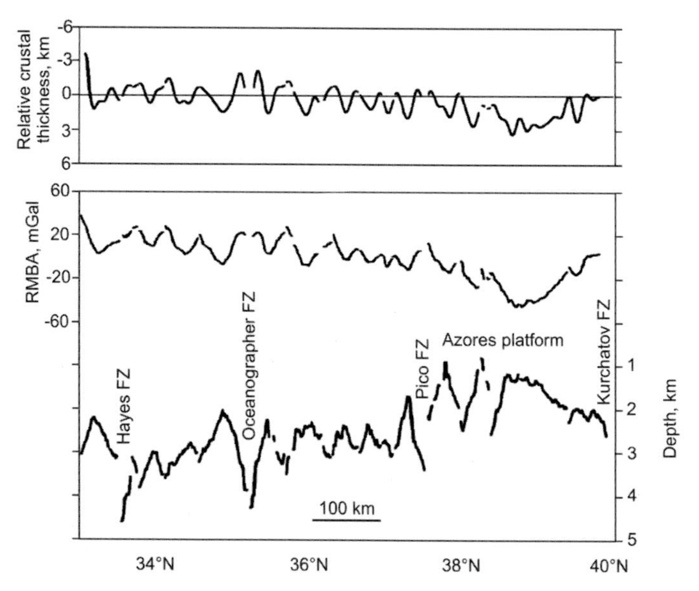


Figure 5.2 Axial depth profile (bottom), residual mantle Bouguer anomaly (middle), and inferred crustal thickness variations (top) for the MAR 33° N to 40° N, after Detrick *et al.* (1995). Segment centres are shallow with low gravity reflecting thick crust, while their ends are deep with high gravity and thin crust.

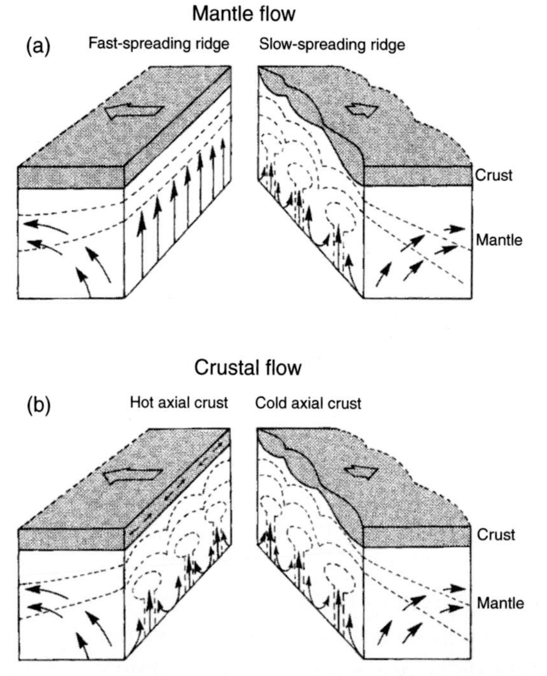


Figure 5.3 Two mechanisms for the variations in along-axis crustal thickness as a function of spreading rate, after Bell and Buck (1992). Black arrows: mantle upwelling; dashed lines: representative isotherms. (a) At fast-spreading ridges, sub-axial mantle flow is sheet-like and uniform, producing constant thickness crust (grey), while at slow spreading, mantle flow is three-dimensional and plume-like, focussing melt at segment centres and generating thicker crust there. (b) Alternatively, 3D upwelling is ubiquitous but at fast-spreading ridges either melt is re-distributed along-axis in a continuous magma chamber or hot, ductile crust flows along-axis (small arrows). Either mechanism could produce a near-uniform crustal thickness. Reprinted by permission from Macmillan Publishers Ltd: *Nature* © 1992.

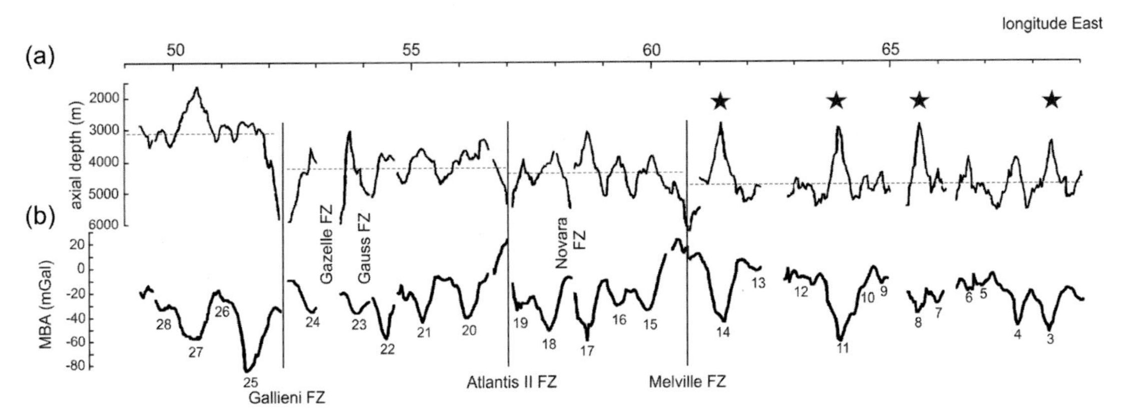


Figure 5.4 Top: axial depth, and bottom: axial mantle Bouguer anomaly, along the Southwest Indian Ridge between 49° E and 69° E, after (Cannat *et al.*, 1999). Horizontal dashed lines show average depth in each super-segment. Segments are numbered sequentially from the Rodrigues triple junction at 70° E, and major transform faults are labelled. West of Melville Fracture Zone the variation of axial depth and gravity is similar to that on slow-spreading ridges; east of the fracture zone, many segments show smaller depth and gravity relief, while a few, marked by stars, display enhanced depth and gravity variations attributed to unusually thick volcanic crust.

sub-axial magma chamber (Lin and Phipps Morgan, 1992) or hot, ductile, lower crust (Bell and Buck, 1992) to allow along-axis redistribution of crustal material (Figure 5.3b).

The 3D upwelling model predicts that most heat will be advected to the segment centre where isotherms, including the brittle–plastic transition, will be shallowest. Thus, where the crust is thickest, the lithosphere will be thinnest. This seems to be supported by the distribution of faulting at slow-spreading segments, where larger faults occur at segment ends, as would be expected for thicker lithosphere (Shaw, 1992; Section 7.3.2). This suggests a positive feedback, because rising melt will be constrained to move along the sloping base of the lithosphere towards the segment centre (Magde *et al.*, 1997).

At ultra-slow-spreading such as parts of the Southwest Indian and Gakkel ridges, the crust is mostly thinner than normal, indicating generally low amounts of melt emplacement. However, there are occasional segments with large central volcanic massifs and thick crust inferred from large negative mantle Bouguer anomalies (Mendel and Sauter, 1997; Michael *et al.*, 2003; Figure 5.4). These have been attributed to along-axis melt focussing (Cannat *et al.*, 1999; Sauter *et al.*, 2004) or intersection of the ridge axis with small areas of unusually fertile mantle (Cochran *et al.*, 2003; see also Section 6.8).

Basaltic rocks characteristic of the upper crust are common along slow-spreading segment centres, while ultramafic rocks characteristic of the upper mantle are more common at segment ends and other areas of high gravity indicative of thin crust. Cannat *et al.* (1995) combined these observations with determinations of lithospheric thickness by seismicity and of crustal thickness by seismic refraction and gravity, to produce the so-called 'plum-pudding' model with thick, continuous crust at melt-rich segment centres but progressively thinner and more discontinuous crust towards segment ends (Figure 5.5). There, small,

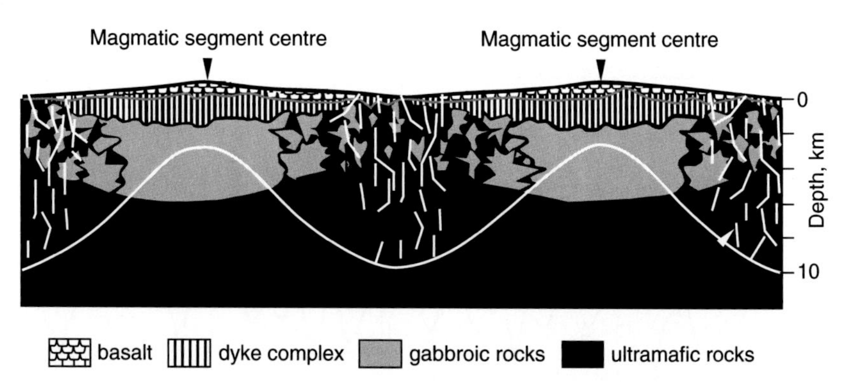


Figure 5.5 Along-axis crustal structure at a slow-spreading ridge, after Cannat *et al.* (1995). Long white line indicates base of lithosphere. Crust thins and becomes discontinuous, while lithosphere thickens, towards segment ends.

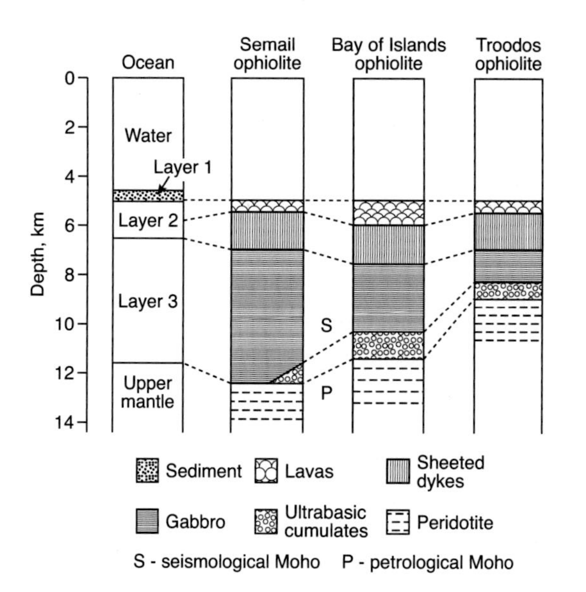


Figure 5.6 Generalised structure of the oceanic crust compared with that of three ophiolites, after Mason (1985).

isolated bodies of gabbro may be embedded in an ultramafic matrix, with a network of strike-slip and oblique-slip faults accommodating tectonic constraints such as ridge offsets.

## 5.3 Seismology and the layered model

Early seismic refraction results suggested that the oceanic crust consisted of a series of uniform layers (Hill, 1957; Raitt, 1963; Figure 2.12). This structure is similar to that seen in the on-land exposures of inferred oceanic crust known as ophiolites (Gass, 1968; Figure 5.6; Section 5.7); it is termed the Penrose model after the conference that adopted it (Penrose,

**Table 5.1** Seismic velocity structure of the oceanic crust, after Bott (1982), Bratt and Purdy (1984), White *et al.* (1992), Fowler (2005), and Kearey *et al.*, 2009)

| Layer | Lithology | Thickness, km | P-wave velocity, km $s^{-1}$ | Velocity gradient, $s^{-1}$ |
|---|---|---|---|---|
| Layer 1 | Sediments | Thin or absent at mid-ocean ridges | 1.6–2.5 | ~2 |
| Layer 2A | Fractured and porous basaltic lava and sediment | 0.2–0.8 | 2.0–5.24 | ~1–2 |
| Layer 2B | Massive basaltic lava | 0.3–1.0 | 4.8–5.5 | ~0 |
| Layer 2C | Massive basalts and dykes | 1.2 | 5.8–6.2 | ~1 |
| Layer 3 | Gabbro | $4.97 \pm 0.90$ | 6.6–7.0 | ~0.1 |

1972). A similar lithological model for the oceanic crust can be deduced independently of ophiolites (Cann, 1970). In the model, a low-velocity layer 1, interpreted as sediments, lies above a moderate-velocity layer 2, while the basal layer 3 has a fairly uniform velocity of about 6.8 km $s^{-1}$ (Table 5.1). The mantle, with a velocity of around 8 km $s^{-1}$, is sometimes referred to as layer 4. The layer 2/3 nomenclature remains common amongst seismologists and geologists. However, it is essentially a seismological terminology, and its extension to lithology relies on interpretation. Where there might be confusion it would be better to refer to the lithological layers as the basaltic (extrusive and sheeted dyke) and gabbroic (or plutonic) layers.

In the 1970s, the 'tau-p' interpretation method was introduced (Bessonova *et al.*, 1974; Kennett and Orcutt, 1976; Kennett *et al.*, 1977). This removed the restriction of using only layered models, allowed the use of continuous velocity gradients and, importantly, enabled the computation of error bounds on the models (Figure 5.7). Further improvements in interpretation (Section 2.7.2) showed that a layered structure remains, but the individual layers tend to be characterised by velocity gradients rather than constant velocities, with large gradients (1–2 $s^{-1}$) in the upper crust and lower gradients (~0.1 $s^{-1}$) in the lower crust (White, 1984; Collier and Singh, 1998; Table 5.1; Figure 5.8; Figure 5.9). The seismic layers are described in detail below, including their likely lithological composition (Figure 5.10).

### 5.3.1 Seismic layer 1 (sediment layer)

Direct sampling of the seabed, particularly by the Deep Sea Drilling Project and its successors, shows that layer 1 is composed of sediments that are progressively consolidated as their depth of burial increases. Oceanic sediments deposited above the carbonate compensation

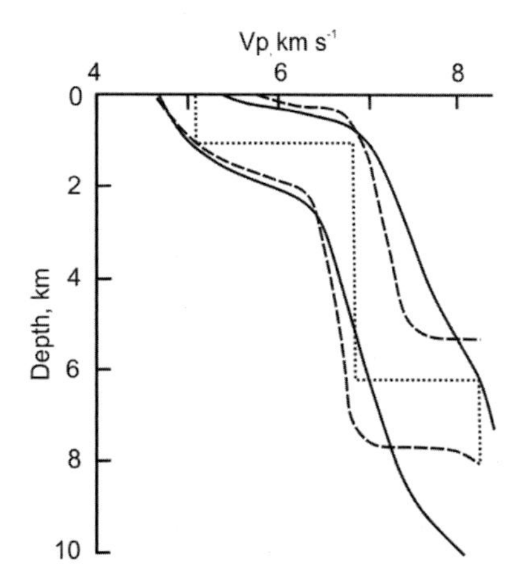


Figure 5.7 Comparison of a layered model of the oceanic crust with the allowable bounds obtained using tau-p inversion, after Kennett and Orcutt (1976). Data are from a split (i.e., fired in opposite directions from a central shot) refraction experiment over 1 Ma oceanic crust on the flank of the EPR. Dotted line shows original layered solution; curved lines indicate the envelopes within which any model compatible with the observations and their error estimates must lie. Continuous line, 'inward' shot; dashed line, 'outward' shot.

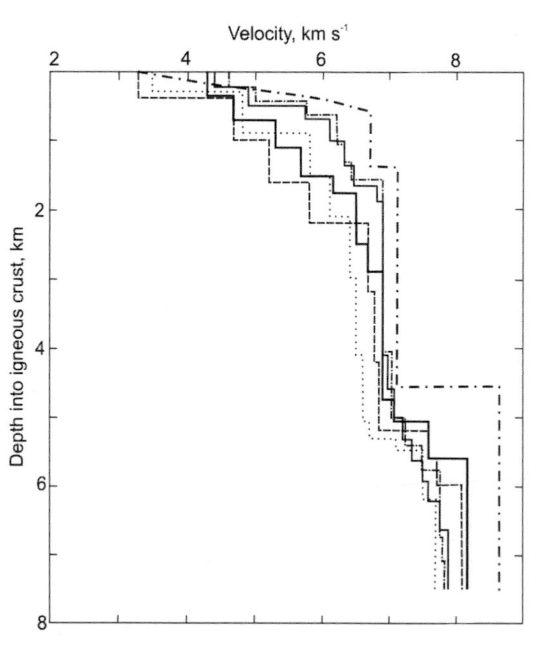


Figure 5.8 P-wave velocity models for north Atlantic and eastern Pacific oceanic crust younger than 20 Ma, after Spudich and Orcutt (1980). Note the steep velocity gradients in the upper crust, low gradients in the lower crust, and increase in velocity to ~8 km $s^{-1}$ (mantle) at ~6 km depth.

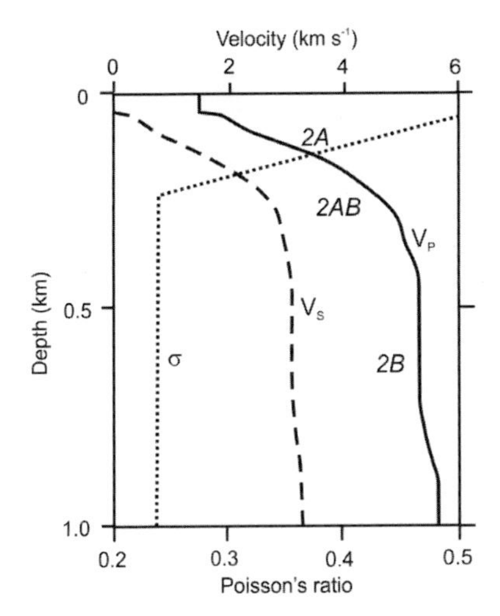


**Figure 5.9** Crustal P-wave and S-wave velocities, after Collier and Singh (1998). The P-wave model was obtained by inversion using a genetic algorithm by Tolstoy *et al.* (1997); S-wave velocities computed from the P-wave curve assuming the Poisson's ratio $\sigma$ shown by the dotted line.

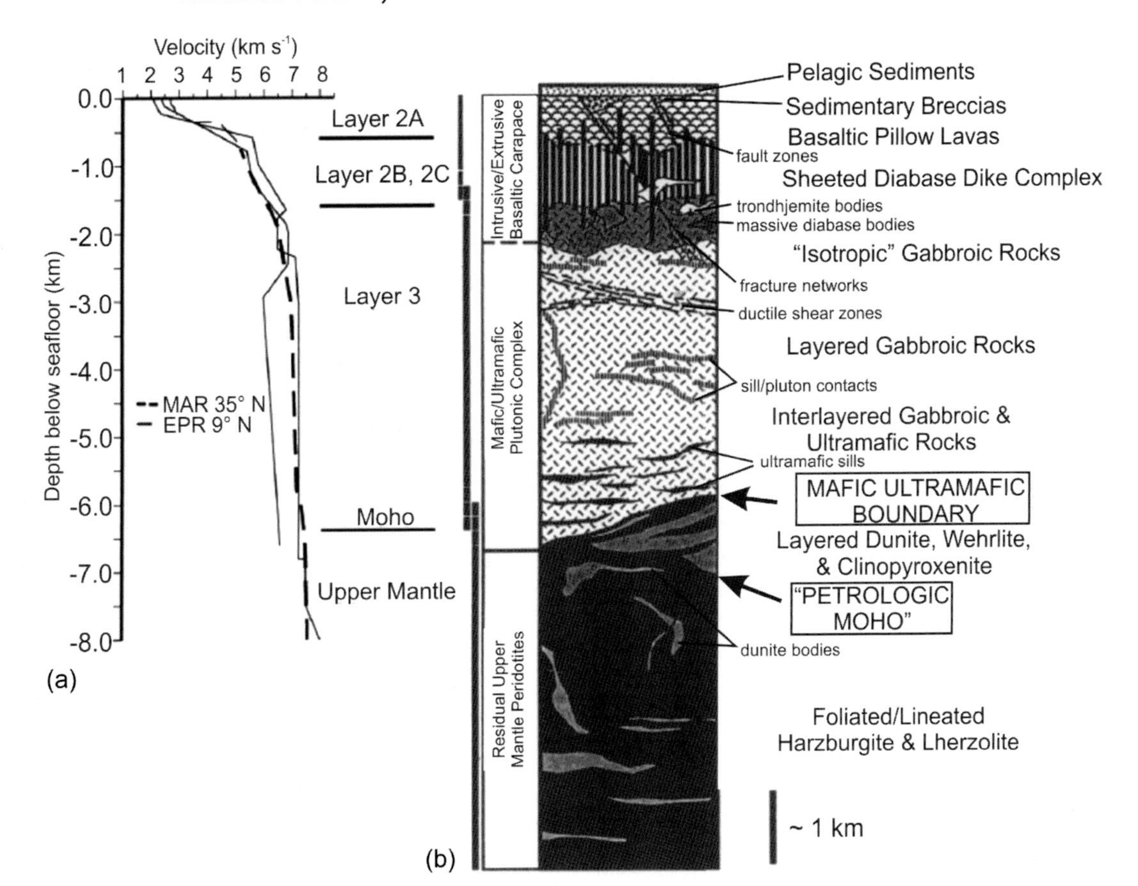


**Figure 5.10** (a) Typical P-wave velocity profiles in oceanic crust, and (b) generalised stratigraphy from ophiolites, matched to the layered oceanic crustal model, after Karson (1998).

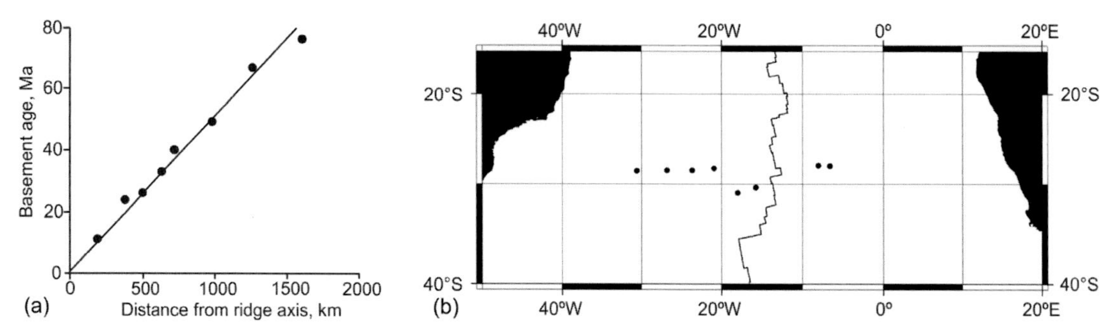


**Figure 5.11** a) Age of sediment immediately overlying igneous basement in DSDP Leg 3 boreholes against distance from ridge axis; b) location of boreholes (dots) and Mid-Atlantic Ridge axis (Maxwell *et al.*, 1970).

depth[1] mostly comprise oozes: either calcareous ooze, consisting of the calcium carbonate exoskeletons of foraminifera, or siliceous ooze comprising the skeletons of radiolaria and diatoms. Siliceous oozes occur principally in areas of high biological productivity such as the polar oceans and equatorial upwelling zones. A minor component of oozes is very fine clay particles, derived from either the finest fraction of water-borne terrigenous sediments or carried by wind off the continents. Below the CCD (which excludes most but not all MORs) and away from high-productivity areas, sediment is composed mostly of clay formed from such fine-grained particles.

Sediment thickness increases with crustal age, i.e. with distance from the MOR axis. An important early test of plate tectonics was the demonstration of the monotonically increasing age of the deepest sediments in the South Atlantic with distance from the ridge axis by DSDP Leg 3 (Maxwell *et al.*, 1970; Figure 5.11).

Typical pelagic sedimentation rates are a few centimetres per thousand years for oozes and considerably less for clays. Because of their youth, most MOR axes therefore have relatively little sediment cover. The inner floors of median valleys at slow-spreading ridges are typically ~10 km wide and <500 ka old, so have up to a few metres of sediment cover (Figure 5.12c, d). Sediment accumulation on young crust is uneven, with ocean currents, tides and earthquakes able to redistribute sediment into low-lying areas and keep steep slopes relatively clear. With increasing age, areal sediment cover increases and outcrops of igneous basement (the extrusive layer) become rarer, until at about 50 Ma sediment cover buries all igneous outcrops and completely inhibits hydrothermal circulation (Stein and Stein, 1992; Chapter 8).

Seismic P-wave velocities in layer 1 range from near the velocity of seawater (1.5 km $s^{-1}$) immediately below the sea floor to some 2.5 km $s^{-1}$ in consolidated sediments at the base of the layer.

[1] The depth at which the rate of solution of calcium carbonate exceeds the rate of supply, so that no solid carbonate is precipitated. It depends on the local pressure, temperature and water chemistry, and in the modern ocean varies between about 4200 m and 5000 m depth.

Figure 5.12 Lavas and sediments near MOR axes: (a) on the EPR, after CYAMEX (1981), with kind permission from Springer Science and Business Media, and (b–d) on the MAR at 45° N, reprinted from Searle *et al.* (2010), © 2010, with permission from Elsevier. (a) Bare sheet flows; (b) bare pillow lavas less than a few thousand years old; (c) pockets of sediment surrounding pillow lavas on crust several thousand years old; (d) approximately 1 m of sediment on ~20 ka crust revealed by fissuring. For colour version, see plates section.

## 5.3.2 Seismic layer 2 (extrusive and dyke layer)

Seismic layer 2 corresponds approximately to a layer of extrusive lavas overlying sheeted dykes (Figure 5.10). Lavas can be observed directly on the sea floor at MOR axes where there is no sediment cover (Figure 5.12a, b), and are commonly sampled by dredge, submersible or ROV. Except for young, shallow lavas, seismic velocities measured by refraction experiments are in agreement with velocities measured on basaltic samples recovered from the sea floor (Fox *et al.*, 1973).

Lava morphologies are variable and are thought to depend mostly on effusion rate and sea-floor slope (Bonatti and Harrison, 1988; Gregg and Fink, 1995; Kennish and Lutz, 1998; Fundis *et al.*, 2010; Section 6.4). Pillow lavas (Figure 5.12b, c) predominate at low effusion rates, moderate slopes and on slow-spreading ridges, while sheet flows (Figure 5.12a) are more common at high effusion rates, lower slopes and on fast-spreading ridges. Occasional thick, extensive, massive lava flows are seen. Ridge volcanism is discussed more fully in Chapter 6.

The seismic structure of layer 2 is complex. A threefold subdivision was proposed by Houtz and Ewing (1976). Although this was confirmed by Bratt and Purdy (1984), Fowler (2005) suggests that a better description is a single layer in which the seismic velocity increases rapidly with depth. The base of layer 2A near ridge axes often produces a clear seismic reflection (Figure 5.16), and the layer 2A/2B subdivision remains in common use.

Layer 2A is the uppermost part of layer 2, with velocities increasing from 2.5 km $s^{-1}$ to over 5 km $s^{-1}$, partly as a function of crustal age (e.g., Carlson, 1998). Sampling shows the layer to consist of basaltic lavas, fractured lavas, and basaltic talus, with sediment often filling the voids. The basalts typically show low-temperature metamorphism (Chapter 8). Detailed seismic experiments on the EPR around 9° N–10° N show layer 2A thickness increasing from 150–200 m on the rise axis to 300–600 m several kilometres off-axis (Harding *et al.*, 1993; Christeson *et al.*, 1994; Vera and Diebold, 1994). This partly reflects the emplacement of significant volumes of lava off-axis, perhaps transported by channels from near-axis vents (Soule *et al.*, 2005; Section 6.5). Smallwood and White (1998) made a detailed seismic reflection study of the axis of the slow-spreading but hotspot-influenced Reykjanes Ridge. They found layer 2A with a mean thickness of $400 \pm 100$ m and a basal velocity increasing from $3.3 \pm 0.3$ km $s^{-1}$ at the ridge axis to $\sim$4.0 km $s^{-1}$ at 1.5 Ma. In mature crust, layer 2A thins to $\sim$100 m though it is somewhat thinner at the fast-spreading EPR than at the slow spreading MAR.

The lower part of layer 2 comprises layers 2B and 2C. It corresponds to the deeper basalts and dykes, and lies (at least at fast-spreading ridges) above the high-level magma sill (Section 5.4.1). Layer 2B has P-wave velocities ranging from $\sim$4.8 km $s^{-1}$ to >6.0 km $s^{-1}$, and ranges from $\sim$300 m thick at ridge axes to 1–2 km in mature crust, inversely with the thinning of layer 2A.

The nature of the layer 2A/2B boundary is still somewhat uncertain. A strong velocity gradient between layers 2A and 2B has been interpreted as marking the transition from lavas to dykes (e.g., Harding *et al.*, 1993; Vera and Diebold, 1994; Smallwood and White, 1998), though not everywhere (Christeson *et al.*, 2010). The near-axis increase in layer 2A thickness at the EPR is then attributed to the growth of the volcanic layer by accumulation of axial eruptives, and has been modelled by stochastic emplacement of dykes and lava flows at the axis (Hooft *et al.*, 1996). At 14° S on the EPR, simultaneous measurement of P-wave and S-wave velocities allowed calculation of the crustal porosity (Collier and Singh, 1998), which falls from >30% in layer 2A to 6%–7% at the layer 2A/2B transition and to 5% within layer 2B. This change is consistent with a transition from lavas to dykes, but also with the closure under pressure, or infilling by sediment or hydrothermal minerals, of numerous gaps and thin cracks within a lava pile (e.g., Carlson, 2011).

Christeson *et al.* (2010) describe detailed seismic studies of the layer 2A/2B boundary in fast-spread crust near Hess Deep and near the intermediate slip rate Blanco transform fault. At the former the boundary lies near the top of the sheeted dyke complex as mapped in the adjacent scarp, while at the latter it projects 600–650 m above the dyke/lava transition seen in the transform wall. This suggests that the boundary does not correspond closely to

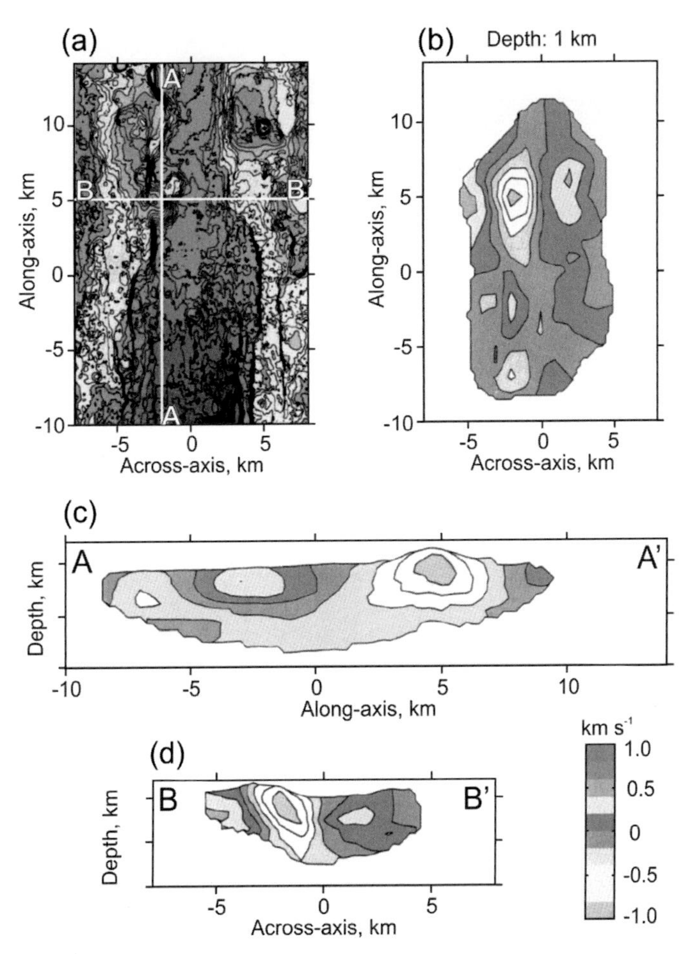


Figure 5.13 Seismic tomographic model of the upper crust at the centre of the OH-1 segment, MAR, 34°50′ N, after Barclay *et al.* (1998). (a) Bathymetry; blue central area is the median valley floor. (b) Horizontal slice, (c) longitudinal vertical section along west side of median valley, and (d) vertical cross-section across the valley, showing departures in P-wave velocity from a 1D reference velocity model. For colour version, see plates section.

a lithological boundary (though it may approximate that in young crust), but to a chemical alteration boundary (see also Section 5.6).

Barclay *et al.* (1998), in a seismic tomographic study of the centre of a magmatically robust segment on the MAR at 35° N (Figure 5.13), failed to find a clear layer 2A/2B boundary. They concluded that the extrusive/dyke transition zone there is significantly thicker than at the EPR. The same study found a greater lateral variability of seismic velocity in the upper crust than at the EPR (Figure 5.14). Barclay *et al.* concluded that in their study area the position of the neo-volcanic zone is unstable over times $\gtrsim$25 ka and can wander over a considerably broader 'crustal accretion zone' on a 1 Ma timescale. This is consistent with the idea that a random emplacement of extrusives is required to match the

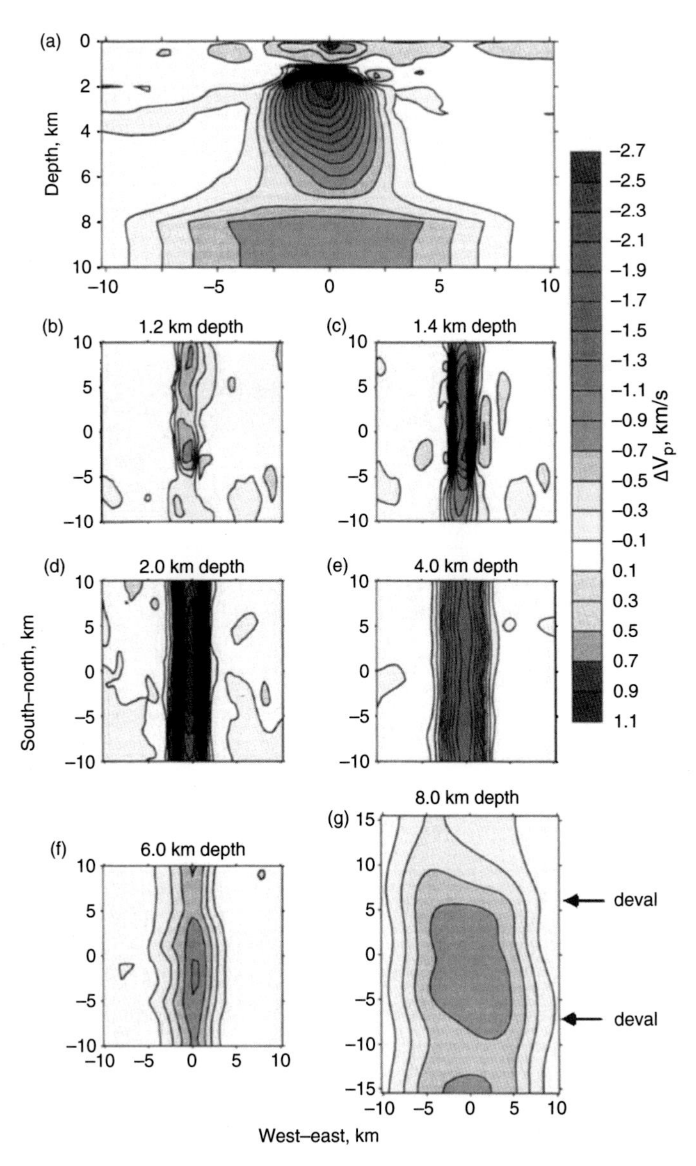


Figure 5.14 Sections through a 3D tomographic model of seismic velocity under the EPR near 9°30′ S, after Dunn *et al.* (2000). (a) Across-axis; (b–g) horizontal slices at increasing depths. Contours show difference in P-wave velocity from a 1D average model. For colour version, see plates section.

observed variations in magnetic anomalies and their polarity transition widths (Macdonald, 1977; Schouten and Denham, 1979; Figure 5.15; see also Section 5.9).

While the layer 2A/layer 2B nomenclature is still common, layer 2C is only rarely used. It probably mostly corresponds to the lower dykes. However, at DSDP hole 504B on the flank of the medium-spreading Costa Rica Rift, comparison between a seismic survey and samples retrieved from the borehole suggested layer 2C comprised a complex transition

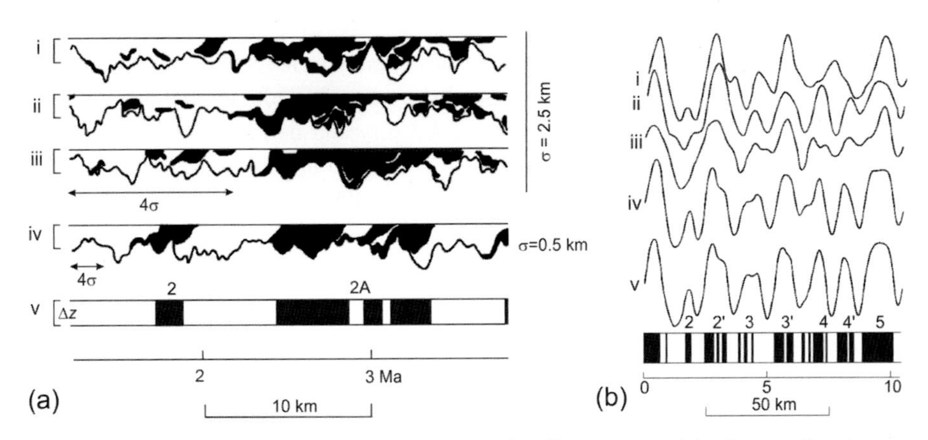


Figure 5.15 Models of the magnetic polarity structure of the upper crust produced by random spatial and temporal emplacement of 1 km-wide lava flows, after Schouten and Denham (1979). (a) Models i, ii, and iii: results of random Gaussian emplacement with 5 km standard deviation; iv: emplacement with 0.5 km standard deviation; v: standard block model (zero width emplacement zone), from 1 to 4 Ma. Average source thickness ($\Delta z$) is 500 m. Black and white represent normal and reverse magnetisation, respectively. (b) The magnetic anomalies predicted by models i–v over 10 Ma.

zone from the base of the extrusives to the top of the sheeted dykes (Bratt and Purdy, 1984). In contrast, at IODP hole 1256D in older, superfast crust, gabbros were encountered within seismic layer 2 (Wilson *et al.*, 2006). The precise nature of the boundary is still unclear.

Layer 2 densities are in the range 2400–2700 kg m$^{-3}$, depending on the porosity of the rock (Blackman *et al.*, 2009).

### 5.3.3 Seismic layer 3 (gabbro layer)

Seismic layer 3 is the deepest crustal layer. It ranges in thickness from <4 km to >6 km and is generally thinnest at the ends of spreading segments and in areas of ultra-slow spreading (Muller *et al.*, 1999; Minshull *et al.*, 2006); it may even be absent from the ends of ultra-slow-spreading segments. The layer has a very low velocity gradient and a uniform velocity of ~6.8–7.2 km s$^{-1}$ (Figure 5.10). This velocity is characteristic of oceanic, but not continental, crust; layer 3 is therefore sometimes termed 'the oceanic layer' (Ewing and Ewing, 1959).

Measured velocities of sea-bed samples are consistent with a gabbroic composition (Fox *et al.*, 1973), as are Poisson's ratio measurements (Kearey *et al.*, 2009). By analogy with ophiolites, seismic layer 3 is usually interpreted as comprising layered and/or isotropic gabbros, representing the plutonic foundation of the oceanic crust (Fox and Stroup, 1981; Figure 5.10). Layered gabbros are thought to be produced by crystal fractionation in a magma chamber (but see Section 5.4.3), isotropic gabbros by solidification of more-or-less uniform magma. Sampling by deep drilling is broadly consistent with this (Section 5.6).

However, Swift and Stephen (1992) suggest that inconsistencies between P-wave velocities seen in refraction experiments and seismic attenuation in gabbro samples from ODP Hole 735B (SWIR) are best explained if the upper part of layer 3 consists of metamorphosed dolerite. In DSDP hole 504B the top of seismic layer 3 corresponds to sheeted dykes (see Figure 5.26). As with layer 2, these examples may reflect modification of initial lithologies (and therefore seismic properties) by chemical alteration of the crust.

Gabbro has similar seismic velocity and density to partially serpentinised peridotite, which is common in exposures of oceanic upper mantle. It is therefore difficult and may be impossible to distinguish geophysically between gabbroic 'lower crust' and peridotitic 'upper mantle' where the latter has been partly serpentinised (Blackman *et al.*, 2009). Indeed, Hess (1962) had proposed that the oceanic crust is formed by hydration of mantle material beginning 5 km below the ocean floor.

Layer 3 gabbros have densities of about 2900 kg $m^{-3}$ (Blackman *et al.*, 2009).

### 5.3.4 Mantle (seismic layer 4)

Although not part of the crust, the uppermost mantle (seismic layer 4) is briefly considered here for completeness. It has typical P-wave velocities of around 8.0–8.2 km $s^{-1}$, producing a strong seismic refraction (and reflection) at the Moho. 'Anomalously' low values of 7.8–8.0 km $s^{-1}$ are found beneath MOR axes and are attributed to high temperatures and the presence of melt in the uppermost mantle (Kearey *et al.*, 2009; Dunn *et al.*, 2000). The upper mantle density is around 3300 kg $m^{-3}$ (Blackman *et al.*, 2009).

Seismic layer 4 velocities match those obtained for samples of ultramafic rocks (e.g., Spudich and Orcutt, 1980). These include the primary upper mantle rock lherzolite and the harzburgite and dunite residues left after removal of basaltic melt (Section 6.2). Such rocks react readily with water to produce the minerals serpentine and talc, and samples of peridotite from the ocean floor are almost always heavily altered in this way.

The Moho is defined seismologically as the depth at which the P-wave velocity reaches 8 km $s^{-1}$. However, in ophiolites, such velocities are also associated with cumulate ultramafic rocks produced by fractional crystallisation at the base of the crust. The top of the mantle is defined petrologically to be where residual mantle material (harzburgite and dunite) appears (Figure 5.10). To avoid confusion, this should be referred to as the 'petrological Moho', with 'seismic Moho' for the velocity boundary. Typically, the petrological Moho in ophiolites occurs tens to hundreds of metres below the seismic Moho (Figures 5.6 and 5.10).

## 5.4 Melt distribution and magma chambers

The distribution of melt at MOR axes is critical to understanding the formation of oceanic crust. It both influences and is influenced by the thermal regime of the crust. Sinha and Evans (2004) reviewed this and its implications for magma distribution and hydrothermal circulation.

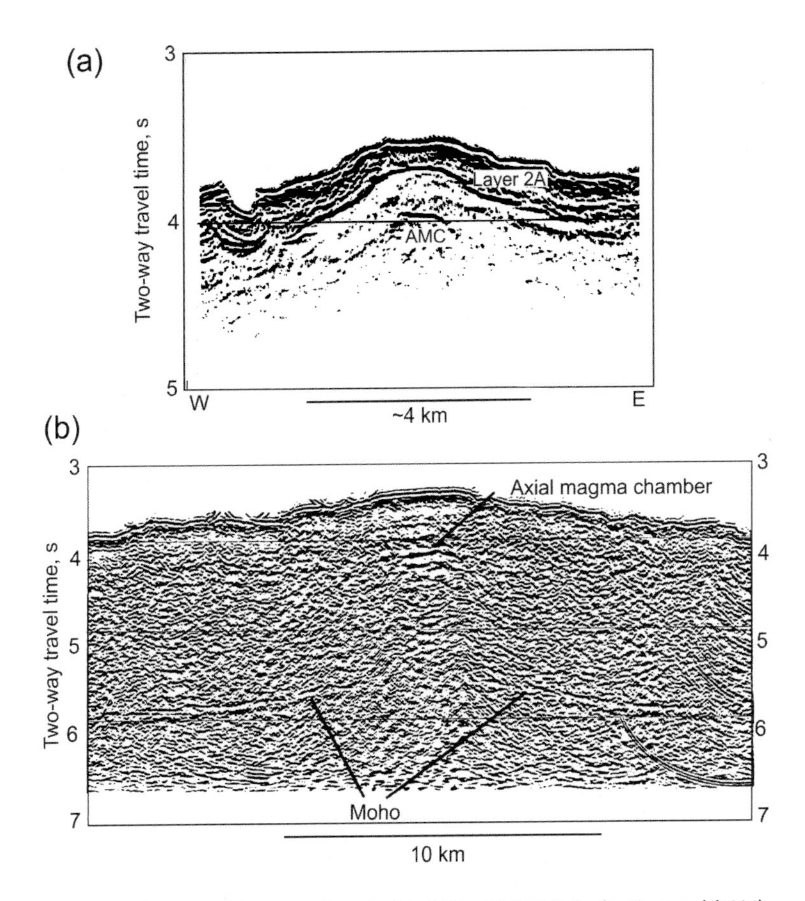


**Figure 5.16** Multi-channel seismic reflection profiles across the axis of the EPR at (a) 14°15′ S, after Kent *et al.* (1994), and (b) 9°30′ N, after Detrick *et al.* (1987). These reveal a narrow reflection from a shallow axial magma chamber (AMC), a continuous reflection across the axis from the base of seismic layer 2A, and a Moho reflection developed within a few kilometres of the axis.

## 5.4.1 Sub-axial magma chambers

Early models of MORs contained large magma chambers several kilometres wide and deep (e.g., Cann, 1974). However, numerous geophysical studies have ruled out such large chambers. The best geophysical evidence for magma chambers comes from multi-channel seismic reflection.

Surveys over fast-spreading ridges show a bright, sub-axial reflection no more than a few kilometres wide, interpreted as coming from the top of a small melt lens (Morton and Sleep, 1985; Detrick *et al.*, 1987; Detrick *et al.*, 1993b; Figure 5.16). Modelling the reflection amplitudes and phase reversals of such signals shows that they are consistent with thin melt layers <50–100 m thick and typically 1–2 km wide (Collier and Sinha, 1990; Kent *et al.*, 1990; Kent *et al.*, 1993). Such axial magma chamber reflectors are more-or-less

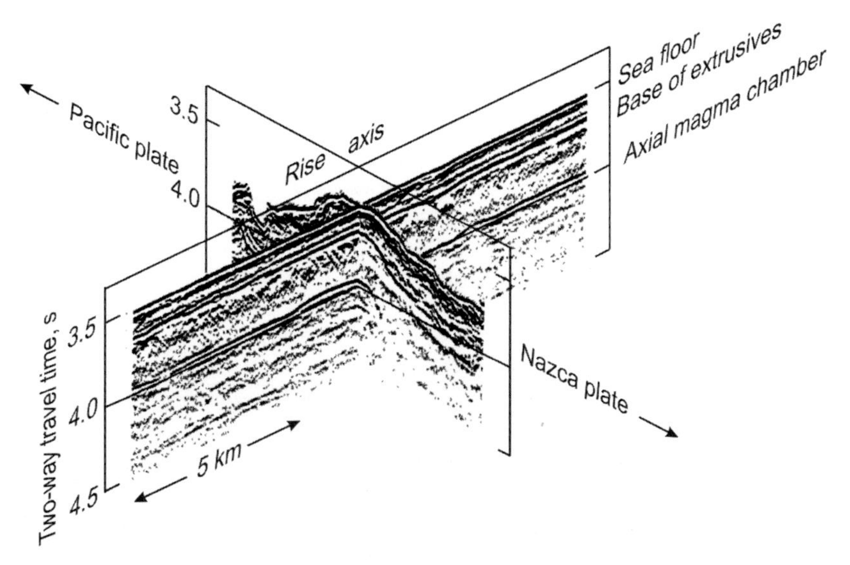


Figure 5.17 Seismic reflection profiles along and across the axis of the EPR at 14°15′ S, after Kent *et al.* (1994), showing along-axis continuity of the axial magma chamber.

continuous along-axis at the EPR (Figure 5.17), though they get smaller and often pinch out at segment ends (Sinton and Detrick, 1992). Similar reflectors have been imaged at the intermediate-spreading Juan de Fuca Ridge (Carbotte *et al.*, 2006), Galapagos Spreading Centre (Detrick *et al.*, 2002) and Southeast Indian Ridge (Baran *et al.*, 2005).

Unambiguous reflections from sub-axial melt lenses have so far been found only in two places on slow-spreading ridges: at a depth of 2.5 km below the southern end of the Reykjanes Ridge at 57°45′N (Navin *et al.*, 1998; Sinha *et al.*, 1998), also confirmed by electromagnetic sounding (MacGregor *et al.*, 1998; Figure 2.21), and at 3 km below the MAR at 37°20′N (Singh *et al.*, 2006a). Both locations are close to mantle hotspots, so may not be typical of slow-spreading ridges. In particular, if their mantle is warmer or more fertile than normal, they are likely to behave more like faster-spreading ridges (e.g., Chen and Morgan, 1990; Searle *et al.*, 1998a; Sections 4.3 and 9.6).

The depth to the top of the axial magma chamber increases with decreasing spreading rate, from occasionally $<1$ km below sea floor at the superfast southern EPR to $\sim 3$ km beneath the Juan de Fuca Ridge and Costa Rica Rift. This is in good agreement with theoretical models of the thermal structure of ridges, which predict a rapid deepening of the axial magma chamber as the spreading rate decreases (Phipps Morgan and Chen, 1993; Chen and Phipps Morgan, 1996; Chen and Lin, 2004; Figure 5.18). Additionally, local variations in mantle melt production and delivery affect the depth (Baran *et al.*, 2005).

### 5.4.2 Other high-temperature and partial melt zones

There is a broader seismic low-velocity zone in the lower crust at the EPR axis (Toomey *et al.*, 1990; Figure 5.14), but the velocities are more consistent with a largely solidified

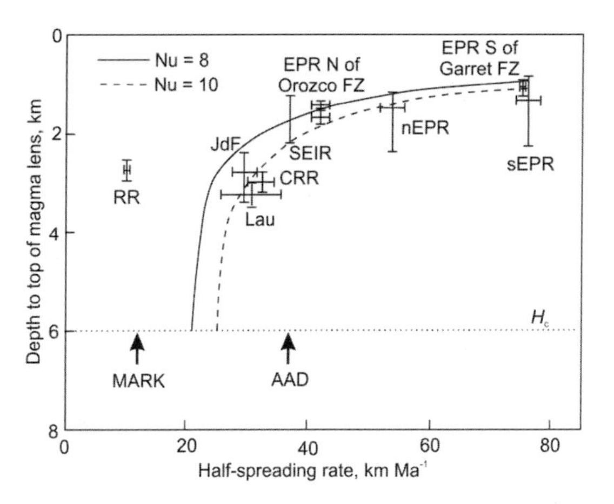


Figure 5.18 Observed and predicted depths to the top of the axial magma chamber, after Chen (2004), with additional data from Baran *et al.* (2005). Curves give predicted depth of melt lens (at the 1200 °C isotherm) for two values of hydrothermal cooling prescribed by the Nusselt number Nu, assuming a mantle temperature of 1350 °C. Dotted line marked $H_C$ is assumed crustal thickness. AAD, Australian-Antarctic Discordance; CRR, Costa Rica Rift; EPR, East Pacific Rise; JdF, Juan de Fuca Ridge; Lau, Lau Spreading Centre; MARK, MAR south of Kane Transform; RR, Reykjanes Ridge; SEIR, Southeast Indian Ridge. Bold arrows indicate two places where no melt lens is observed, at the slow-spreading MARK area and the anomalously cold AAD.

gabbro body than with molten basaltic magma. Such low-velocity zones may be very steep sided, indicating a narrow (~4 km) hot axial region bounded by strong temperature gradients (Dunn *et al.*, 2000). Such structures place important constraints on hydrothermal flow (Chapter 8).

Broad, low-velocity regions are also observed in the lower crust on parts of the MAR (Purdy and Detrick, 1986; Hooft *et al.*, 2000), and some at least have been interpreted as indicating partially molten zones (Hosford *et al.*, 2001). Such regions, which also tend to correlate with topographic highs and low gravity, may represent zones of hot rock that possess small pockets of melt (Sinton and Detrick, 1992). A detailed study at 35° N on the MAR imaged pipe-like low-velocity regions in the lower crust that connect into an axial low-velocity zone at 3 km depth, together with high-amplitude, low-velocity anomalies in the upper crust directly underlying sea floor volcanoes (Magde *et al.*, 2000; Barclay *et al.*, 1998; Figure 5.19). These are interpreted as the signature of a crustal melt delivery system, with focussed melt from the mantle rising and pooling at the brittle–plastic transition whence it supplies sea floor eruptions.

### 5.4.3 Models of melt distribution

Sinton and Detrick (1992) reviewed the geophysical and petrological evidence and experimental constraints for axial magma chambers and developed a model that reconciles them

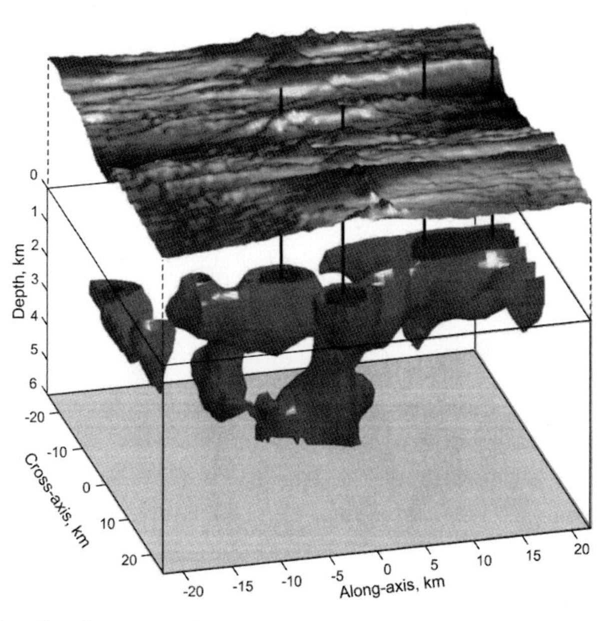


**Figure 5.19** Three-dimensional view of sea floor topography overlying seismic low-velocity anomalies at MAR 35° N, after Magde *et al.* (2000). Reprinted with permission from Elsevier, © 2000. The top 1 km of the model has been removed for clarity. Dark and light red surfaces: boundaries of $-0.5$ km s$^{-1}$ and $-0.3$ km s$^{-1}$ velocity anomalies, respectively, derived from tomographic inversion. Vertical black lines reference two sea floor volcanoes (centre) and the axial volcanic ridge (right). For colour version, see plates section.

(Figure 5.20). They noted that recent petrological and geochemical results are consistent with a small, sill-like magma chamber. Importantly, through most of the temperature interval between solidus and liquidus, magma will have a high crystal content and therefore high viscosity; for about half of this range it will behave rheologically like a solid, and not be able to flow or erupt. In the Sinton and Detrick model, large bodies of low-viscosity melt are confined to a narrow, high-level sill. This is surrounded by a partially solidified crystal mush zone containing connected melt with >25% crystallinity and high viscosity. The mush zone is surrounded by a largely crystalline (60%–80%) transition to completely solidified gabbro. In this model, low-spreading-rate or low-magma-supply ridges lack large, continuous magma lenses, but contain small melt pockets within the mush zone that only become eruptible when mobilised by fresh injections of magma; such refreshment may be coupled to tectonic extensional events.

While the broad outlines of this model are largely accepted, there remains considerable debate concerning the details of melt distribution in the crust and how it is delivered to the volcanic layer. Models such as Figure 5.20b imply that, since all melt is delivered via a high-level magma chamber, the lower crust must be formed by material that has crystallised high up and subsequently been transported deeper by plastic flow – the so-called 'gabbro glacier' model (Quick and Delinger, 1993; Figure 5.21a). An alternative, suggested by Kelemen and Aharonov (1998) largely on the basis of observations at the Oman ophiolite, is that multiple sills exist at permeability barriers within and at the base of the lower crust

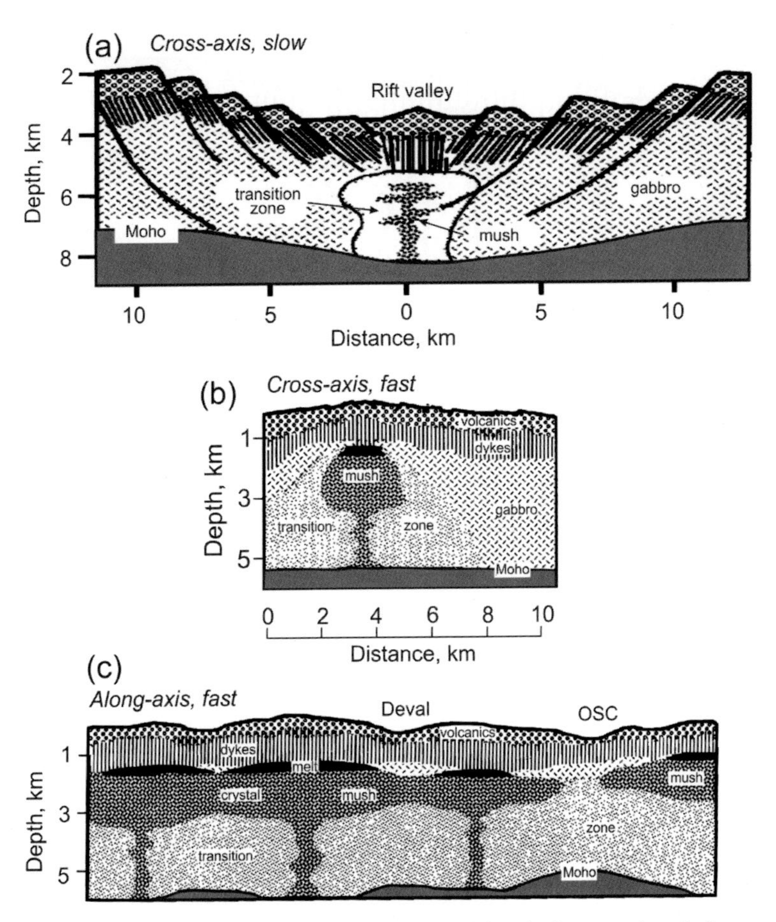


**Figure 5.20** Schematic models of crustal melt distribution, after Sinton and Detrick (1992), all at approximately the same scale. Black: melt lenses; stipple: partially solidified crystal mush zone associated with crustal low-velocity zone. Also shown are a largely crystalline transition zone and solid gabbro.

(Figure 5.21b). Melt is periodically extracted from these lenses, probably via fractures. The model can explain observed compositional layering in gabbros as a result of pressure variations associated with this fracture formation. Seismic techniques are beginning to detect such deep melt pockets (Canales *et al.*, 2009).

Further evidence for deep melt comes from the compliance method (Crawford *et al.*, 1991; Section 2.9). Compliance measurements have revealed low shear-wave velocities throughout the crust under the EPR between 9° N and 10° N (Crawford *et al.*, 1999; Crawford and Webb, 2002). In addition to the shallow, seismically imaged sub-axial melt lens, there must be another at 5.4 km depth, with the intervening crust containing between 2.5% and 18% melt. There must also be isolated melt bodies near the Moho, and a significant amount of melt extending up to 10 km laterally from the ridge axis.

Direct sampling of lower oceanic crust is beginning to provide evidence relevant to this debate, as discussed below.

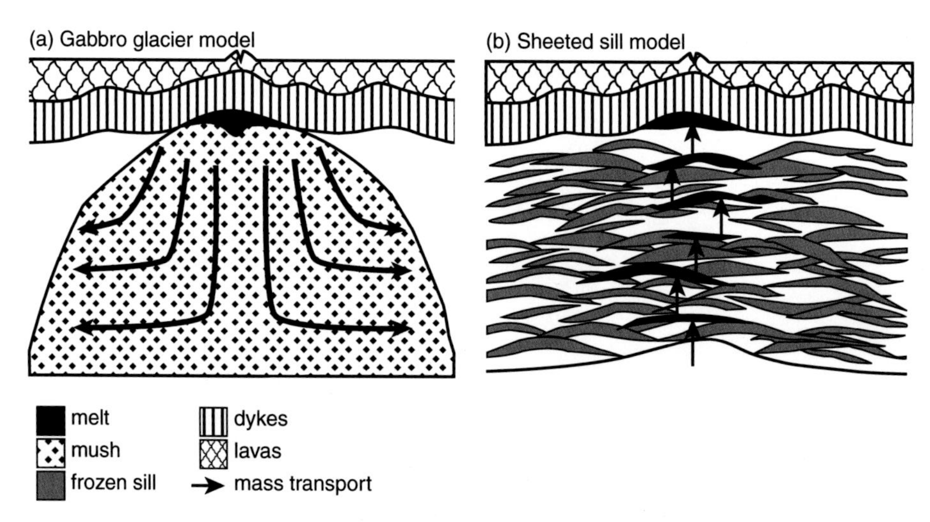


Figure 5.21 End-member models for the formation of lower oceanic crust, after Perk *et al.* (2007), with kind permission from Springer Science and Business Media. (a) Single, high-level magma sill with crystallised melt being transported downwards and outwards by plastic flow; (b) multiple magma sills throughout the lower crust that solidify to produce layer 3 without further movement.

## 5.5 Shallow crustal sampling

Early sampling of the ocean floor by dredging recovered the same lithologies – basalts, dolerites, gabbros, peridotites, and their metamorphosed products – that are found in ophiolites. However, dredging is very imprecise, and the subsequent use of manned submersibles and ROVs, coupled with improvements in underwater navigation, have greatly improved sampling precision (Sections 2.13–2.15).

Slow- and medium-slipping transform faults offer high fault-scarps that may expose deep crustal sections (so-called 'tectonic windows'). Figure 5.22 shows two sections up the south wall of the slow-slipping Vema transform fault, based on a synthesis of results from five manned submersible dives (Auzende *et al.*, 1989). The dives reveal, going up the scarp, a sequence of rocks corresponding to the ophiolite sequence: serpentinised peridotite, gabbro, sheeted dykes, basalts and indurated sediment, fitting the presumed composition of the oceanic layers. More detail is revealed by a study at Blanco transform fault on the medium-spreading Juan de Fuca Ridge, where Karson *et al.* (2002a) synthesised the results of 25 dives up the transform wall (Figure 5.23). These confirm the layering, with 1000–1500 m thick lavas overlying a 500 m thick sheeted dyke layer. Off-axis lavas dip toward the ridge axis and their feeder dykes dip away (Figure 5.23c), implying that young crust warps and subsides rapidly to make space for the thick lava pile, as predicted by Cann (1974).

However, transform faults occur at the ends of spreading segments where crust is thin and may be distorted by transform-parallel normal faulting, so they may not reveal typical crust (Francheteau *et al.*, 1976). Tectonic windows provided by ridges that have propagated (Section 4.8) into pre-existing crust should reveal more typical sections. Karson *et al.*

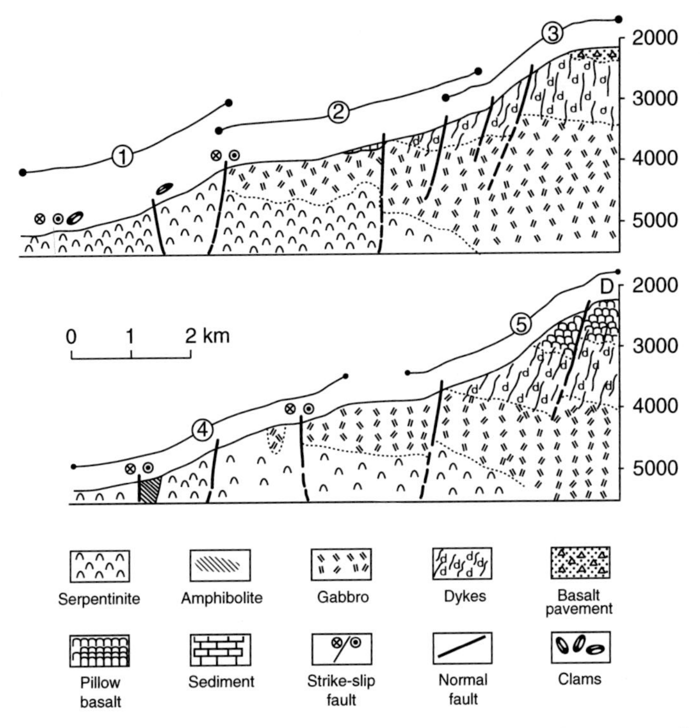


Figure 5.22 Two sections up the southern wall of Vema transform, synthesised from submersible observations and sampling, after Auzende *et al.* (1989). Numbered lines above the sections indicate the positions of the dives. Reprinted by permission from Macmillan Publishers Ltd: *Nature* © 1989.

(2002b) sampled Hess Deep, where the Cocos–Nazca Spreading Centre is propagating into crust produced at the fast-spreading EPR. They found a similar structure to that at Blanco transform, but with a thinner lava (200–800 m) and thicker dyke section (~1000 m). There are significant variations in the thickness and structure of the observed lava and sheeted dyke units, attributed to temporal fluctuations in magma supply accompanied by substantial brittle deformation. Examples of the lithologies exposed at Hess Deep are shown in Figure 5.24.

Pito Deep lies at the corner of Easter microplate, where the microplate's East Rift propagates into crust spread from the EPR (Figure 4.29). Here an ROV recovered a substantial lower crustal gabbro section, overlain by a complete upper crustal section of sheeted dykes and lavas (Perk *et al.*, 2007). The uppermost gabbros contain substantial primitive cumulates, in contrast to Hess Deep where the uppermost gabbros are much more evolved. Perk *et al.* (2007) argue this may be explained by the melt at Pito Deep having been rapidly transported to a shallow magma sill with little or no crystallisation en route; subsequently the crystallised gabbro sank to the lower crust via the gabbro–glacier mechanism (Figure 5.21a). In contrast, at Hess Deep at least some melt crystallised at intermediate depths, allowing that reaching the shallowest sill to be more evolved – the multiple sill model (Figure 5.21b). Such differences are ascribed to either spatial (e.g., mid-segment versus end-segment) or temporal variations in the thermal structure of the crust.

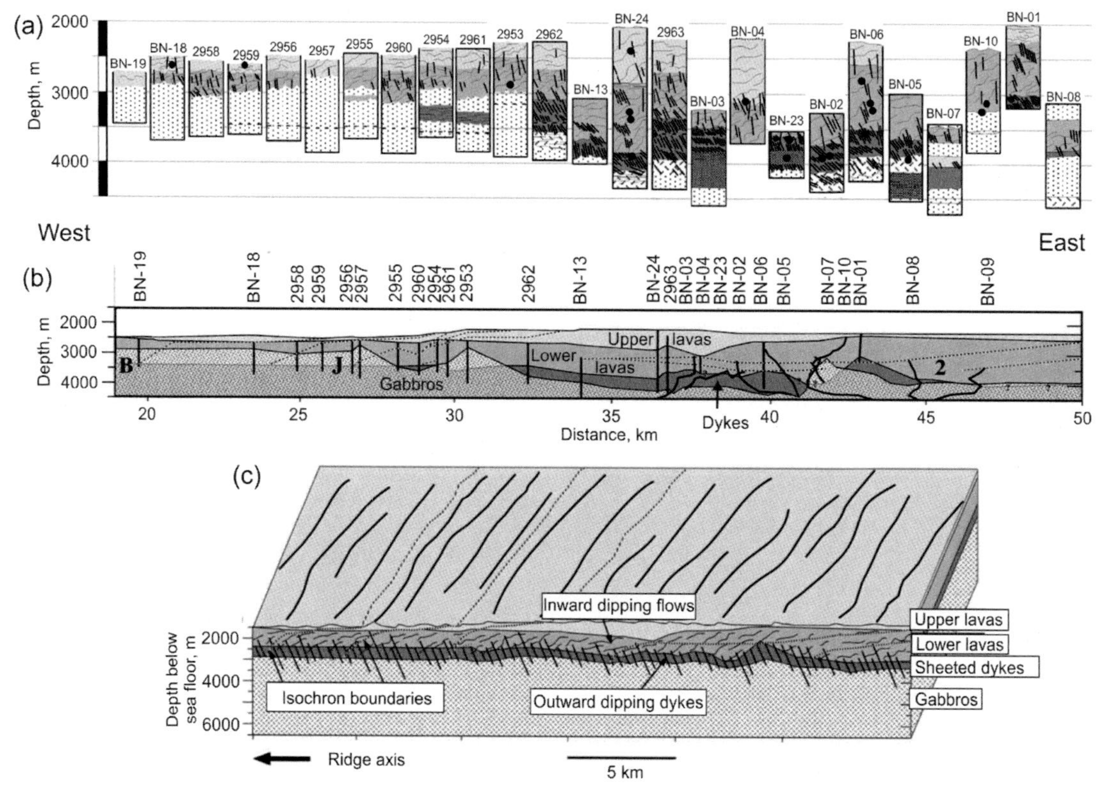


Figure 5.23 Crustal structure exposed in the north wall of Blanco transform, after Karson *et al.* (2002a). (a) Stacked dive transects, labelled at top with dive numbers. Light grey: upper basaltic lavas; mid grey: fractured lower basaltic lavas; dark grey: sheeted dyke complex; stipple: gabbros; short black lines: dykes; dark shading: areas of massive-looking sedimentary rock. (b) Synthetic vertical section based on dive transects; shading as in (a); solid black lines and numbers: dive transects in their correct relative positions; dotted lines: magnetic isochron boundaries: B: Brunhes (0.78 Ma); J: Jaramillo (0.99–1.07 Ma); 2: chron 2 (1.77–1.95 Ma). (c) Block diagram showing orientations of lavas and dykes.

## 5.6 Deep sampling: ocean drilling

The Deep Sea Drilling Project had its origins in the abortive Mohole project (Riedel *et al.*, 1961), and its successors retain the ambition to sample right through the oceanic crust to the mantle. Meanwhile a number of deep boreholes have penetrated significant sections of crust and provided some direct evidence for its deeper composition and structure.

The IODP hole 1256D is sited in 15 Ma crust formed at the superfast ($\sim$212 km Ma$^{-1}$ full rate) EPR. It currently extends 1522 m below the sea floor, and has recovered samples from a continuous section of crust to the top of the plutonic (gabbro) layer (Wilson *et al.*, 2006;

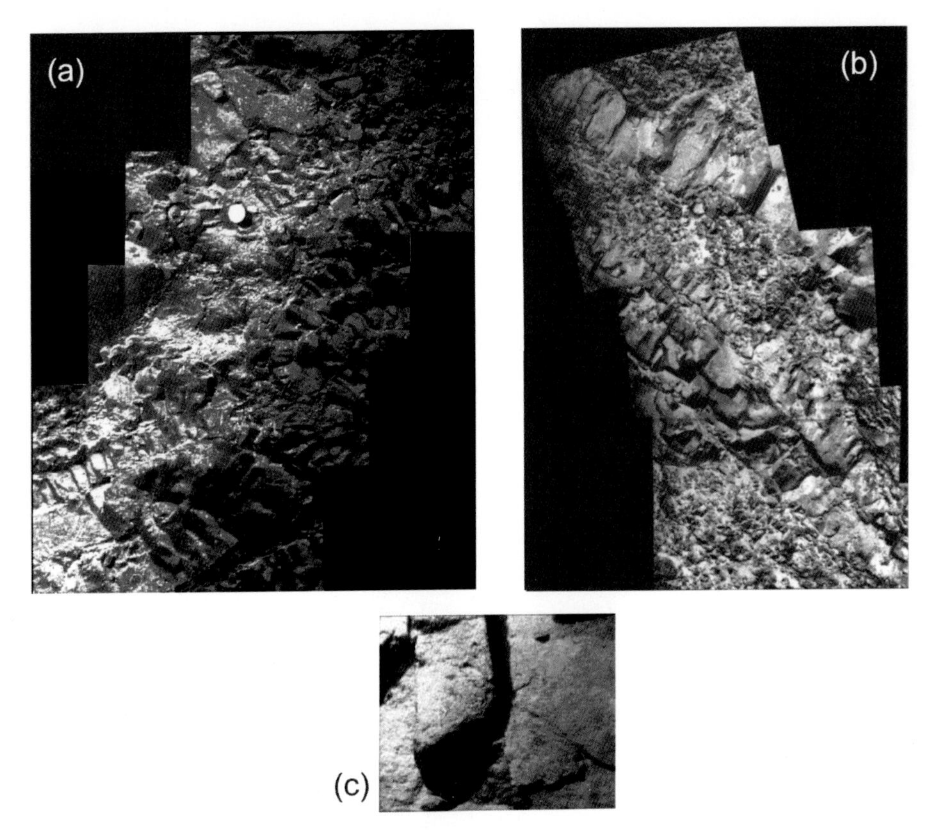


Figure 5.24 Examples of crustal lithologies of fast-spread EPR crust, outcropping at Hess Deep, after Karson *et al.* (2002b). (a) Alvin video mosaic showing pillow, lobate and tabular lavas; image width ~3 m. Note radial jointing in pillow at centre right. (b) Argo II digital video mosaic showing dykes; image width ~3 m. (c) Alvin video frame showing gabbro; image width ~1 m.

Teagle *et al.*, 2012; Figure 5.25). An extrusive layer that is 750 m thick comprises mostly sheet flows and massive lava flows, including a 75 m thick lava pond, with only 2% pillow lavas. Of this, 284 m is inferred to have been emplaced off-axis. A transition zone that is 60 m thick leads to 350 m of untilted sheeted dykes. The top of the plutonic layer is marked by gabbro and trondhjemite dykes 1160 m below the top of the extrusives. The gabbro is evolved, implying the existence of deeper, fractionating melt bodies. Regional seismic refraction experiments put the seismic layer 2/3 transition between 1200 m and 1500 m below the top of the extrusives, although velocities from borehole logging and samples are <6.5 km $s^{-1}$, typical of layer 2, down to and including the gabbros (Wilson *et al.*, 2006). This suggests, as elsewhere, that porosity and alteration may be more important than lithology in determining seismic velocity, especially in older crust.

DSDP hole 504B penetrated 1600 m into intermediate-spreading (74 km $Ma^{-1}$ full rate) crust on the flank of the Costa Rica Rift (Anderson *et al.*, 1982; Becker *et al.*, 1989). It encountered 572 m of extrusives, a 209 m thick transition zone of extrusives and dykes,

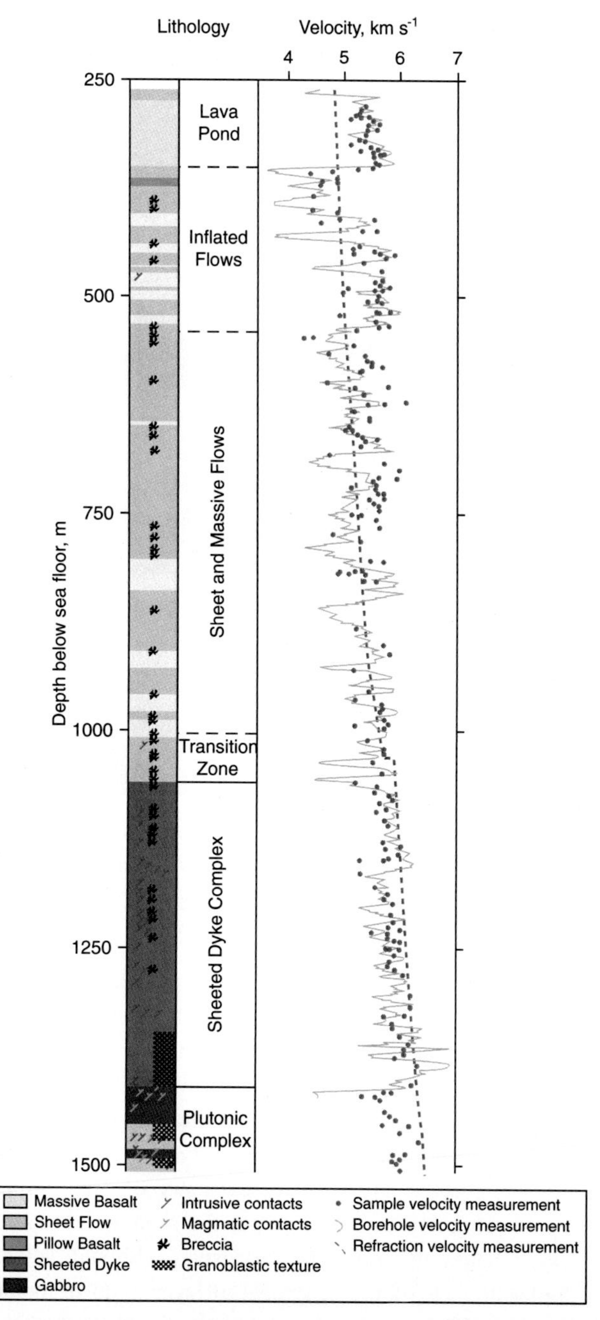


**Figure 5.25** Summary lithological column from IODP hole 1256D drilled into superfast-spread crust, after Wilson *et al.* (2006). Left: lithologies; right: regional seismic refraction model (red dashed line), wire-line logged seismic velocity (green line) and velocity measurements on cored samples (blue dots). Top of column is at base of sedimentary layer. For colour version, see plates section. From Wilson *et al.* (2006). Reprinted with permission from AAAS.

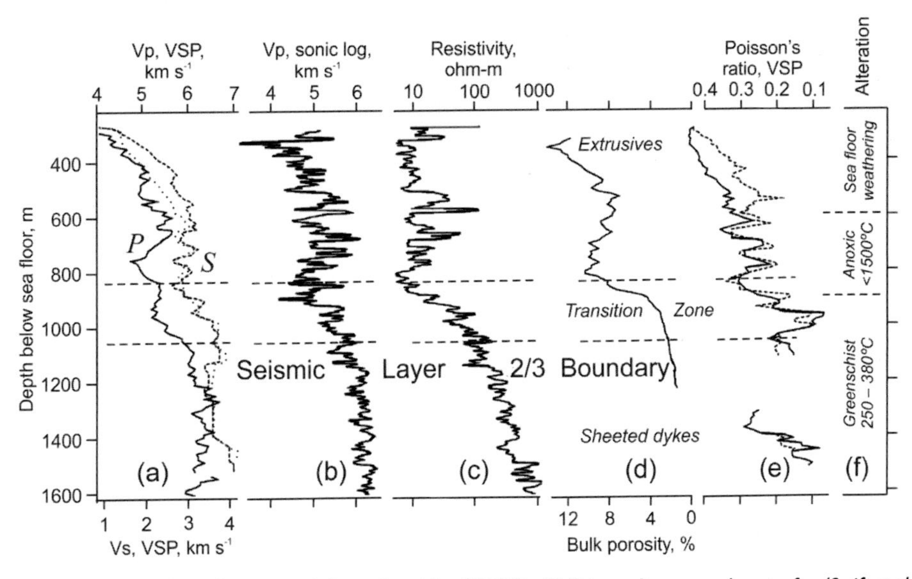


**Figure 5.26** Seismic velocity and physical properties in layers 2 and 3 at DSDP Site 504B in medium-spread crust, after (Swift *et al.*, 1998). (a) Seismic velocities inferred from a vertical seismic profile in the borehole. Two S-wave profiles reflect differing assumptions in the modelling. (b) P-wave velocity measured by logging tool. (c) Electrical resistivity measured by logging tool. (d) Bulk porosity inferred from resistivity logging. (e) Poisson's ratio inferred from vertical seismic profile. (f) Mineral alteration regimes. Horizontal dashed lines indicate major lithological and mineral alteration boundaries.

and 1056 m of sheeted dykes, without reaching the base of the dyke layer. Swift *et al.* (1998) used vertical seismic profiles obtained in the hole to compare the lithological and seismic structure (Figure 5.26). The changes in seismic velocity, particularly near the top and bottom of the extrusive/dyke transition zone, mostly correspond to observed changes in measured physical properties, reflecting changes in porosity related to different lithologies. The regional seismic layer 2/3 boundary, at 1150–1200 m below the top of the extrusives and some 100 m below the top of the sheeted dykes, is marked by a decrease in compressional velocity gradient, and corresponds to the deepest appearance of extrusive lavas in the borehole. Thus it does not correspond precisely to a lithological boundary, and again may be controlled by changes in porosity and chemical alteration (Detrick *et al.*, 1994). Low values of Poisson's ratio in the extrusive/dyke transition zone indicate the presence of large-aspect-ratio cracks, suggested to be caused by rapid expansion of circulating seawater following intrusion of dykes.

Two boreholes have penetrated deep into slowly spread crust: IODP hole 1309D on the slow-spread (22.7 km $Ma^{-1}$) MAR (Ildefonse *et al.*, 2007), and ODP hole 735B into ultra-slowly-spread crust (14 km $Ma^{-1}$ full rate) on the SWIR (Dick *et al.*, 1991, 2000; Figure 5.27). Both sites are at the ends of spreading segments, adjacent to major transform faults, and in the exhumed footwalls of OCCs (Section 7.4), so may not be representative of all slowly spread crust. Each hole recovered samples from a significant depth range within the plutonic complex.

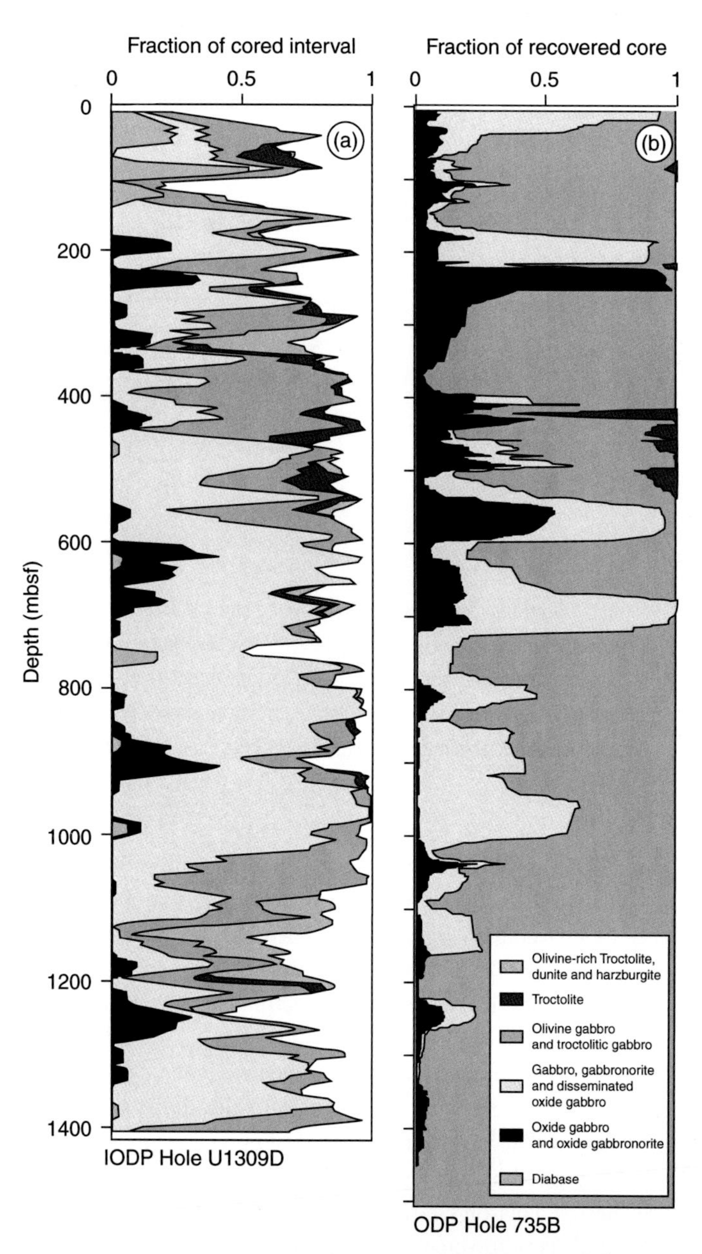


**Figure 5.27** Summary lithological columns from (a) IODP hole 1309D, slow-spreading MAR, and (b) ODP hole 735B, ultra-slow-spreading SWIR, after Ildefonse *et al.* (2007). The extrusive and sheeted dyke sections are absent and there is negligible sediment cover. For colour version, see plates section.

Hole 1309D penetrated 1415 m through predominantly gabbro and olivine gabbro, with minor amounts of more olivine-rich rocks and some oxide gabbro (Blackman *et al.*, 2006; Ildefonse *et al.*, 2007; Blackman *et al.*, 2011). The gabbros generally intrude more olivine-rich rocks, and are themselves intruded by oxide gabbros. These rocks are all interpreted as cumulates produced by crystal fractionation from a common parental magma. The intrusive units range from a few centimetres to ~100–200 m in thickness. There are a few thin intervals of residual mantle peridotite (harzburgite). Most measured seismic velocities were ~6.5 km $s^{-1}$ or less, and none were >7 km $s^{-1}$. A borehole seismic experiment revealed velocities of 5.6–6.0 km $s^{-1}$. A nearby seismic refraction experiment with shots and receivers on the sea floor can be modelled with maximum velocities of 6.0–6.5 km $s^{-1}$ (Collins *et al.*, 2009).

Hole 735B penetrated 1508 m through various gabbros (Dick *et al.*, 2000; Ildefonse *et al.*, 2007), which are interpreted to have crystallised at the margins of small magma chambers or sills (Dick *et al.*, 1991). They reveal a strongly heterogeneous lower oceanic crust very different from the gabbros sampled at fast-spreading ridges and most ophiolites (Dick *et al.*, 2000). A nearby seismic study revealed a thin seismic layer 2 with velocity 5.8 km $s^{-1}$ increasing to 6.5 km $s^{-1}$ at 1.4 km depth (roughly the depth of the bottom of the hole 735B), overlying a layer 3 that is approximately 4.2 km thick with velocity increasing from 6.5 to 6.9 km $s^{-1}$ (Muller *et al.*, 1997; Muller *et al.*, 2000).

## 5.7 Ophiolites

Ophiolites are believed to be examples of ancient oceanic crust that are exposed on land (Gass, 1968; Moores, 1982). They are mostly found in collisional mountain belts, and mark the last remains of a subducted oceanic plate. They are thought to be emplaced by the process of obduction, whereby the crustal and uppermost mantle of the subducting lithosphere is separated from the deeper mantle, elevated, and thrust over the otherwise overriding plate. Some of the best examples are the Troodos ophiolite in Cyprus (e.g., Gass and Masson Smith, 1963; Gass *et al.*, 1984; Gass, 1990; Robertson and Xenophontos, 1993), Bay of Islands ophiolite in Newfoundland (e.g., Malpas, 1978; Girardeau and Nicolas, 1981; Salisbury and Christensen, 1978; Casey and Karson, 1981), and the Semail ophiolite in Oman and the United Arab Emirates (e.g., Lippard *et al.*, 1986; Boudier and Nicolas, 1988; Macleod and Rothery, 1992; MacLeod and Yaouancq, 2000; Kelemen *et al.*, 2000; Nicolas *et al.*, 2000).

The association of ophiolites with oceanic crust is strengthened by the comparison of physical properties (Vine and Smith, 1990) and particularly their seismic velocity structure (Salisbury and Christensen, 1978). However, several features suggest they are not typical of all oceanic crust. Age relations often suggest ophiolites were obducted shortly after crustal formation, arguing for formation in a marginal sea (e.g., a back-arc basin or young ocean basin such as the Red Sea). Geochemical relations frequently suggest a supra-subduction zone setting (Robinson *et al.*, 1983; Pearce *et al.*, 1984; Elthon, 1991), also consistent with a back-arc basin. Spreading rate estimates suggest ophiolites are mainly derived from medium- or fast-spreading ridges (e.g., Nicolas *et al.*, 2000). High-resolution observations in ocean floor tectonic windows reveal details that differ significantly from the traditional ophiolite model (Karson *et al.*, 2002a). Finally, the very fact of ophiolite emplacement onto continental crust indicates something unusual about these examples. Thus, ophiolites are probably not perfect analogues for 'normal' oceanic crust, so direct inferences concerning the properties and processes of oceanic crust should be made with caution. Nevertheless, comparisons between oceanic crust and ophiolites have been very fruitful in the study of each (e.g., Nicolas, 1989; Malpas *et al.*, 1990; Parson *et al.*, 1992; Dilek *et al.*, 2000).

Ophiolites classically contain some or all of the following components (Penrose, 1972): a basal ultramafic complex composed of lherzolite, harzburgite and dunite; a gabbroic complex, usually including a thick, basal unit of layered gabbros and cumulate peridotites and pyroxenites overlain by isotropic gabbros; a mafic sheeted dyke complex consisting entirely of parallel, juxtaposed and inter- and intra-intruded doleritic dykes; a mafic volcanic complex, usually containing pillow basalts; all topped by pelagic sediments (Figure 5.6). At Troodos, the pillow lavas are about 1 km thick and the sheeted dyke layer 2–4 km thick (Gass, 1968), although there are significant 'transition zones' of mixed dykes and lavas (Hall *et al.*, 1987). The sheeted dykes are underlain by gabbros and plagiogranites, which both cut and are cut by dykes in a zone ~200 m thick (Baragar *et al.*, 1990). Within the gabbros and plagiogranites, structural relations and mineral chemistry supports their crystallisation in and emplacement from multiple magma chambers at various depths (Malpas, 1990). There may be significant differences at other ophiolites (Figure 5.6).

## 5.8 Departures from the layered crust model

Substantial departures from the layered 'Penrose' model of oceanic crust occur where the melt supply is reduced. This tends to occur where spreading rate decreases, towards the ends of spreading segments, and over anomalously cool or infertile mantle.

### 5.8.1 'Plum-pudding' model

An important example occurs towards the ends of slow-spreading segments, where the mafic crust becomes discontinuous and there are believed to be small pockets of gabbro embedded in peridotite, overlain by patchy, thin or non-existent sheeted dyke and extrusive layers (Cannat *et al.*, 1995; Section 5.2; Figure 5.5). These areas of discontinuous crust have come to be known informally as 'plum-pudding' structures.

### 5.8.2 Oceanic core complexes

'Normal' slow-spreading segments exhibit intermittent outcrops of lavas and peridotite at their ends, along with large-offset normal faults (Shaw, 1992). In places these developments become so extreme that a new mode of spreading and crustal structure occurs via detachment faulting and production of oceanic core complexes (OCCs; Figure 7.23).

Oceanic core complexes are domed massifs where crustal generation occurs by exhumation of deep material directly onto the sea floor along large-scale normal or 'detachment' faults; they thus represent a highly asymmetric mode of accretion (Cann *et al.*, 1997; Tucholke *et al.*, 1998; Escartin and Canales, 2011). Oceanic core complexes are thought to form where the melt supply falls to a critical level of about half the crustal accretion rate (Tucholke *et al.*, 2008), at the ends of slow-spreading segments and in some areas of ultra-slow spreading (Cannat *et al.*, 2006). Escartin *et al.* (2008) suggest such asymmetric accretion occurs along almost half the length of the MAR between 12.5° N and 35.6° N. The tectonic development of OCCs is discussed in Section 7.4.

Many studies have investigated the structure of OCCs. They are invariably associated with high gravity anomalies, indicative of positive density anomalies within them, and are mostly surrounded by thinner crust than on the conjugate plate (Tucholke *et al.*, 1998; Fujiwara *et al.*, 2003; Searle *et al.*, 2003; Cannat *et al.*, 2006; Blackman *et al.*, 2008; Blackman *et al.*, 2009). Their domal surfaces typically expose serpentinised peridotite and/or gabbro (Tucholke *et al.*, 1998). Where they have been drilled, these gabbros are emplaced as a series of thin units, not as a single pulse of magma (Blackman *et al.*, 2011; Section 5.6). Some core complexes apparently expose mostly peridotite in their domes, suggesting direct exhumation of mantle rocks (MacLeod *et al.*, 2009); others appear to expose gabbros mainly in those parts proximal to the ridge axis, suggesting that renewed magmatism may be associated with the termination of slip on the detachment fault (MacLeod *et al.*, 2002).

Canales *et al.* (2008) have analysed geophysical and particularly seismic data from three well-studied Atlantic core complexes, and find a heterogeneous velocity structure (Figure 5.28). By considering the geophysical characteristics and comparison with available lithological samples, they suggest that high velocities indicate gabbroic rock while low velocities represent serpentinised peridotite. (Although fresh peridotite would be expected to have a faster velocity than gabbro, serpentinisation can lower the velocity substantially). Both Kane and Atlantis core complexes show low-velocity serpentinite under their older parts, and high-velocity gabbro under their younger parts, but Dante's Domes are characterised by high velocity all along the spreading-parallel line (Figure 5.28). Similarly, Atlantis and Dante's Domes show considerable variation along their N–S (axis-parallel) sections whereas Kane appears relatively homogeneous in the along-axis section. Further work is needed to fully understand the structure and lithological development of these features.

It is generally accepted that the corrugated domes of OCCs represent the exposed surfaces of detachment faults (Escartin and Canales, 2011), but the subsurface geometry of such faults is poorly known. Smith *et al.* (2008) have suggested that closely spaced core complexes might represent multiple exposures of a single detachment fault and be linked subsurface along the spreading direction. MacLeod *et al.* (2009), following Tucholke *et al.* (1998), have emphasised the independent development of each complex, while Reston and Ranero (2011) propose that detachments may be linked both along and across the ridge-axis direction. So far the only constraints on the subsurface geometry comes from the TAG area of the MAR. There, deMartin *et al.* (2007) connected the shallow-dipping sea floor exposure of a detachment, via a boundary between two tomographic velocity anomalies, to a cluster of earthquake hypocentres in the lower crust, to infer a downward-steepening fault (Figure 7.24). While numerical models (Buck *et al.*, 2005) also suggest that detachment faults steepen with depth, so far no deep detachments have been directly imaged.

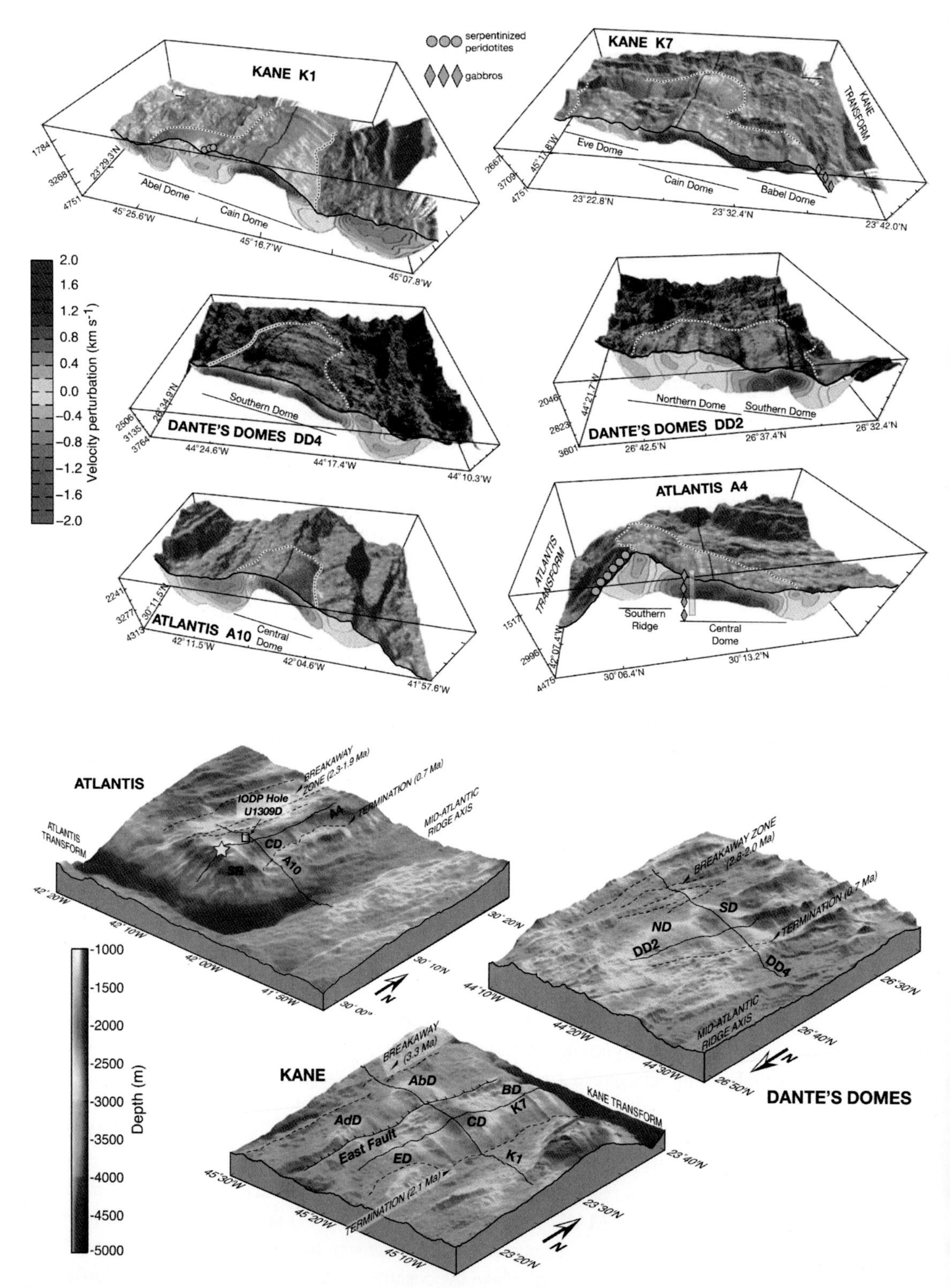


**Figure 5.28** Topography and tomographic seismic velocity models of three Atlantic oceanic core complexes: Atlantis Massif, Dante's Domes and Kane OCC, after Canales *et al.* (2008). Top: sea floor topography shown in shaded relief (grey) above coloured 2D tomographic sections showing velocity anomalies in shades of red to blue. Spreading-parallel sections (E–W) are shown on the left and axis-parallel ones (N–S) on the right. Bottom: coloured topographic images showing locations of seismic sections (black lines). Dotted lines mark edges of smooth, corrugated detachment surfaces. Yellow star: Lost City hydrothermal field. For colour version, see plates section.

### 5.8.3 'Smooth' sea floor

A further, recently discovered mode of sea floor spreading, producing so-called 'smooth' sea floor, occurs at ultra-slow spreading rates (Cannat *et al.*, 2006). Here there is little sign of sea floor volcanism, but detachment faults exhume mantle directly on to the sea floor, where it is rapidly serpentinised and forms smooth-sided, axis-parallel ridges (Dick *et al.*, 2003; Michael *et al.*, 2003; Sauter *et al.*, 2013; Section 7.5). In such areas, the 'crust' on both plates may consist largely of serpentinised peridotite, as once hypothesised by Hess (1962) for all oceanic crust.

## 5.9 Crustal magnetisation

Knowing the magnetic structure of the oceanic crust is important both for understanding overall crustal structure, for the proper interpretation of marine magnetic anomalies, and for their use to date the crust. It was reviewed by Gee and Kent (2007).

Magnetisation is carried in naturally magnetic minerals such as magnetite and titano-magnetite, which are minor constituents of crustal rocks such as basalt. These minerals carry a small induced magnetisation that is proportional and parallel to the ambient Earth's magnetic field, with a constant of proportionality (the magnetic susceptibility) ranging from $\sim 10^{-4}$ to $10^{-1}$. However, they typically also acquire a stronger, stable, thermo-remanent magnetisation as they cool below the magnetic blocking temperature (~500 °C) following eruption, and sometimes a chemical remanent magnetisation following alteration, especially serpentinisation (e.g., Dunlop and Prévot, 1982). The ratio of remanent to induced magnetisation (the Königsberger ratio) is high in oceanic basalts, in the range 1–160 (Fowler, 2005), so that often the induced component can be ignored.

Young oceanic basalts can have natural remanent magnetisations up to many tens of amps per metre (A $m^{-1}$), so the extrusive layer is commonly considered to be the only, or major, source of marine magnetic anomalies. A shallow magnetic source layer ~500 m thick and with magnetisation ~10–20 A $m^{-1}$ is often assumed in modelling these anomalies. However, there is evidence of considerable variation on this pattern, particularly near ridge axes (Tivey and Johnson, 1993). Very young basalts typically have higher magnetisations, and older ones have considerably lower magnetisations (Tivey and Johnson, 1987; Hussenoeder *et al.*, 1996). Lineated magnetic anomalies are observed to extend, though somewhat discontinuously, across several OCCs and, since the extrusive layer is absent in these structures, lower crustal or mantle rocks there must carry a significant magnetisation acquired at or near the time of sea floor generation (Tucholke *et al.*, 2001; Fujiwara *et al.*, 2003; Searle *et al.*, 2003; Okino *et al.*, 2004; Blackman *et al.*, 2009).

The magnetic structure of individual crustal components is discussed below; the results are summarised in Table 5.2.

**Table 5.2** Magnetisations of typical oceanic rocks

| Rock type | Magnetisation ($A\ m^{-1}$) | Susceptibility | Reference |
|---|---|---|---|
| Young basalt | 40–42 | | Johnson and Tivey (1995) |
| Basalts | 2.1–2.7 | | Lowrie (1997); Harrison (1976) |
| Basalts | 5.24–5.4 | 0.02–0.04 | Pariso and Johnson (1991) |
| Dykes | 1.6 | 0.02 | Pariso and Johnson (1991) |
| Gabbro | 2.5 | | Pariso and Johnson (1993b) |
| Olivine gabbro | 1–2 | | Pariso and Johnson (1993b) |
| Moderately serpentinised peridotite | <5 | <0.05 | Oufi *et al.* (2002) |
| Strongly serpentinised peridotite | 4–10 | 0.07 | Oufi *et al.* (2002) |

## 5.9.1 The basaltic layer

Most oceanic basalt samples have magnetisations $<10\ A\ m^{-1}$. Irving *et al.* (1970) reported values of a few $A\ m^{-1}$ from the median valley of the MAR. Lowrie (1974) and Harrison (1976) found mean magnetisations of 2.4–3.7 $A\ m^{-1}$ from 50 pillow basalt samples from Deep Sea Drilling Project cores around the world, while Johnson and Tivey (1995) report values $<10\ A\ m^{-1}$ for Ocean Drilling Program samples older than 1 Ma. However, very young basalts have magnetisations up to many tens of amperes per metre. Samples from two recently erupted lava flows on the Juan de Fuca Ridge yielded magnetisations of 40 $A\ m^{-1}$ and 42 $A\ m^{-1}$, while modelling their observed anomalies suggested magnetisations $>50\ A\ m^{-1}$. Samples from the Juan de Fuca Ridge with ages of 100–1000 years have magnetisations of 17–23 $A\ m^{-1}$ (Johnson and Tivey, 1995).

A uniform magnetic source layer that is 500 m thick is geologically unrealistic. A variable source thickness corresponding to the seismically determined thickness of layer 2A is consistent with both the measured intensity of magnetisation and the intensity inferred from anomaly inversions. On the Juan de Fuca Ridge, a simple inversion of the magnetic anomaly assuming a constant source layer that is 500 m thick yielded magnetisations $\sim\pm 6\ A\ m^{-1}$; however, allowing the source thickness to vary between 100 m and 800 m could match both the 60 $A\ m^{-1}$ magnetisations determined from rock samples and the seismically determined layer 2A thickness (Tivey and Johnson, 1993). On the Reykjanes Ridge, Lee and Searle (2000) obtained a moderately good fit to observed anomalies by using a constant 19 $A\ m^{-1}$ magnetisation and the seismically determined (Smallwood and White, 1998) layer 2A thickness. Williams *et al.* (2008) found that the amplitude of the central anomaly magnetic high correlated with seismic layer 2A thickness.

It is now generally recognised that the magnetisation of young basalts diminishes over timescales of $10^4$–$10^6$ years (Johnson and Atwater, 1977; Macdonald, 1977; Hussenoeder *et al.*, 1996). This produces a positive anomaly superimposed upon the Brunhes anomaly referred to as the Central Anomaly Magnetic High (Tivey and Johnson, 1987; Schouten *et al.*, 1999). It is thought to reflect either oxidation of titanomagnetite to titanomaghaemite (Johnson and Merrill, 1973; Prévot and Grommé, 1975) or a recent increase in the strength of the Earth's axial magnetic dipole (Gee *et al.*, 1996).

In addition to these effects, basalts with a high Fe–Ti content tend to have higher magnetisations (Vogt and deBoer, 1976). The axial magnetic anomaly along both the EPR and the MAR is higher at segment ends than segment centres, which may be at least partially explained by higher proportions of Fe and Ti in more differentiated basalts at segment ends (Carbotte and Macdonald, 1992; Gac *et al.*, 2003).

### 5.9.2 Dykes

The sheeted dyke complex generally has low magnetisation, and normally makes only a minor contribution to oceanic magnetic anomalies. Tivey (1996) measured the vertical magnetic structure in the northern escarpment of Blanco Fracture Zone, and consistently found a large-amplitude anomaly over the transition between lavas and dykes, indicating that the former are more strongly magnetised and confirming that they constitute the main crustal magnetic source, at least for crust <2 Ma.

Samples cored at ODP site 504B show that significant changes in mineral alteration and magnetisation occur near the lithological boundaries between extrusive basalts, the basalt–dyke transition zone, and the sheeted dykes (Pariso and Johnson, 1991). Away from the lava–dyke transition, dykes had a mean magnetisation of 1.6 A $m^{-1}$ and mean susceptibility of 0.02. It was concluded that here the dykes were responsible for a 'significant contribution' to the overlying magnetic anomalies.

### 5.9.3 Gabbros

Several authors have investigated the magnetic properties of the gabbros drilled at ODP site 735B. Pariso and Johnson (1993a) studied the magnetic petrology there, and found that primary, igneous magnetite is common only in Fe–Ti oxide-rich gabbros at about 2% by volume. Secondary magnetite, produced by high-temperature exsolution and hydrous alteration, occurred in all the gabbros studied and was the most important magnetic phase in most. The results strongly suggested that layer 3 rocks are capable of recording magnetic field reversals. Pariso and Johnson (1993b) found a mean effective remanent magnetisation of 2.5 A $m^{-1}$ in the drilled crustal section, while the olivine gabbros that constitute 60% of the section had an average effective magnetisation between 1 and 2 A $m^{-1}$. These values are comparable to or slightly lower than the values for off-axis extrusives, and suggest that the gabbroic section could contribute 25%–75% of the sea-surface magnetic anomaly. Worm (2001) also studied the hole 735B gabbros, and concluded that they would be an ideal source for marine magnetic anomalies. Allerton and Tivey (2001) sampled the gabbros exposed at the sea floor around hole 735B and found palaeomagnetic evidence for the outcrop of a normal polarity magnetic lineation that correlated with observed deep-towed and sea-surface magnetic anomalies. However, in the TAG region of the MAR, gabbros and dykes exposed along a detachment surface are associated with a zone of low magnetisation (Tivey *et al.*, 2003).

Slow cooling of the lower crust means that gabbros acquire their thermo-remanent magnetisation later than the overlying lavas, resulting in dipping isotherms and/or downward variation in polarity (Gee and Meurer, 2002; Meurer and Gee, 2002; Morris *et al.*, 2009).

### 5.9.4 Serpentinite

There have been many studies of the magnetisation of serpentinised peridotite, showing it may be responsible for a significant component of magnetic anomalies.

Oufi *et al.* (2002) compiled new and published data on the magnetic properties of 245 serpentinised abyssal peridotites from seven DSDP and ODP sites. They found that the magnetic susceptibility is modest in partially serpentinised peridotites, but increases rapidly for levels of serpentisation over 75%. Some samples had natural remanent magnetisations comparable to those of basalts, while others had low NRM even where the susceptibility was high; these differences were attributed to differences in the effective magnetic grain size related to serpentine textures. Oufi *et al.* proposed that the most likely average magnetisation and susceptibility for extensively (>75%) serpentinised peridotites are 4–10 A $m^{-1}$ and ~0.07, whereas for moderately (<75%) serpentinised peridotites the corresponding values would be <5 A $m^{-1}$ and < 0.05.

Nazarova (1994) studied serpentinised peridotites from five ODP sites on the MAR and found that all samples were characterized by natural remanent magnetisation averaging about 3.5 A $m^{-1}$, comparable with that of altered oceanic basalts. The average Königsberger ratio was about 2.

In the Australian–Antarctic Discordance, terrains containing OCCs are characterised by higher magnetisation than adjacent volcanic terrains, probably as a result of serpentinisation of peridotites exposed at detachment surfaces (Okino *et al.*, 2004).

Magnetisations at segment ends on the MAR at 20° N–24° N have a positive bias, which Pockalny *et al.* (1995) attributed to the preferential emplacement of serpentinised peridotite there. Gac *et al.* (2003) suggested the same explanation for higher magnetisations at the ends of large-offset segments, where the effect of enhanced Fe–Ti content is inadequate (Section 5.9.1).

Marine magnetic anomalies often exhibit an anomalous skewness, over and above that caused by the obliqueness of ridges (Section 2.3.2). This suggests that reversal boundaries dip toward the ridge axis, as suggested above for gabbros. The effect has been successfully modelled assuming moderate hydrothermal cooling of the whole crust and serpentinisation at 200–300 °C (Dyment *et al.*, 1997).

At St. Paul Fracture Zone (equatorial Atlantic), the central normal polarity zone is 40% wider than predicted for the known plate motions, an effect attributed to prolonged acquisition of chemical remanent magnetisation during serpentinisation of peridotites (Sichler and Hekinian, 2002).

## 5.10 Summary

The oceanic crust has a basic layered structure, similar to that seen in ophiolites. The lithological layers correlate approximately with those seen in the seismic velocity structure. The shallowest, layer 1, is composed of sediments deposited on top of the igneous crust. It is absent at the ridge crest and thickens away from it. P-wave velocity is $<2.5$ km s$^{-1}$. Seismic layer 2 is about 2 km thick and has P-wave velocity increasing steeply from 2.5 km s$^{-1}$ to 6.5 km s$^{-1}$. A three-part subdivision has been suggested. A low-velocity layer 2A is identified with young, fractured and porous basalts; this thins with age, and the underlying layer 2B thickens, reflecting the closing of pore spaces as the basalts age, and perhaps some off-axis emplacement of lavas. Layer 2C may correlate approximately with the sheeted dyke layer. However, in the few places where the seismic layer 2/3 boundary has been sampled, it lies either within the sheeted dykes or within the gabbros, suggesting the effects of alteration with age may overlay primary lithological control. Seismic layer 3, sometimes called the oceanic layer, is some 4–6 km thick and has an almost uniform P-wave velocity of 6.8–7.2 km s$^{-1}$. It is interpreted as comprising mostly the plutonic gabbros of the lower crust. At the base of layer 3 the P-wave velocity increases fairly sharply to $\sim$8 km s$^{-1}$ marking the seismological Moho. However, such high velocities may be associated with ultramafic cumulate rocks that are genetically part of the crust. A transition to mantle rocks (peridotites) typically occurs somewhat deeper than the seismological Moho, and is called the petrological Moho.

The lithological interpretations of seismic structure are broadly confirmed by shallow sea floor sampling using dredges, ROVs and manned submersibles, often making use of tectonic scarps to provide windows into the crust. Deeper sampling has been accomplished in a few places by the Deep Sea Drilling Project and its successor programmes.

An important feature of the structure of MORs is the location and distribution of melt in the lithosphere. Seismic studies show the existence of shallow bodies of connected melt (magma chambers) beneath the axes of fast- and intermediate-spreading ridges; they are small and lens shaped, no more than $\sim$100 m thick and a few kilometres wide. Similar axial magma chambers are rare at slow-spreading ridges, but have been imaged under parts of the MAR that may be influenced by the Azores and Iceland hotspots. Recent work has detected some deeper and slightly off-axis melt lenses. Two end-member models of melt delivery and crustal construction are the gabbro glacier and multiple sill models. The former has a single, shallow axial magma chamber, requiring that solidified melt (gabbro) must move plastically down and away from the axial chamber to form the lower crust. The multiple sill model has melt lenses throughout the crust allowing lower crustal gabbros to form *in situ*. Combinations of these two models are also possible.

The relatively simple layered model of crustal structure breaks down at slower spreading rates, at segment ends and at low levels of melt delivery. At segment ends the amount of melt delivered to the crust is reduced, resulting in isolated areas of basaltic rocks and gabbros surrounded by mantle peridotite. If the melt delivery falls below a critical level, lithospheric separation is accommodated by long-lived detachment faulting rather than magmatic accretion. This leads to a highly asymmetric mode of accretion, with one plate comprising relatively normal magmatic crust, while the other forms by direct extrusion of ductile mantle (possibly intruded locally by gabbro) onto the sea floor. At ultra-slow spreading rates, yet another type of crust is produced whereby mantle is continuously extruded onto the sea floor via detachment faults of alternating polarity, so both plates consist largely of peridotite.

Crustal rocks provide the source of most marine magnetic anomalies. Basaltic lavas generally have the highest magnetisations, up to tens of A $m^{-1}$ but diminishing with age. Dykes generally have low magnetisations. Gabbros can carry intermediate levels, and serpentinised peridotite can have very variable magnetisation depending on the degree of serpentinisation.

# 6 Volcanism

## 6.1 Introduction

Much of the oceanic crust is built from the products of mantle melting. Approximately 20 km$^3$ of melt are produced per year around the 65 000 km long global ridge system, creating some 3 km$^2$ of new crust with an average thickness of about 7 km. As the lithospheric plates separate at a ridge, plastic asthenospheric mantle is drawn up and rises adiabatically, i.e. too fast to allow for significant conductive cooling. Thus the geotherm under the ridge is raised until it intersects the solidus, when partial melting begins. (Alternatively, one may think of the melting point for a batch of mantle being progressively reduced as the pressure falls.) Once melt has been produced it rises buoyantly, eventually accumulating in or below the crust. From there some melt may erupt onto the sea floor, while the residue cools and solidifies to form the gabbroic lower crust (Chapter 5). Mantle melting was reviewed by Langmuir *et al.* (1992).

## 6.2 Mantle melting

Most ridges are not situated directly above hot, rising columns of convecting mantle; their positions are largely uncorrelated with the underlying mantle convection (McKenzie, 1967; McKenzie and Bickle, 1988). This means that ductile asthenosphere of normal mantle temperature is drawn up passively by the separating plates at ridges (Cann, 1970; Figure 6.1). This model successfully predicts both the thickness and composition of the oceanic crust (McKenzie and Bickle, 1988).

Far from plate boundaries, the oceanic geotherm is as shown in Figure 6.2a. The sub-lithospheric mantle convects vigorously, so heat transport, by convection, is relatively efficient, resulting in a geotherm where temperature varies only slowly with depth. Heat conduction in the mantle is poor, so this part of the geotherm is close to the adiabatic gradient. In contrast, the lithosphere is rigid and transports heat by less efficient conduction, resulting in a steep geotherm. (Radiative heat transfer is negligible.) The two parts of the geotherm intersect at the base of the lithosphere (one definition of which is the Earth's outermost thermal boundary layer), forming a sharp bend or 'knee' in the geotherm. As the knee approaches the solidus, it causes the mantle to become more plastic, giving rise to the widespread seismic low-velocity zone in the upper mantle – Section 3.3); however,

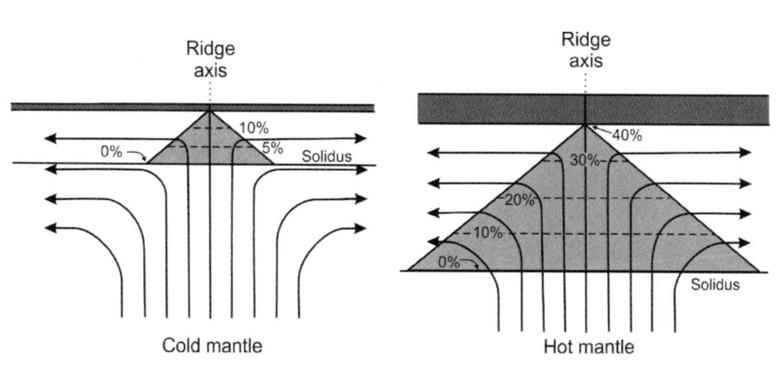


Figure 6.1 Cartoon of passive mantle upwelling and melting at MORs for cold (left) and hot mantle (right), after Langmuir *et al.* (1992). Dark grey: crust. Light grey triangular region: zone of partial melting, which roughly defines the base of the axial lithosphere. Arrows: asthenospheric trajectories. Horizontal dashed lines: extent of melting. 'Solidus' is the point at which partial melting begins, and is approximately the base of the lithosphere far from the ridge axis. Mantle begins to melt at the solidus and melts progressively as it rises, so that material reaching the apex of the melting triangle has suffered the greatest extent of partial melting.

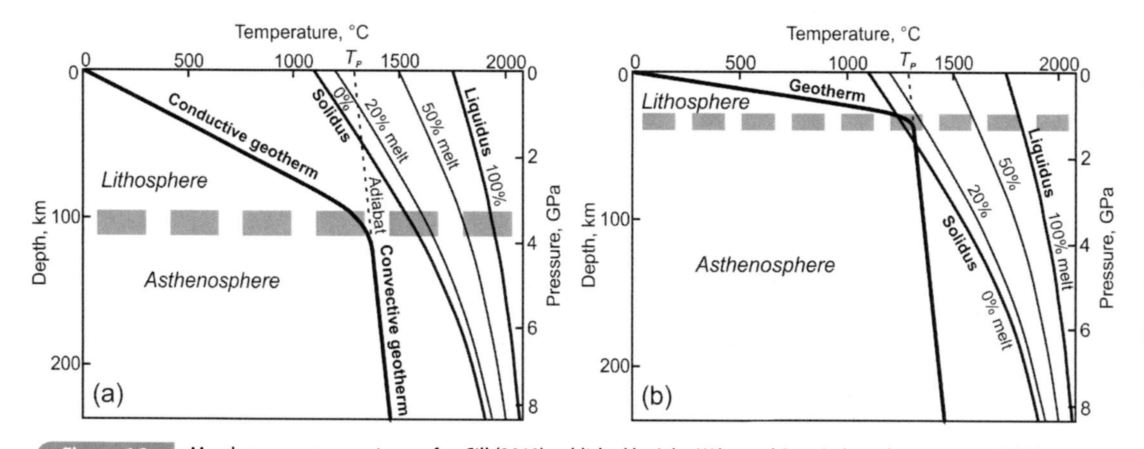


Figure 6.2 Mantle temperature regimes, after Gill (2010) published by John Wiley and Sons Ltd., with permission. Bold lines: geotherm; medium thickness lines: solidus and liquidus; light lines: degrees of melting at intermediate temperatures; broad, dashed grey line: gradational lithosphere/asthenosphere boundary. (a) Far from a MOR, heat is efficiently transported through the plastic asthenosphere by convection, giving a low-gradient geotherm that follows the adiabat, while heat transport through the lithosphere is by conduction, producing a steep geotherm. (b) Under MORs the separating plates draw up ductile asthenosphere, modifying the geotherm so that it intersects the solidus and the mantle begins to melt. Projecting the adiabat to the surface (dashed line) gives the mantle potential temperature, $T_P$.

unless the geotherm reaches the solidus, no melting will occur. The upward projection of the adiabat intersects the Earth's surface at a temperature called the mantle potential temperature, $T_p$. McKenzie and Bickle (1988) determined the normal value of $T_p$ to be 1280 °C, although a slightly higher value of 1300 °C is generally recognised today (Gill, 2010).

**Table 6.1** Chemical and mineral compositions of typical peridotites. Data from Best and Christiansen (2001) and Gill (2010)

| Spinel lherzolite | | Garnet lherzolite | | Harzburgite | |
|---|---|---|---|---|---|
| Oxide | Weight% | Oxide | Weight% | Oxide | Weight% |
| $SiO_2$ | 44.22 | $SiO_2$ | 45 | $SiO_2$ | 43.59 |
| $TiO_2$ | 0.09 | $TiO_2$ | 0.08 | $TiO_2$ | 0.03 |
| $Al_2O_3$ | 2.28 | $Al_2O_3$ | 1.31 | $Al_2O_3$ | 1.27 |
| $Cr_2O_3$ | 0.39 | $Cr_2O_3$ | 0.38 | $Cr_2O_3$ | 0.43 |
| $FeO_{total}$ | 8.47 | $FeO_{total}$ | 6.97 | $FeO_{total}$ | 5.71 |
| MnO | 0.14 | MnO | 0.13 | MnO | 0.07 |
| MgO | 41.6 | MgO | 44.86 | MgO | 48.39 |
| NiO | 0.27 | NiO | 0.29 | NiO | 0.31 |
| CaO | 2.16 | CaO | 0.77 | CaO | 0.21 |
| $Na_2O$ | 0.24 | $Na_2O$ | 0.09 | $Na_2O$ | 0.06 |
| $K_2O$ | 0.05 | $K_2O$ | 0.1 | $K_2O$ | 0.07 |
| $P_2O_5$ | 0.06 | $P_2O_5$ | 0.01 | | |
| Normative mineral[a] | Weight% | Normative mineral[a] | Weight% | Normative mineral[a] | Weight% |
| Olivine | 62 | Olivine | 68 | Olivine | 68 |
| Orthopyroxene | 24 | Orthopyroxene | 25 | Orthopyroxene | 32 |
| Clinopyroxene | 12 | Clinopyroxene | 2 | | |
| Spinel | 2 | Garnet | 5 | | |

[a] 'Normative minerals' provide a theoretical mineral content calculated from the actual chemical analysis.

Beneath MORs, the separation of the lithospheric plates draws up material from the asthenosphere. This material rises adiabatically, and the thin axial lithosphere brings the knee closer to the solidus (Figure 6.2b). When it reaches the solidus, partial melting begins. The sloping base of the lithosphere defines a relatively broad, roughly triangular melting zone (Figure 6.1). Melting begins where the geotherm first intersects the solidus, and continues as long as the asthenosphere rises adiabatically. The amount of melting is approximately 1%–2% per kilobar (100 kPa) of pressure release (Langmuir *et al.*, 1992; Figure 6.1), with more melt being produced for greater $T_p$. If all the extracted melt is delivered to the crust, then the crustal thickness is equal to the total extent of melting times the height of the melting column. A 60 km high column undergoing a total of 12% melting will produce a 7 km thick crust.

The mantle comprises rocks of the peridotite family, for which typical compositions are given in Table 6.1.[1] As well as oxygen, their main constituents are Si and Mg, with minor amounts of Fe, Ca, Al and other elements. These elements are arranged into crystalline minerals, whose detailed composition and crystal structure depend on environmental conditions such as temperature and pressure. Prior to melting, the mantle comprises

[1] By convention, rock analyses are given as weight% of the element oxides. Iron can exist in two oxidation states, $FeO_2$ and $Fe_2O_3$, and is often quoted as total iron, the sum of these two.

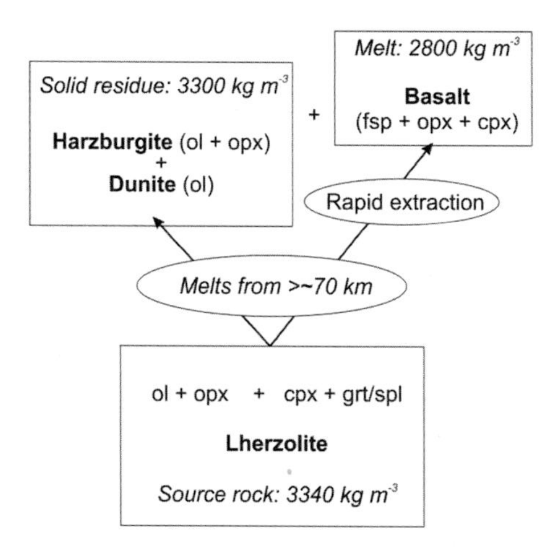


Figure 6.3 Simplified cartoon of the process of mantle melting. Rock names in bold, with constituent minerals in brackets: cpx, clinopyroxene; fsp, feldspar; grt, garnet; ol, olivine; opx, orthopyroxene; spl, spinel.

lherzolite, a rock composed of olivine ($(Mg,Fe)_2SiO_4$), orthopyroxene ($(Mg,Fe)SiO_3$), clinopyroxene ($MgCaSi_2O_6$), and minor amounts of either spinel ($MgAl_2O_4$) or garnet ($Mg_3Al_2Si_3O_{12}$).[2] Garnet is stable at pressures greater than about 1.6–2.3 GPa (~45–70 km depth), depending on temperature, while spinel is stable at shallower depth (Gill, 2010). Note that MOR lavas are very dry, suggesting that the upper mantle beneath ridges contains only small amounts (~100–200 ppm) of water (Bell and Rossman, 1992).

The result of melting lherzolite is summarised in very simple terms in Figure 6.3. Essentially, the mantle beneath ridges begins melting at depths ~70 km or more, and continues melting to ~10 km depth at the ridge axis, producing a melt of basaltic composition. This is rapidly transported through mantle conduits and eventually erupts as MOR basalt, or MORB (Kelemen *et al.*, 1995). Typical MORB has higher Si, Al and Ca content and much lower Mg content than peridotite, reflected in the dominance of feldspar minerals (Table 6.2). The residual, unmelted mantle forms the rock harzburgite, which has higher Mg content than lherzolite and no clinopyroxene, garnet or spinel (Table 6.1). The migrating magma may dissolve pyroxene from surrounding lherzolite, leaving residual dunite, while some molten olivine may crystallise in the uppermost mantle as dunite veins. Both melt and residue are less dense than primitive mantle; this enhances the buoyancy of the rising mantle, and helps focus the mantle flow to the ridge axis (Sparks and Parmentier, 1991).

Because rocks are not simple chemical compounds but combinations of several minerals, they do not melt in a simple way. Not only does the degree of melting vary (Figures 6.1 and 6.2), but so does the precise composition of the melt. This is determined by partition coefficients, which describe the proportion of each element or mineral that will reside in the molten or solid phase at a given temperature, pressure and composition, and are

[2] The chemical formulae given here for minerals are a guide only: proportions and elements may vary. For example, olivine can have a composition anywhere between $Mg_2SiO_4$ and $Fe_2SiO_4$ with variable Fe/Mg ratio; garnet can have a variety of compositions in which Fe, Mn, Ca and Cr can substitute for some or all of the Mg and Al.

**Table 6.2** Average chemical and normative compositions of N-MORB, after Best and Christiansen (2001, Table 13.1)

| Oxide | Weight% | Normative[a] mineral | Type | Formula | Weight% |
|---|---|---|---|---|---|
| $SiO_2$ | 49.93 | Orthoclase | Potassium feldspar | $KAlSi_3O_8$ | 1.00 |
| $TiO_2$ | 1.51 | Albite | Plagioclase feldspar | $(Na,Ca)AlSi_3O_8$ | 22.06 |
| $Al_2O_3$ | 15.90 | Anorthite | Calcium feldspar | $CaAl_2Si_2O_8$ | 31.14 |
| FeO | 10.43 | Diopside | Clinopyroxene | $MgCaSi_2O_6$ | 21.04 |
| MnO | 0.17 | Hypersthene? | Orthopyroxene | $(Mg,Fe)SiO_3$ | 15.55 |
| MgO | 7.56 | Olivine | Magnesium–iron silicate | $(Mg,Fe)_2SiO_4$ | 2.47 |
| CaO | 11.62 | Magnetite | Oxide mineral | $Fe_3O_4$ | 3.89 |
| $Na_2O$ | 2.61 | Ilmenite? | Oxide mineral | $FeTiO_3$ | 2.87 |
| $K_2O$ | 0.17 | Apatite | Phosphate mineral | $Ca_5(PO_4)_3(F,Cl,OH)$ | 0.19 |
| $P_2O_5$ | 0.08 | | | | |

[a] 'Normative minerals' provide a theoretical mineral content calculated from the actual chemical analysis.

determined experimentally. All MORBs tend to have similar major element chemistry (Table 6.2), although there is minor variation among $Na_2O$, FeO, $TiO_2$ and $SiO_2$ contents and $Ca/Al_2O_3$ ratios that correlate with variations in axial depth and crustal thickness (Langmuir *et al.*, 1992). Langmuir *et al.* (1992) ascribe the control of these variations to global variations in mantle temperature, but Niu and O'Hara (2007) find the global variation in $T_p$ is $<50\,°C$ and instead attribute the variation to varying mantle composition. The chemical systematics of MORB suggest that the melting process is mostly fractional melting, whereby melt is rapidly and efficiently removed from the melting region and does not remain in contact with the residual solids (Kelemen *et al.*, 1995). The melts do not undergo significant low-pressure equilibration during their ascent, so they presumably move relatively freely along wide conduits towards the surface (Langmuir *et al.*, 1992).

Most MORBs have similar major element compositions, and are referred to as N-MORB (normal MORB). However, there are small but significant differences between typical MORB from the EPR and the MAR: EPR basalts are slightly more enriched in 'incompatible' trace elements (those that tend to remain in the melt as minerals crystallise) compared to those from the MAR (Table 6.3 and Figure 6.4). Another class of MORB, referred to as E-MORB (enriched MORB), has very different trace element compositions and is significantly enriched in the more incompatible elements (Figure 6.4). Even more highly enriched basalts are associated with oceanic islands, and referred to as ocean island basalt (OIB, Figure 6.4). Such islands are associated with mantle hotspots that are thought to tap a mantle reservoir largely unaffected by MOR processes. Hence, OIB is considered to represent this less depleted mantle reservoir, whereas N-MORB is considered to be sourced from shallow mantle that has been depleted by previous cycles of mantle melting. E-MORB, which tends to be found nearer hotpots (at least in the Atlantic) is thought to derive from a mixture of the two source types.

Studies of mantle peridotites dredged from the ocean floor tend to confirm the results from basalt geochemistry (Dick *et al.*, 1984). Analyses demonstrate that these peridotites are probably the residues of variable degrees of mantle melting, and there is a good

**Table 6.3** Trace element compositions of typical MORBs, in ppm, after (Gill, 2010)

| Element | Primitive mantle | N-MORB, MAR | N-MORB, EPR | E-MORB | OIB tholeiite |
|---|---|---|---|---|---|
| Rb | 0.635 | 0.9 | 0.8 | 5.04 | 8.4 |
| Ba | 6.989 | 13.5 | 10 | 57 | 125 |
| Th | 0.085 | 0.191 | 0.171 | 0.6 | 1.14 |
| Nb | 0.713 | 2.5 | 3.1 | 8.3 | 24 |
| La | 0.687 | 3.08 | 4.06 | 6.3 | 14.6 |
| Ce | 1.775 | 8.71 | 12.3 | 15 | 36.5 |
| Sr | 21.1 | 126 | 142 | 155 | 368 |
| Nd | 1.354 | 7.91 | 11.3 | 9 | |
| Sm | 0.444 | 2.53 | 3.92 | 2.6 | 5.55 |
| Zr | 11.2 | 78.3 | 100 | 73 | 141 |
| Gd | 0.596 | 3.51 | 5.23 | 2.97 | |
| Y | 4.55 | 26.9 | 37 | 22 | 22 |
| Yb | 0.493 | 2.53 | 3.67 | 2.37 | 1.9 |
| Lu | 0.074 | 0.4 | 0.57 | 0.354 | 0.27 |

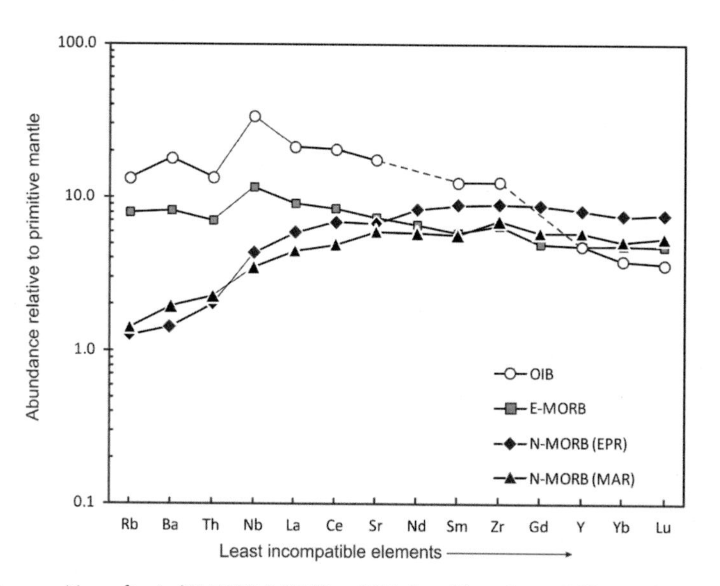


**Figure 6.4** Trace element compositions of typical N-MORB, E-MORB and OIB plotted from data in Table 6.3. Element abundances are relative to primitive mantle composition; elements are plotted in order of decreasing incompatibility.

correlation between compositions of peridotites and those of spatially associated basalts. Peridotites from the vicinity of hotspots have the most depleted compositions, implying that the associated basaltic melts should be enriched, such as E-MORB and OIB, and that the hotspots should be characterised by high mantle temperatures. Peridotites remote from hotspots are less depleted.

## 6.3 Melt delivery to the crust

As stated above, most melt is thought to be rapidly removed from the melting region and efficiently channelled towards the surface. Given that the melting regime may be many tens of kilometres wide, and relatively little MOR volcanism is observed outside a narrow axial zone only a few kilometres across, there must be an effective mechanism for focussing the rising melt. Buck and Su (1989) modelled mantle flow driven by plate separation and compositional buoyancy, and showed that the degree of focussing depends strongly on the amount of partial melt present, which determines the viscosity of the mantle. Buoyant melt focussing also leads to a 3D pattern of mantle upwelling, especially at slow-spreading ridges (Sparks *et al.*, 1993), producing strong along-axis variations in crustal thickness, as noted in Chapter 5. Melt focussing may also occur as melt flows up the sloping base of the lithosphere (Sparks and Parmentier, 1991; Magde *et al.*, 1997) and is aided by the interaction of stresses around dykes (Ito and Martel, 2002).

Some melt may be directly erupted onto the sea floor without fractionating at intermediate depth, as discussed below. However, most melt is thought to pond at some intermediate depth within (or near the base of) the crust (e.g., Smith and Cann, 1992; Section 5.4). Thence it may undergo fractional crystallisation, possibly via several cycles (O'Neill and Jenner, 2012), and may form ultramafic cumulate rocks such as those between the seismic and petrological Mohos. The remaining melt either erupts onto the sea floor or freezes *in situ* to form the gabbroic lower crust (e.g., Carbotte and Scheirer, 2004).

## 6.4 Lava morphologies

A variety of lava flow morphologies is found at MORs, including ropy, folded, whorled and jumbled sheet flows, lobate flows, pillow lavas and elongated pillows (Fox *et al.*, 1988; Perfit and Chadwick, 1998; Figure 6.5). Theory and laboratory experiments suggest that the variation from sheet flows to pillows reflects a decreasing effusion rate for a given viscosity and slope (Gregg and Fink, 1995; Table 6.4, Figure 6.6 and Figure 6.7). Increased cooling rate and decreased sea floor slope also favour pillows over sheet flows, though Gregg and Smith (2003) found sheet flows off-axis only on slopes $\leq 15°$.

Pillow lavas (Figure 6.5a) are roughly spherical or cylindrical bodies approximately 1 m across. They advance one pillow at a time, new ones breaking out from the surface of existing ones. Their surfaces are often corrugated where the cooling skin is scored by irregularities around the vent. Pillows become increasingly elongated over steeper slopes, and can reach aspect ratios of 20:1 when flowing over near-vertical scarps (Figure 6.5b, c). They typically flow only a few tens to hundreds of metres from their source, though they may form on the cooling flanks and extremities of more extensive sheet flows.

Somewhat higher effusion rates favour lobate flows (Figure 6.5d). These also advance one lobe at a time, but fast enough to allow individual lobes to coalesce back into an interconnected flow. Lobate flows are often more extensive, and individual lobes are generally

**Table 6.4** Ranges of effusion rates inferred for submarine lava flows of given types, assuming a lava viscosity of $10^2$ Pa s, after Griffiths and Fink (1992)

| Lava type | Minimum rate, $m^3\ s^{-1}$ | Maximum rate, $m^3\ s^{-1}$ |
|---|---|---|
| Pillows | Not defined | 1 |
| Lobate sheet flows | 1 | 100 |
| Ropey sheet flows | 100 | 3000 |
| Jumbled sheet flows | 3000 | Not defined |

**Figure 6.5** Lava flow morphologies seen on MORs. Images (a–e) are on the MAR, 45° N, from the author; (f) and (h) are from Axial Seamount, Juan de Fuca Ridge, 46° N, after Fox *et al.* (1988); (g) is from the EPR, 21° N, after CYAMEX (1981); (i) is from the Galapagos Ridge, 86° W, after Ballard *et al.* (1979). All images are ~2–3 m across. (a) Pillow lavas; (b) elongate pillows; (c) highly elongate pillows draping a 'haystack'; (d) flattened lobate flow; (e) collapsed lobate flow; (f) jumbled sheet flow; (g) folded pahoehoe; (h) ropey sheet flow; (i) lava pillar. For colour version, see plates section. Images (f) and (h) © 1988, with permission from Elsevier; (g) reproduced with kind permission of Springer Science and Business Media.

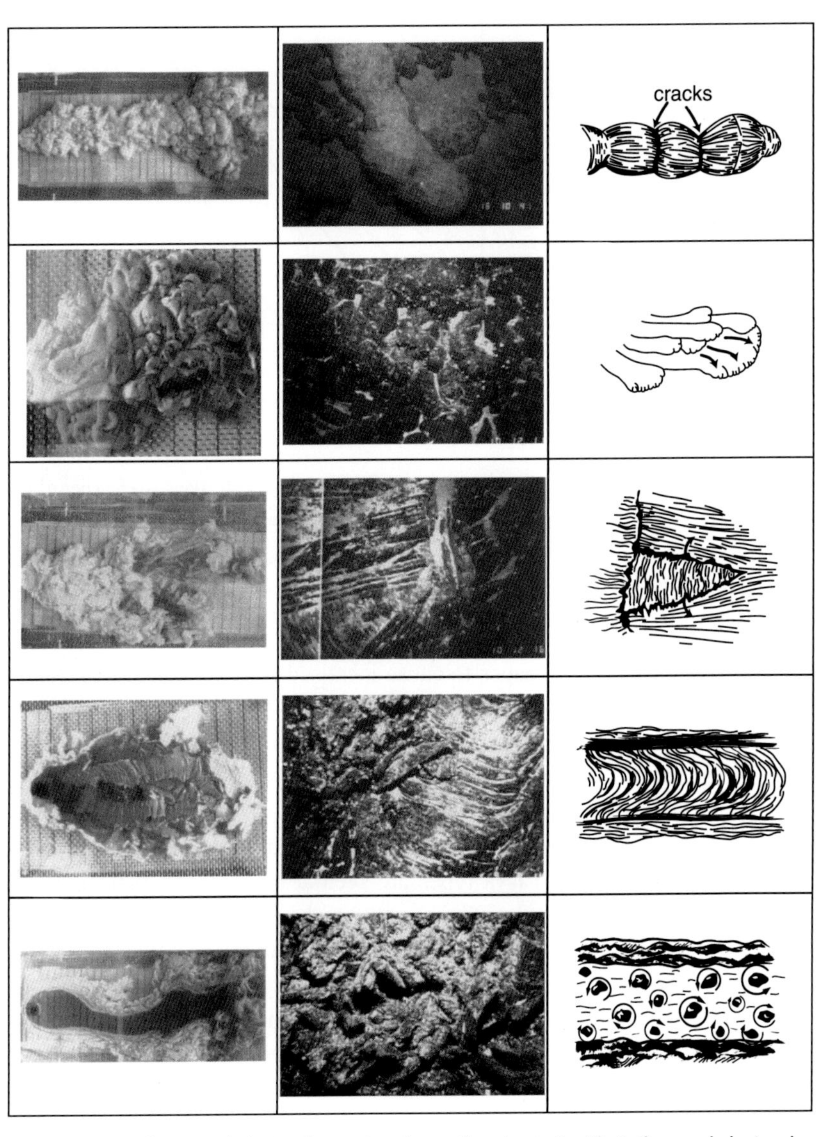


Figure 6.6 Examples of various lava flow morphologies observed on the sea floor (centre), with similar morphologies observed in solidified molten wax during laboratory experiments (left), and sketches of the morphological formation (right), after Gregg and Fink (1995). Lines in left-hand photos are 2 cm apart; solid wax is lighter. Sea floor photos are about 5 m across. Flow is from left to right in left- and right-hand panels. Top represents high-viscosity, low flow rate eruptions, grading to low-viscosity, high flow rate at bottom.

broader and flatter, than pillows. Lobate flows often inflate vertically, and may subsequently collapse if some underlying lava drains away, producing 'trapdoors' or 'skylights' in the flow surface (Figure 6.5e). If collapse occurs on a larger scale, the remains of supporting 'lava pillars' may be seen, often with drain-back marks on them (Ballard *et al.,* 1979; Francheteau *et al.,* 1979; Figure 6.5i). They are thought to form when water trapped

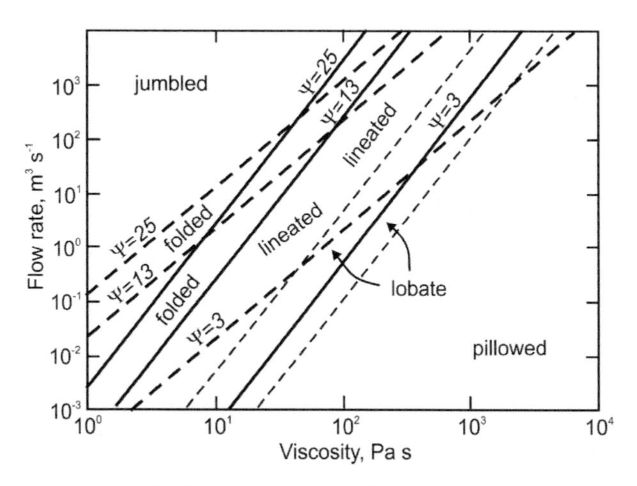


Figure 6.7 Calculated morphology of submarine basalts emplaced on a 10° slope as a function of effusion rate and viscosity, after Gregg and Fink (1995). Dashed lines are for linear sources (fissures), solid lines for point sources (cones). $\Psi$ is the non-dimensional ratio of time to form a solid crust on the flow surface, to a characteristic timescale for horizontal advection. As sea floor slope increases, lines of constant $\Psi$ move down and to the right.

beneath a flow escapes to cause localised freezing of lava. Extensive areas of ponded and possibly collapsed flows are often referred to as 'lava lakes'.

Sheet flows (Figure 6.5f–h) form by rapid effusion onto gentle slopes, with lava flowing relatively freely to produce wide, continuous sheets. The pressure of lava flowing underneath may cause the lava crust to buckle into low, fractured mounds called tumuli. Sheet flows up to ~20 km long have been observed near MORs (Macdonald *et al.,* 1989).

Although most lava types are seen at most ridges, there is a clear tendency for those representing higher effusion rates to be more prevalent at faster spreading, with fast-spreading ridges displaying >70% sheet flows and slow-spreading ridges >80% pillow basalts (Figure 6.8).

Several examples of explosive volcanism are known. Sohn *et al.* (2008) found unconsolidated pyroclastics covering >10 km$^2$ at 4000 m depth on the ultra-slow-spreading Gakkel Ridge. The volume change associated with boiling of seawater is highly sensitive to pressure and becomes negligible above the critical pressure of 2.98 Mpa (*c.* 3000 m depth, Figure 8.10), so steam-based explosive volcanism should be rare at ridges. Clague *et al.* (2009) report numerous pyroclastic deposits on the Juan de Fuca and Gorda Ridges, EPR, Fiji back-arc basin and various seamounts, many of them deeper than 3000 m. They find high effusion rate eruptions produce sheet flows with abundant pyroclastics, while low effusion rates produce pillows with few pyroclastics, suggesting that high effusion rates deliver higher magmatic gas contents. Subsequently, Helo *et al.* (2011) found $CO_2$ concentrations of 0.9% in crystals from Axial Seamount, Juan de Fuca Ridge, sufficient to drive explosive eruptions. This suggests the upper mantle is more variably enriched in $CO_2$ than previously thought.

The rest of this chapter discusses the volcanic structures and processes that are observed on the sea floor, many of which were reviewed by Perfit and Chadwick (1998). The

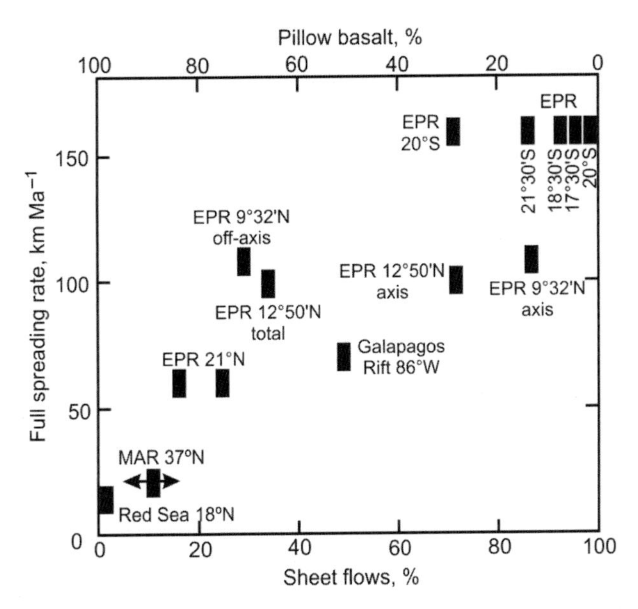


Figure 6.8 Proportion of main lava type (abscissa) as a function of spreading rate (ordinate), after Perfit and Chadwick (1998).

expressions of sea floor volcanism change markedly with spreading rate, so in the following sections they are classified by spreading rate. However, it should be remembered that in reality there is a continuum of both spreading rates and volcanic processes.

## 6.5 Fast-spreading ridges

Fast and superfast spreading (Table 1.1) currently occurs only on the EPR south of Orozco Transform. Eruptions are frequent, small volume and relatively homogeneous (Perfit and Chadwick, 1998). They are estimated to occur about once every 3–14 years with average volume ~0.001–0.01 km$^3$ (Macdonald, 1998; Perfit and Chadwick, 1998; Sinton *et al.*, 2002). The volcanic structures and inferred magma sources and pathways beneath a section of the northern EPR are summarised in Figure 6.9. Particular components of this structure are described in the following sections.

### 6.5.1 The volcanic axial high

The axes of fast-spreading MORs are characterised by shallow ridges (sometimes called 'axial highs' or 'crestal highs') a few kilometres wide and a few hundred metres high where most eruptive activity is focussed. These ridges, which are like an elongated shield volcano (Lonsdale, 1977), are essentially continuous along whole spreading segments (Macdonald *et al.*, 1984; Searle, 1984; Carbotte and Macdonald, 1994a). They vary in cross-sectional area and shape (Figure 6.10), from ~8 km wide and 250 m high at 13° N on the EPR to 15 km

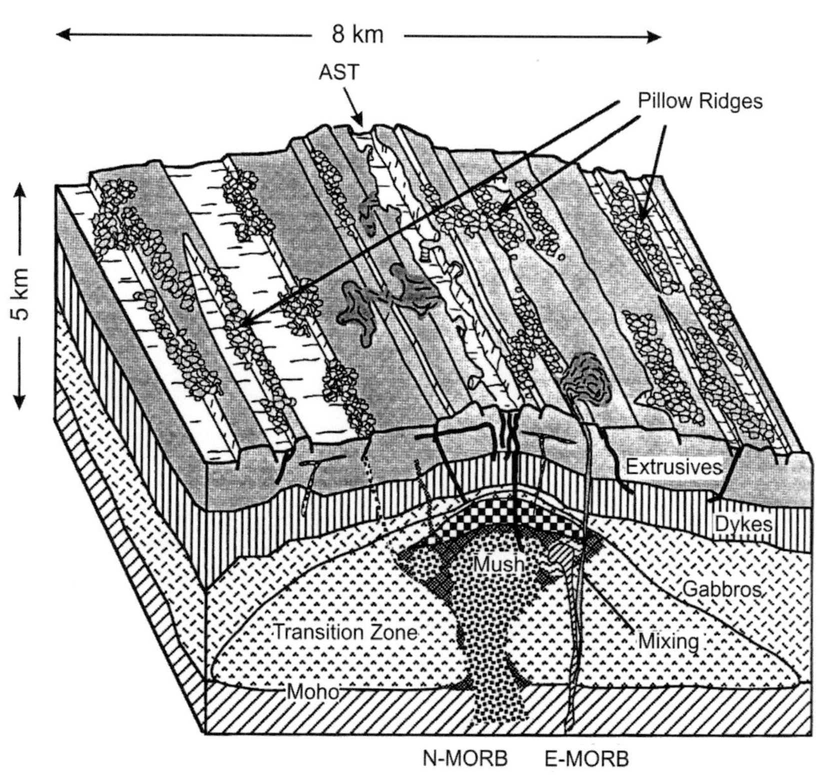


Figure 6.9 Volcanic structure of the fast-spreading EPR near 9°30′ N, after Perfit and Chadwick (1998). On-axis eruptions are mainly associated with the axial summit trough (AST); off-axis pillow lavas with fissures and faults.

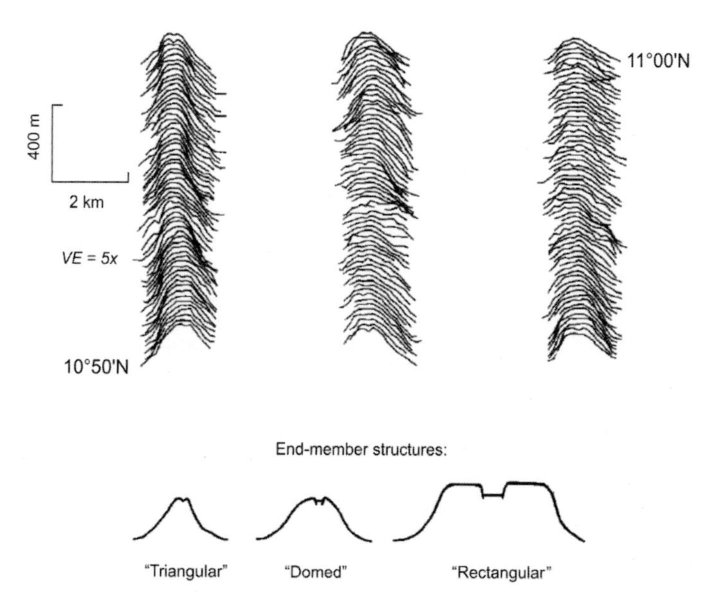


Figure 6.10 Stacked, 100-m-spaced profiles across the EPR axis near 11° N, reflecting variation in cross-sectional shape, after Macdonald *et al.* (1984). Profiles are from single bathymetry pings to preserve maximum resolution.

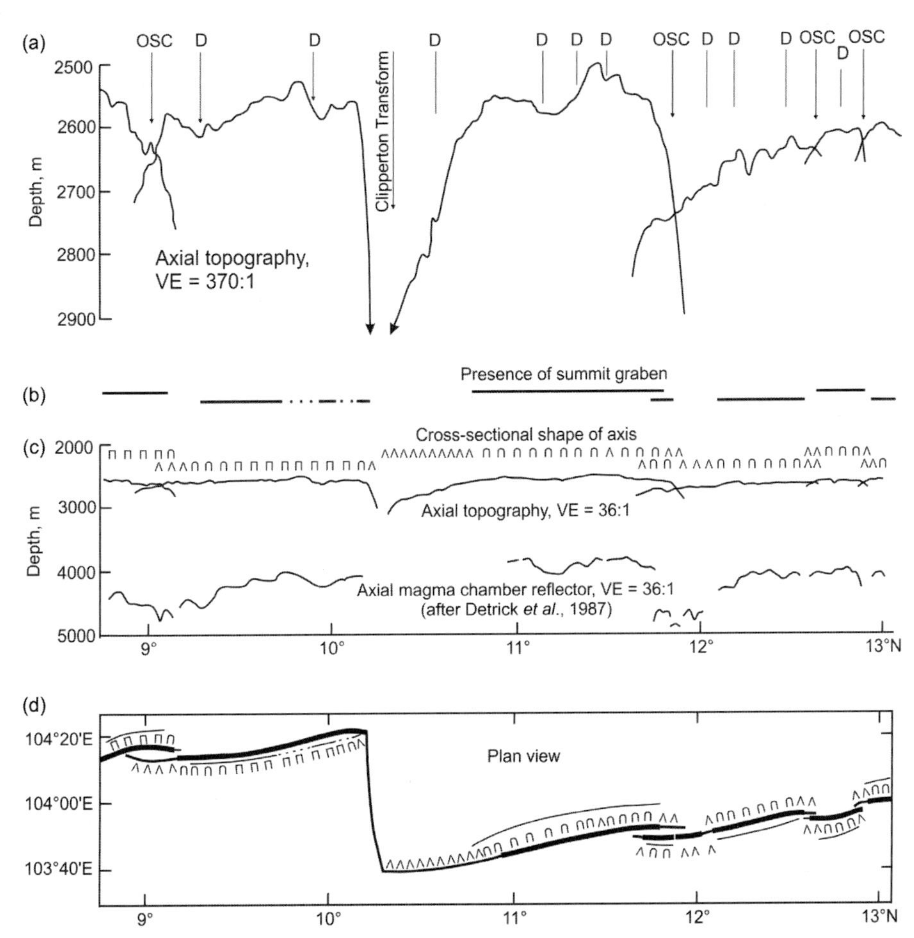


Figure 6.11 Correlations between the shape of the axial ridge, axial depth profile, presence of axial summit graben, and depth to axial magma chamber along part of the fast-spreading northern EPR, after Macdonald and Fox (1988). © 1988, with permission from Elsevier. (a) Highly exaggerated axial depth profile showing segment boundaries at transform faults, overlapping spreading centres (OSC) and DEVALS (D). (b) Horizontal bars indicate presence of axial summit graben. (c) Top, symbols indicate approximate cross-sectional shape and inferred magmatic state of axial ridge (triangles: narrow and deflated; hoops: domed and intermediate; square: broad, inflated); upper curve, axial depth profile at reduced vertical exaggeration; lower curve, depth to top of axial magma chamber at same scale. (d) Plan view of ridge axis: broad lines, spreading axis with axial magma chamber; intermediate lines, spreading axis with no AMC; thin continuous lines (offset from spreading axis for clarity), axial summit graben present; thin dotted lines, recent lava flows but no axial summit graben.

wide and 300–400 m high on the superfast EPR at 20° S (Francheteau and Ballard, 1983). Macdonald and Fox (1988) recognise three end-member shapes – triangular, domed and rectangular – related to the presence and depth of the axial magma chamber (Figure 6.11). Narrow, triangular sections occur where the AMC is deep or absent, and are interpreted as magmatically deflated; broad, rectangular ridges are associated with shallow AMC and are interpreted as being magmatically inflated, while domed cross-sections are intermediate.

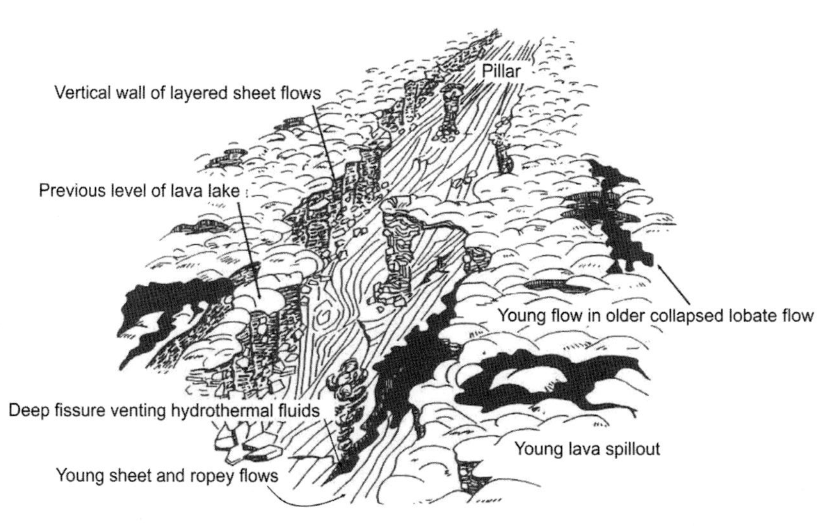


Figure 6.12 Composite sketch of the axial summit trough along the EPR near 9°45′ N, after Perfit and Chadwick (1998). Width of view ~100 m. Sheet flows dominate the floor of the AST, with lobate flows more common on the flanks.

## 6.5.2 The axial summit trough and neo-volcanic zone

The axial ridge has, along most of its crest, a small valley ≤2 km wide and 100 m deep (Figures 6.9 and 6.10). This has been called the 'axial summit graben' (Francheteau and Ballard, 1983), and 'axial summit caldera' (Haymon *et al.*, 1991). Fornari *et al.* (1998) distinguished between narrow (<200 m wide, <~15 m deep) 'axial summit collapse troughs' formed by collapse of lava flow surfaces above primary eruptive fissure zones, and larger (>~300 m wide) tectonically formed graben. Often the broad axial graben contains a narrow collapse trough within it. The more generic 'axial summit trough' (Perfit and Chadwick, 1998; Fornari *et al.*, 2004) is used here.

The neo-volcanic zone (NVZ) may be 500 m to 1 km wide (Macdonald *et al.*, 1984) but is often less than 250 m (Perfit and Chadwick, 1998). While it usually contains an axial summit trough, it can also be recognised where there is no axial trough. The NVZ is mostly characterised by sheet flows, which may show evidence of along-axis flow. Lava lakes are common, especially in the axial summit trough (AST), whose floor frequently contains collapse pits with lava pillars (Figure 6.12). Some places with an inflated axial ridge lack the expected AST, perhaps because it has filled to the brim with recent lava (e.g., Soule *et al.*, 2009).

Some parts of the axis, especially but not exclusively near segment ends (Cormier *et al.*, 2003), contain broad fields of circular lobate or pillow mounds averaging 20 m high and 200 m in diameter (White *et al.*, 2002), implying lower magma supply and lower effusion rates there. Fast-spreading neo-volcanic zones have been said to contain no point-source volcanoes or seamounts (Scheirer and Macdonald, 1995); however, the pillow mounds appear virtually identical to similar features on intermediate and slow-spreading ridges, sometimes called volcanic 'hummocks' (Section 6.7) and interpreted as originating from

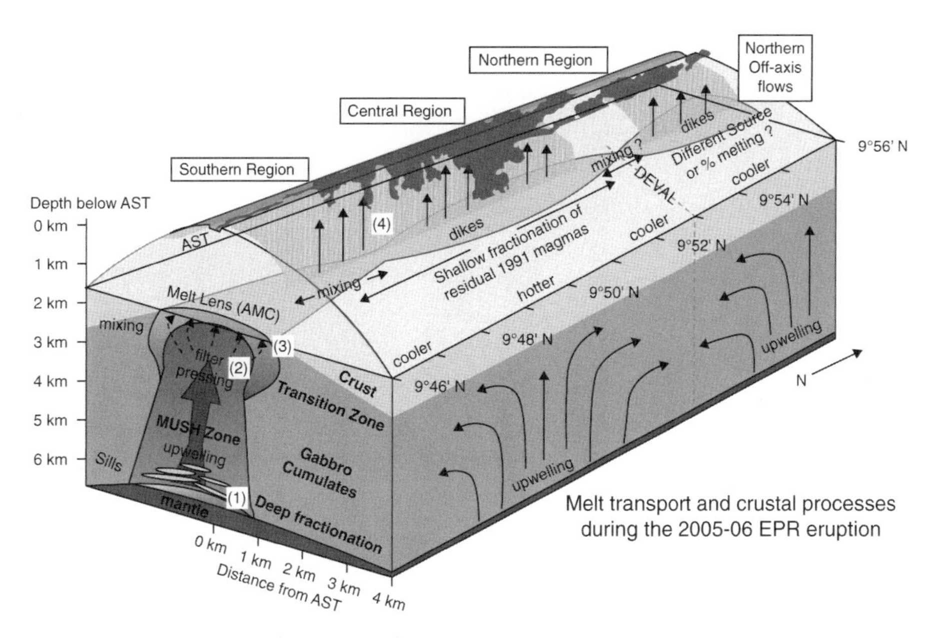


**Figure 6.13** Magmatic model of the 2005–2006 EPR eruption at 9°50′ N, after Goss *et al.* (2010). AMC: axial magma chamber; AST: axial summit trough. Dark grey shows extent of the eruption. Wide red arrow and curved black arrows show directions of upwelling melt from lower crustal sills (1), through crystal mush zone (2), to axial magma chamber (3), and subsequent eruption through dykes (vertical black arrows) to the surface (4). Long horizontal arrows indicate along-axis fractionation within AMC. For colour version, see plates section.

point sources (Head *et al.*, 1996). Fornari *et al.* (1998) suggest that a sequence from broad lava mounds free of faults, through a narrow volcanic collapse trough, to a broad, faulted axial graben, reflects a volcano–tectonic cycle.

## 6.5.3 Evidence from recent eruptions

In 1991–1992 an eruption occurred at the magmatically robust 9°50′ N section of the EPR (Haymon *et al.*, 1993; Gregg *et al.*, 1996), providing a baseline for further studies. In 2006 in the same area, intense seismic activity indicated a dyking event, while water-column anomalies and subsequent camera tows showed that a new eruption had occurred (Tolstoy *et al.*, 2006). Subsequent detailed studies imaged the new lava flow and lava morphology (Fundis *et al.*, 2010) and investigated its geochemistry (Goss *et al.*, 2010). The flow was characterised by sheet and lobate lavas in its interior and pillow lavas at its edges. Lavas were found to be sourced both from fissures within the AST (and then distributed down numerous lava channels) and from fissures some 600 m off-axis (Soule *et al.*, 2007). The flow volume was estimated at $\sim 22 \times 10^6$ m$^3$. The geochemical studies showed the lava was more evolved towards its ends, suggesting fractional crystallisation and magma mixing within the crystal mush zone and axial magma chamber prior to eruption. Many lavas were estimated to have partially crystallised within the uppermost 6 km of the mantle before migrating up and along-axis in the crystal mush zone (Figure 6.13).

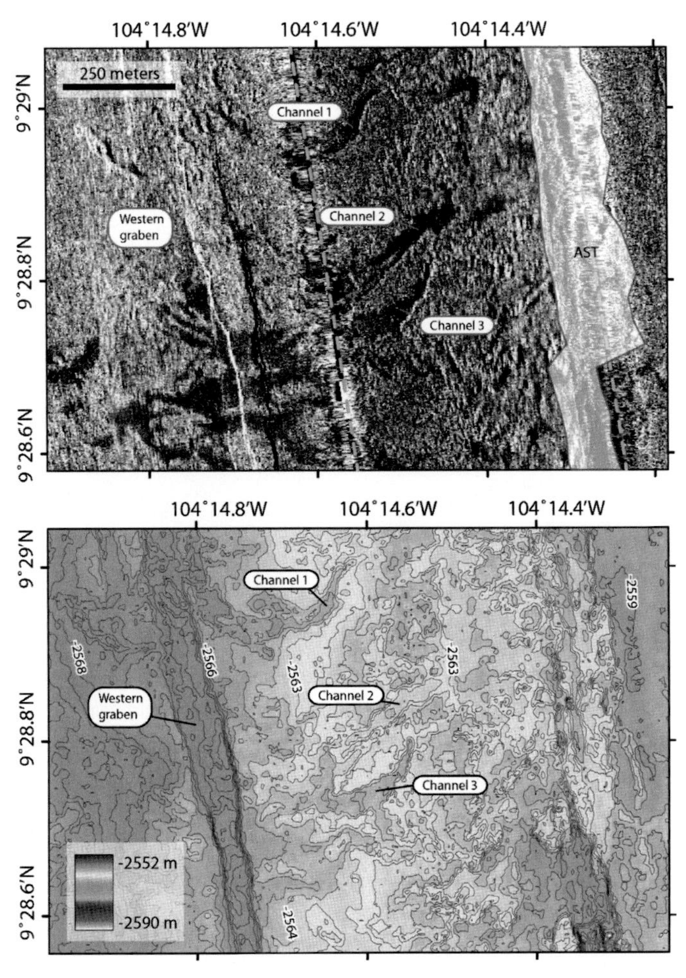


Figure 6.14 Lava channels on the flank of the EPR, after Soule *et al.* (2005). Top, acoustic backscatter intensity from deep-towed side-scan sonar. Dark areas (low backscatter) are floors of lava channels composed of smooth sheet flows. Blue lines mark centres (nadir) of side-scan swathes. Red lines outline axial summit trough (AST). Bottom, microbathymetry of same area (contour interval 1 m) from Imagenex echosounder mounted on an AUV. For colour version, see plates section.

## 6.5.4 The summit plateau and flanks

The neo-volcanic zone is flanked by a zone up to 500 m wide characterised by lobate lava flows that have overflowed the axial summit trough. Repeated flooding and collapse has created a highly permeable zone where the sea floor is underlain by a network of connected tubes and channels (Perfit and Chadwick, 1998).

Beyond the axial zone and summit plateau are the gently sloping flanks of the axial high. These flanks frequently display plan-view lobate shapes, reflecting individual lava flow fronts (Francheteau and Ballard, 1983; Fornari *et al.*, 2004; Figure 6.14). Lobate flows predominate, and pillow lavas are relatively rare, but may form along the edges of more fluid flows, and on pillow ridges (Section 6.5.5).

Well-defined lava channels are seen on ridge sections with elevated magma budgets. They are up to 50 m wide and a few metres deep, spaced ~500–1000 m along-axis, and extend to ~3 km off-axis, where they may end in lobate lava fans (Cormier *et al.*, 2003; Fornari *et al.*, 2004; Soule *et al.*, 2005, Figure 6.14). Their interiors contain lineated or flat sheet flows, with more brecciated lava at their margins. Detailed side-scan sonar images and lava geochemistry indicate that they originate at the rim of the axial summit trough. Such channelised lava transport may be an important mechanism for increasing the off-axis thickness of seismic layer 2A (Section 5.3.2).

### 6.5.5 Seamounts and other off-axis eruptions

In addition to the lavas channelled from the AST, some lavas appear to erupted up to 4 km off-axis, as indicated by their thin sediment cover, fresh appearance, young radiometric ages and distinct geochemistry (Hekinian *et al.*, 1989; Reynolds *et al.*, 1992; Goldstein *et al.*, 1994; Perfit *et al.*, 1994; Sims *et al.*, 2003). These lavas mostly comprise pillow mounds and pillow ridges up to ~1 km long and 30 m high, and occur more than 1.5 km from the axis (Fornari *et al.*, 2004; Figure 6.9). Near 9°30′ N on the EPR, all axial eruptions are N-MORB and some off-axis eruptions are E-MORB (Perfit and Chadwick, 1998); at 12.0° N–12.5° N, lavas transitional between N-MORB and E-MORB are also erupted on-axis (Reynolds *et al.*, 1992)

In addition to these near-axis features, larger, circular seamounts up to ~1 km high are found on the EPR flanks. They appear to develop 5–15 km off-axis, and often form chains, sometimes with kinks in them, oriented between the relative and absolute plate velocity vectors (Scheirer and Macdonald, 1995). Larger seamounts tend to occur on faster-spreading ridges. These seamounts have a chemistry distinctive from that of the near-axis lavas, suggesting that their magmas may be highly heterogeneous and not usually stored in crustal magna chambers (Perfit and Chadwick, 1998).

Turner *et al.* (2011) investigated samples, mostly dredged from fault scarps, from three transects up to 50 km from the EPR axis, at 9°30′ N, 10°30′ N, and 11°20′ N. They found evidence from $^{230}Th/^{238}U$ ratios implying eruption tens of kilometres from the rise axis. Sohn and Sims (2005) suggested that bending stresses that formed as the lithosphere evolved (see Section 7.3.4) can promote off-axis volcanism, by increasing the pressure in lower crustal magma sills and simultaneously reducing stress and opening faults and fissures in the upper crust.

## 6.6 Intermediate-spreading ridges

Intermediate-rate ridges spread at between 50 and 90 km $Ma^{-1}$ full rate. Examples include the Gorda, Juan de Fuca and Endeavour Ridges, the EPR north of Orozco Transform, the

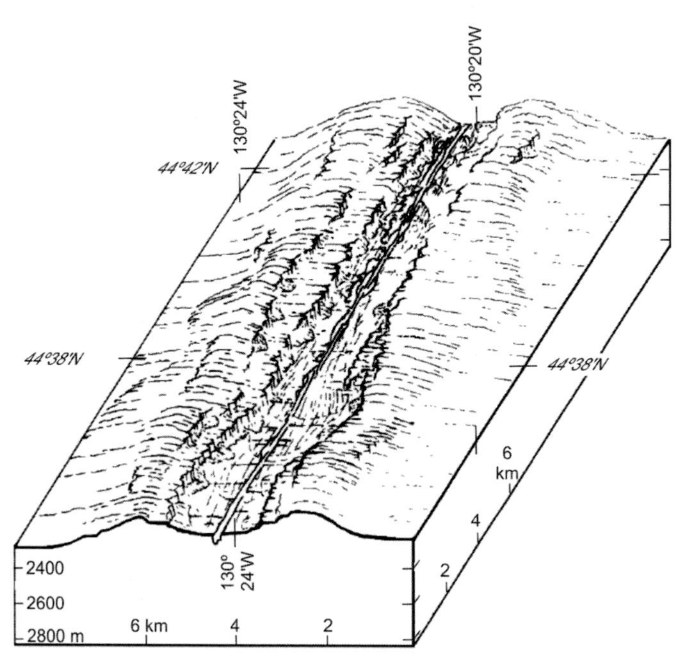


Figure 6.15 Physiographic sketch of the axial part of the southern Juan de Fuca Ridge (Cleft Segment), after Normark *et al.* (1987).

Central and Southeast Indian Ridges, and the Cocos–Nazca Spreading Centre. The Juan de Fuca Ridge has been particularly well studied because of its proximity to western North America. Continuous hydroacoustic monitoring there has built up an invaluable record of seismic activity (Dziak *et al.*, 2011) and led to the detection and rapid investigation of several active eruptions.

Intermediate-spreading ridges occupy a critical spreading rate below which a quasi-steady-state magma chamber, as seen under fast-spreading ridges, may no longer exist (Phipps Morgan and Chen, 1993). Small changes may thus lead to the presence or absence of such a chamber, which can have a marked effect on the ridge morphology, melt supply and tectonic style (Perfit and Chadwick, 1998). The Juan de Fuca Ridge is at least in part underlain by an axial magma chamber (Canales *et al.*, 2006).

## 6.6.1 Axial volcanism

The axial morphology at intermediate-spreading rates is itself intermediate between those of fast and slow-spreading ridges (Perfit and Chadwick, 1998). On average there is an almost flat cross-axial profile, modified by small-scale faulting (Chapter 7) and volcanic constructs. The morphology can be quite variable between ridge segments, ranging from a subdued axial high to a shallow median valley a few kilometres wide (Figure 6.15). The neo-volcanic zone is somewhat broader, at around 1 km, than at fast-spreading ridges, and the sites of even temporally close eruptions can migrate hundreds of metres across axis. Eruptions

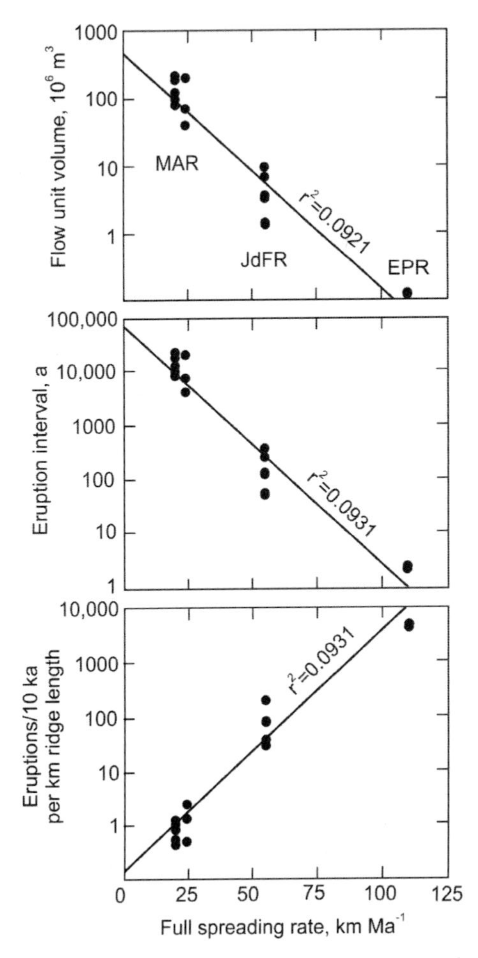


**Figure 6.16** Plots of temporal and spatial intervals and volumes of eruptions against spreading rate, after Perfit and Chadwick (1998).

are more voluminous but less frequent than on fast-spreading ridges (Figure 6.16). The morphology of volcanic constructs tends to be more variable than at fast-spreading ridges.

Intermediate ridges generally display a smaller proportion of sheet flows and a greater proportion of pillow basalt (Figure 6.8), but each lithology is similar to its equivalent from faster-spreading ridges. Pillow mounds are common (Embley and Wilson, 1992; Embley and Chadwick, 1994), but may be clustered into linear ridges up to tens of metres high, hundreds of metres wide and several kilometres long (Perfit and Chadwick, 1998). Where well developed, such pillow ridges form an 'AVR' (Ballard *et al.*, 1982; Figure 6.17) similar to, but smaller than, those that are common on slow-spreading ridges (Section 6.7.3). Major eruptive episodes may begin with broad sheet flow eruptions and end in more spatially restricted eruptions of pillow basalts forming mounds and ridges (Van Andel and Ballard, 1979; Embley and Chadwick, 1994).

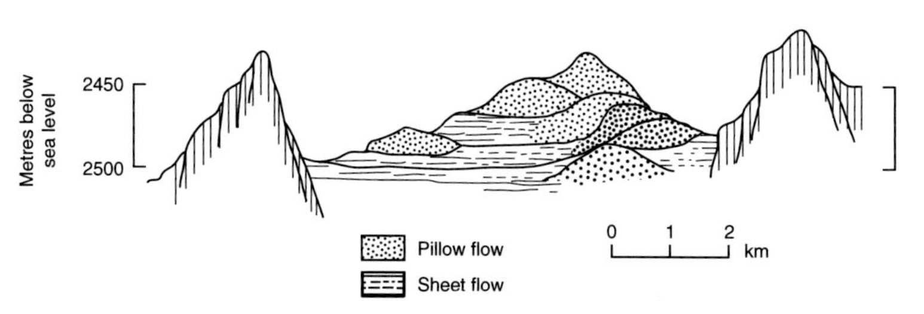


Figure 6.17 Cross-section of an axial volcanic ridge on the Cocos–Nazca Spreading Centre, 86° W, after Ballard *et al.* (1982). Vertical exaggeration 5.5. Vertical shading indicates older sea floor.

The concept of volcano–tectonic cycles is well developed for intermediate-spreading ridges, based on morphological, volcanic and structural evidence (Van Andel and Ballard, 1979; Kappel and Ryan, 1986). In this model, periods of inflation and volcanic construction alternate with periods of stretching and faulting. For example, on the southern Juan de Fuca Ridge the young pillow basalts are associated with dyking in an inflated volcanic ridge and come from a poorly fractionated melt source; by contrast, older sheet flows, which are now in an area of tectonic extension, were fed from a more fractionated source (Smith *et al.*, 1994).

## 6.6.2 Recent eruptions

A number of recent and ongoing eruptions have been observed on the Juan de Fuca and Gorda Ridges, which have allowed detailed follow-up studies to be carried out.

In 1996 an eruptive episode was detected seismically on the northern Gorda Ridge, and a 'rapid response' effort was mobilised to study it only two months after the eruption (Chadwick *et al.*, 1998). New bathymetry, side-scan and video imagery, compared with earlier bathymetry, indicated that the flow was 2600 m long, 400 m across and up to 75 m thick, with a volume of $18 \times 10^6$ $m^3$ (Figure 6.18). Such repeat mapping surveys have indicated possible sites of other recent eruptions, so far not confirmed by other data.

On the southern Juan de Fuca Ridge a series of lava flows erupted in the mid 1980s from the 100 m wide, 15–25 m deep axial cleft (Normark *et al.*, 1987). These include a young sheet flow around the northern end of the cleft (Embley and Chadwick, 1994) and a series of pillow mounds and pillow ridges extending northwards (Chadwick and Embley, 1994; see Figure 8.17). These recent, axial lavas are more chemically primitive than off-axis ones (Stakes *et al.*, 2006).

An important study area is Axial Volcano (or Axial Seamount), a large central volcano on the Juan de Fuca Ridge where a series of eruptions occurred in 1998. They were detected hydro-acoustically (Dziak and Fox, 1999), and were followed by repeat bathymetry surveys that detected the lava flows, leading to an estimate of $18–76 \times 10^6$ $m^3$ for the total volume erupted (Embley *et al.*, 1999). Further deformation has been monitored and modelled

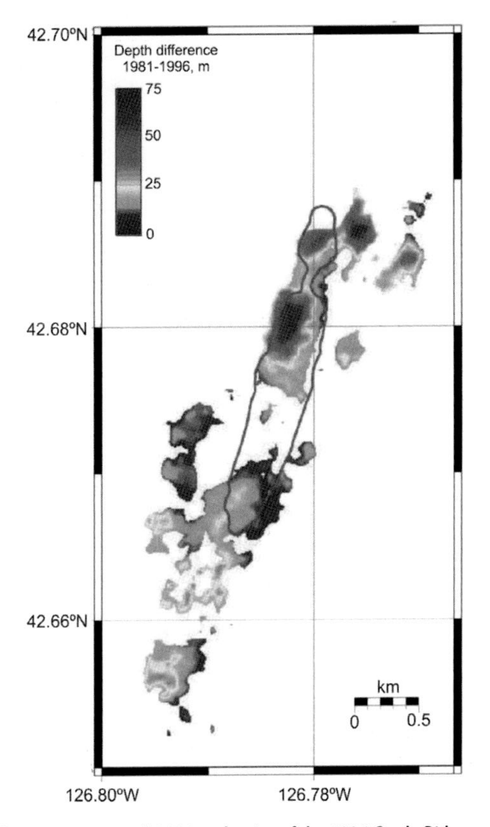


Figure 6.18 Depth differences observed between 1981 and 1996 at the site of the 1996 Gorda Ridge eruption, after Chadwick *et al.* (1998). Red line indicates mapped outline of 1996 eruption. For colour version, see plates section. © 1998, with permission from Elsevier.

using acoustic extensometers (Chadwick *et al.*, 1999) and sea floor pressure measurements (Chadwick *et al.*, 2006; Nooner and Chadwick, 2009), providing the first measurements of volcanic inflation at a submarine volcano.

### 6.6.3 Off-axis volcanism

Many of the extensive lava flows on the flanks of the southern Juan de Fuca Ridge appear to emanate from ridge-parallel faults and fissures in the walls of the median valley, and have an evolved, heterogeneous chemistry consistent with an origin from the cooler flanks of the axial magma chamber or mush zone (Stakes *et al.*, 2006). Anomalously young radiometric ages have been found on the Gorda Ridge, implying off-axis eruption ~1 km from the ridge axis (Cooper *et al.*, 2003). There is limited evidence of large off-axis flows (Davis, 1982).

In addition, chains of seamounts occur mainly to the west of the Juan de Fuca Ridge, which appear to have similar characteristics to those formed off the axis of the EPR (Section 6.5.5; Scheirer and Macdonald, 1995; Clague *et al.*, 2000b).

## 6.7 Slow-spreading ridges

Slow-spreading ridges, with a full spreading rate between 20 and 50 km $Ma^{-1}$, are exemplified by the MAR, Carlsberg Ridge and Gulf of Aden. Here the ridge morphology is distinctly different from that seen at intermediate and fast spreading rates, with a substantial median valley several kilometres deep and tens of kilometres wide. Volcanism is mostly confined to the floor of this valley, which ranges from just a few kilometres to 12 or more kilometres wide. Most spreading ridge segments contain an AVR (Section 6.7.3), which is more-or-less coincident with the neo-volcanic zone (Parson *et al.*, 1993; Sempéré *et al.*, 1993). The AVRs are composed of hundreds to thousands of individual pillow mounds and small volcanoes making up a 'hummocky' terrain (Smith and Cann, 1990; Yeo *et al.*, 2012). The median valley floor outside the AVR may consist of lower relief hummocky terrain and/or smoother, flatter terrain, which may comprise lobate and sheet flow lavas.

### 6.7.1 Hummocky and smooth volcanic terrains

Hummocky volcanic terrain has been revealed most clearly by side-scan sonars operating at around 30 kHz frequency (e.g., Kong *et al.*, 1988/89; Smith *et al.*, 1995b; Briais *et al.*, 2000). It consists of clusters of small volcanic edifices, called 'hummocks' or 'rounded mounds' (Smith *et al.*, 1995a), similar to the pillow mounds of intermediate and fast ridges. Many hummocks build on and overlap each other (Figure 6.19). Recent detailed studies using video imaging and high-resolution bathymetry have revealed that individual hummocks range from less than 30 m to over 300 m diameter, with heights ranging from $<5$ m to $>140$ m (Yeo *et al.*, 2012). They are mostly dome-shaped, with steep, $\sim 45°$ flanks, occasional flatter or more pointed summits, and a range of height to diameter ratios (Figure 6.20). They are composed entirely of pillow lavas and lobate lavas, usually with elongated pillows on their steeper flanks; flatter hummocks have more lobate flows near the summit suggesting higher effusion rate or lower effective viscosity. Individual hummocks are thought to be produced by single, continuous eruptions that might last from several hours to several weeks (Yeo *et al.*, 2012).

While hummocky terrain is mostly found in AVRs, patches of it occur elsewhere on the median valley floor and walls (e.g., Cann and Smith, 2005; Searle *et al.*, 2010). Usually these off-axis exposures appear degraded, probably by partial sediment cover, and may be parts of older AVRs now partially buried and/or tectonically dismembered. Occasionally, young-looking hummocky terrain occurs at the edges of the medial valley floor, and reflects either off-axis eruptions or possibly the recent shifting of the main volcanic axis to a marginal site (e.g., Mallows and Searle, 2012).

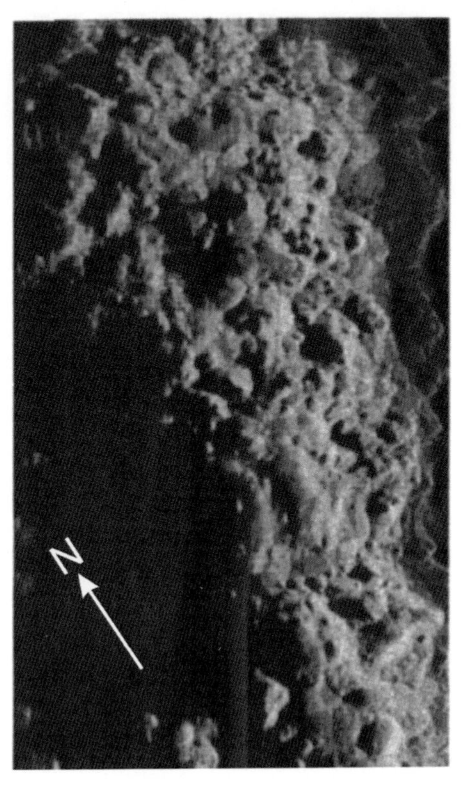


**Figure 6.19** Hummocky volcanic terrain imaged by the TOBI side-scan sonar on the Reykjanes Ridge axis at 58°20′ N, after Smith *et al.* (1995a). Light tones indicate high acoustic backscatter. Image is 5 km by 3 km. Vehicle track (zero range) is along the right side of image, with insonification towards the left. The crest of an axial volcanic ridge runs from bottom right to top left, throwing the west flank of the AVR into shadow. Eastern flank is composed of hundreds of volcanic 'hummocks', ranging from tens to hundreds of metres in diameter.

Often the median valley contains small areas of flat sea floor outside the AVR. These areas appear smooth on side-scan sonar (Figure 6.21), with variable degrees of acoustic backscatter probably reflecting varying sediment cover and age (Cann and Smith, 2005; Searle *et al.*, 2010). They are often interpreted as sheet flows, though even thin sediment cover makes flow morphology difficult to identify. Video observations of such flows exposed by fissuring at 45° N on the MAR showed they are mostly lobate flows (Yeo and Searle, 2013), while similar sonar features at 25° N on the MAR were shown by camera imaging to be all pillow lavas (Cann and Smith, 2005). In the latter area, about half the segment was floored by smooth flows, with hummocky flows restricted to the southern half but propagating northwards (Figure 6.22). However, most slow-spreading segments have a much smaller proportion of smooth to hummocky flows. Younger flows generally occur nearer the axis of the median valley (Cann and Smith, 2005; Searle *et al.*, 2010; Figure 6.22).

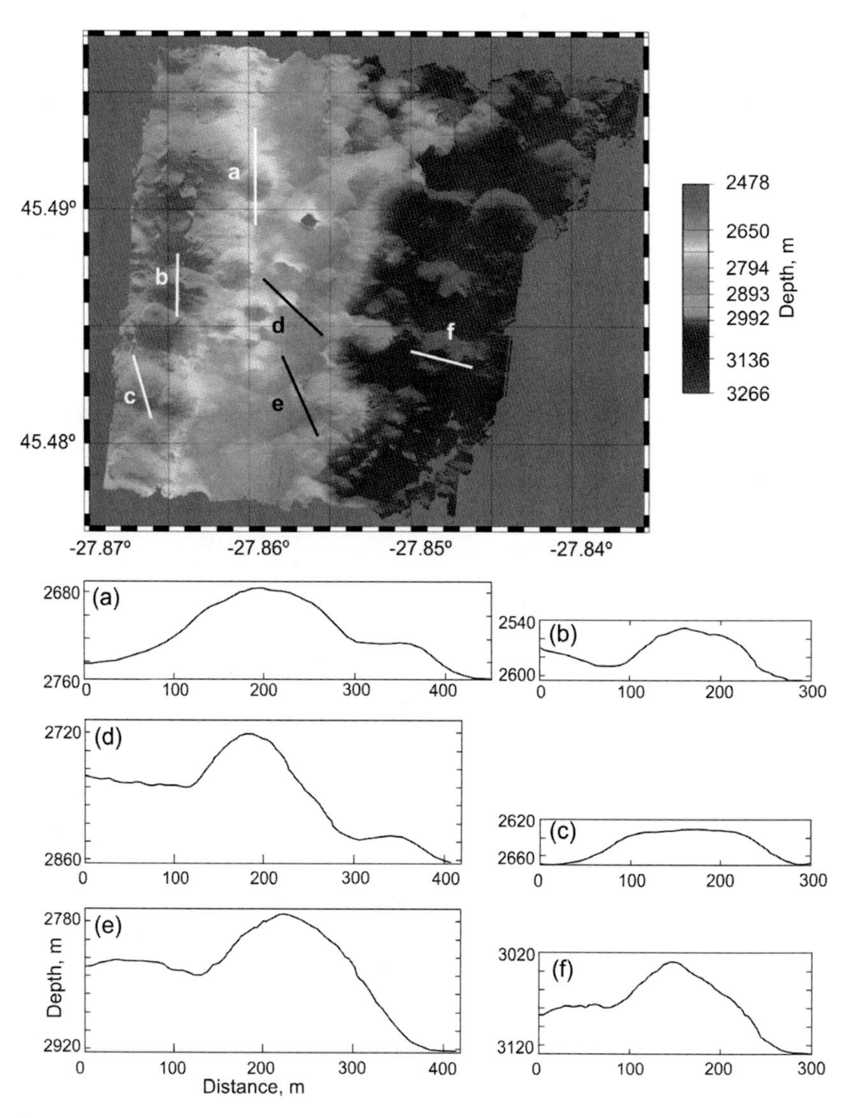


Figure 6.20 Top: high-resolution (~1 m) shaded-relief bathymetric map of part of the axial volcanic ridge flank at 45.5° N on the MAR, illuminated from the north. The AVR crest is at the left and the foot of its eastern flank on the right. Lettered lines: positions of profiles shown below. Bottom: vertical profiles (no vertical exaggeration) over six individual volcanic hummocks. For colour version of top section, see plates section.

## 6.7.2 Volcano morphologies

While volcanic hummocks may occur in apparently chaotic, unstructured distributions (Figure 6.19), often they cluster to form larger, distinct hummocky volcanoes or linear hummocky ridges (Figure 6.23a, b). Small shield volcanoes (Figure 6.23c), lava terraces

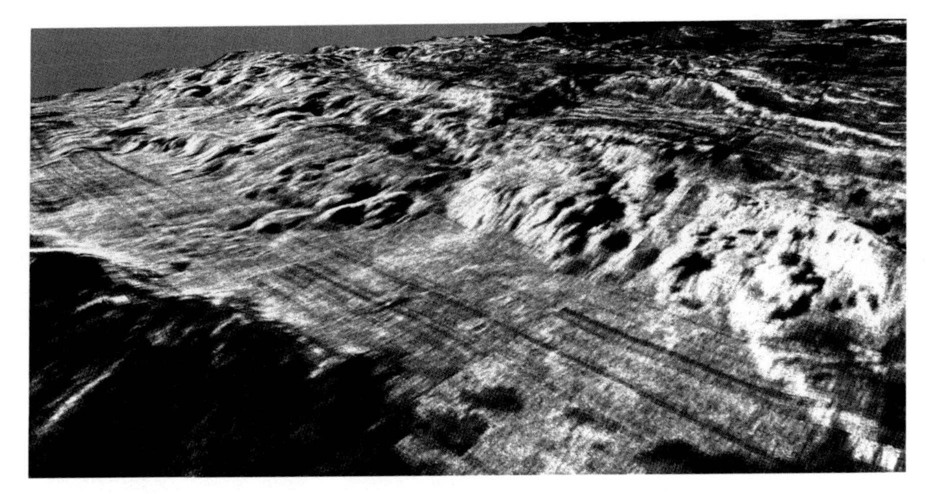

**Figure 6.21** Oblique view of the MAR median valley floor at 45°40′ N, looking southwest. TOBI side-scan sonar, insonified from the south, is draped over EM120 multi-beam bathymetry with no vertical exaggeration. Image width approximately 6 km at mid-range. Smooth sheet-flow and lobate lavas (relatively uniform, mid-grey) are in foreground, with AVR composed of volcanic hummocks in middle ground. Top right has older, faulted median valley floor. Dark area at bottom left is thickly sedimented sea floor on top of fault block in median valley wall. Mottled appearance of smooth lavas may reflect underlying, almost totally buried hummocky terrain. Dark stripes are acoustic artefacts.

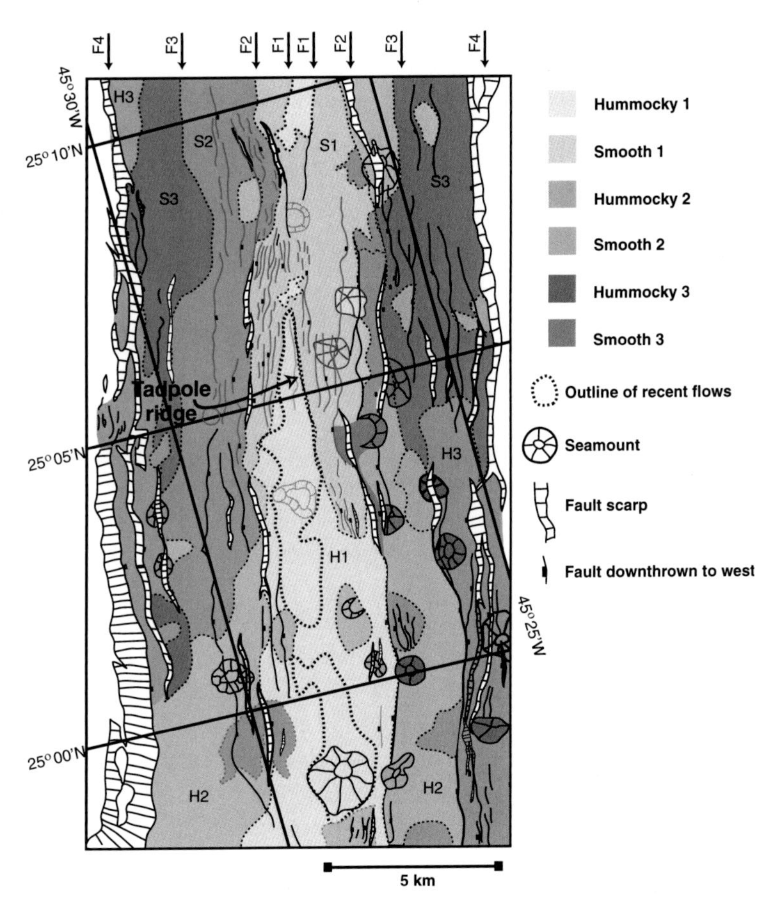


**Figure 6.22** Geological interpretation of a TOBI side-scan sonar survey of the median valley floor on the MAR at 25° N, after Cann and Smith (2005). Inferred ages of faults (F) and hummocky and smooth sea floor are labelled 1 (youngest) to 4 (oldest). For colour version, see plates section.

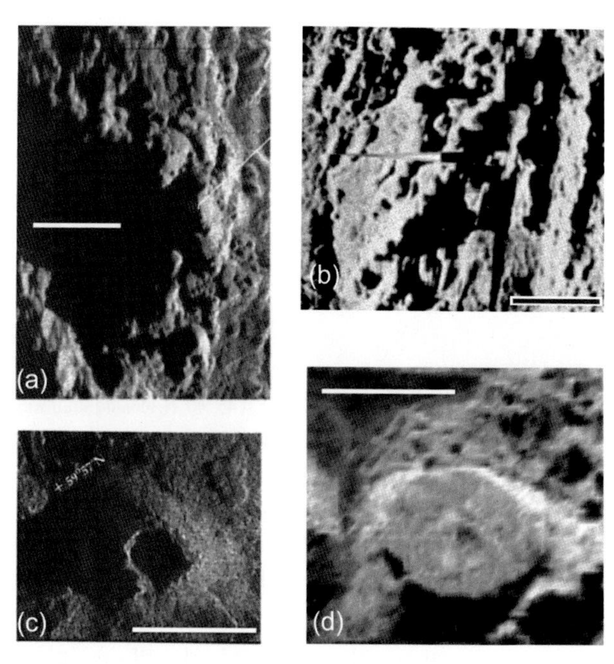


**Figure 6.23** TOBI side-scan sonar images of volcanic forms on the slow-spreading MAR. Insonification is from the right for a and c, from the left for b and from above for d. The scale bar associated with each image is 1 km long. (a) Hummocky volcano (half in shadow), 25°45′ N, after Smith *et al.* (1995b); (b) Hummocky ridges (most prominent one trending top right to bottom left), 23°56′ N; (c) smooth-sided, cratered seamount, 59°57′ N, after Smith *et al.* (1995b); (d) flat-topped seamount, 28°58′ N. Images (a) and (c) © 1995, with permission from Elsevier.

and flat-topped seamounts (Figure 6.23d) also occur, possibly formed by somewhat different mechanisms. Smith *et al.* (1995b) have given detailed descriptions of these volcano morphologies.

Hummocky ridges consist of lines of hummocks, usually one hummock wide and containing ~10 hummocks in a row (Figure 6.23b), and are very common. They are believed to form from fissure eruptions where flow becomes unstable, breaking up into a line of small point sources (Head *et al.*, 1996; Wylie *et al.*, 1999; Figure 6.25). However, some small alignments of volcanic cones have been found to have unrelated magma sources (Bryan *et al.*, 1994).

Hummocky volcanoes (sometimes called ‘hummocky shields’ or ‘hummocky mounds’) consist of many individual hummocks piled together to form an edifice up to several hundred metres high and several kilometres in diameter (Smith *et al.*, 1995b; Figure 6.23a). They may be the commonest type of central (as opposed to linear) seamount on the MAR (Smith *et al.* 1995b). It is not yet clear whether their individual hummocks are related, e.g. as separate phases of a single eruption, or as separate eruptions from a common body of magma. At least one segment, at 45° N on the MAR, contains very few hummocky volcanoes (Searle *et al.*, 2010).

Flat-topped volcanoes (Figure 6.23d) apparently form a distinct class of volcanoes. They occur at all spreading rates (Clague *et al.*, 2000a), but are a particular feature of slow-spreading ridges. There they are more common at segment ends and are relatively rare on AVRs (Briais *et al.*, 2000); many may be emplaced off-axis. They are significantly larger than the other volcano types, with diameters ~600–2000 m. Unlike hummocks they have a characteristic height-to-diameter ratio of about 1:10. They have steep sides formed of pillow lavas and flat tops comprising sheet or lobate flows that appear smooth on side-scan sonar. They frequently have one or more craters in their summit plateaus. Flat-topped seamounts tend to have more primitive chemistry than hummocky volcanoes, and may tap less-differentiated magma and/or rise rapidly from the lower crust or upper mantle with little intermediate storage or differentiation in crustal magma chambers (Lawson *et al.*, 1996; Murton *et al.*, 2013). Clague *et al.* (2000a) proposed that flat-topped volcanoes are formed by long-lived eruptions producing a lava lake and surrounding levee. As lava spills over the levee, it solidifies into pillows, reinforcing the levee and building it upwards and outwards, while the top of the enclosed lava lake freezes to form the seamount summit.

Possibly related to flat-topped seamounts are 'smooth seamounts' or 'smooth shields' (Smith *et al.*, 1995b; Head *et al.*, 1996; Figure 6.23c). They have smooth-looking flanks in side-scan sonar images, but are generally topped by a crater without a surrounding summit plateau. Although Head *et al.* (1996) proposed they grow by prolonged eruption from axial dykes, their mode of formation and relation to other volcano morphologies remain unclear.

Lava terraces were identified on the flanks of the AVR by Smith and Cann (1999), who proposed that they form by a similar mechanism to flat-topped volcanoes, but are fed laterally by lava transported in tubes from the ridge axis. Serocki volcano at 22°55′ N on the MAR appears similar. Side-scan imagery shows it as a cratered, flat-topped volcano, but it has been interpreted as a tube-fed 'megatumulus' or inflated lava delta (Humphris *et al.*, 1990; Bryan *et al.*, 1994).

Smith and Cann (1990) measured the distribution of seamounts on the MAR between 24° N and 30° N. The seamounts (probably mostly hummocky and smooth shield volcanoes) were interpreted from multi-beam bathymetry maps as circular features >50 m high. The density was 80 seamounts per 1000 km$^2$. Elsewhere, estimated densities range from 7 per 1000 km$^2$ (MAR near 26° S; Batiza *et al.*, 1989), through 14–52 per 1000 km$^2$ (Gakkel Ridge; Cochran, 2008) to 310 per 1000 km$^2$ (Reykjanes Ridge; Magde and Smith, 1995). Individual hummocks are much more common, at some 6000 per 1000 km$^2$ over the whole median valley floor, or 22 000 per 1000 km$^2$ over the AVR alone (Yeo, 2012).

### 6.7.3 Axial volcanic ridges

Most segments on the MAR and other slow-spreading ridges display an AVR (Figure 6.24). AVRs are tens of kilometres long, a few kilometres wide and a few hundred metres high. There is usually one AVR in each spreading segment and they tend to extend the length of the segment; however, occasionally there is only a single, short ridge (Smith and Cann, 1999; Cann and Smith, 2005), and sometimes there are several ridges in the same segment (Ballard and Van Andel, 1977). AVRs consist of thousands of individual volcanic

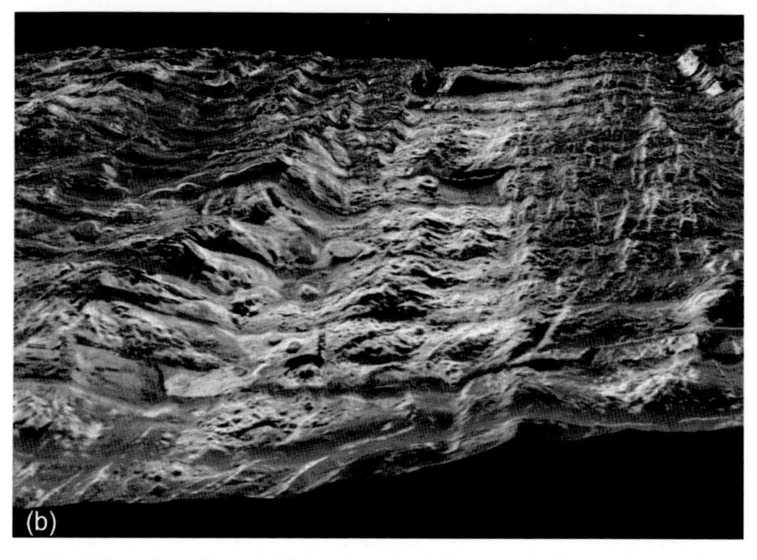


Figure 6.24 Oblique views of the MAR median valley at 29° N, looking north. (a) Topography, from GeoMapApp (http://www.geomapapp.org; Ryan *et al.*, 2009). Depths range from approximately 2000 m (grey) to 4000 m (purple). (b) TOBI side-scan sonar mosaic of same area draped over bathymetry; no vertical exaggeration. Side-scan is insonified from the north. Images approximately 25 km wide at mid-range, 30 km front to back. Axial volcanic ridge runs N–S along centre of each image (pale blue in (a) composed mostly of hummocky and a few smooth and some flat-topped volcanoes). Major fault scarps bounding the median valley flank the AVR. Note AVR is slightly oblique to median valley, crossing from centre of valley (foreground) to western valley wall in far distance. For colour version, see plates section.

hummocks, often organised into hummocky ridges and hummocky and smooth volcanoes (Head *et al.*, 1996; Figure 6.25); they sometimes include other volcanic forms such as flat-topped volcanoes and lava terraces. Although sheet flows often partially surround AVRs, they are rare within them.

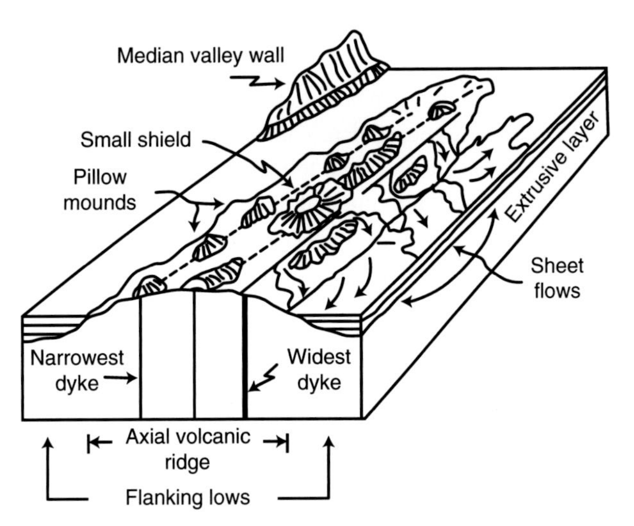


**Figure 6.25** Diagram showing the main elements of the axial volcanic ridge and median valley floor of a slow spreading MOR, after Head *et al.* (1996). Narrow dykes ($<$ 1 m wide) feed fissure eruptions that break down to produce a series of small, point-source volcanic hummocks. Larger dykes permit more voluminous eruptions that may additionally produce small shield (hummocky and smooth) volcanoes. The largest dykes ($\sim$3 m wide) feed eruptions that flow down the AVR flanks, either on the sea floor or through lava tubes, feeding lava terraces and sheet flows. Note, however, that although sheet flows are widely recognised on the median valley floor, their origin and possible connection to AVRs is still debated; there is evidence that at least some have geochemical sources distinct from AVR lavas (Murton *et al.*, 2013).

Note that AVRs are usually aligned approximately normal to the spreading direction, since their feeding fissures follow this direction (Section 7.2). However, the overall trend of slow spreading MORs, and thus of their median valley wall faults, is often slightly oblique to the spreading normal. Thus AVRs may lie oblique to the valley trend and cross it from one side to the other (Figure 6.24).

Magnetic polarity transitions inferred from high-resolution magnetic anomalies are a few kilometres wide (Macdonald, 1977), implying that this is the range over which dike intrusions and lava flows spread from the axis. Since this is so similar to the width of AVRs, it is likely that most of the uppermost crust is emplaced by volcanism at AVRs, though surrounding smooth flows must contribute some. The AVRs usually contain the youngest lavas, as inferred from sediment thickness and acoustic reflectivity, although some smooth lava flows have similar sediment cover to the adjacent AVR.

The median valley floor mostly lies within the most recent, 800 ka Brunhes–Matuyama magnetic reversal. Consequently, dating of axial volcanism and AVRs has been hampered by the lack of a suitable method, though recently magnetic and radiometric methods have been developed. The Central Anomaly Magnetic High (Section 5.9.1) lies within the Brunhes normal chron and reflects the youngest sea floor (Tivey and Johnson, 1987). The width of the CAMH is consistent with AVR ages of a few hundred ka or less (Hussenoeder *et al.*, 1996). Samples from the AVR at 45° N on the MAR have been dated to $\sim$10 ka using magnetic palaeointensity observations (Searle *et al.*, 2010), while U-series dating of axial lavas from the MAR at 23° N has revealed ages of 10–20 ka (Sturm *et al.*, 2000).

Generally, AVRs have been thought to be built episodically (Parson *et al.*, 1993; Cann and Smith, 2005). A geophysically imaged melt lens under an AVR at the southern Reykjanes Ridge contains enough melt to supply 20 000 years of crustal growth, but would solidify after 1500 years, implying that melt delivery must be highly episodic (Sinha *et al.*, 1998). In some models, initial high-volume, high-effusion rate eruptions of sheet flows cover the median valley floor and erase most previous morphology, and are followed by lower-volume, low-effusion rate eruptions that build a new AVR (e.g., Cann and Smith, 2005). Alternately, the AVR may be rifted to one side or other of the ridge axis by subsequent faulting and intrusion; occasionally, the AVR may be split down the middle and the two sides rifted apart. Aspects of all these models may apply from time to time. However, the near-ubiquitous presence of single AVRs on slow-spreading ridges implies they may be near-steady-state features, being continually built from within and broken down by faulting and rifting on their flanks (Murton *et al.*, 2013). This issue is not yet resolved.

### 6.7.4 Off-axis volcanism

Some (or most) flat-topped volcanoes may be emplaced off-axis (Section 6.7.2). Some smooth lavas have geochemistry that is similar to that of neighbouring, flat-topped volcanoes and distinct from that of the adjacent AVR, suggesting that they too may have been erupted off-axis (Murton *et al.*, 2013). Individual smooth, off-axis lava flows are sometimes recognised from their relatively high backscatter and lobate outlines. They have been imaged by side-scan sonar on the flanks of the median valley of the MAR at 13° N (Mallows and Searle, 2012).

## 6.8 Ultra-slow-spreading ridges

The style of volcanism changes again at ridges with very slow spreading rates (<20 km $Ma^{-1}$) that spread highly obliquely to yield a small component of normal opening, or where the mantle is unusually cold (Mendel *et al.*, 1997; Dick *et al.*, 2003; Cannat *et al.*, 2006). Such ridges display similar volcanic terrains to those seen at slow-spreading ridges, but in different proportions and arrangements, at both large and small scale.

At a large scale, volcanic centres appear to become more focussed but more widely spaced along the ridge, as at the eastern end of the SWIR (Southwest Indian Ridge) over a mantle cold-spot (Cannat *et al.*, 1999; Figure 5.4). Here, many segments display lower axial relief and smaller gravity anomalies than typical slow-spreading segments, while a few, separated by ~200 km along-axis, display very high relief and large anomalies implying thick crust. Cannat *et al.* (1999) proposed that this reflects a change in the way axial topography is formed: melt delivery is focussed away from the low-relief segments towards a few, widely

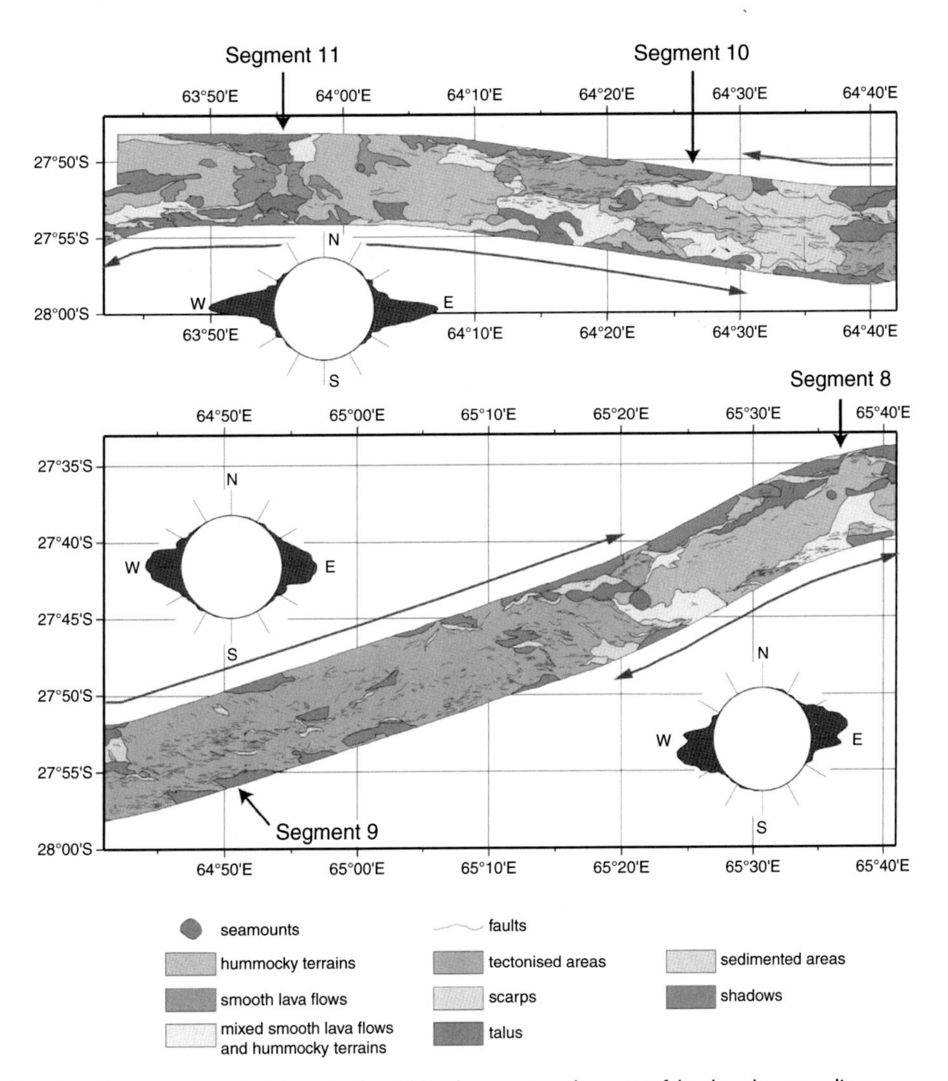


Figure 6.26 Volcanic and tectonised terrains interpreted from TOBI side-scan sonar along part of the ultra-slow-spreading Southwest Indian Ridge axis, after Sauter *et al.* (2004). Rose diagrams show distribution of fault directions. Note highly tectonised segment 9 which entirely lacks recent volcanism, compared with nearby highly focussed volcanic segments 8 and 11. See also Figure 5.4. For colour version, see plates section.

spaced, high-relief ones, where it feeds excess volcanism and builds a thick crust largely by eruptions directly onto the sea floor.

Figure 6.26 shows an example of the along-axis variability of volcanic terrain on the SWIR. The volcanically robust segment 11 is dominated by a large volcanic massif covered in smooth lava flows at the segment centre, with extensive hummocky lavas throughout the rest of the segment. By contrast, segment 9 is entirely devoid of recent volcanism and is characterised throughout by tectonised terrain. Some segments of the SWIR appear to

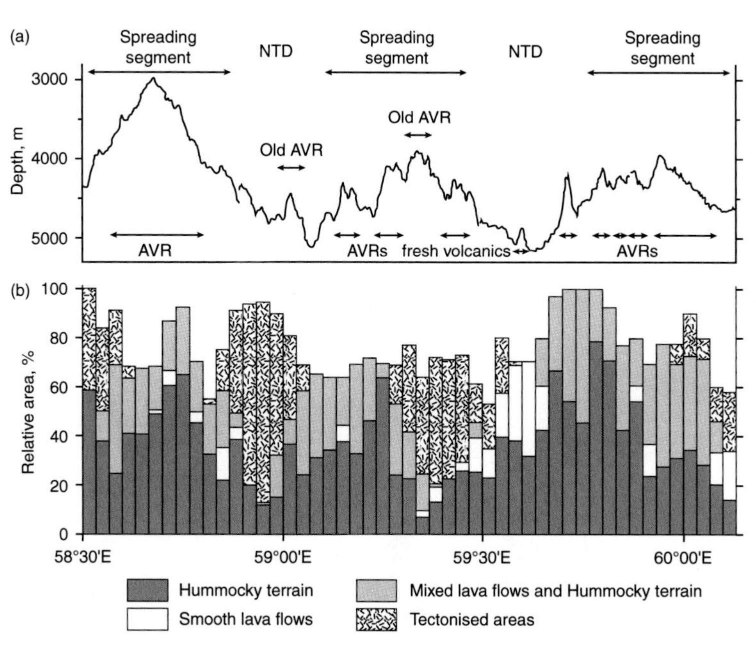


**Figure 6.27** (a) Axial depth profile, and (b), relative areas of lava types, along part of the ultra-slow Southwest Indian Ridge after Sauter *et al.* (2002). The same lava types are seen as at other ridges, but the distribution and proportions are different, for example, smooth lava flows in NTDs and greater areas of tectonised terrain. © 2002, with permission from Elsevier.

have been completely devoid of volcanism for millions of years (Cannat *et al.*, 2006; Sauter *et al.*, 2013; Section 7.5).

In contrast to these 'focussed' volcanic terrains, parts of the ultra-slow SWIR show an 'unfocussed' volcanic terrain. Here, apparently fresh volcanic terrains are found not only in substantial AVRs at segment centres, but also at segment ends and in non-transform offsets (Figure 6.27), where eruptions may be controlled by the local opening of fractures in response to magma pressure (Sauter *et al.*, 2002). Similar patterns of focussed volcanism and intervening amagmatic zones occur on the ultra-slow-spreading Gakkel Ridge (Cochran *et al.*, 2003; Michael *et al.*, 2003).

Seamounts on the Gakkel Ridge were detected automatically from bathymetry by Cochran (2008). He found smaller seamount heights and densities about half those of the MAR (Section 6.7.2). Seamount densities on the Gakkel Ridge fall as the spreading rate falls, and seamounts tend to be clustered in the widely spaced magmatic centres and on the edges of the median valley or in the valley walls.

There is evidence of off-axis eruptions at several ultra-slow ridges. Standish and Sims (2010), using U-series radiometric dating, found anomalously young lavas (<8 ka) at 10 km off-axis on the SWIR, where the predicted age based on a uniform spreading rate is 1.5 Ma. They suggest these young lavas have been emplaced along faults that outcrop off-axis (Figure 6.28). Lavas apparently erupted from ridge-flanking normal faults several kilometres off-axis and several hundred metres above the ridge axis have been imaged in the Mid-Cayman Spreading Centre (Searle, 2012).

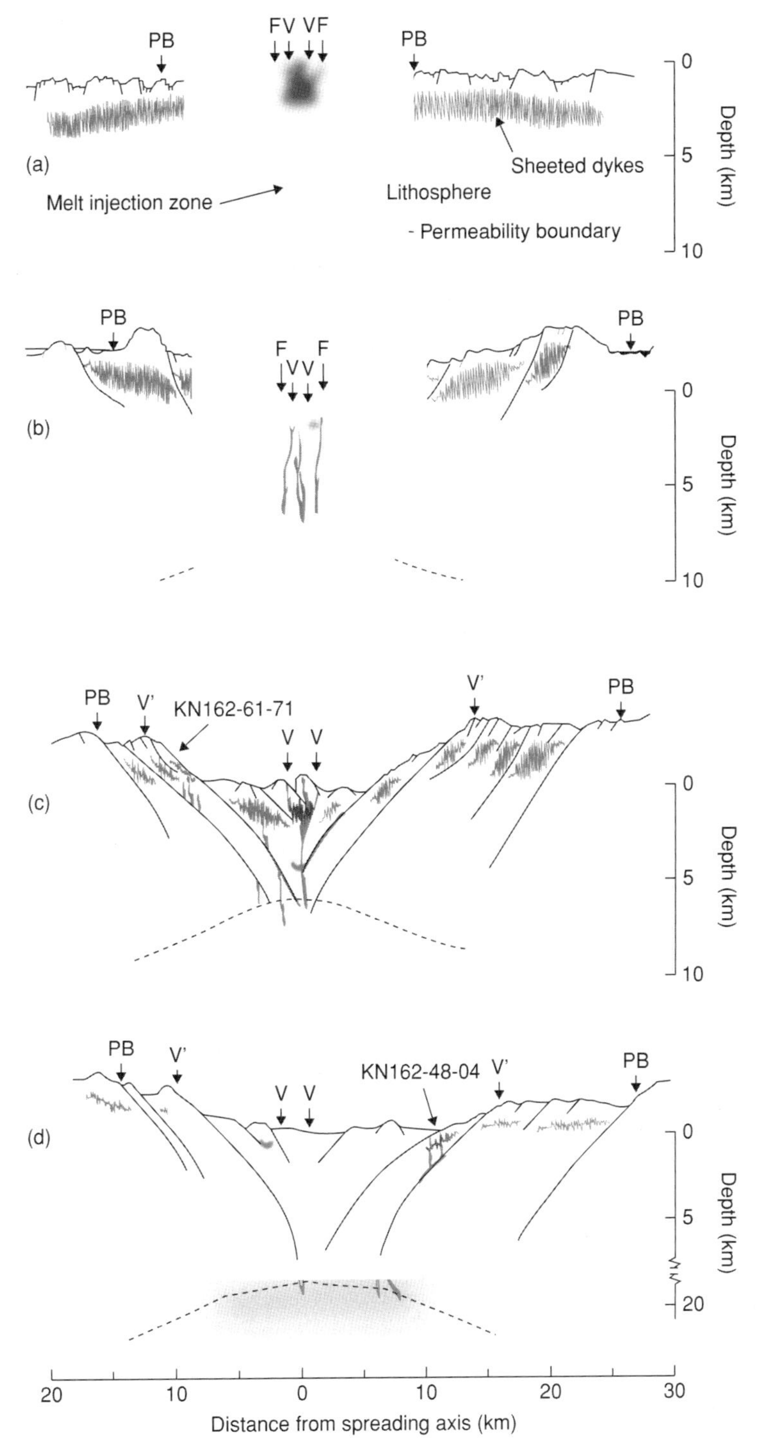


**Figure 6.28** Cartoons showing the cross-axis distribution of volcanism at different spreading rates, after Standish and Sims (2010). (a) Fast-spreading (55 km $Ma^{-1}$ half-rate) EPR, 3° S; (b) slow-spreading (12 km $Ma^{-1}$) MAR, 37° N; (c) magmatic segment of the ultra-slow-spreading (7 km $Ma^{-1}$) Southwest Indian Ridge, 14.5° E; (d) amagmatic segment of the Southwest Indian Ridge, 12.5° E. V: neo-volcanic zone; F: zone of fissuring; PB: plate boundary zone defined by active faulting; V': crustal accretion zone; red shading: intensity and width of melt injection zone; grey shading: thickness of sheeted dykes; dashed line: notional permeability boundary to upwelling melt. For colour version, see plates section. Reprinted by permission from Macmillan Publishers Ltd: *Nature Geoscience* © 2010.

## 6.9 Summary

As tectonic plates diverge the asthenosphere wells up adiabatically and undergoes partial melting in a triangular melting zone, starting at a depth of ~70 km or more and continuing to the base of the lithosphere. Degrees of partial melting may reach ~40%, but average degree of melting is ~10%. The melt composition is basaltic, containing mainly Si, Al, Ca, Fe, Mg and O, with a much lower Mg content than its mantle source.

Melt rises until it reaches a barrier – probably the plastic–brittle transition – and then pools in magma chambers, where it is stored at pressure until released by fracturing of the overlying rock. It then rises up the fractures and may solidify *in situ*, forming dykes, or erupt onto the sea floor.

The type of lava flow depends on effusion rate, viscosity and sea floor slope. High effusion rates produce fluid, relatively flat-lying sheet flows, whereas lower effusion rates lead to progressively rounder lavas, ending at pillow lavas for low effusion rates. The type of volcanism varies with spreading rate.

At fast spreading, sheet and lobate flows dominate; they are erupted from axial fissures and may flow several kilometres down the rise flanks, often in channels. Pillow mounds and pillow ridges form intermittently where effusion is slower. Chains of seamounts occasionally form at or near the axis.

Slow-spreading ridges show a wide variety of volcanic structures. Most common are small, ~100 m diameter, cone- or dome-like volcanic hummocks. They are thought to have pipe-like feeders, which are thought to derive from fissure eruptions that degenerate into lines of point sources. Hummocks often aggregate into linear hummocky ridges or larger, complex hummocky volcanoes. Smoother shield and flat-topped volcanoes also occur. Many volcanoes are grouped into AVRs tens of kilometres long and a few kilometres wide, usually one to a spreading segment. AVRs are often surrounded by smooth lava flows and flat-topped volcanoes. The precise relations between these different volcanic features remain unclear.

Some parts of ultra-slow ridges contain segments volcanologically similar to slow-spreading ridges. Elsewhere there are widely spaced segments with highly focussed volcanism building thick crust by extensive sea floor eruptions, interspersed among segments almost devoid of volcanism. Some segments have remained avolcanic for millions of years.

At all spreading rates, some lavas erupt up to a few kilometres, or even tens of kilometres, off-axis. There is increasing evidence that off-axis volcanism at very slow spreading rates is focussed along major normal fault scarps.

# 7 Tectonism

## 7.1 Introduction

Earthquake focal mechanisms show that MORs are everywhere under tension (Huang *et al.*, 1986). As a result, as soon as young lithosphere is created by magmatic extrusion, intrusion and accretion, the forces of plate separation cause it to fracture (some of these fractures allowing emplacement of new melt – Chapter 6). This fracturing results in a wide range of tectonic features, from tensional fissures less than 1 m wide, through normal faults with horizontal and vertical offsets of tens to hundreds of metres, to detachment faults that cut right through the lithosphere and have offsets of several kilometres. Many of the characteristics of these fractures depend on the spreading rate through its effect on the thermal structure and hence rheology of the lithosphere.

## 7.2 Fissures

The smallest scale of fracturing occurs right at the ridge axis and consists of tensional fissures. The distributions of fissures have been mapped, amongst others, by Crane (1987) using deep-towed side-scan sonar and Haymon *et al.* (1991) using a towed camera system. Many aspects of fissuring were reviewed by Wright (1998).

Most fissures have widths ~0.5–3.0 m and lengths from ~10 m to >500 m (Figure 7.1 and Figure 7.2). They have vertical, parallel sides and no vertical offsets, so are known in rock mechanics as Mode I cracks, produced by pure tension (Figure 7.3). The only stress is horizontal extension normal to the ridge axis.

Fissures on the fast-spreading EPR are generally found within an axial zone some 2 km wide (Figure 7.2), but on slow-spreading ridges they can occur throughout the median valley floor. Some fissures occur within and at the crests of AVRs (e.g., Searle *et al.*, 2010; Yeo *et al.*, 2012). Others typically surround the bases of AVRs, sometimes with a sharp cut-off indicating that recent AVR volcanism post-dates fissure formation (e.g., Lawson *et al.*, 1996; Figure 7.4). Offsets across fissures generally increase away from the AVR axis (Searle *et al.*, 1998b).

The opening of fissures under tension allows magma to rise through the upper crust and erupt onto the sea floor, though such fissure eruptions may in time evolve into a number of

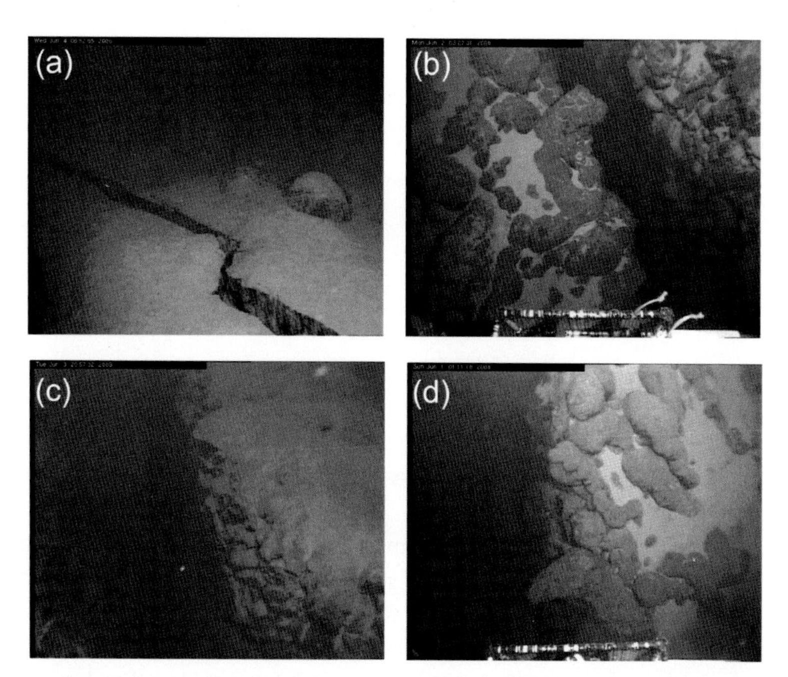


**Figure 7.1** Fissures from the MAR axis near 45°30′ N. (a) Narrow extensional fissure (offset ~7 cm) showing sharp, vertical, parallel sides; (b) 25 cm wide fissure; (c) ~1 m wide extensional fissure showing fractured pillow lavas in its walls. (d) Multiple lava buds (largest ~30 cm) extruding from side of eruptive fissure. For colour version, see plates section.

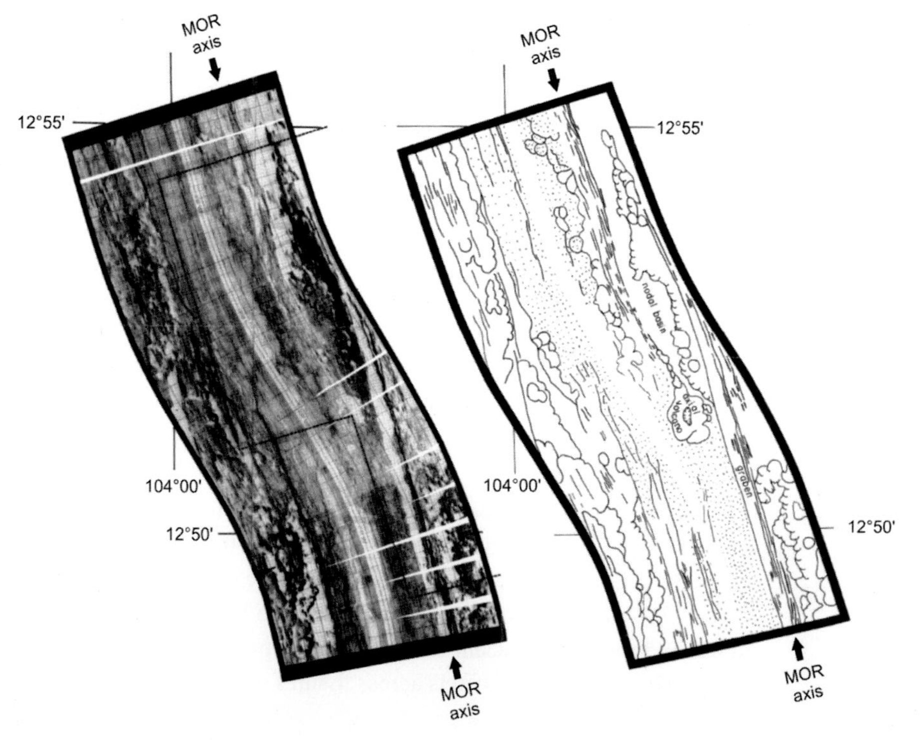


**Figure 7.2** SeaMARC I deep-towed side-scan sonar image (left) and geological interpretation (right), showing field of closely-spaced fissures (fine lines) along the EPR axis, after Crane (1987). Insonification is outwards from sinuous vehicle track along centre of image. Dark tones represent high backscatter.

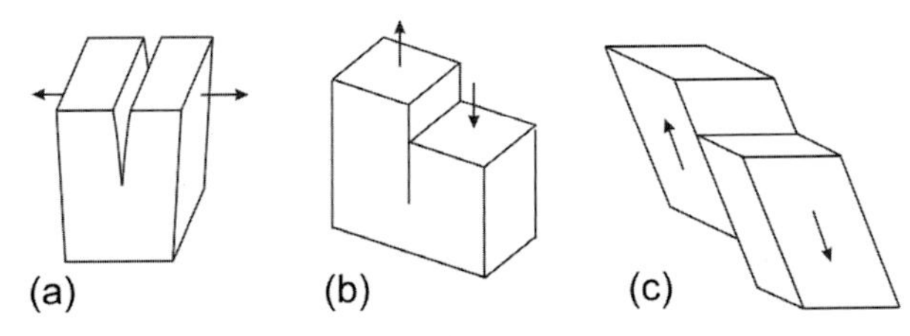


**Figure 7.3** (a) Mode I crack (tensile); (b) mode II crack (in-plane shear); (c) normal fault. Arrows indicate relative motion.

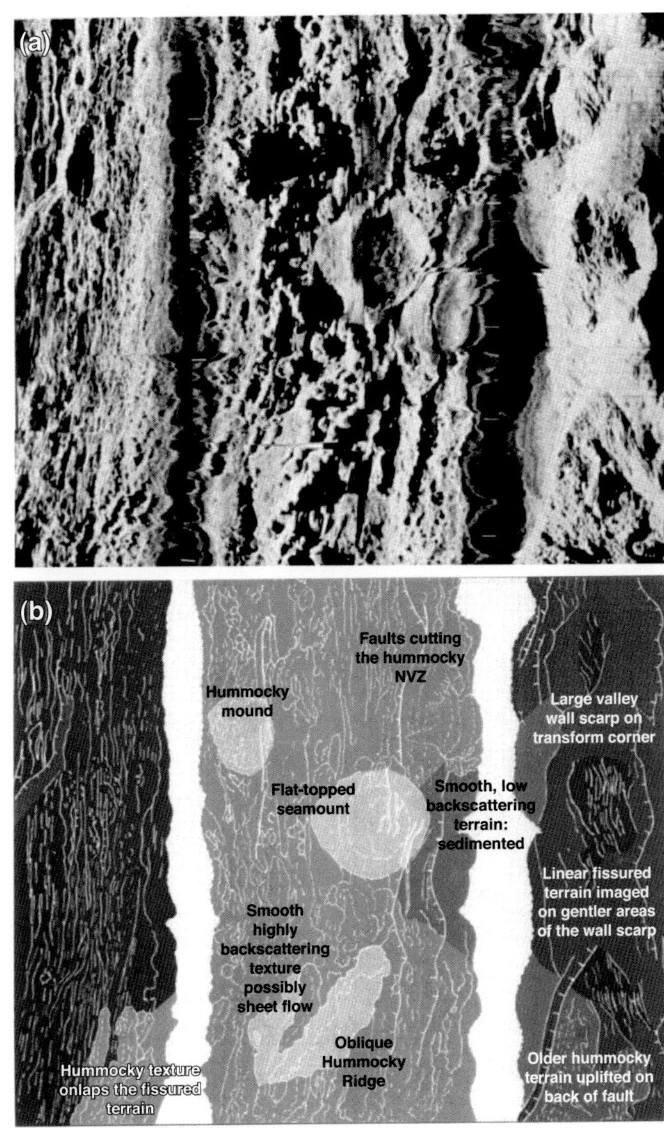


**Figure 7.4** (a) TOBI deep-towed side-scan image, and (b) interpretation, of part of the MAR median valley floor at 23°57′ N, after Lawson *et al*. (1996). Light tones represent high backscatter. White, vertical bands in interpretation are regions of poor signal in side-scan nadir. Insonification directions are outwards from these bands. Axial volcanic ridge runs up centre of figure. Eastern median valley wall fault is at far right; zone of intense fissuring flanks the AVR to the left.

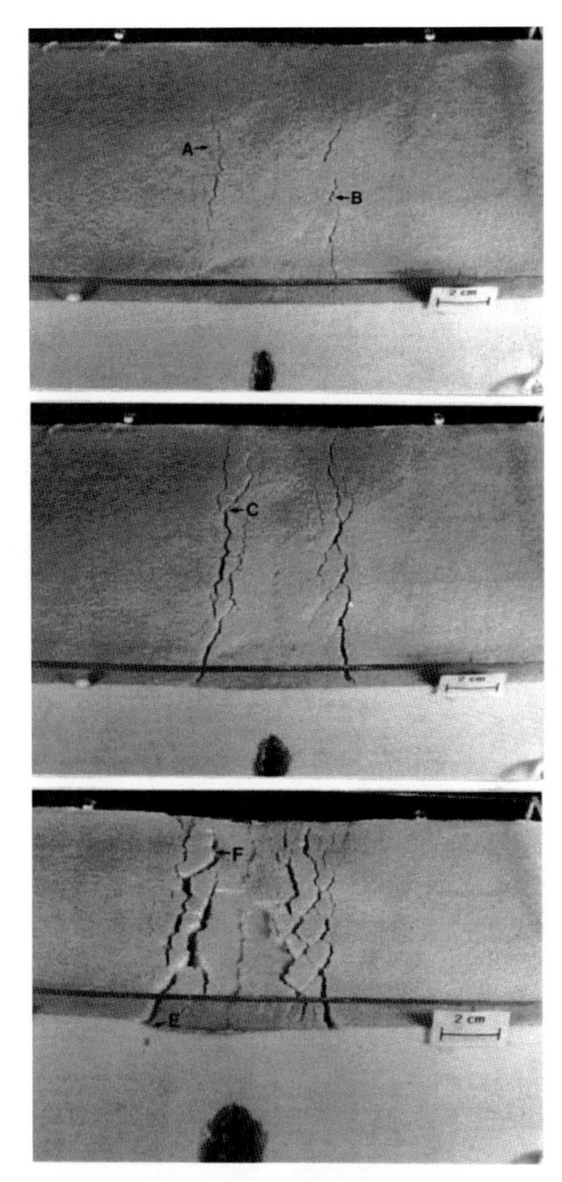


Figure 7.5 Oblique views of an analogue model showing three stages of fissure and fault formation above a dyke, after Mastin and Pollard (1988). Vertical front of box is seen in foreground of each panel, with dyke (dark mass) at bottom. Top: in first stage, small fissures (A, B) grow in the surface to either side of the buried dyke. Middle: additional fissures (C) grow and link with the earlier ones. Bottom: some fissures develop dip-slip movement to form small faults and graben.

point-source eruptions (Chapter 6). Because the resultant lava flows often cover the feeding fissures, such fissures are rarely observed. Consequently, many mapped fissures, especially off-axis, may represent purely tectonic fissuring of previously created crust. However, the magma in eruptive fissures may sometimes fail to reach the surface or, having done so, may partially withdraw, leaving an open fissure at the surface (e.g., Figure 7.1d). If a dyke intrudes the crust but fails to reach the sea floor (i.e. there is no eruption), it creates a stress field at the surface that can produce fissures or graben (Pollard *et al.*, 1983; Mastin and Pollard, 1988; Figure 7.5).

Fissures open along the direction of the least compressive stress (conventionally called $\sigma_3$), and strike perpendicular to that direction. Thus, fissures are usually normal to the spreading direction. However, $\sigma_3$ is deflected near ridge offsets (Alexander and Macdonald, 1996), so fissure (and fault) directions there depart from spreading-normal (see Section 7.3). Shifting patterns of crustal stress can produce a superposition of differently orientated fissures; such cross-cutting sets may provide foci for hydrothermal activity (Kleinrock and Humphris, 1996; Section 8.3.3).

## 7.3 Normal faults

As lithospheric extension continues, fissures begin to develop vertical offsets to form faults. The young faults grow by lengthening and linking with neighbouring faults (Cowie *et al.*, 1993a; Searle *et al.*, 1998b; Figure 7.5). At the same time they penetrate deeper into the thickening brittle crust, so the increasing rock overburden increases the vertical stress. Both bending stresses and buoyancy stresses arising from the magma lens and surrounding hot rock contribute additional stresses (Bohnenstiehl and Carbotte, 2001; Shah and Buck, 2001; see also Chapter 3). In general there is a tri-axial stress system, in which the least compressive stress ($\sigma_3$) is tensile and normal to the ridge axis, the most compressive stress ($\sigma_1$) is compressional and vertically down (the overburden of rock), and the intermediate stress ($\sigma_2$) is zero and parallel to the ridge. This stress regime produces normal faults with theoretical dips of 60° (Anderson, 1951; Figure 7.6a). The block below the fault plane is the footwall; that above the fault is the hanging wall. In strike-slip faults, one of the horizontal stress components is greater than the overburden stress; then $\sigma_2$ is vertical while $\sigma_1$ and $\sigma_3$ are horizontal and at 60° to the vertical plane of the fault (Figure 7.6b).

Spreading rate has a major influence on the rheology of the lithosphere (Searle and Escartín, 2004), particularly the thickness of the brittle layer (Carbotte and Macdonald, 1994b), and hence on the properties of faults. Table 7.1, which combines reported fault statistics from studies at a range of spreading rates and using a number of different measurement systems, shows some clear trends. Mean fault offsets increase with decreasing spreading rate, from 10–20 m at the fast-spreading EPR to several hundred metres at the slow-spreading MAR. At the same time, the proportion of outward facing faults decreases dramatically, from ~60% at super-fast to ~6% at slow spreading. Carbotte and Macdonald (1994b) suggest the predominance of inward facing faults at slow spreading rates may reflect the rapid increase there of lithospheric strength with age, giving a steeply outward dipping base to the lithosphere. Finally, there is a suggestion that fault spacing increases somewhat with decreasing spreading rate.

**Table 7.1** Faulting parameters

| Area | Full rate[a], km $Ma^{-1}$ | Dip direction | % of faults | Mean spacing, km | Mean offset, m | Strain, %[d] | Instrument | Source |
|---|---|---|---|---|---|---|---|---|
| MAR 40° N | 23 | In | | 1.9 | 147[b] | | GLORIA | Searle (1984) |
| MAR 40° N | 23 | Out | | >10.0 | | | GLORIA | Searle (1984) |
| MAR 29° N | 23 | In | 94 | 1.7 | 193[c] | 11.0 | TOBI | Escartin *et al.* (1999) |
| MAR 29° N | 23 | Out | 6 | 4.6 | 270[c] | 1.0 | TOBI | Escartin *et al.* (1999) |
| Explorer Ridge | ~50 | In | >~50 | | 12.1[b,f] | 5–8 | ABE | Deschamps *et al.* (2007) |
| Explorer Ridge | ~50 | Out | <~50 | | 8.5[b,f] | 3–4 | ABE | Deschamps *et al.* (2007) |
| Explorer Ridge | ~50 | Both | | 0.035 | | 3.7–18.4 | ABE | Deschamps *et al.* (2007) |
| Coc-Naz 86° W | 62 | In | | 1.7 | ~50[b] | | GLORIA | Searle (1984) |
| Coc-Naz 86° W | 62 | Out | | ~8.0 | <50[b] | | GLORIA | Searle (1984) |
| **Coc-Naz 85° W[g]** | **64** | **Both** | **100** | **1.5** | | **4.1** | **SeaMARC II** | **Carbotte & Macdonald (1994b)** |
| Coc-Naz 85° W | 64 | In | 92 | 1.6 | ~50[b] | | SeaMARC II | Carbotte & Macdonald (1994b) |
| Coc-Naz 85° W | 64 | Out | 8 | 5.3 | <50[b] | | SeaMARC II | Carbotte & Macdonald (1994b) |
| EPR 13° N–15° N | 83–90 | Both | | | | 4.8–15.5 | SeaMARC II | Cowie *et al.* (1993b) |
| EPR 9.4° N–10° N | 102 | In | 68[e] | 0.53[e] | 11[b,e,f] | 1.6[e,f] | High-resolution sonars | Escartin *et al.* (2007) |
| EPR 9.4° N–10° N | 102 | Out | 32[e] | 1.11[e] | 7[b,e,f] | 0.5[e,f] | High-resolution sonars | Escartin *et al.* (2007) |
| EPR 9.3° N–9.8° N | 102 | Both | 100 | ~2.0 | | | SeaBeam, SeaMARC II | Alexander & Macdonald (1996) |
| EPR 9.3° N–9.8° N | 102 | In | 53 | ~2.0 | 60–80 | 4.2 | SeaBeam, SeaMARC II | Alexander & Macdonald (1996) |
| EPR 9.3° N–9.8° N | 102 | Out | 47 | ~2.0 | 60–80 | 4.3 | SeaBeam, SeaMARC II | Alexander & Macdonald (1996) |
| **EPR 8.5° N–10° N** | **100–106** | **Both** | **100** | | | **3.2** | **SeaMARC II** | **Carbotte & Macdonald (1994b)** |
| EPR 8.5° N–10 N | 100–106 | In | 60 | 2.3 | 41[b] | | SeaMARC II | Carbotte & Macdonald (1994b) |
| EPR 8.5° N–10° N | 100–106 | Out | 40 | 3.8 | 36[b] | | SeaMARC II | Carbotte & Macdonald (1994b) |
| EPR 8.5° N | 106 | Both | 100 | | | 5.2 | SeaMARC II | Carbotte & Macdonald (1994b) |
| EPR 3° S | 130 | Both | | | | 3.4–5.5 | GLORIA | Cowie *et al.* (1993b) |
| EPR 3° S–4° S | 131 | In | 73 | 1.7 | 38[b] | | GLORIA | Searle (1984) |
| EPR 3° S–4° S | 131 | Out | 27 | 2.6 | 30[b] | | GLORIA | Searle (1984) |

| **EPR 18° S–19° S** | **147** | **Both** | **100** | | | **4.8** | **SeaMARC II** | **Carbotte & Macdonald (1994b)** |
|---|---|---|---|---|---|---|---|---|
| EPR 18° S–19° S | 147 | In | 41 | 1.7 | 20.0[b] | | SeaMARC II | Carbotte & Macdonald (1994b) |
| EPR 18° S–19° S | 147 | Out | 59 | 1.9 | 15.0[b] | | SeaMARC II | Carbotte & Macdonald (1994b) |
| EPR 19°30′ S (east) | 147 | In | 45 | 0.60 | 22.3[b] | | SeaBeam, Deep-Tow | Bicknell *et al.* (1988) |
| EPR 19°30′ S (east) | 147 | Out | 55 | 0.49 | 14.0[b] | | SeaBeam, Deep-Tow | Bicknell *et al.* (1988) |
| EPR 19°30′ S (west) | 147 | In | 51 | 0.55 | 12.0[b] | | SeaBeam, Deep-Tow | Bicknell *et al.* (1988) |
| EPR 19°30′ S (west) | 147 | Out | 49 | 0.57 | 17.7[b] | | SeaBeam, Deep-Tow | Bicknell *et al.* (1988) |
| EPR 19°30′ S (east) | 147 | Both | | | | 6.4 | SeaBeam, Deep-Tow | Bicknell *et al.* (1988) |
| EPR 19°30′ S (west) | 147 | Both | | | | 4.2 | SeaBeam, Deep-Tow | Bicknell *et al.* (1988) |
| EPR 19°30′ S (1E) | 147 | In | 38 | 0.857 | 21.5[b] | *2.5* | Deep-Tow | Bohnenstiehl & Carbotte (2001) |
| EPR 19°30′ S (1E) | 147 | Out | 62 | 0.517 | 11.0[b] | *2.1* | Deep-Tow | Bohnenstiehl & Carbotte (2001) |
| EPR 19°30′ S (1W) | 147 | In | 45 | 0.963 | 24.5[b] | *2.6* | Deep-Tow | Bohnenstiehl & Carbotte (2001) |
| EPR 19°30′ S (1W) | 147 | Out | 55 | 0.720 | 16.5[b] | *2.2* | Deep-Tow | Bohnenstiehl & Carbotte (2001) |
| EPR 19°30′ S (2E) | 147 | In | 38 | 1.129 | 25.0[b] | *2.2* | Deep-Tow | Bohnenstiehl & Carbotte (2001) |
| EPR 19°30′ S (2E) | 147 | Out | 62 | 0.678 | 13.0[b] | *1.9* | Deep-Tow | Bohnenstiehl & Carbotte (2001) |
| EPR 19°30′ S (2W) | 147 | In | 36 | 1.039 | 18.0[b] | *1.7* | Deep-Tow | Bohnenstiehl & Carbotte (2001) |
| EPR 19°30′ S (2W) | 147 | Out | 64 | 0.572 | 10.5[a] | *1.8* | Deep-Tow | Bohnenstiehl & Carbotte (2001) |

[a] Calculated from NUVEL-1A poles (DeMets *et al.*, 1990; DeMets *et al.*, 1994).
[b] Measured as scarp height (vertical offset).
[c] Measured as fault heave (horizontal offset).
[d] Total strain in plain type, strain on *either* inward or outward facing faults in italics.
[e] Data for 'old'-looking sea floor that had not been repaved by recent lava flows.
[f] Axial zone only.
[g] Locations in bold have been determined to follow an exponential distribution, with parameters as given in Table 7.2.

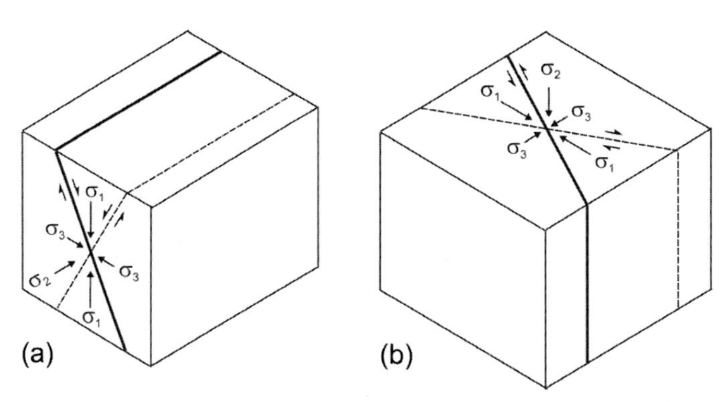


Figure 7.6 Stress patterns associated with (a) normal faulting, and (b) strike-slip (transform) faulting, according to Andersonian theory (see text). Bold lines: fault planes; dashed lines: alternative locations. Here, $\sigma_1$, $\sigma_2$ and $\sigma_3$ are the most, intermediate and least compressive stresses, respectively, and are mutually orthogonal. Fault planes form at 60° to $\sigma_3$.

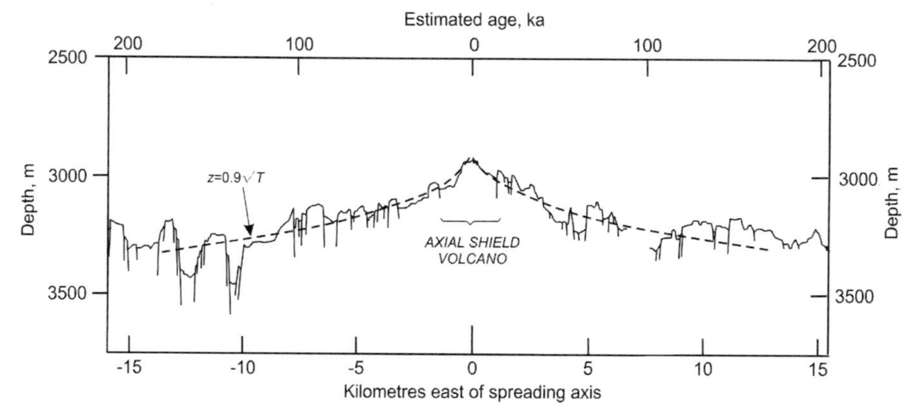


Figure 7.7 High-resolution topographic profile across the EPR axis at 3°20′ S from deep-towed observations, after Lonsdale (1977), with kind permission from Springer Science and Business Media. Inferred faults are drawn extending into the subsurface. Dashed line gives the best fit square root of age curve.

### 7.3.1 Fast-spreading ridges

At fast-spreading ridges (and hotspot-dominated ridges such as Reykjanes that partially mimic them), the axial lithosphere is thin and weak and can support only small-offset faulting. Detailed observations of sea floor topography, initially by deep-towed instruments and later by surface-towed side-scan and multi-beam echosounding, revealed a pattern of small, ridge-parallel normal faults (e.g., Lonsdale, 1977; Searle, 1984; Alexander and Macdonald, 1996; Crowder and Macdonald, 2000; Bohnenstiehl and Carbotte, 2001; Figures 7.7–7.9). These faults are up to tens of kilometres long and may extend the whole length of a spreading segment. Both inward and outward facing faults have throws of 10–20 m, forming a series of small horsts and graben (Figure 7.7).

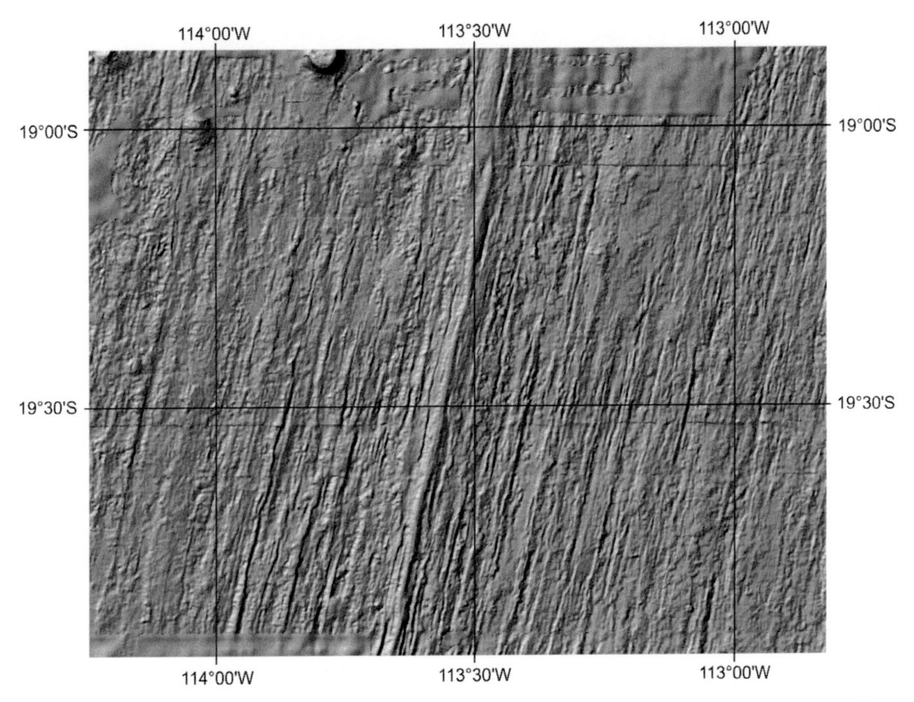


Figure 7.8 Shaded relief topography of the EPR around 19°30′ S, illuminated from the NW, showing inward and outward facing normal fault scarps. EPR axial high runs NNE–SSW at 113°30′ W. Image produced using GeoMapApp (http://www.geomapapp.org; Ryan *et al.*, 2009).

Carbotte and Macdonald (1994b) and Bohnenstiehl and Carbotte (2001) showed that faulting at the super-fast southern EPR is best characterised by two series of faults: the first are wider spaced and higher, comprising inward facing 'master' faults, while the second consists of smaller but more frequent antithetic outward facing faults (Figure 7.10). This pattern is consistent with faults nucleating above a shallow, internally pressured magma sill, but is strongly dependent on the sill depth, perhaps explaining differences from the slightly slower spreading northern EPR where the magma sill is somewhat deeper.

Both inward and outward facing faults begin to develop within about 2 km of the ridge axis, though the innermost are always inward facing. Along most of the northern and southern EPR, outward facing faults may grow to their maximum offsets within 7 km of the axis, but continue to lengthen beyond that; inward dipping faults continue to grow in both offset and length to at least 30 km off axis (Alexander and Macdonald, 1996). Crowder and Macdonald (2000) determined that active faulting at the EPR 8.5° N–10° N continues to ~45 km (0.8 Ma) off-axis, producing a zone of active faulting up to 90 km wide.

Fault parameters are linked to local magma supply as well as overall spreading rate. For example, in the ~60 km long survey of Alexander and Macdonald (1996), fault scarp height increased from N to S in the direction of inferred decreased magma supply towards the segment end; a similar effect was found by Carbotte and Macdonald (1994b) near transform faults and on the eastern Cocos–Nazca Spreading Centre.

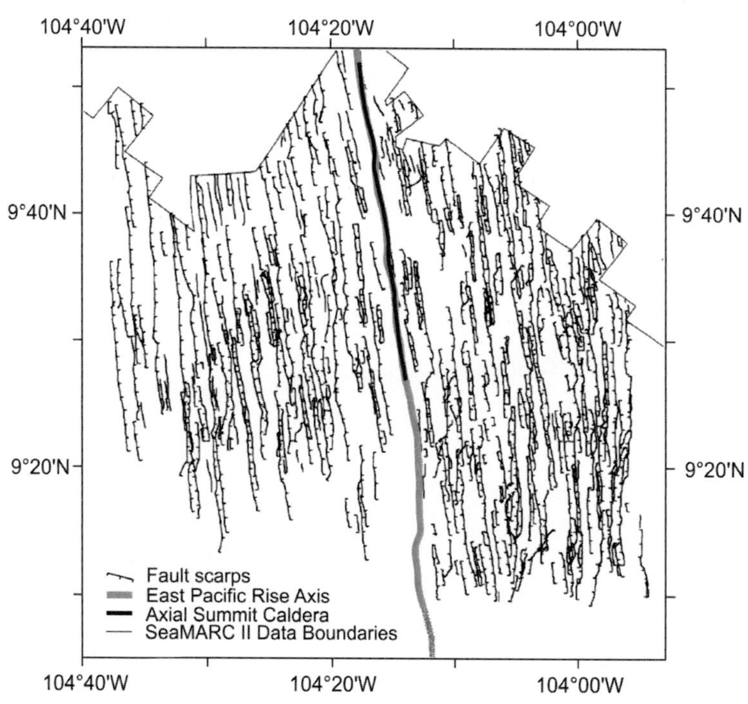


**Figure 7.9** Map of faults imaged on the EPR crest at 9°30′ N using the 12 kHz SeaMARC II side-scan sonar, after Alexander and Macdonald (1996), with kind permission from Springer Science and Business Media. Tick marks are on the downthrown sides.

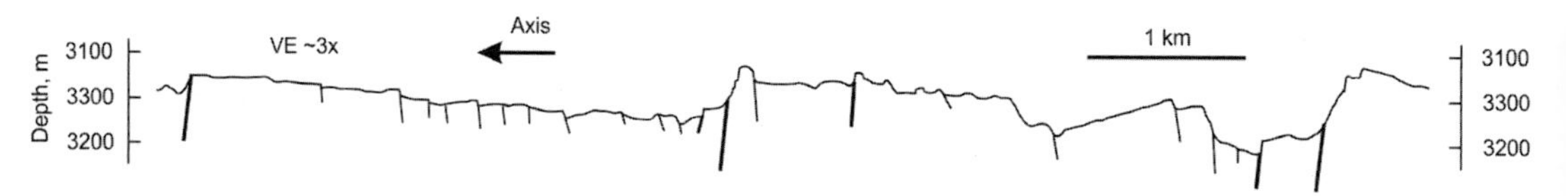


**Figure 7.10** Sea floor depth profile and inferred faults on the east flank of the super-fast EPR near 19°30′ S, after Bohnenstiehl and Carbotte (2001). Inward facing 'master' faults are shown by heavy lines, with sea floor profile and outward facing antithetic faults in fine lines.

In a few places where fault distributions have been examined, spacings and lengths approximately follow exponential distributions (Carbotte and Macdonald, 1994b; Table 7.2). Cowie *et al.* (1994) suggest that these apparent distributions arise because small packets of faults are formed independently in a narrow (few kilometres wide) fault generation zone, and only with high-resolution bathymetry, as used at 19° S on the EPR, is the fractal nature revealed. However, fault parameters on the southern EPR fit neither exponential, normal, nor power law distributions (Bohnenstiehl and Carbotte, 2001).

There is an intimate interplay between faulting and volcanism on the EPR (Macdonald *et al.*, 1996). Lavas erupted near the rise axis can flow many kilometres off-axis, becoming dammed behind some inward facing faults and overflowing some outward facing ones. Escartin *et al.* (2007) find that low-relief volcanic growth faults strongly control the

**Table 7.2** Parameters of fault populations fitting an exponential distribution, after Carbotte & Macdonald (1994b)

| Area | Full rate[a], km Ma$^{-1}$ | Characteristic spacing, km | Characteristic throw, m | Characteristic length, km | Instrument |
|---|---|---|---|---|---|
| Cocos–Nazca 85° W | 64 | 1.2 | | 5.8 | SeaMARC II |
| East Pacific Rise 8.5° N–10° N | 100–106 | 1.1 | 81 | 5.3 | SeaMARC II |
| East Pacific Rise 18° S–19° S | 147 | 0.9 | | 4.5 | SeaMARC II |

[a] Calculated from NUVEL-1A poles (DeMets *et al.*, 1990; DeMets *et al.*, 1994).

development of oceanic layer 2A. They estimate that at least 2% of plate separation at the EPR is accommodated by brittle fracturing, mostly producing inward facing faults, in an axial zone that is 4 km wide. However, as much again may be hidden by subsequent eruptions. See also Figure 7.17. Further deformation up to 40 km off-axis is accommodated by lithospheric flexure due to thermal subsidence, producing roughly half the inward facing faults that accommodate ~50% of the strain. Most estimates of brittle strain at the EPR are in the range 3%–6% (Table 7.1).

## 7.3.2 Slower spreading rates

Faulting parameters change as spreading rates lessen. The relatively few studies of fault statistics at intermediate rates show a decrease in the number of outward facing faults and a modest increase in fault offset and length (Searle, 1984; Carbotte and Macdonald, 1994b; Mitchell and Searle, 1999; Table 7.1 and Figure 7.11).

These changes continue at slow spreading rates. Fault throws increase by a factor of ~5 to some 200 m, with the largest faults reaching offsets >1 km. Almost 95% of faults are inward facing, and the proportion of plate divergence taken up by tectonic strain increases from ~4% on fast-spreading ridges to >10% at slow-spreading ones (Escartin *et al.*, 1999). The average fault length is ~10 km, almost twice that at the EPR. Fault scarps typically dip ~30°, although detailed investigation shows that such scarps often comprise several steeper-dipping (45°–60°) faults separated by narrow terraces.

At slow-spreading ridges the style of faulting varies systematically throughout a spreading segment (Karson *et al.*, 1987; Shaw and Lin, 1993; Searle *et al.*, 1998b). At segment centres, faults on both sides of the median valley have relatively small throws and close spacings; at segment ends, those on the inside corners have larger offsets and wider spacings than those on the outside corners, which more nearly resemble those at the segment centre (Figures 7.12 and 7.13). Thus segment centres are quite symmetrical and segment ends strongly asymmetrical. These changes correlate with inferred changes in lithospheric thickness and strength (Chapters 3 and 5), with thicker lithosphere (colder) and thinner crust (less melt emplacement) towards segment ends (Escartin and Lin, 1995; Figure 7.14). The very large faults at inside corners may achieve offsets of well over 1 km. Many dip at low angles and may cut right through the lithosphere so are detachment faults (Dick *et al.*, 1981; Section 7.4.1).

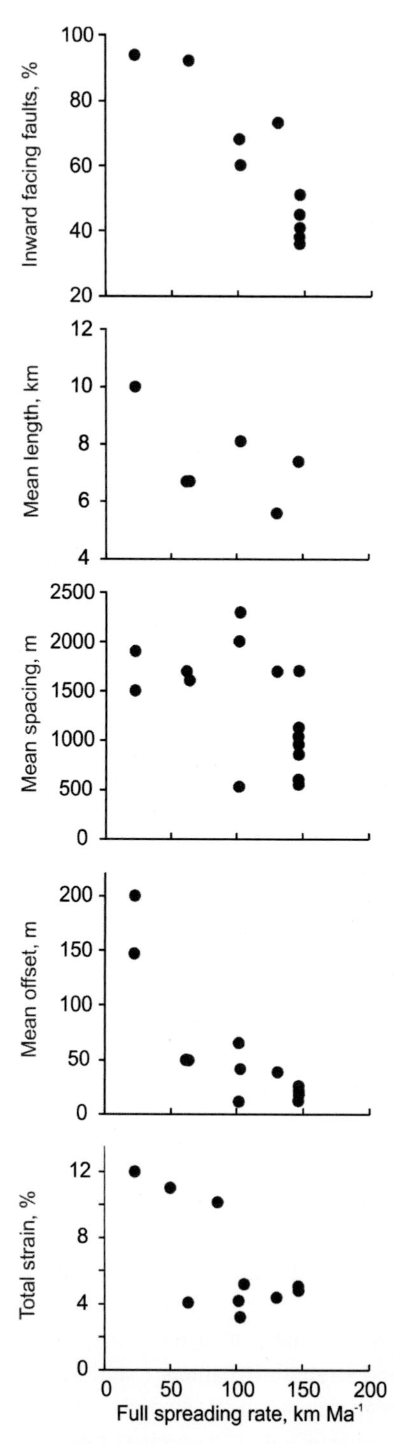


Figure 7.11 Faulting parameters from Table 7.1 plotted against spreading rate. Length, spacing and offset are for inward facing faults only.

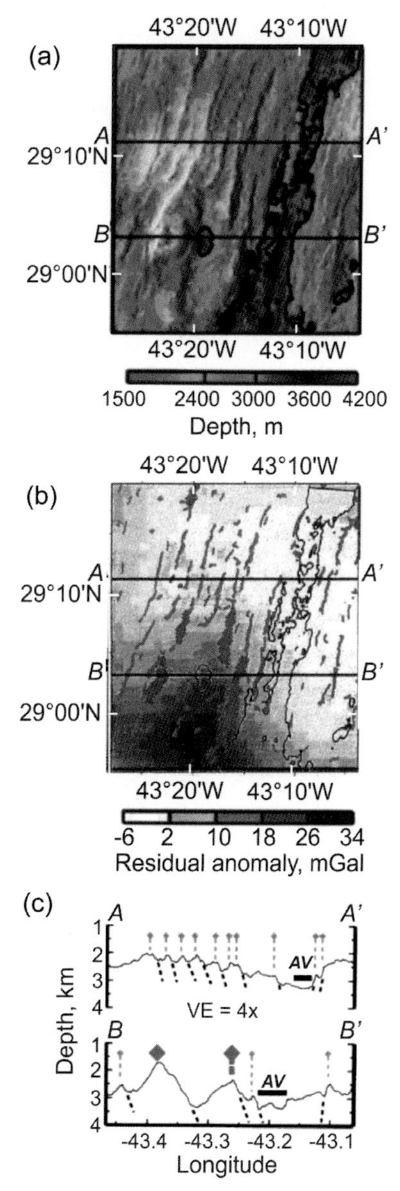


**Figure 7.12** Fault pattern in a single spreading segment of the MAR, 29° N, after Shaw and Lin (1993). (a) Bathymetry. (b) Scarps (colour) detected automatically from the curvature of bathymetry, superimposed on residual mantle Bouguer anomaly (grey). Blue points determined using a 0.45 km radius window, highlighting small scarps; red points used a 2.2 km window showing largest scarps. (c) Interpreted topographic sections along segment centre and segment end. Red and blue points: scarps highlighted in (b); dashed lines: inferred faults; black bars: position of median valley floor. For colour version, see plates section.

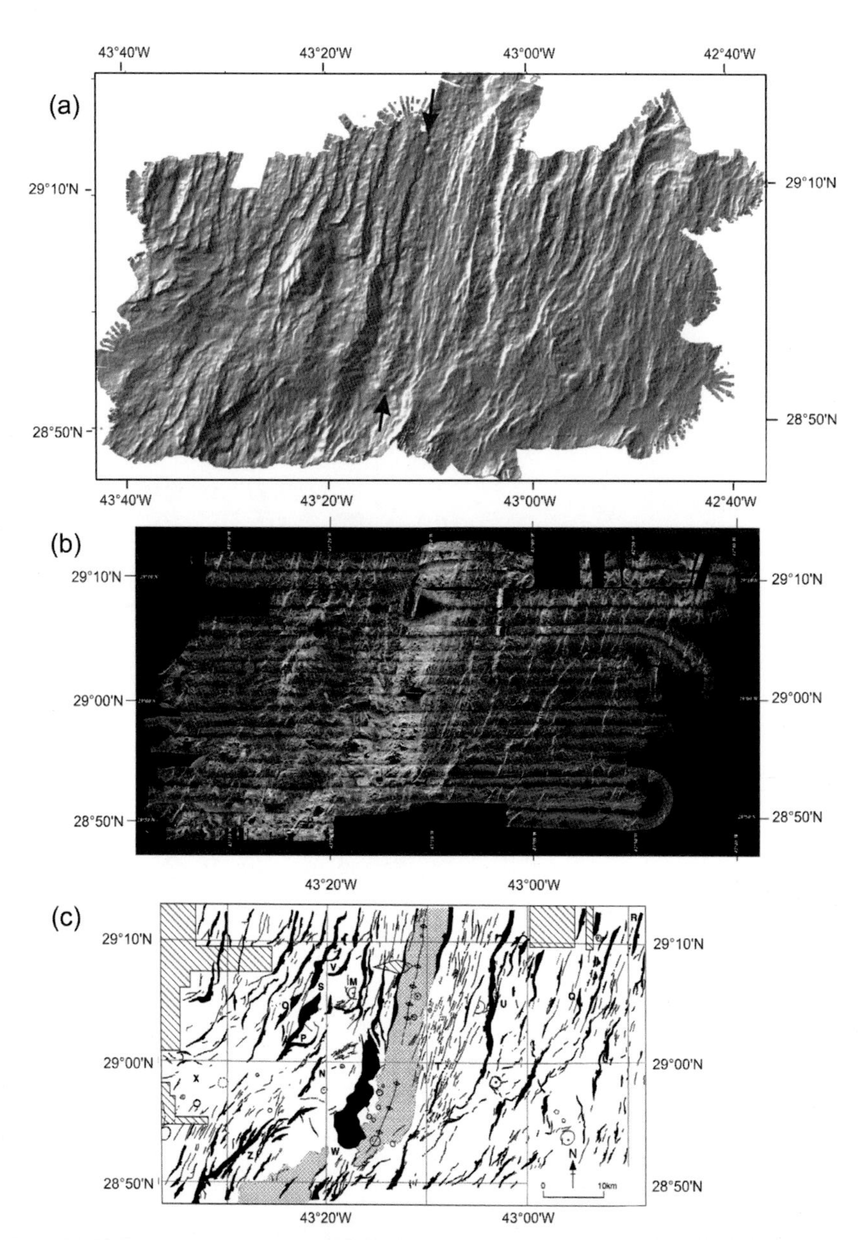


**Figure 7.13** Faulting at the MAR around 29° N. (a) Shaded relief bathymetry, illuminated from the NW, emphasising prevalent inward facing faults. Ridge axis trends NNE between black arrows. (b) Deep-towed side-scan sonar mosaic at same scale, after Escartin *et al.* (1999). Insonification is to the south and light tones indicate high backscatter. West-facing faults, mostly east of the axis, produce sharp, narrow reflections. The high-backscatter zone through the centre of image is the neo-volcanic zone. (c) Fault patterns (black) deduced from the side-scan images, after Searle *et al.* (1998b), © 1998, with permission from Elsevier; stipple: neo-volcanic zone; diagonal shading: areas of no coverage. Note growth of faults off-axis by lengthening, broadening, and linking with neighbouring faults.

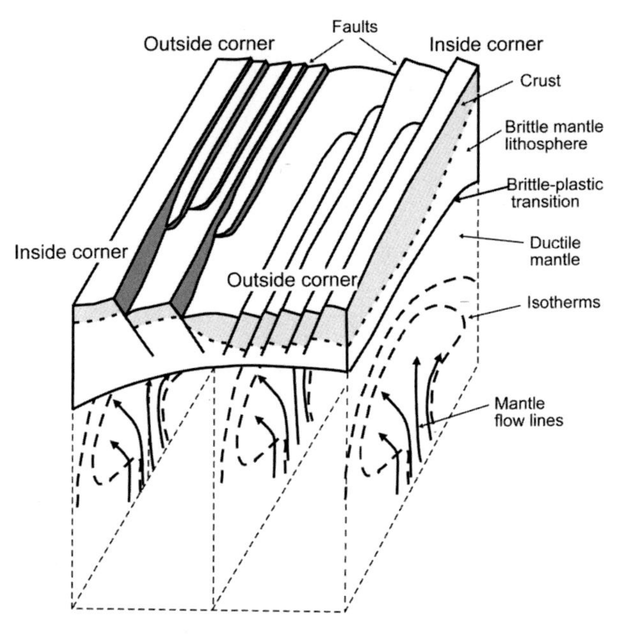


Figure 7.14 Schematic diagram of lithospheric and crustal thickness variations and resultant faulting pattern in a slow-spreading segment, after Searle and Escartín (2004). Light grey: crust; short-dashed line: base of crust; continuous line: base of lithosphere; long-dashed lines: shallow isotherms resulting from focussed mantle upwelling and/or melt emplacement (arrows) at segment centre.

### 7.3.3 Strength of faults

The strength of rocks depends on their rheology, which is controlled by a number of factors including composition, strain rate, fluid content and temperature. The rheology of the oceanic lithosphere was reviewed by Searle and Escartín (2004). Most rocks behave elastically (and deform brittlely when the elastic limit is exceeded) at low temperatures and pressures and high strain rates, with strain being proportional to applied stress. Anderson's theory of faulting (Anderson, 1951) leads to a relationship between fault dip $\theta$ and the coefficient of static friction $\mu$:

$$\tan 2\theta = -\frac{1}{\mu} \tag{7.1}$$

(Turcotte and Schubert, 1982). Thus the dip ranges from 45° for $\mu = 0$ to 67.5° for $\mu = 1.0$. This relationship, with a value of $\mu = 0.85$, is known as Byerlee's Law (Byerlee, 1978).

Brittle behaviour gives way to plastic deformation at higher temperatures and pressures and lower strain rates. There is an approximate characteristic temperature for the brittle–plastic transition, which is typically about 700 °C–750 °C for crustal rocks. Above this temperature, rocks deform plastically by solid state creep according to a power law:

$$\dot{\varepsilon} = A\sigma^n d^m e^{-Q/RT}, \tag{7.2}$$

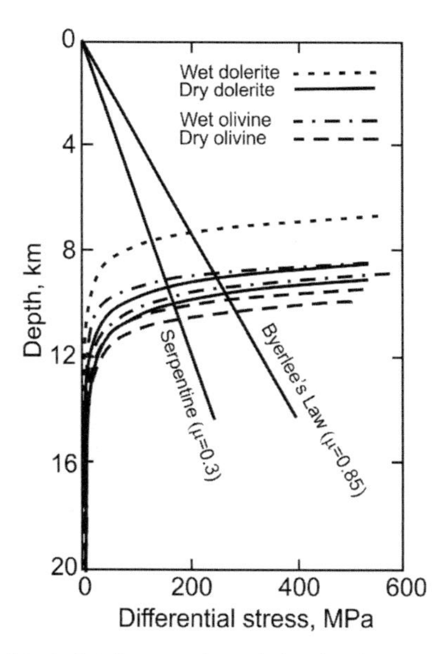


**Figure 7.15** Plot of predicted frictional strength in the brittle regime (straight lines) and measured power-law rheology for wet and dry dolerite and olivine (curves) after Searle and Escartín (2004). Doubled curves for olivine and dry dolerite are from different studies. Dry dolerite and/or olivine are considered most applicable to oceanic lithosphere at MORs. Byerlee's law with coefficient of friction $\mu = 0.85$ is probably appropriate for unaltered oceanic rocks, but lithosphere and faults containing extensive serpentinised peridotite or talc is better described by $\mu \leq 0.3$.

where $\dot{\varepsilon}$ is the strain rate, $d$ is grain size, $R$ the universal gas constant, $T$ is temperature, and $A$, $n$, $m$ and $Q$ are experimentally determined constants specific to the material (Goetze, 1978). Two creep mechanisms are relevant: dislocation creep, which occurs at lower temperature, higher stress and larger grain size and has $n = 3$, and diffusion creep, which occurs at higher temperature, lower stress and smaller grain size, for which $n \sim 1$. In both mechanisms the strain rate depends strongly and non-linearly on temperature, so rock strength rapidly falls with increasing temperature and thus depth.

Combining Equations (7.1) and (7.2) with a specified temperature–depth function leads to a general profile of strength against depth such as those shown in Figure 7.15. The strength increases linearly with depth in the brittle regime, then falls rapidly with increasing depth in the underlying plastic regime. Strain accrues when this strength envelope is exceeded: either brittle faulting or plastic deformation depending on the regime.

A wet dolerite rheology was used in early models of oceanic lithosphere, and results in a weak, plastic lower crust decoupling a strong upper crust from the mantle. Such models reproduced some significant features of MORs, such as the transition from axial high to median valley with decreasing spreading rate (Chen and Morgan, 1990b; Figure 4.9). Nevertheless, a dry dolerite rheology is now known to be more appropriate (Escartin *et al.*, 1997; Searle and Escartín, 2004). This has a brittle–plastic transition at a higher temperature, typical of the upper mantle, and is less likely to produce a weak lower crust. With this rheology, crustal faulting more closely follows Byerlee's Law.

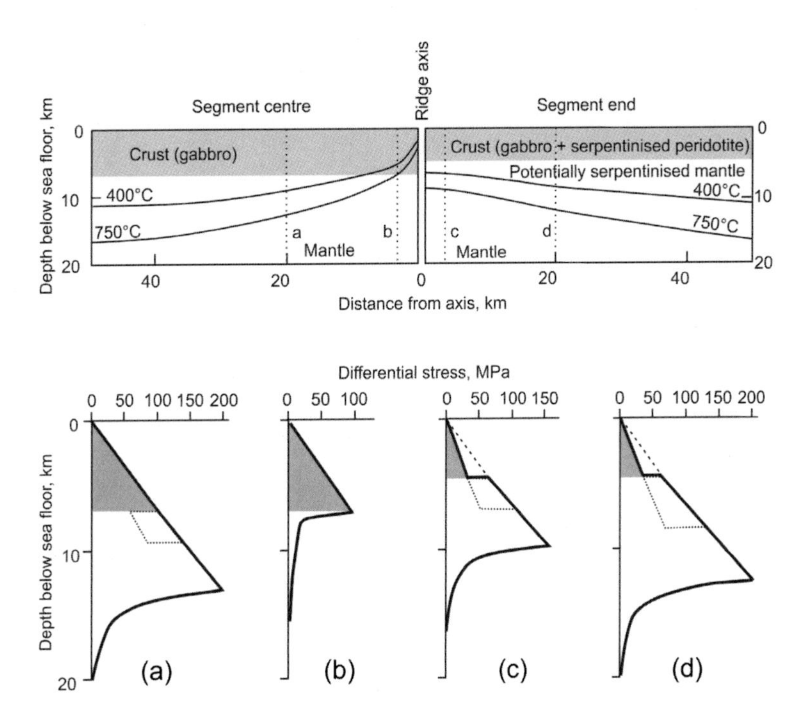


Figure 7.16 Proposed variations in lithospheric strength as a result of serpentinisation, after Escartin *et al.* (1997). Reprinted with permission from Elsevier, © 1997. Top: cross-sections of a slow-spreading ridge axis at segment centre and segment end, showing 400 °C isotherm (approximate stability limit of serpentine) and 750 °C isotherm (approximate temperature of brittle-plastic transition). Grey shading: crust; dotted lines: positions of the strength profiles illustrated below. Bottom: strength vs. depth profiles assuming the thermal structure shown above and a lithosphere that follows Byerlee's Law for brittle deformation at temperature <750 °C and the power-law rheology of dry dolerite above 750 °C. Coefficient of friction $m = 0.85$ is assumed at segment centre (a, b); at segment end, $\mu = 0.3$ for $T < 400$ °C and $\mu = 0.85$ for 400 °C $< T <$ 750 °C (c, d). Dashed lines in (c) and (d) show strength for $\mu = 0.85$ for comparison. Serpentinisation of upper mantle may also weaken lithosphere (dotted lines in a, c, d).

An important factor in controlling rheology is the degree of alteration of the rocks. Olivine is converted to serpentine in the presence of water (Section 8.5.2) and is stable at temperatures <~400 °C. Serpentine is much weaker than dolerite, with $\mu = 0.3$ (Escartin *et al.*, 1997). Another weak mineral commonly produced by hydration of olivine and serpentine is talc, which is even weaker than serpentine (Moore and Lockner, 2007; Moore and Rymer, 2007). As discussed in Chapter 5, the 'crust' at the ends of slow-spreading segments may comprise large proportions of serpentinised peridotite. Thus, although the lithosphere there is thick and isotherms are depressed relative to segment centres, the presence of serpentine significantly weakens the crust (Figure 7.16). In contrast, serpentine is rare at segment centres, so faulting there occurs at the higher friction typical of Byerlee's Law ($\mu = 0.85$).

One model for fault development suggests they grow until the stress required to create a new fault is less than that needed to maintain slip on an increasingly large and strong

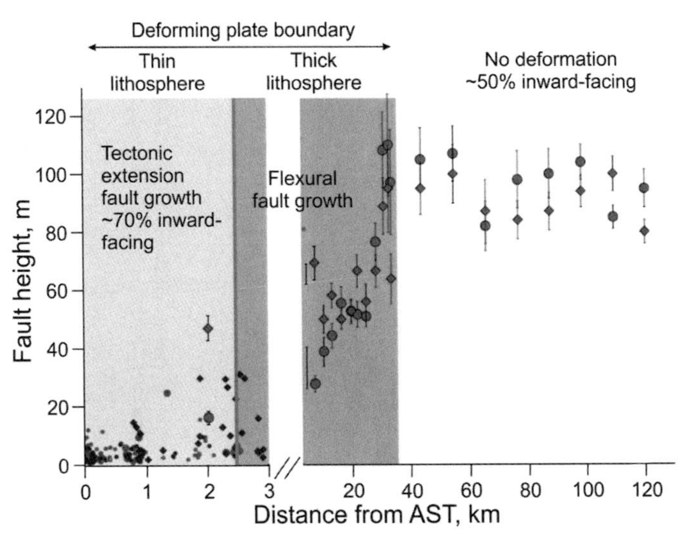


Figure 7.17 The development of faulting on the EPR 9°–10° N, after Escartin *et al.* (2007). Red symbols: data from Alexander and Macdonald (1996); diamonds: inward facing faults; circles: outward facing faults. Note change in horizontal scale at 3 km. Main mechanism for fault growth in innermost 2.5 km is tectonic extension; beyond that, lithospheric flexure. For colour version, see plates section.

existing one (Shaw and Lin, 1993). In this model, weak faults will grow larger than strong ones, perhaps partly explaining the preponderance of large faults at segment ends. Escartin *et al.* (1997) show that such an effect can allow for fault heaves of over 1.5 km at segment ends, compared to about 0.6 km at segment centres.

### 7.3.4 Fault growth and development

Once initiated, fissures and faults continue to be influenced by the tensional stress field at the ridge axis, which causes them both to increase their horizontal and vertical offsets and to extend along their lengths (Figure 7.13c). Although such development may continue up to tens of kilometres from the ridge axis, especially at fast spreading rates, the majority of the development takes place within just a few kilometres of fault formation, within the inner floor of the median valley at slow-spreading ridges and on the flanks of the axial high at fast-spreading ones.

At fast-spreading ridges, bending stresses may play a significant part in controlling faulting (Shah and Buck, 2003). The axial lithosphere is created with a concave-up curvature as a result of axial buoyancy, and gradually unbends as it cools. This places the top of the lithosphere into tension, creating faults. The effect may extend from a few kilometres off-axis to as much as 35 km away, depending on the initial width of the axial high (Figure 7.17).

As faults grow and lengthen they may approach one another. When this happens, the force promoting propagation increases and changes direction, causing the faults to overlap, hook toward each other, and eventually join (Pollard and Aydin, 1984). Strain progressively accumulates onto fewer, longer faults with greater offsets. This pattern is clearly seen in

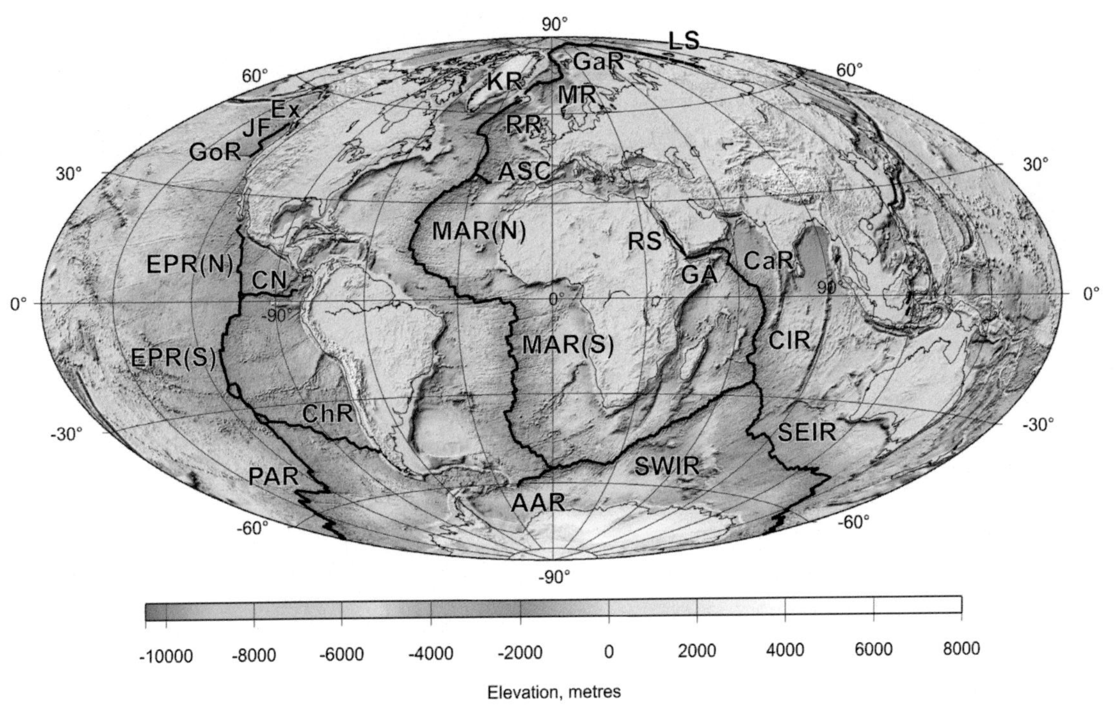


**Figure 1.1** World topography with mid-ocean ridges superimposed (heavy black lines). Back-arc spreading centres have been omitted for clarity. AAR, American–Antarctic Ridge; ASC, Azores Spreading Centre; CaR, Carlsberg Ridge; ChR, Chile Rise; CIR, Central Indian Ridge; CN, Cocos-Nazca Spreading Centre; EPR, East Pacific Rise (north and south); Ex, Explorer Ridge; GA, Gulf of Aden; GoR, Gorda Ridge; GaR, Gakkel Ridge; JF, Juan de Fuca Ridge; KR, Kolbeinsey Ridge; LS, Laptev Sea Rift; MAR, Mid-Atlantic Ridge (north and south); MR, Mohns Ridge; PAR, Pacific–Antarctic Rise; RR, Reykjanes Ridge; RS, Red Sea; SEIR, Southeast Indian Ridge; SWIR, Southwest Indian Ridge.

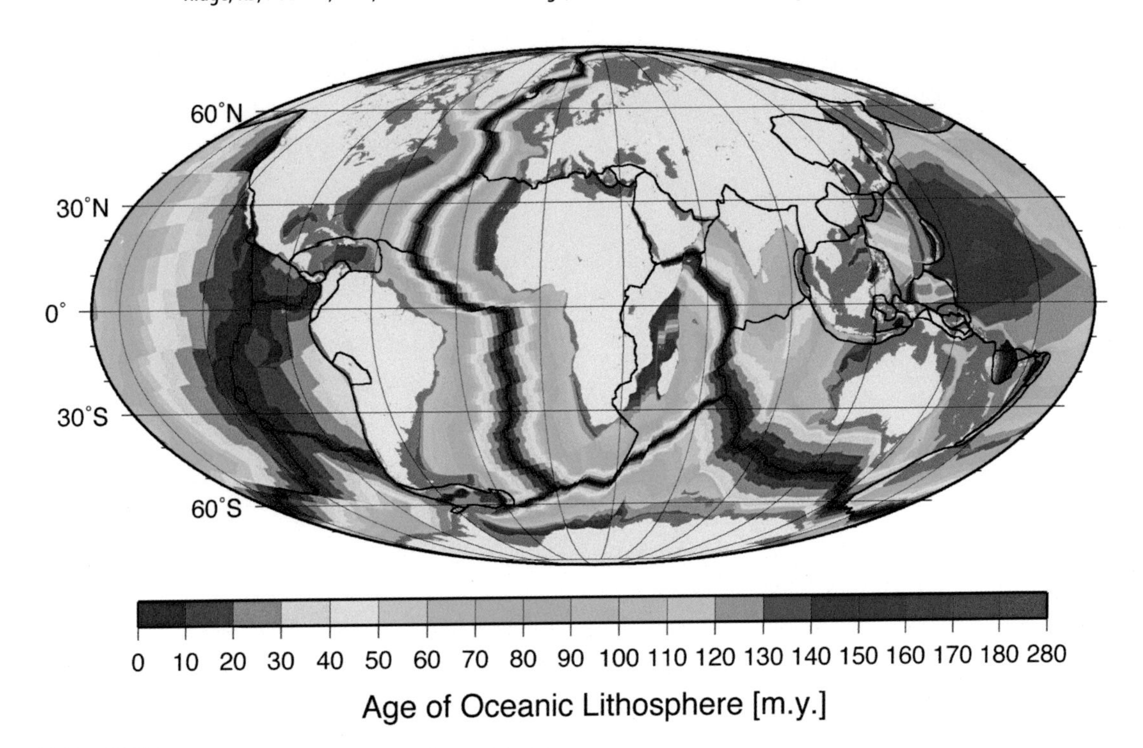


**Figure 1.8** Age of the ocean floor, after Müller *et al.* (2008).

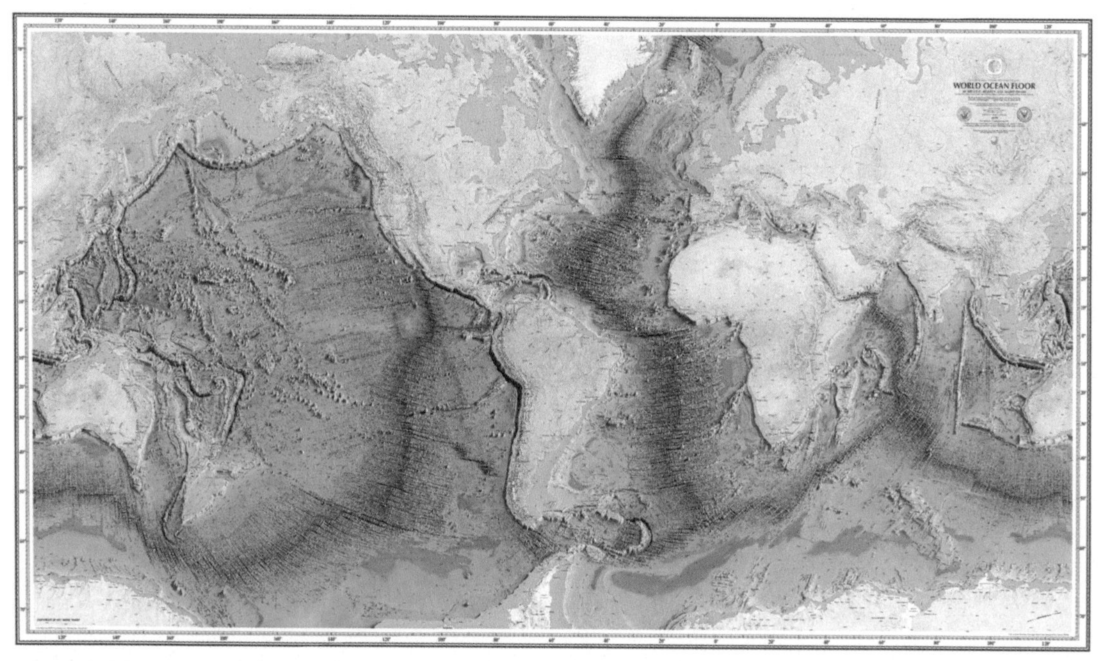


Figure 2.3 World Ocean Floor Panorama by Bruce C. Heezen and Marie Tharp (1977). Copyright by Marie Tharp 1977/2003. Reproduced by permission of Marie Tharp Maps, LLC 8 Edward Street, Sparkill, New York 10976.

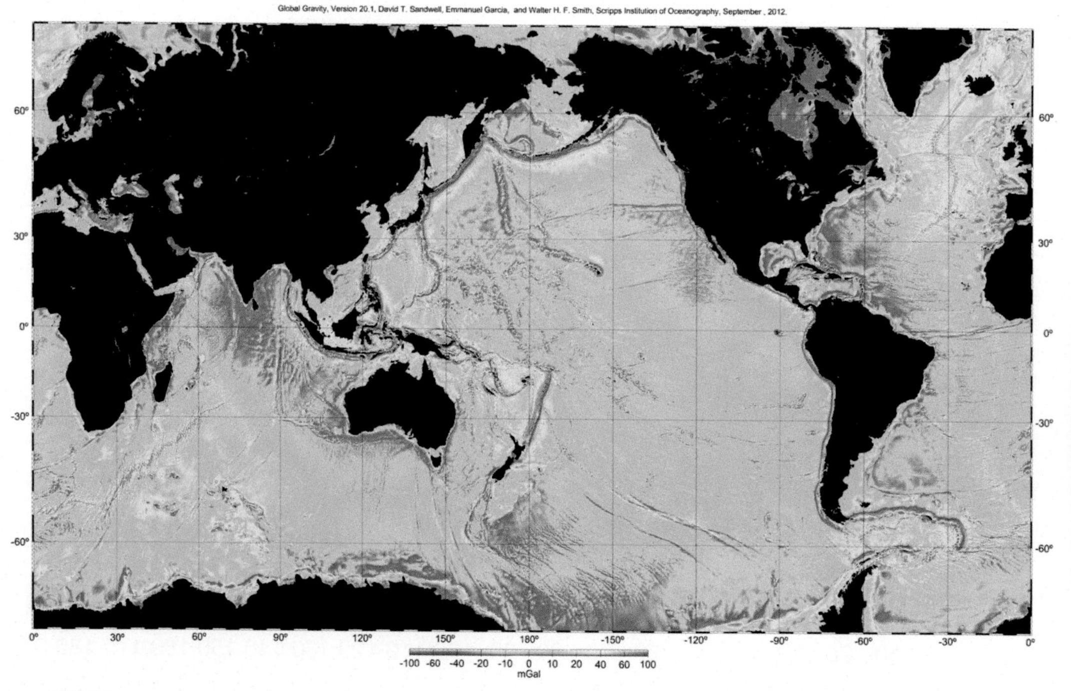


Figure 2.9 Global free-air gravity field obtained from satellite altimetry, after Sandwell and Smith (2009). Image courtesy of David Sandwell.

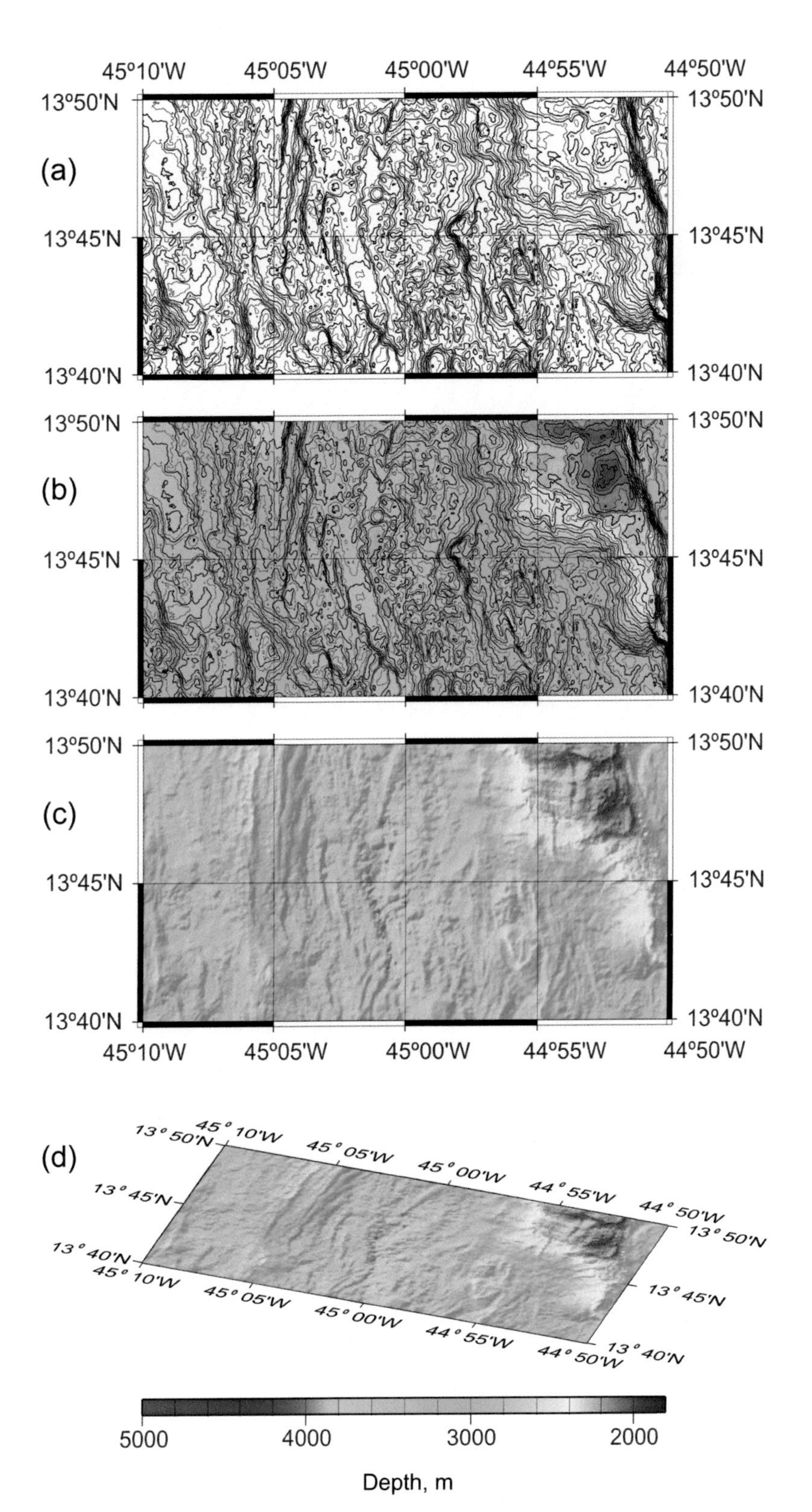


Figure 2.5 Examples of depth visualisation using multi-beam sonar, showing part of the MAR axis: (a) simple contour plot, with 50 m contour interval; (b) the same contours with colour-coded depth; (c) colour-coded shaded relief image with illumination from the NW; (d) same data as (c) displayed in an oblique view.

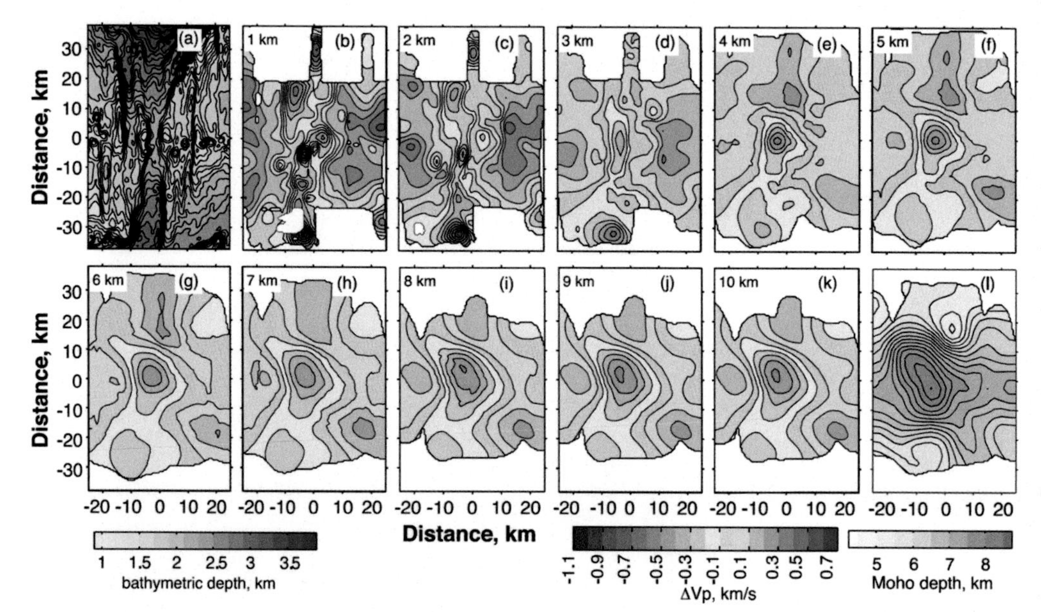


**Figure 2.14** Three-dimensional (3D) seismic tomographic model of a segment of the MAR near 35° N, after Dunn *et al.* (2005). (a) Bathymetry of the segment. (b) to (k) Two-dimensional (2D) horizontal sections through the model at 1 km depth increments, showing computed departure of P-wave velocity from a simple one-dimensional (1D) starting model. (l) Crustal thickness inferred from the model.

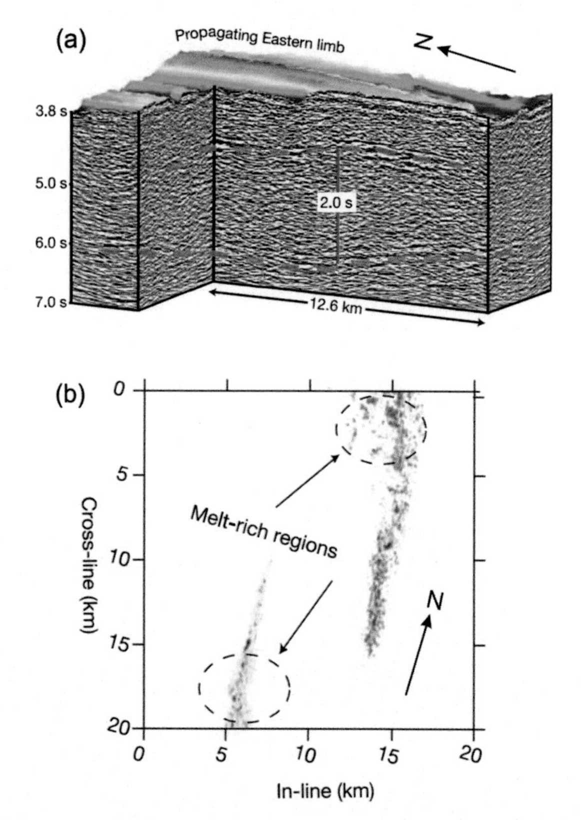


**Figure 2.18** Three-dimensional (3D) seismic image of the EPR overlapping spreading centre near the 9° N, after Singh *et al.* (2006b). (a) Section through the data volume. Coloured upper surface is sea floor topography. Dashed orange lines indicate reflections from axial magma chamber (upper line) and base of crust (lower line). (b) Amplitude map of magma chamber reflections, emphasising positions of melt-rich regions. Reprinted by permission from Macmillan Publishers Ltd: *Nature* (Singh, S. C., Harding, A. J., Kent, G. M. *et al.*, vol. **442**, pp. 287–290), copyright 2006.

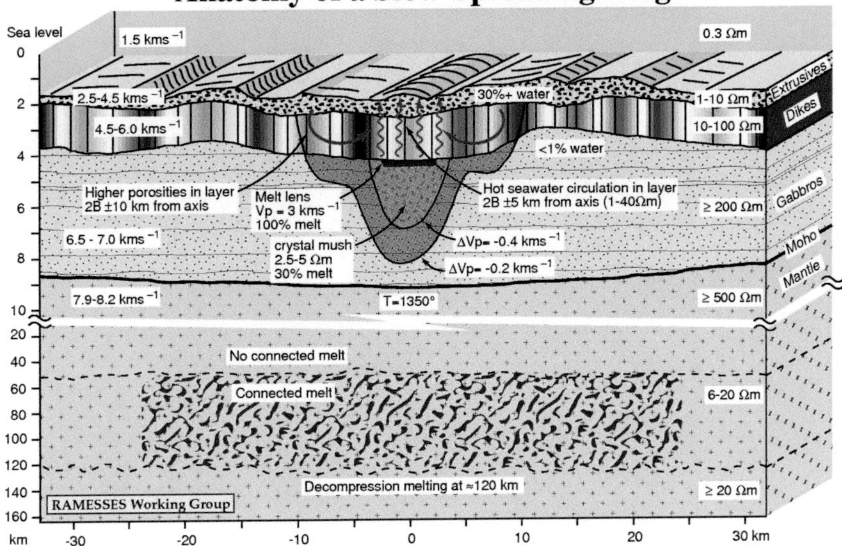


Figure 2.21 Lithospheric structure and melt distribution beneath the Reykjanes Ridge from a combination of controlled-source electromagnetic sounding, seismic refraction and reflection and other geophysical data, after Sinha *et al.* (1998, Figure 5). Note change of depth scale at 10 km.

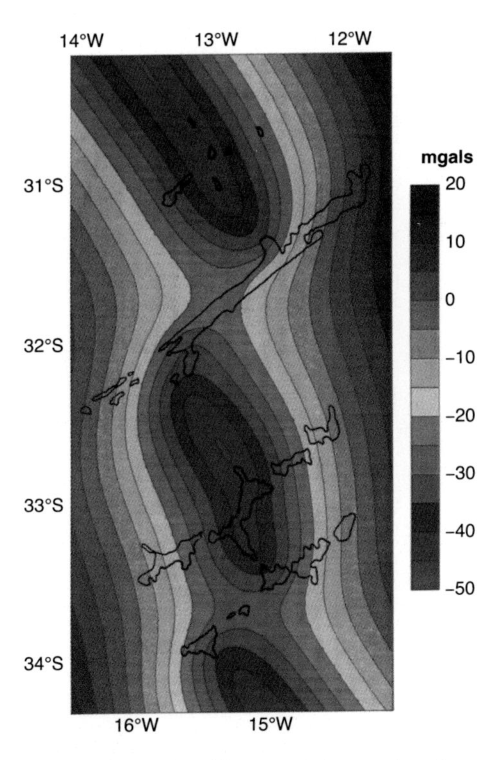


Figure 3.12 Predicted gravitational effect of the thermally induced density variations resulting from passive mantle upwelling under part of the southern MAR, after Kuo and Forsyth (1988, Figure 8) with kind permission from Springer Science and Business Media. Bold contours indicate 3600 m depth and partially outline the NW–SE spreading centres and their SW–NE transform offsets and associated fracture zones.

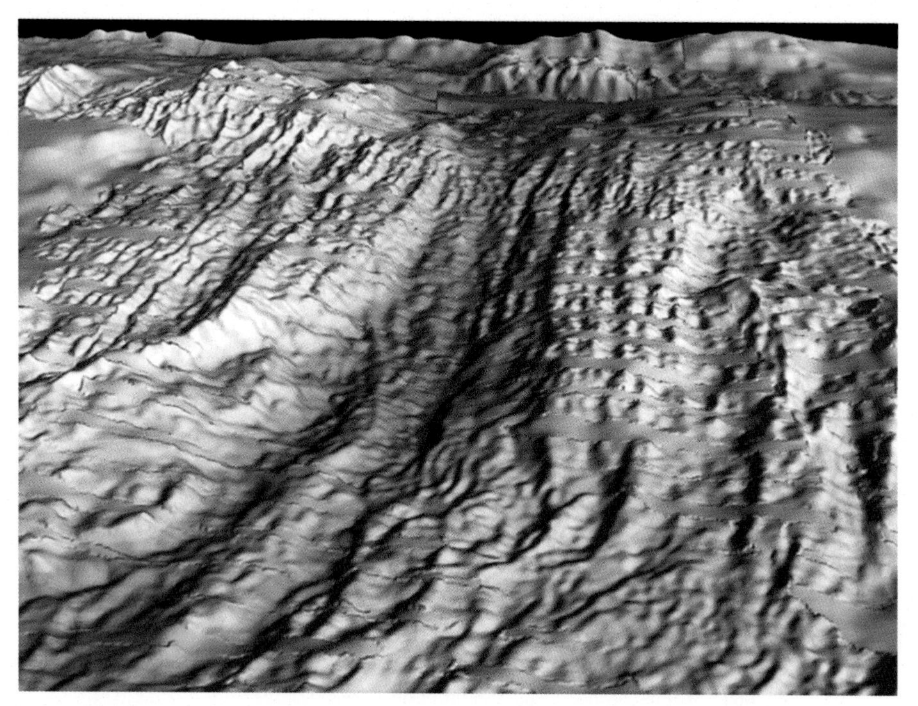

**Figure 4.5** Shaded relief image of an ~100 km length of the MAR median valley at 24° N, looking south towards the Kane transform fault, seen in the distance. Vertical exaggeration 2×. Width of image ~20 km in foreground, 40 km at middle distance. Depth ranges from about 2600 m (white) to 5000 m (purple). Image produced using GeoMapApp (http://www.geomapapp.org; Ryan *et al.*, 2009).

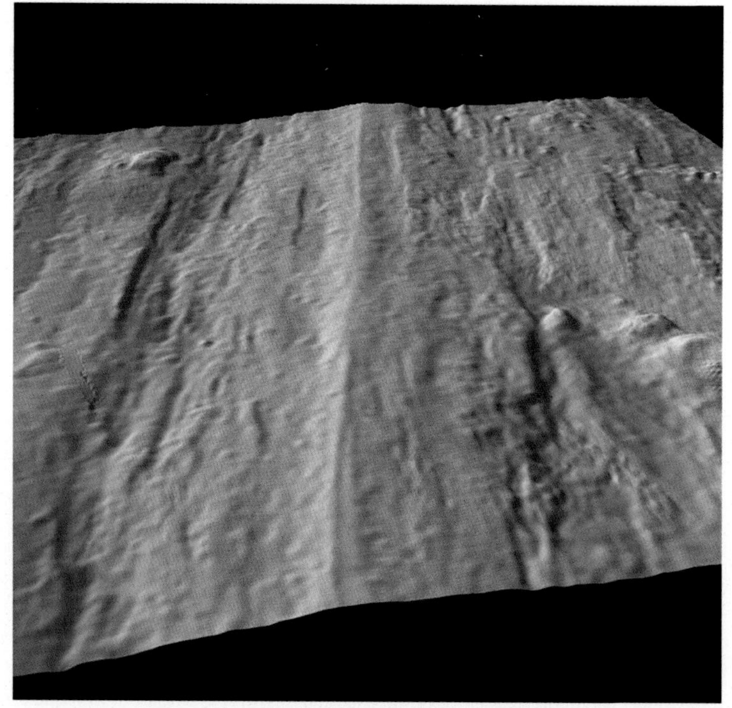

**Figure 4.6** Shaded relief image of an ~60 km length of the EPR axial high at 17°30′ S, looking north. Vertical exaggeration 2×. Width of image ~35 km. Depth ranges from about 2600 m (pale grey) to 3100 m (purple). The crestal high can be seen as a narrow ridge in the centre of the pale grey strip. Image produced using GeoMapApp (http://www.geomapapp.org; Ryan *et al.*, 2009).

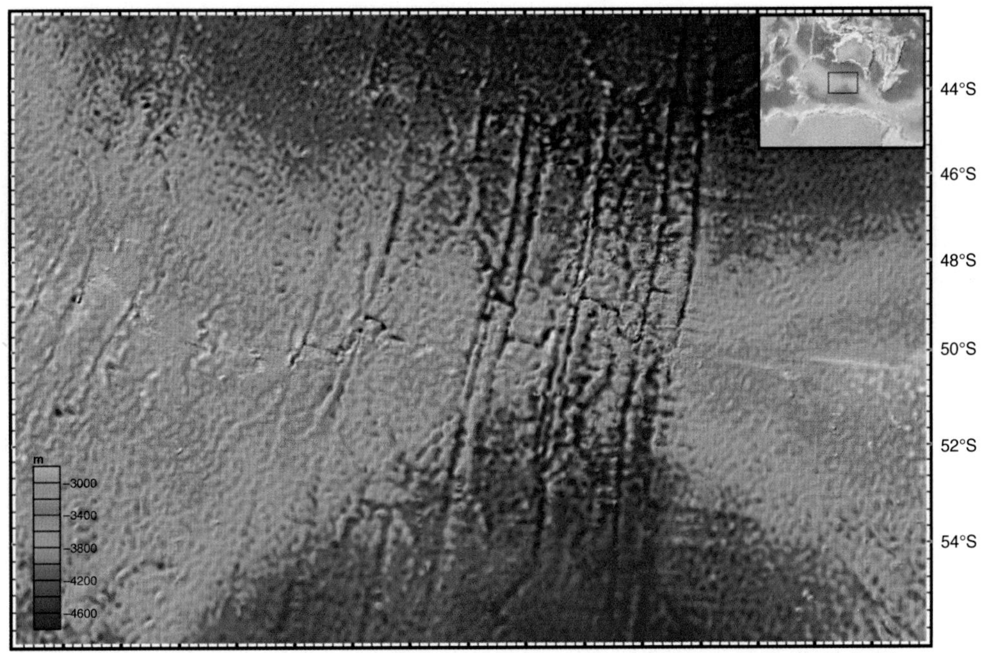


**Figure 4.8** Bathymetry of the Australia–Antarctic Discordance, showing changes in along-axis character from axial highs to median valleys. Inset shows location of main figure. Image produced using GeoMapApp (http://www.geomapapp.org; Ryan *et al.*, 2009).

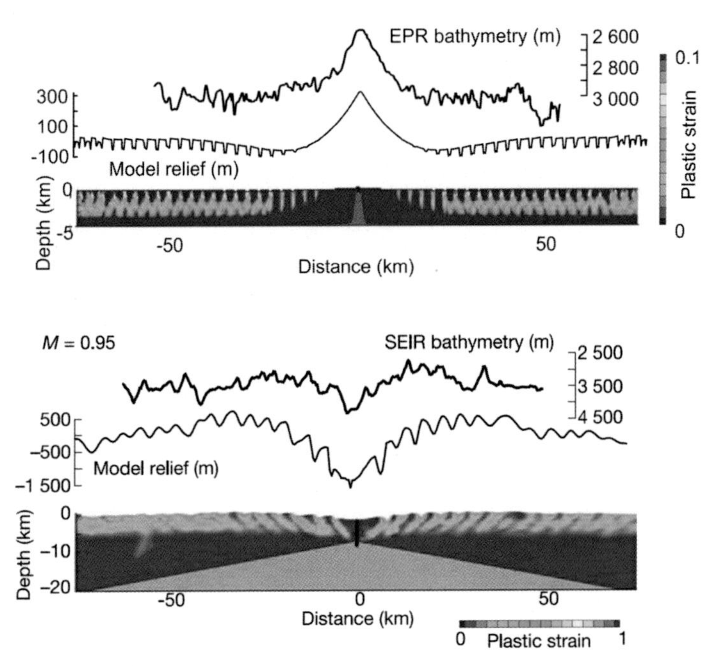


**Figure 4.10** Numerical models of an elastic–plastic–viscous lithosphere overlying a weak asthenosphere for fast (top) and slow (bottom) MORs, after Buck *et al.* (2005). Coloured sections show computed strain in vertical, ridge-normal cross-sections. Thin black curves show sea floor topography computed from the models, compared with actual topographic profiles (bold curves). The fast-spreading case is dominated by axial buoyancy, while the slow-spreading case is dominated by stretching. Note the two sets of profiles have different vertical exaggerations. Reprinted by permission from Macmillan Publishers Ltd: *Nature* (Buck *et al.*, vol. **434**, pp. 719–723), copyright 2005.

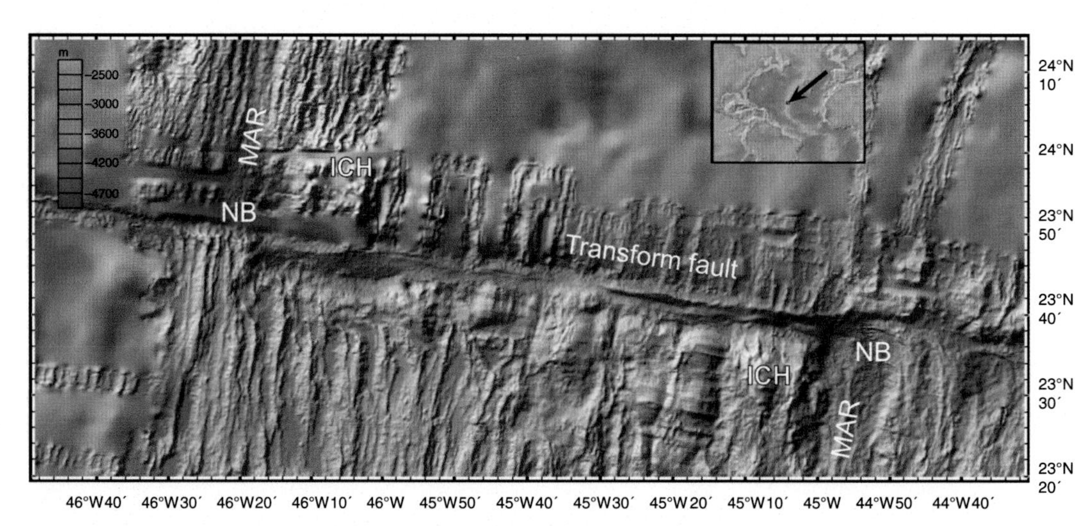


**Figure 4.12** Shaded relief image of the slow-slipping, 150 km offset Kane transform fault, MAR. Inset shows location of main figure. ICH, Inside Corner High; MAR, MAR spreading axis; NB, Nodal Basin. Image produced using GeoMapApp (http://www.geomapapp.org; Ryan *et al.*, 2009).

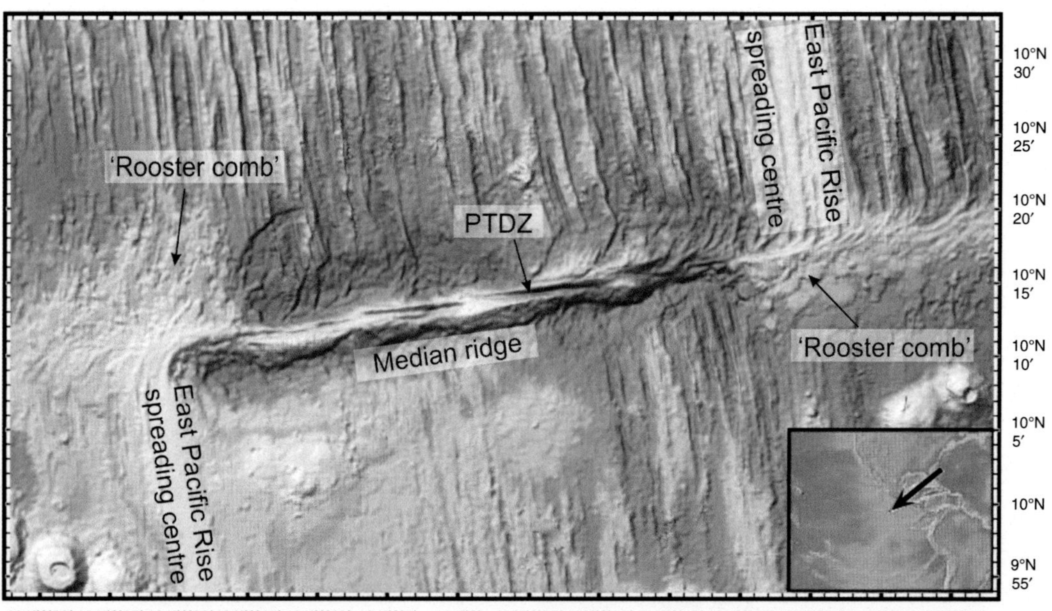


**Figure 4.13** Shaded relief image of the fast slipping, 85 km offset Clipperton transform fault, EPR. Note prominent median ridge with superimposed scarps and troughs marking the principal transform displacement zone (PTDZ). Shallow 'rooster comb' areas opposite active spreading centres are thought to reflect reheating of older lithosphere. Inset shows location of main figure. Image produced with GeoMapApp (http://www.geomapapp.org; Ryan *et al.*, 2009).

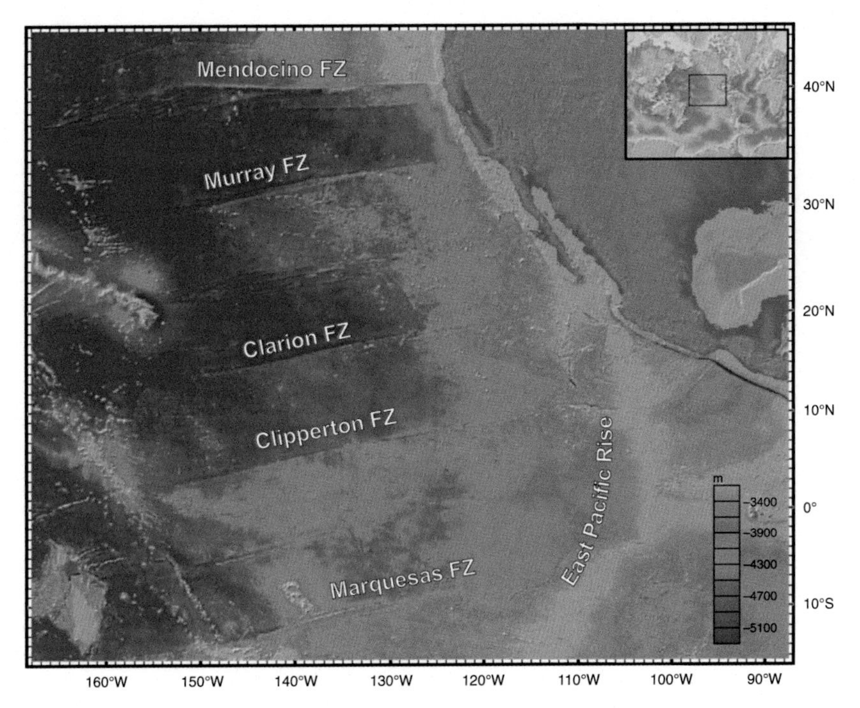


**Figure 4.17** Bathymetry of the north-eastern Pacific Ocean, showing the great Pacific fracture zones with their large topographic scarps. Inset shows location of main figure. Image produced using GeoMapApp (http://www.geomapapp.org; Ryan *et al.*, 2009).

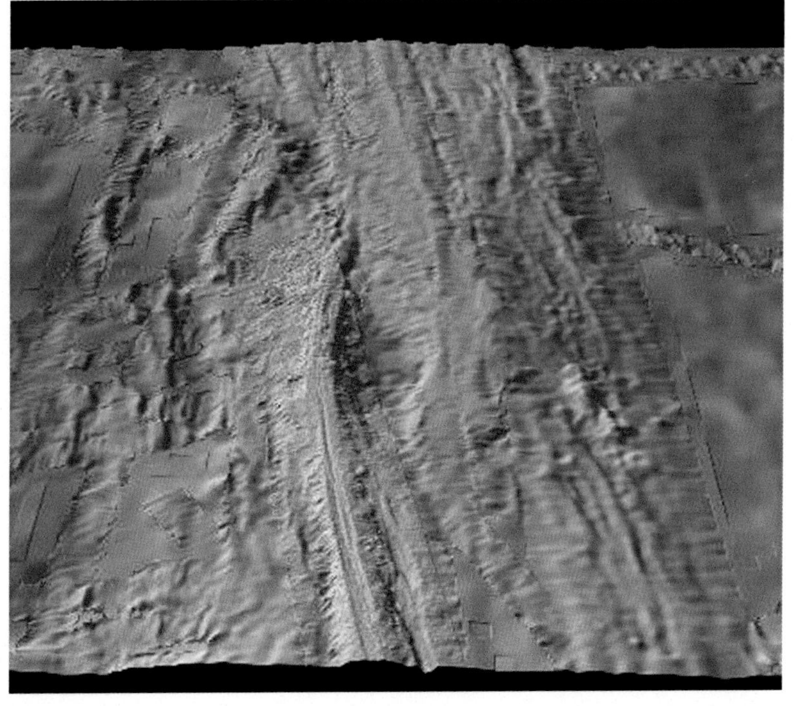

**Figure 4.19** Oblique, shaded relief image of an overlapping spreading centre, EPR 11°45′ N, viewed towards the north. Depths range from approximately 2600 m (grey) to 3100 m (purple). Image is approximately 50 km wide at middle distance, and E–W gap between overlapping spreading centres is approximately 10 km at the widest point. Image produced using GeoMapApp (http://www.geomapapp.org; Ryan *et al.*, 2009).

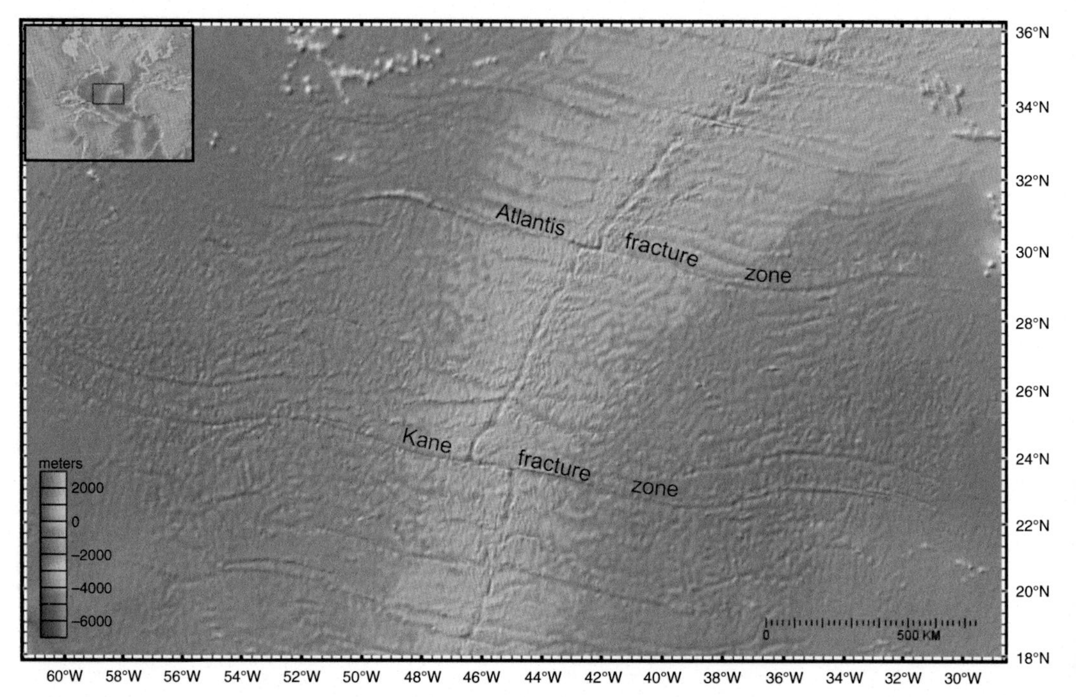


**Figure 4.21** Shaded relief bathymetry of the central North Atlantic, showing off-axis traces of transform faults and non-transform offsets. Fracture zone traces show an early history of WNW–ESE spreading, followed by a period of WSW–ENE, then reverting to WNW–ESE. Traces of some but not all non-transform offsets approximately follow these trends, while others migrate along-axis. Image produced using GeoMapApp (http://www.geomapapp.org; Ryan *et al.*, 2009).

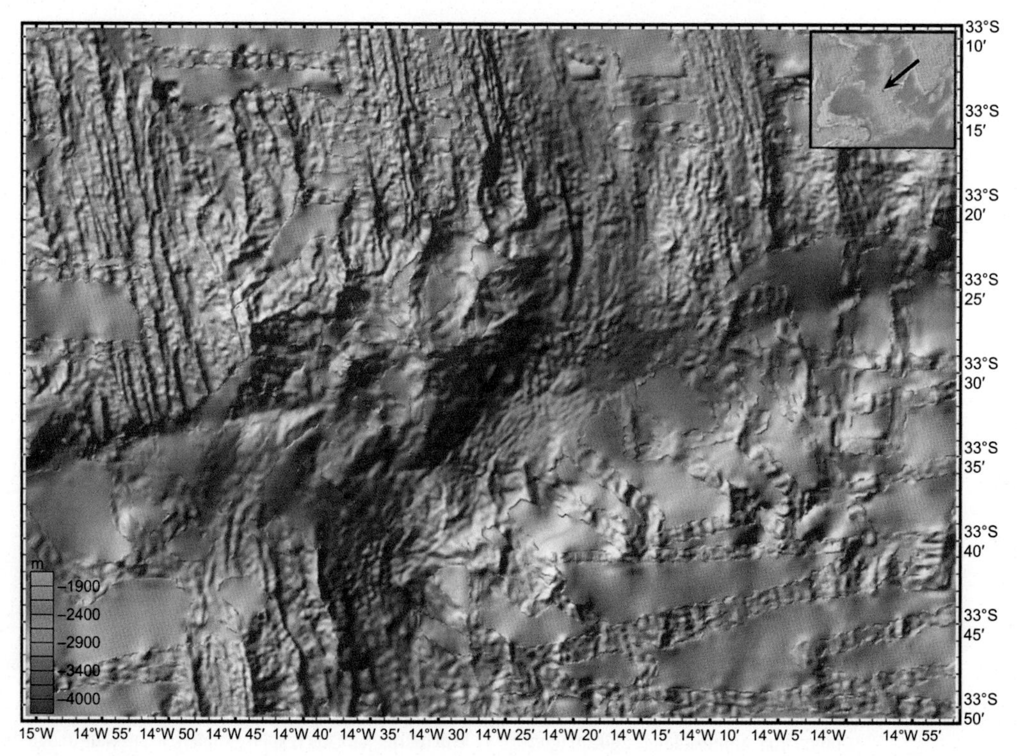


**Figure 4.23** Bathymetry of a non-transform offset: MAR at 33°30′ S. Inset shows location. Image is approximately 120 km wide. Spreading centre at 14°35′ W in the south is offset approximately 30 km to the NE at 14°20′ W. Image produced using GeoMapApp (http://www.geomapapp.org; Ryan *et al.*, 2009).

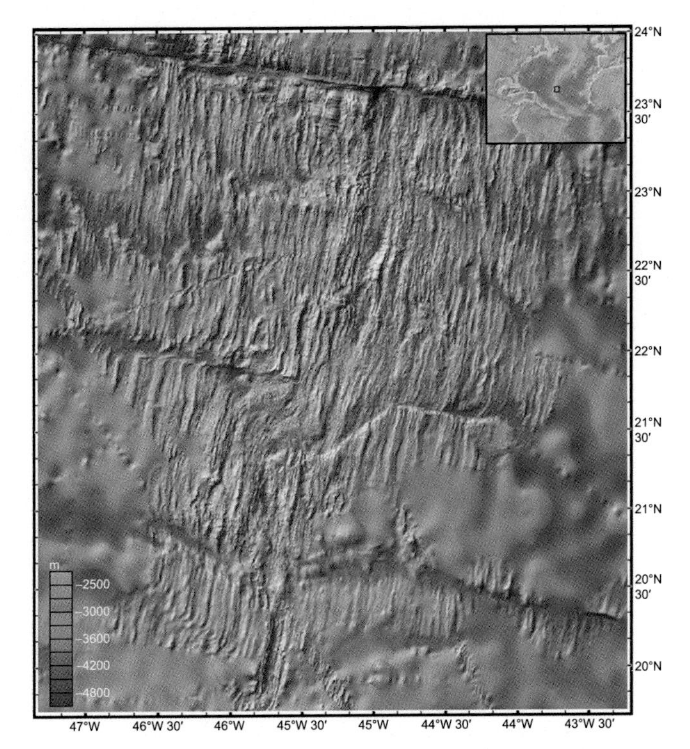


**Figure 4.24** Shaded relief bathymetry of MAR 20° N–24° N, showing oblique traces of expanding segments and migrating non-transform offsets. Note that V-shaped wakes imply migration of offsets away from a centre near 22°30′ N, and that an erstwhile transform near 21°30′ N has recently converted to a southward-migrating non-transform offset. Image width approximately 400 km. Image produced using GeoMapApp (http://www.geomapapp.org; Ryan *et al.*, 2009).

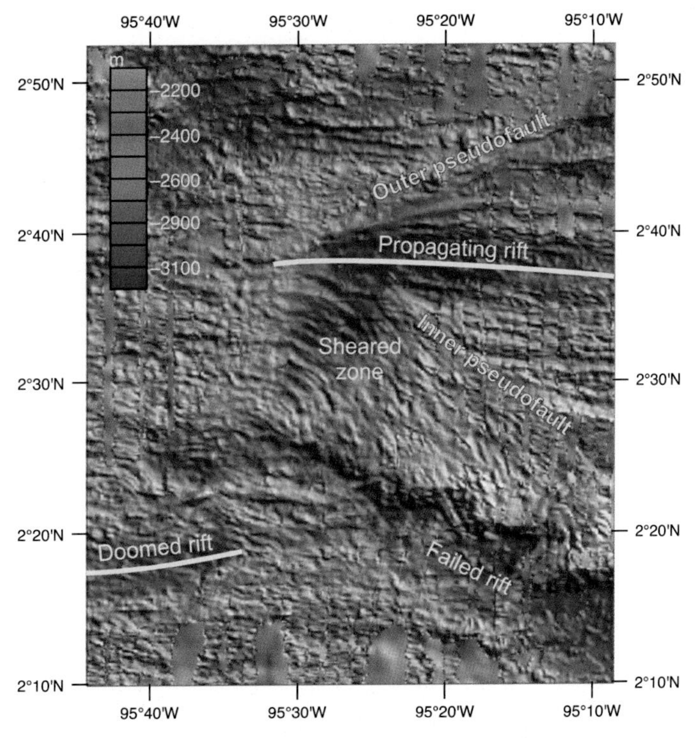


**Figure 4.28** Shaded relief bathymetry of the propagating rift at 95° W on the Cocos–Nazca Spreading Centre, illuminated from NE. Yellow lines mark the propagating and doomed rift axes. Image produced using GeoMapApp (http://www.geomapapp.org; Ryan *et al.*, 2009).

**Figure 5.12** Lavas and sediments near MOR axes: (a) on the EPR, after CYAMEX (1981), with kind permission from Springer Science and Business Media, and (b–d) on the MAR at 45° N, reprinted from Searle *et al.* (2010), copyright 2010, with permission from Elsevier. (a) Bare sheet flows; (b) bare pillow lavas less than a few thousand years old; (c) pockets of sediment surrounding pillow lavas on crust several thousand years old; (d) approximately 1 m of sediment on ~20 ka crust revealed by fissuring.

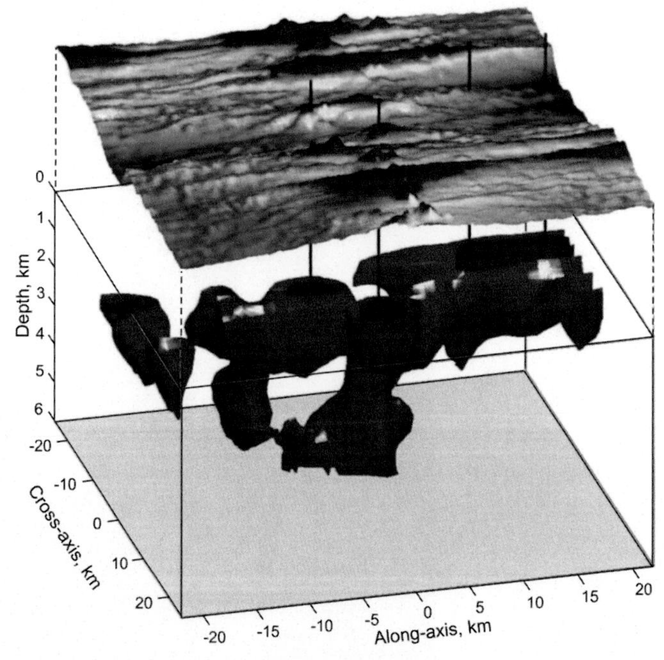


**Figure 5.19** Three-dimensional view of sea floor topography overlying seismic low-velocity anomalies at MAR 35° N, after Magde *et al.* (2000). Reprinted from *Earth and Planetary Science Letters*, vol. **175**, pp. 55–67, copyright 2000, with permission from Elsevier. The top 1 km of the model has been removed for clarity. Dark and light red surfaces: boundaries of $-0.5$ km s$^{-1}$ and $-0.3$ km s$^{-1}$ velocity anomalies, respectively, derived from tomographic inversion. Vertical black lines reference two sea floor volcanoes (centre) and the axial volcanic ridge (right).

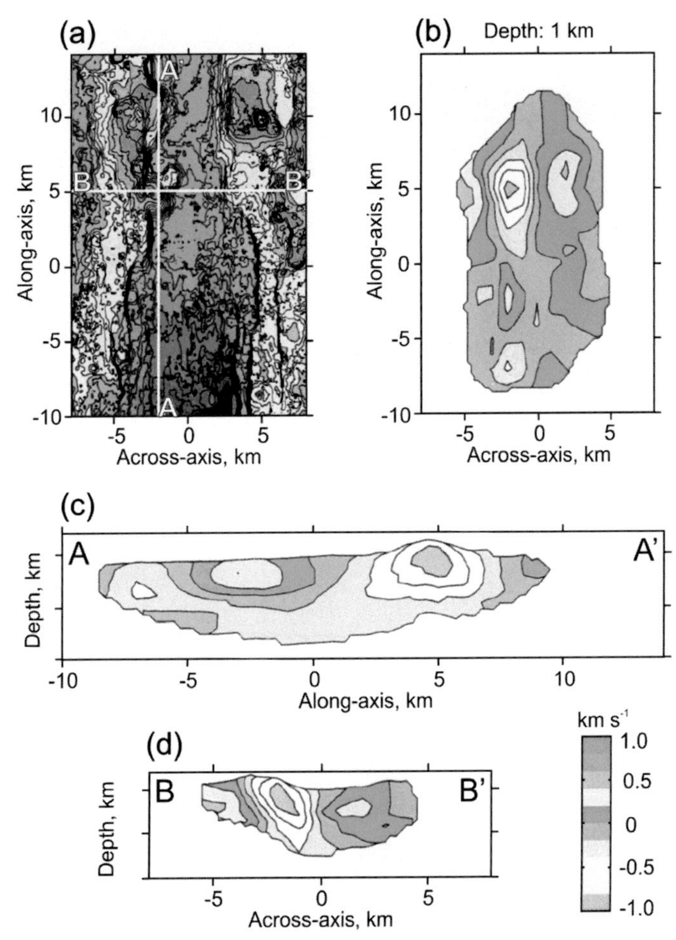


**Figure 5.13** Seismic tomographic model of the upper crust at the centre of the OH-1 segment, MAR, 34°50′ N, after Barclay *et al.* (1998). (a) Bathymetry; blue central area is the median valley floor. (b) Horizontal slice, (c) longitudinal vertical section along west side of median valley, and (d) vertical cross-section across the valley, showing departures in P-wave velocity from a 1D reference velocity model.

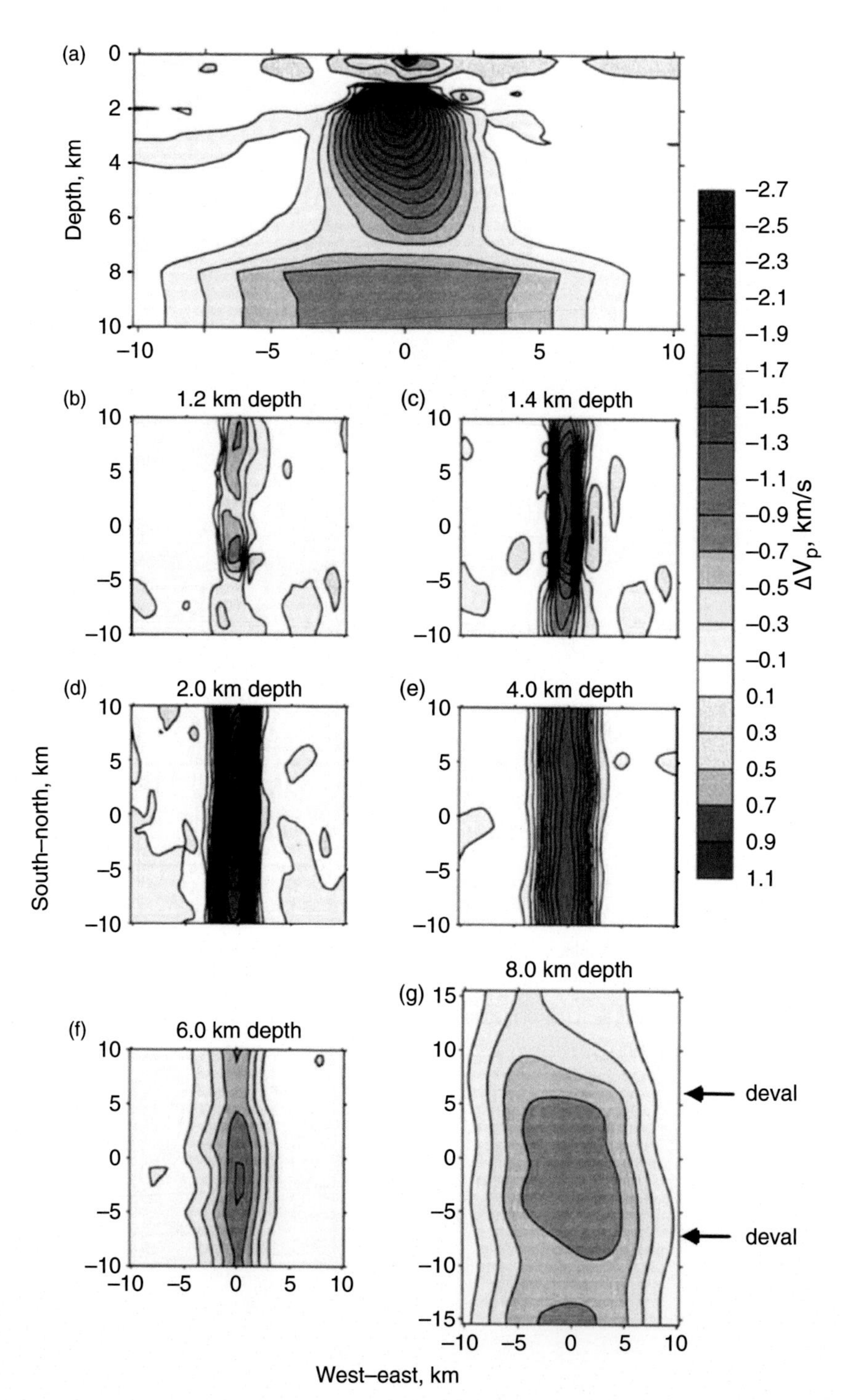


**Figure 5.14** Sections through a 3D tomographic model of seismic velocity under the EPR near 9°30′ S, after Dunn *et al.* (2000). (a) Across-axis; (b–g) horizontal slices at increasing depths. Contours show difference in P-wave velocity from a 1D average model.

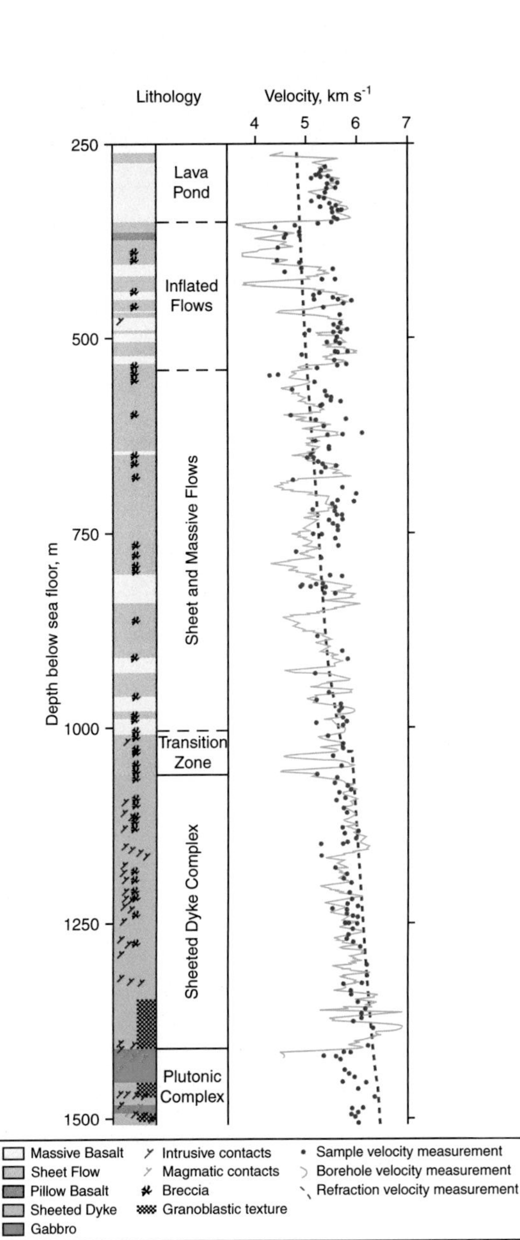


**Figure 5.25** Summary lithological column from IODP hole 1256D drilled into superfast-spread crust, after Wilson *et al.* (2006). Left: lithologies; right: regional seismic refraction model (red dashed line), wire-line logged seismic velocity (green line) and velocity measurements on cored samples (blue dots). Top of column is at base of sedimentary layer. From Wilson, D. S., Teagle, D. A. H., Alt, J. C. *et al.* (2006), Drilling to gabbro in intact ocean crust, *Science*, vol. **312**, pp. 1016–1020. Reprinted with permission from AAAS.

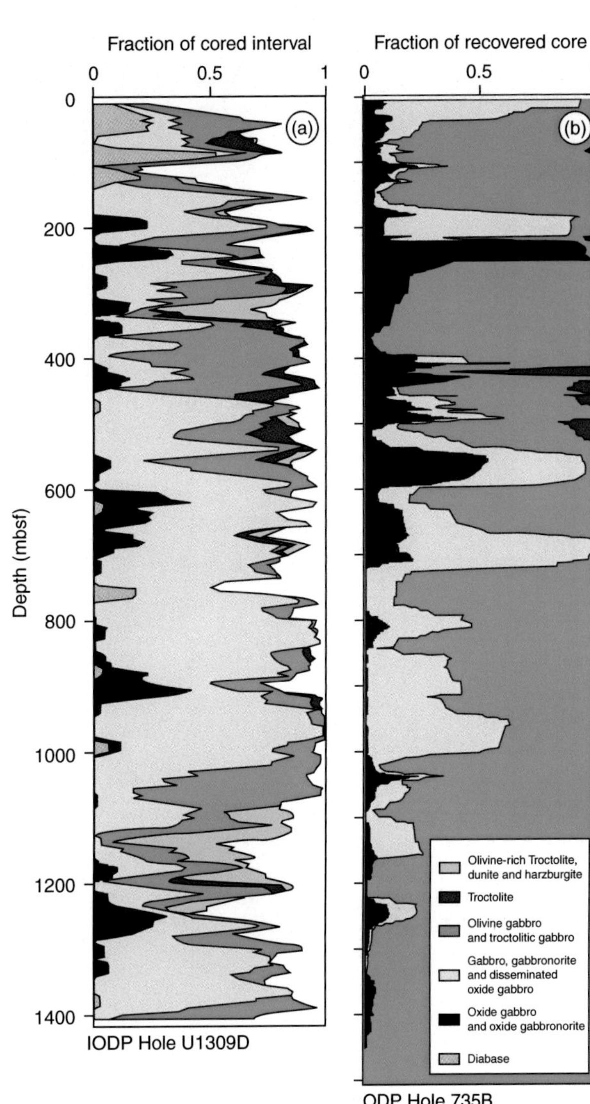


**Figure 5.27** Summary lithological columns from (a) IODP hole 1309D, slow-spreading MAR, and (b) ODP hole 735B, ultra-slow-spreading SWIR, after Ildefonse *et al.* (2007). The extrusive and sheeted dyke sections are absent and there is negligible sediment cover.

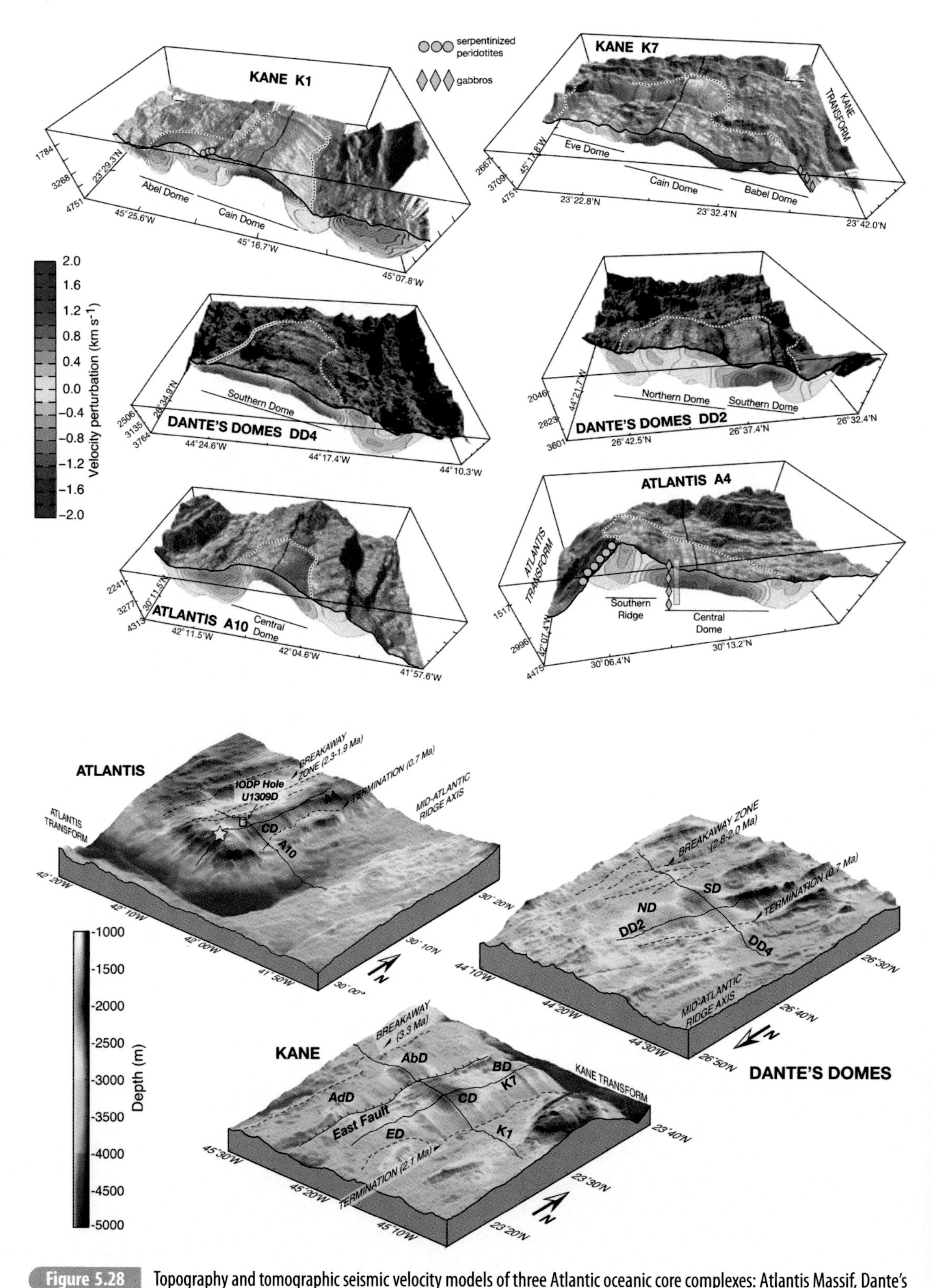


**Figure 5.28** Topography and tomographic seismic velocity models of three Atlantic oceanic core complexes: Atlantis Massif, Dante's Domes and Kane OCC, after Canales *et al.* (2008). Top: sea floor topography shown in shaded relief (grey) above coloured 2D tomographic sections showing velocity anomalies in shades of red to blue. Spreading-parallel sections (E–W) are shown on the left and axis-parallel ones (N–S) on the right. Bottom: coloured topographic images showing locations of seismic sections (black lines). Dotted lines mark edges of smooth, corrugated detachment surfaces. Yellow star: Lost City hydrothermal field.

**Figure 6.5** Lava flow morphologies seen on MORs. Images (a–e) are on the MAR, 45° N, from the author; (f) and (h) are from Axial Seamount, Juan de Fuca Ridge, 46° N, after Fox *et al.* (1988); (g) is from the EPR, 21° N, after CYAMEX (1981); (i) is from the Galapagos Ridge, 86° W, after Ballard *et al.* (1979). All images are ~2–3 m across. (a) Pillow lavas; (b) elongate pillows; (c) highly elongate pillows draping a 'haystack'; (d) flattened lobate flow; (e) collapsed lobate flow; (f) jumbled sheet flow; (g) folded pahoehoe; (h) ropey sheet flow; (i) lava pillar. Images (f) and (h) reprinted from *Marine Geology*, vol. **78**, pp. 199–216, copyright 1988, with permission from Elsevier; (g) reproduced with kind permission of Springer Science and Business Media.

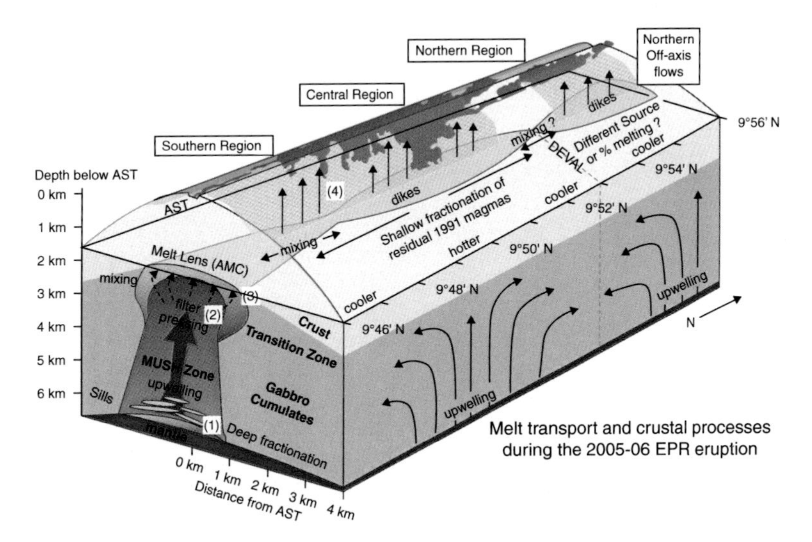


**Figure 6.13** Magmatic model of the 2005–2006 EPR eruption at 9°50′ N, after Goss *et al.* (2010). AMC: axial magma chamber; AST: axial summit trough. Dark grey shows extent of the eruption. Wide red arrow and curved black arrows show directions of upwelling melt from lower crustal sills (1), through crystal mush zone (2), to axial magma chamber (3), and subsequent eruption through dykes (vertical black arrows) to the surface (4). Long horizontal arrows indicate along-axis fractionation within AMC.

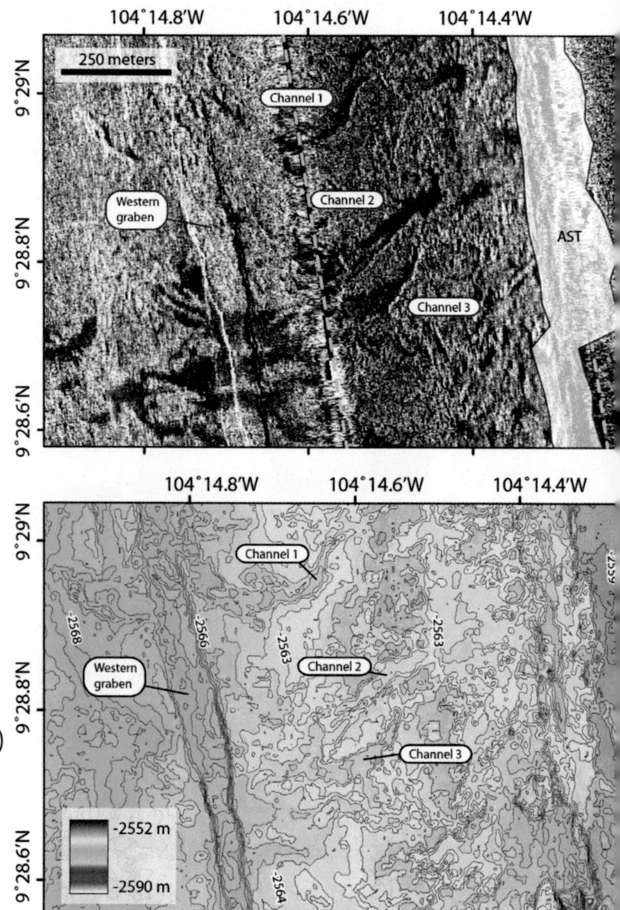


**Figure 6.14** Lava channels on the flank of the EPR, after Soule *et al.* (2005). Top, acoustic backscatter intensity from deep-towed side-scan sonar. Dark areas (low backscatter) are floors of lava channels composed of smooth sheet flows. Blue lines mark centres (nadir) of side-scan swathes. Red lines outline axial summit trough (AST). Bottom, microbathymetry of same area (contour interval 1 m) from Imagenex echosounder mounted on an AUV.

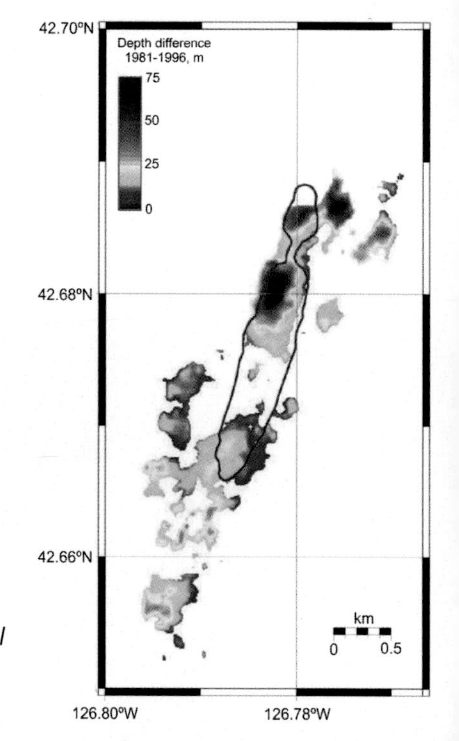


**Figure 6.18** Depth differences observed between 1981 and 1996 at the site of the 1996 Gorda Ridge eruption, after Chadwick *et al.* (1998). Red line indicates mapped outline of 1996 eruption. Reprinted from *Deep-Sea Research Part I – Topical Studies in Oceanography*, vol. **45**, pp. 2547–2569, copyright 1998, with permission from Elsevier.

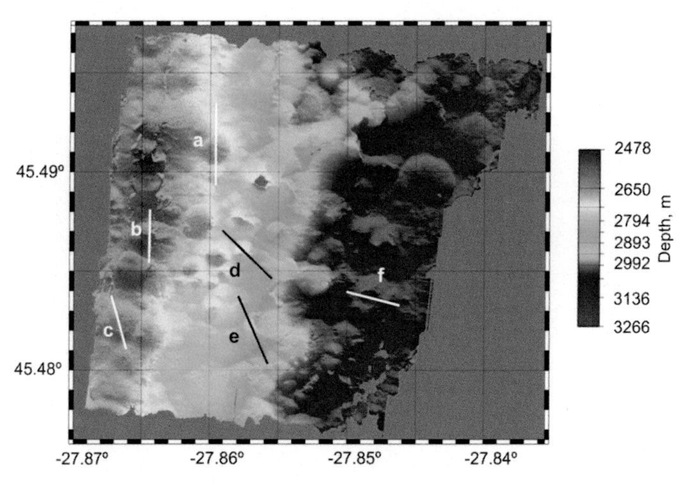


Figure 6.20 Top: high-resolution (~1 m) shaded-relief bathymetric map of part of the axial volcanic ridge flank at 45.5° N on the MAR, illuminated from the north. The AVR crest is at the left and the foot of its eastern flank on the right. Lettered lines: positions of profiles shown in text Figure 6.20.

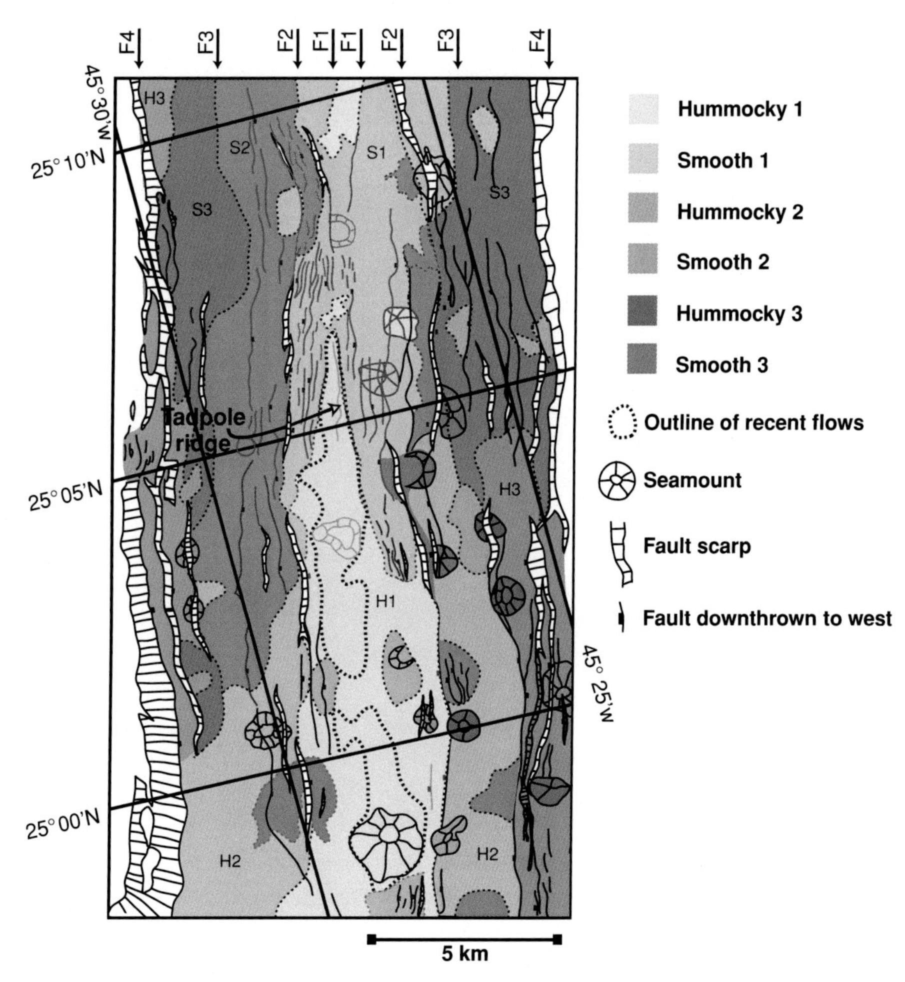


Figure 6.22 Geological interpretation of a TOBI side-scan sonar survey of the median valley floor on the MAR at 25° N, after Cann and Smith (2005). Inferred ages of faults (F) and hummocky and smooth sea floor are labelled 1 (youngest) to 4 (oldest).

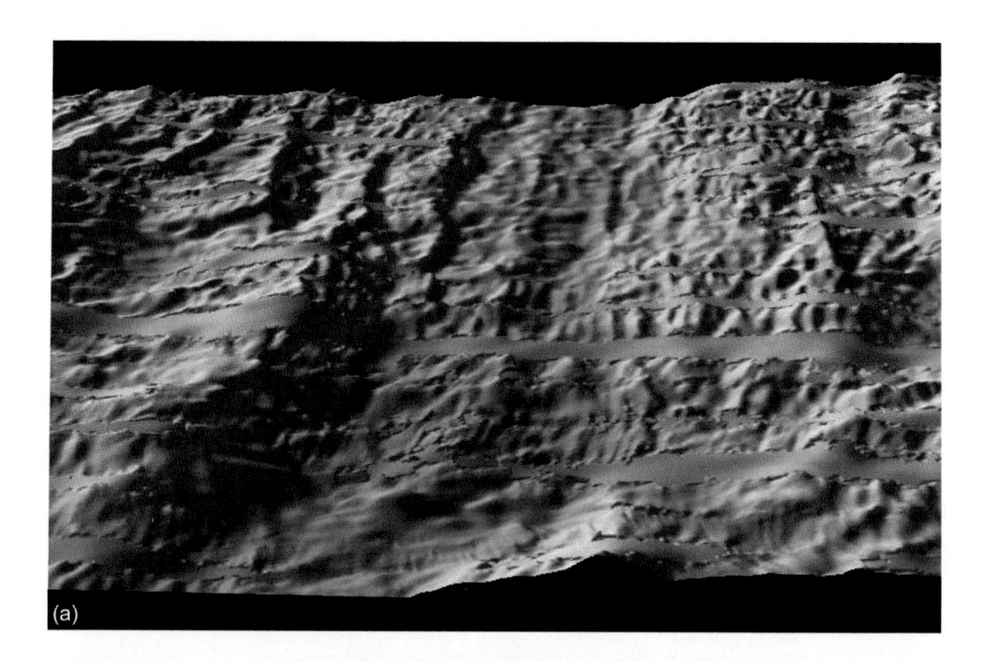


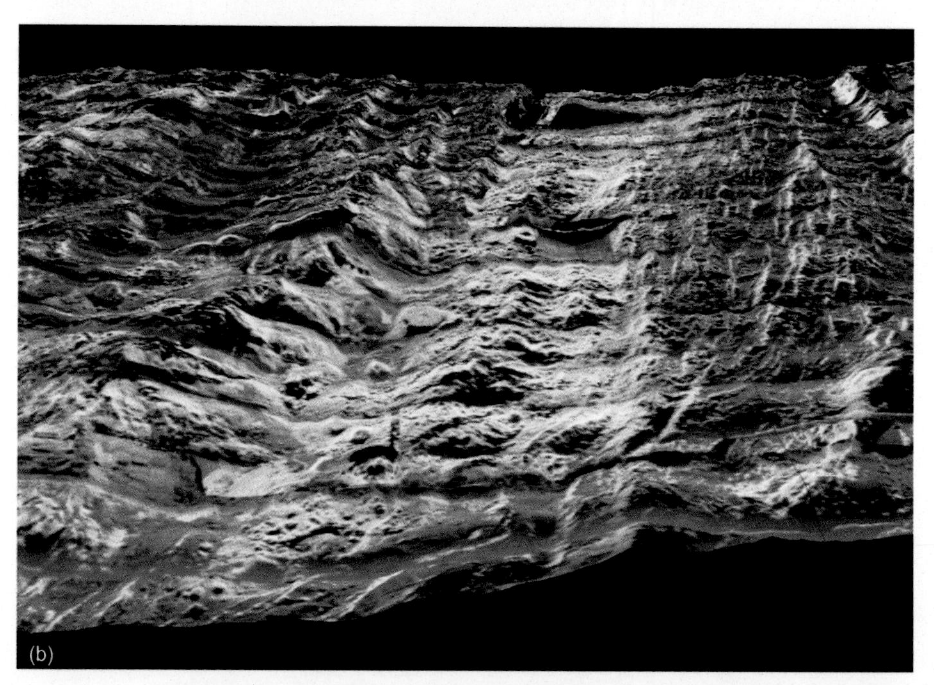


**Figure 6.24** Oblique views of the MAR median valley at 29° N, looking north. (a) Topography, from GeoMapApp (http://www.geomapapp.org; Ryan *et al.*, 2009). Depths range from approximately 2000 m (grey) to 4000 m (purple). (b) TOBI side-scan sonar mosaic of same area draped over bathymetry; no vertical exaggeration. Side-scan is insonified from the north. Images approximately 25 km wide at mid-range, 30 km front to back. Axial volcanic ridge runs N–S along centre of each image (pale blue in (a) composed mostly of hummocky and a few smooth and some flat-topped volcanoes. Major fault scarps bounding the median valley flank the AVR. Note AVR is slightly oblique to median valley, crossing from centre of valley (foreground) to western valley wall in far distance.

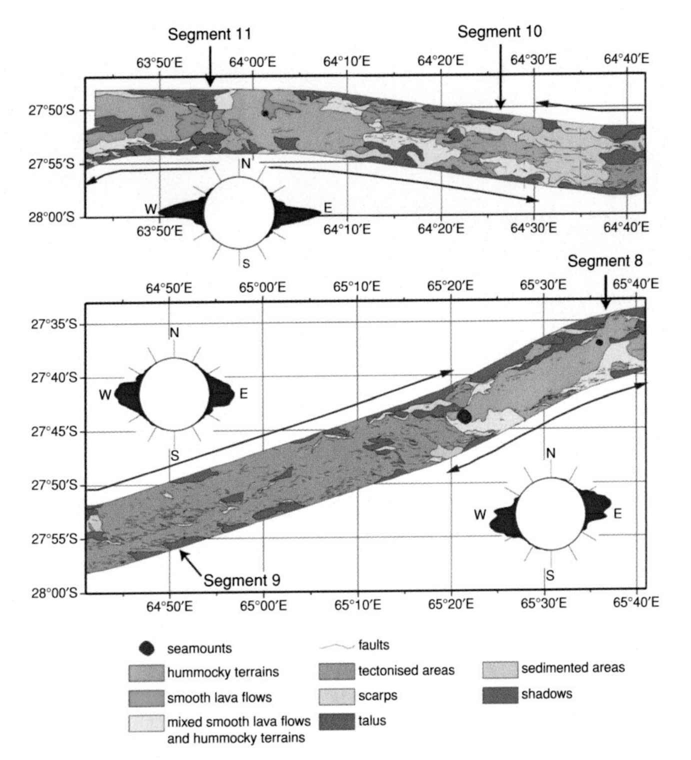


**Figure 6.26** Volcanic and tectonised terrains interpreted from TOBI side-scan sonar along part of the ultra-slow-spreading Southwest Indian Ridge axis, after Sauter *et al.* (2004). Rose diagrams show distribution of fault directions. Note highly tectonised segment 9 which entirely lacks recent volcanism, compared with nearby highly focussed volcanic segments 8 and 11. See also Figure 5.4.

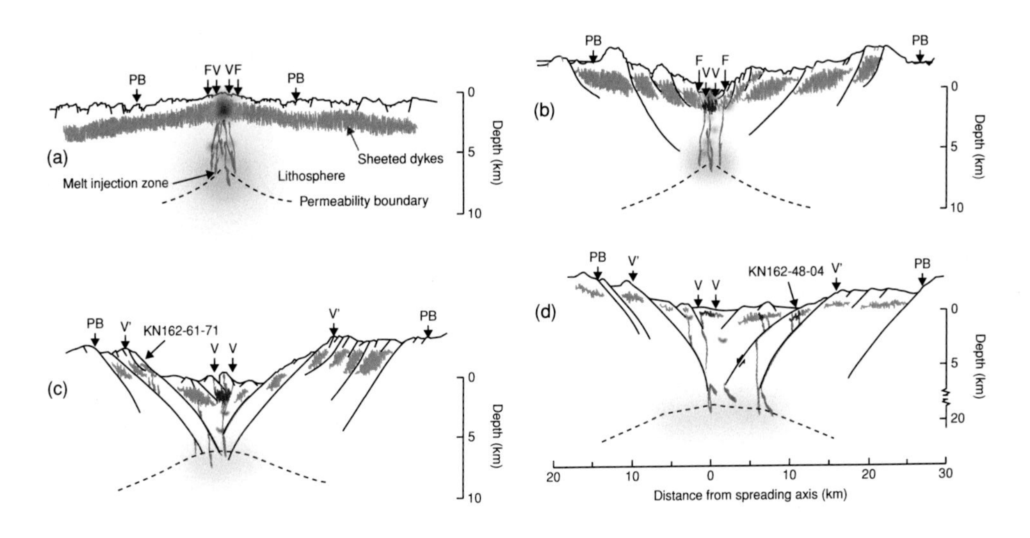


**Figure 6.28** Cartoons showing the cross-axis distribution of volcanism at different spreading rates, after Standish and Sims (2010). (a) Fast-spreading (55 km Ma$^{-1}$ half-rate) EPR, 3° S; (b) slow-spreading (12 km Ma$^{-1}$) MAR, 37° N; (c) magmatic segment of the ultra-slow-spreading (7 km Ma$^{-1}$) Southwest Indian Ridge, 14.5° E; (d) amagmatic segment of the Southwest Indian Ridge, 12.5° E. V: neo-volcanic zone; F: zone of fissuring; PB: plate boundary zone defined by active faulting; V′: crustal accretion zone; red shading: intensity and width of melt injection zone; grey shading: thickness of sheeted dykes; dashed line: notional permeability boundary to upwelling melt. Reprinted by permission from Macmillan Publishers Ltd: *Nature Geoscience*, vol. **3**, pp. 286–292, copyright 2010.

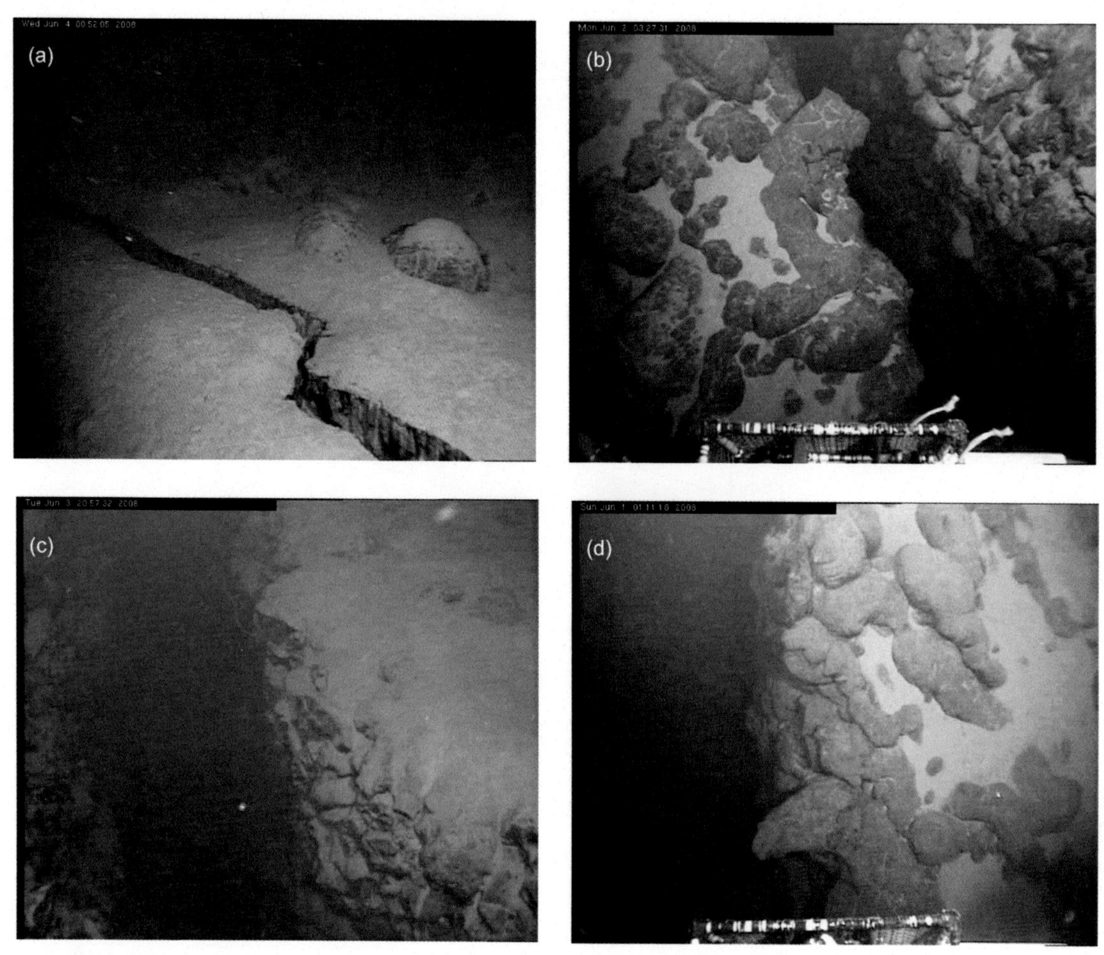


**Figure 7.1** Fissures from the MAR axis near 45°30′ N. (a) Narrow extensional fissure (offset ~7 cm) showing sharp, vertical, parallel sides; (b) 25 cm wide fissure; (c) ~1 m wide extensional fissure showing fractured pillow lavas in its walls. (d) Multiple lava buds (largest ~30 cm) extruding from side of eruptive fissure.

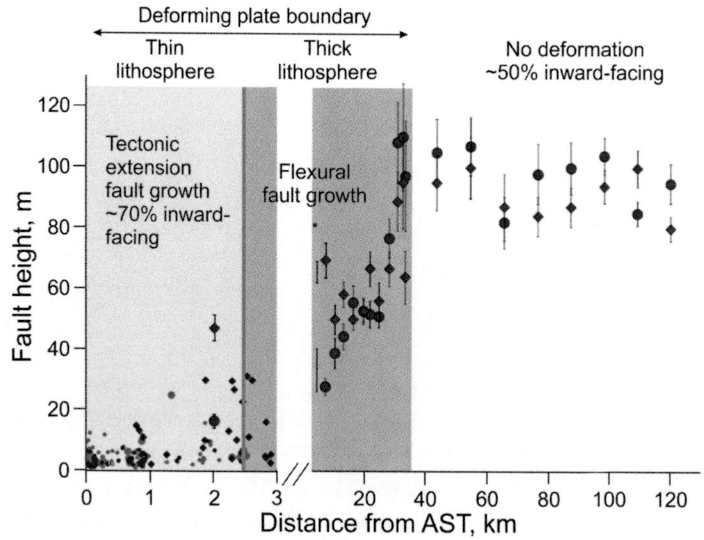


**Figure 7.17** The development of faulting on the EPR 9°–10° N, after Escartin *et al.* (2007). Red symbols: data from Alexander and Macdonald (1996); diamonds: inward facing faults; circles: outward facing faults. Note change in horizontal scale at 3 km. Main mechanism for fault growth in innermost 2.5 km is tectonic extension; beyond that, lithospheric flexure.

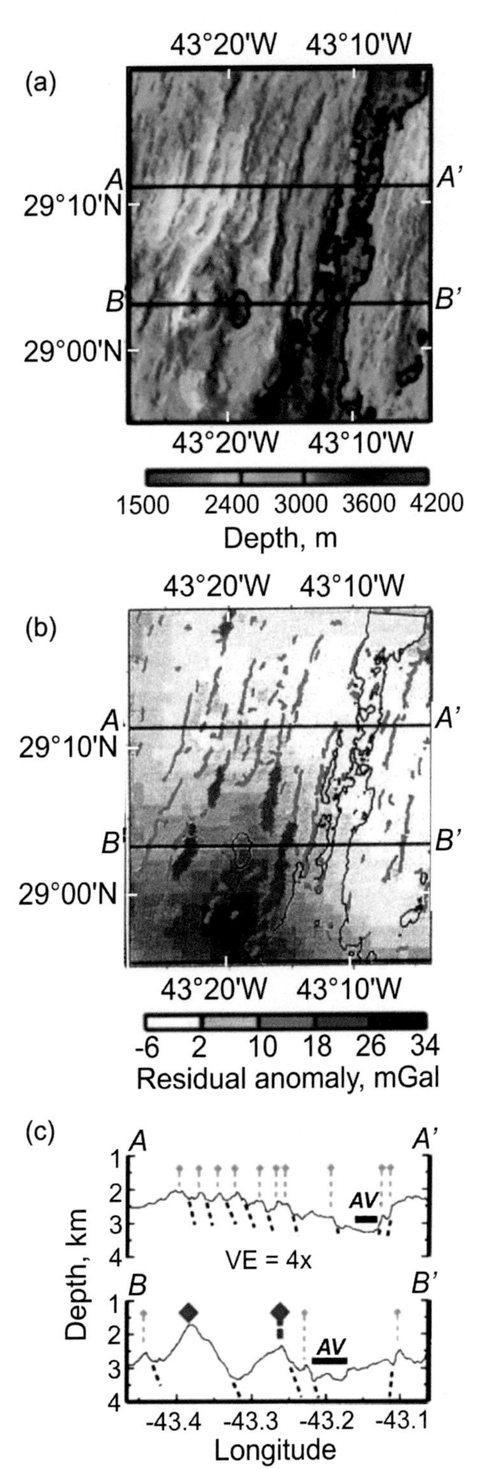


**Figure 7.12** Fault pattern in a single spreading segment of the MAR, 29° N, after Shaw and Lin (1993). (a) Bathymetry. (b) Scarps (colour) detected automatically from the curvature of bathymetry, superimposed on residual mantle Bouguer anomaly (grey). Blue points determined using a 0.45 km radius window, highlighting small scarps; red points used a 2.2 km window showing largest scarps. (c) Interpreted topographic sections along segment centre and segment end. Red and blue points: scarps highlighted in (b); dashed lines: inferred faults; black bars: position of median valley floor.

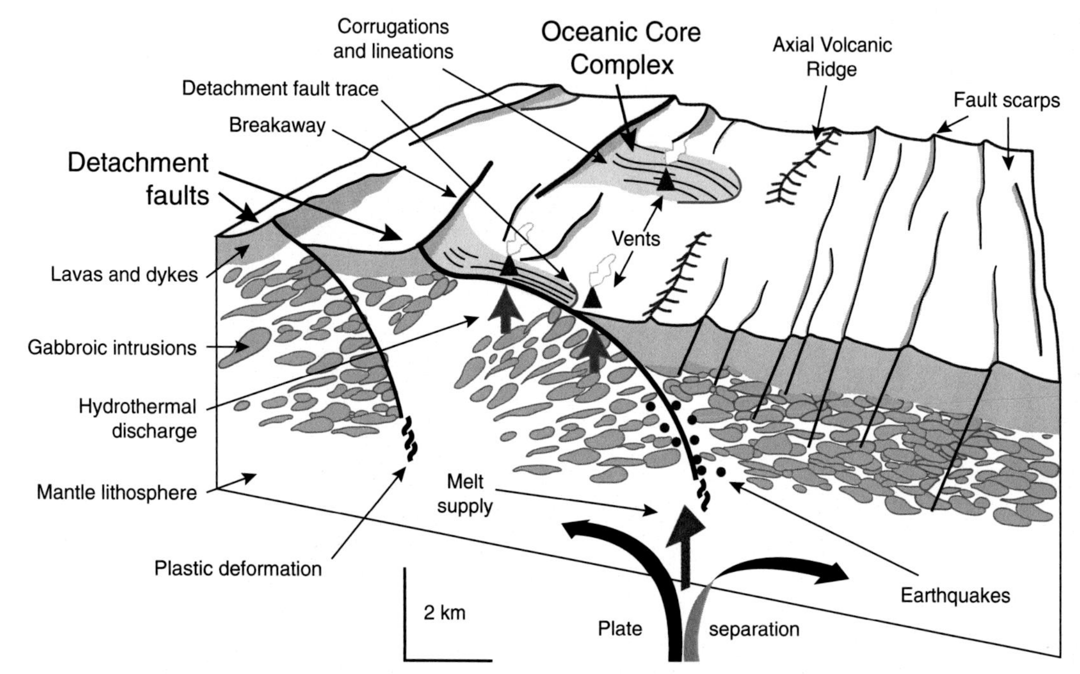


**Figure 7.23** Diagram of oceanic detachment faults, oceanic core complexes and associated structures, after Escartin and Canales (2011). Black lines indicate faults: bold, curved lines are detachments, straight lines are smaller normal faults. Orange domes are oceanic core complexes. Red triangles mark hydrothermal vents; red dots, earthquakes. Proposed subsurface lithology comprises basaltic lavas and dykes (pink), variously serpentinised peridotite (white), and intruded gabbro bodies (blue), though such structures await confirmation. The 'detachment fault trace' where the fault intersects the sea floor is sometimes referred to as the 'termination' (Tucholke *et al.*, 1998) or 'emergence zone' (MacLeod *et al.*, 2009).

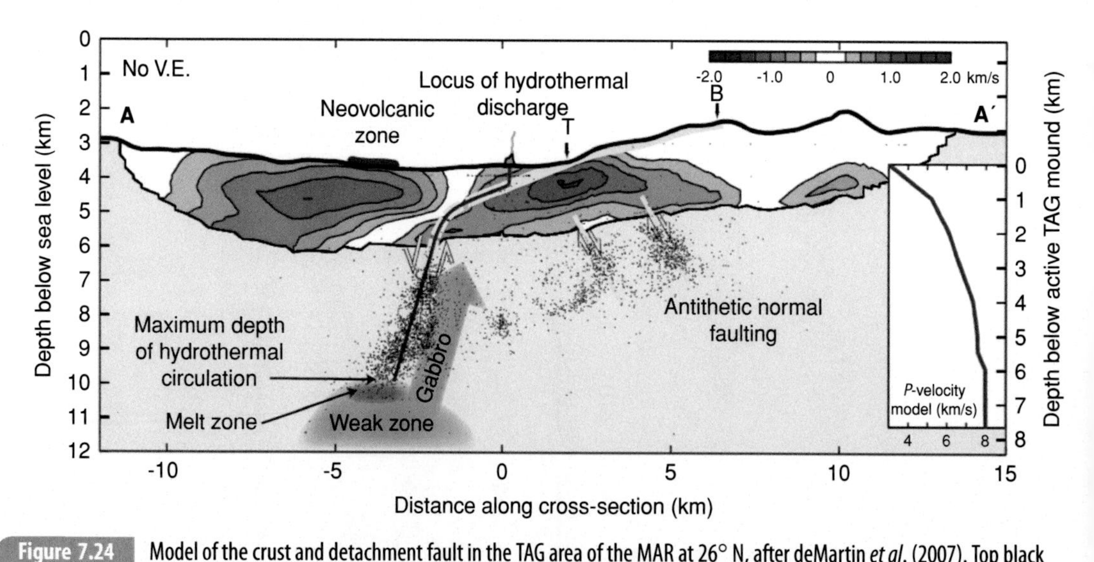


**Figure 7.24** Model of the crust and detachment fault in the TAG area of the MAR at 26° N, after deMartin *et al.* (2007). Top black line is the across-axis sea floor profile, with neo-volcanic zone marked in red. Red and blue contoured areas show departures from the assumed 1D seismic velocity profile (inset) resulting from a seismic tomographic model, interpreted as volcanic crust (red) and gabbro or serpentinised peridotite (blue). Thin black line is base of tomographic model. Small dots are earthquake hypocentres projected onto plane of figure. Yellow lines indicate faults, including detachment, with arrows showing sense of offset. Detachment drawn to connect steeply dipping band of hypocentres at depth, low-angle boundary between tomographic velocity anomalies, and outcropping fault surface. No vertical exaggeration.

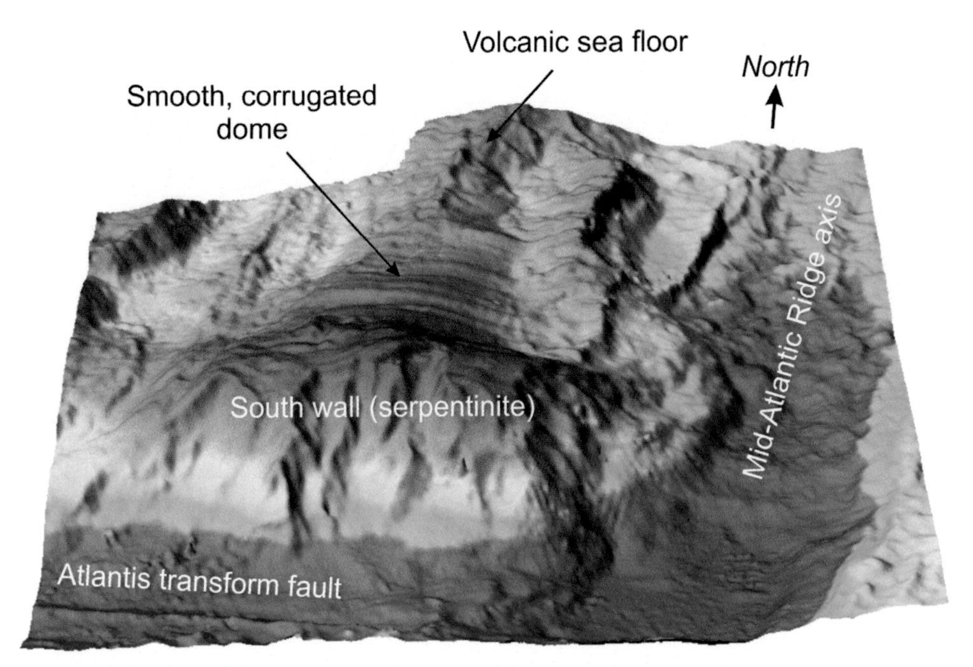


**Figure 7.25** Bathymetry of the Atlantis Massif oceanic core complex in the inside corner junction of the MAR with Atlantis transform fault, after Cann *et al.* (1997). The core complex is characterised by a smooth, corrugated dome exposing the detachment fault, which dips gently to the east and west. The corrugations are aligned along the spreading direction. The complex is surrounded by volcanic sea floor to the west, north and east, and by a wall of serpentinised peridotite flanking the transform to the south. Reprinted by permission from Macmillan Publishers Ltd: *Nature*, vol. **385**, pp. 329–332, copyright 1997.

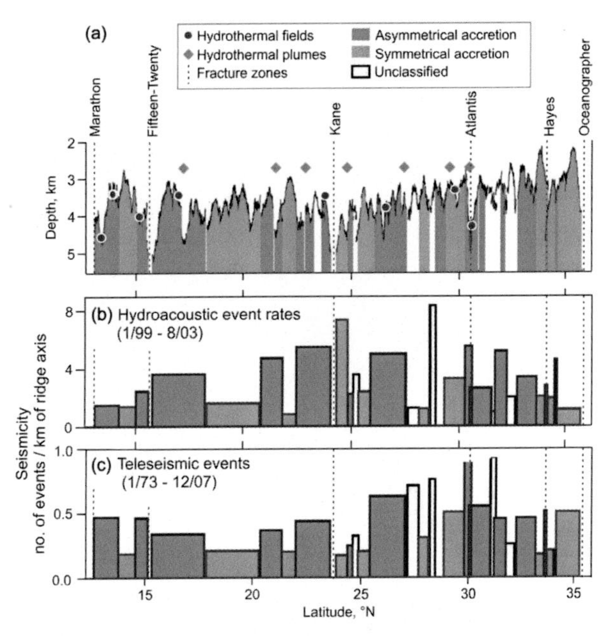


**Figure 7.28** Classification of the MAR, 12.5° N–35.5° N into 'symmetric' (red) and 'asymmetric' (blue) styles of lithospheric accretion, after Escartin *et al.* (2008). (a) Axial depth profile, where segment ends are characterised by localised deeps. (b) Seismicity indicated by hydroacoustic monitoring from 1999 to 2003 using a regional array of moored acoustic detectors. (c) Seismicity from 1973 to 2007 based on global teleseismic recording stations. Note that segments undergoing asymmetric accretion generally show higher hydrothermal and seismic activity than those accreting symmetrically. Reprinted by permission from Macmillan Publishers Ltd: *Nature*, vol. **455**, pp. 790–795, copyright 2008.

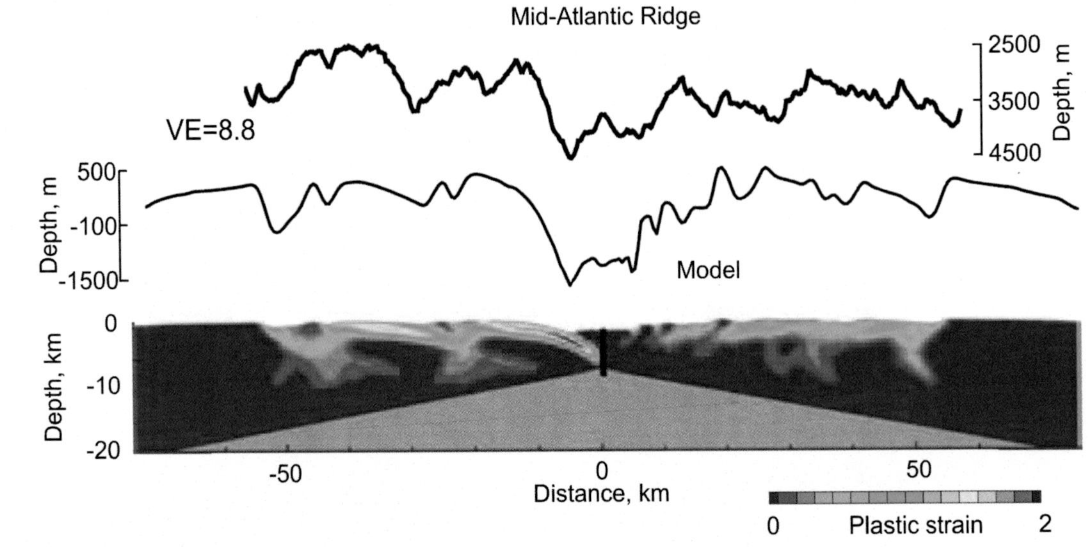


Figure 7.32 Numerical model of MOR deformation in a 'stretching-dominated' ridge with 50% of plate separation taken up by magmatic intrusion, after Buck *et al.* (2005). The lower (coloured) panel shows the degree of plastic strain developed; the curve immediately above this shows the modelled sea floor topography, and the upper curve shows a real topographic profile across the MAR and Kane OCC near 23°25′ N at the same scale. Dark blue represents brittle lithosphere and pink represents weak asthenosphere. Reprinted by permission from Macmillan Publishers Ltd: *Nature*, vol. **434**, pp. 719–723, copyright 2005.

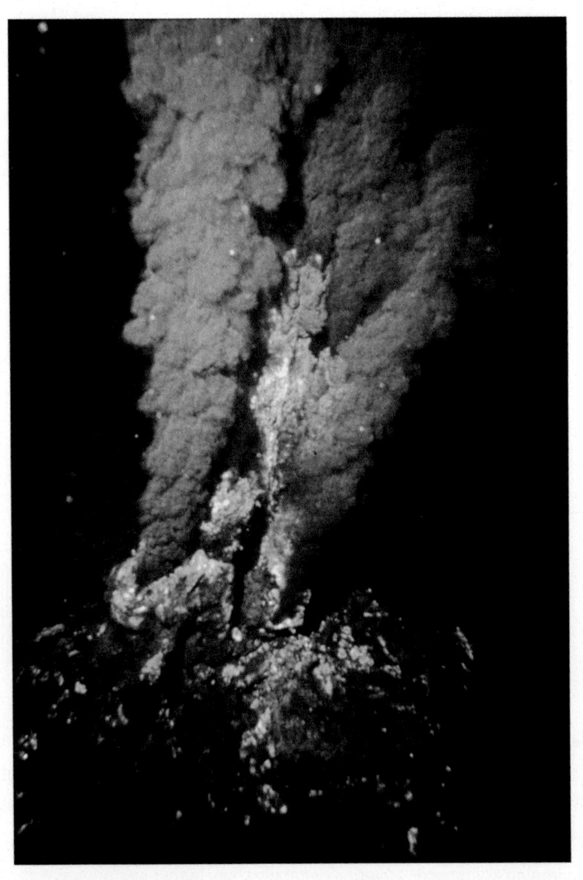

Figure 8.1 Black smoker from the MAR. Image from Wikimedia Commons, courtesy Peter Rona, National Oceanic and Atmospheric Administration. Vent chimneys are visible at foot of image.

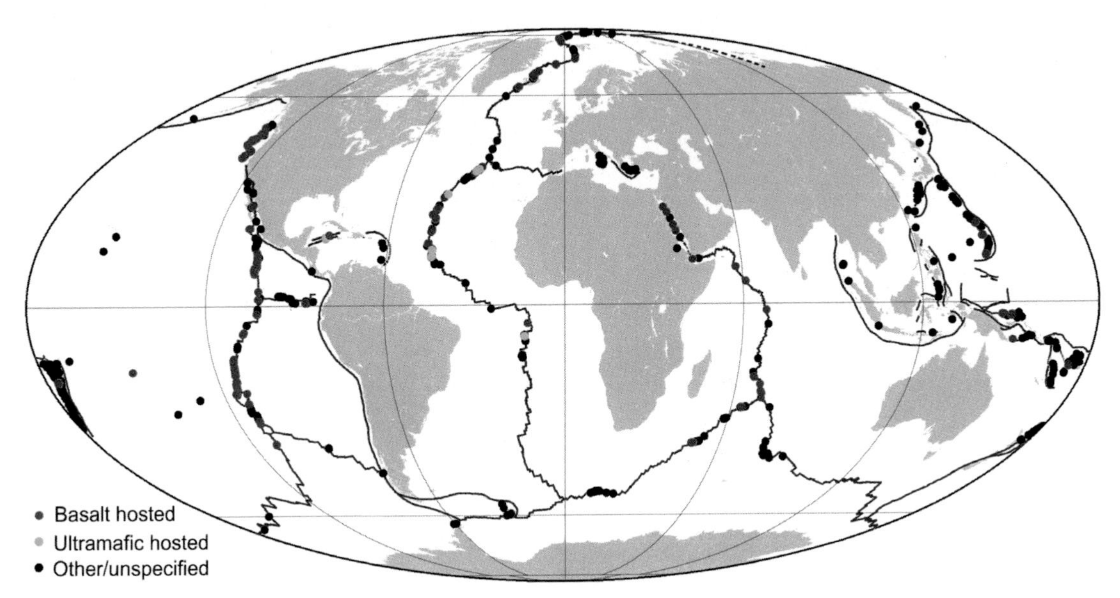


**Figure 8.2** Global distribution of plate boundaries and hydrothermal vent fields, based on the data in the InterRidge vent database (http://www.interridge.org/irvents/).

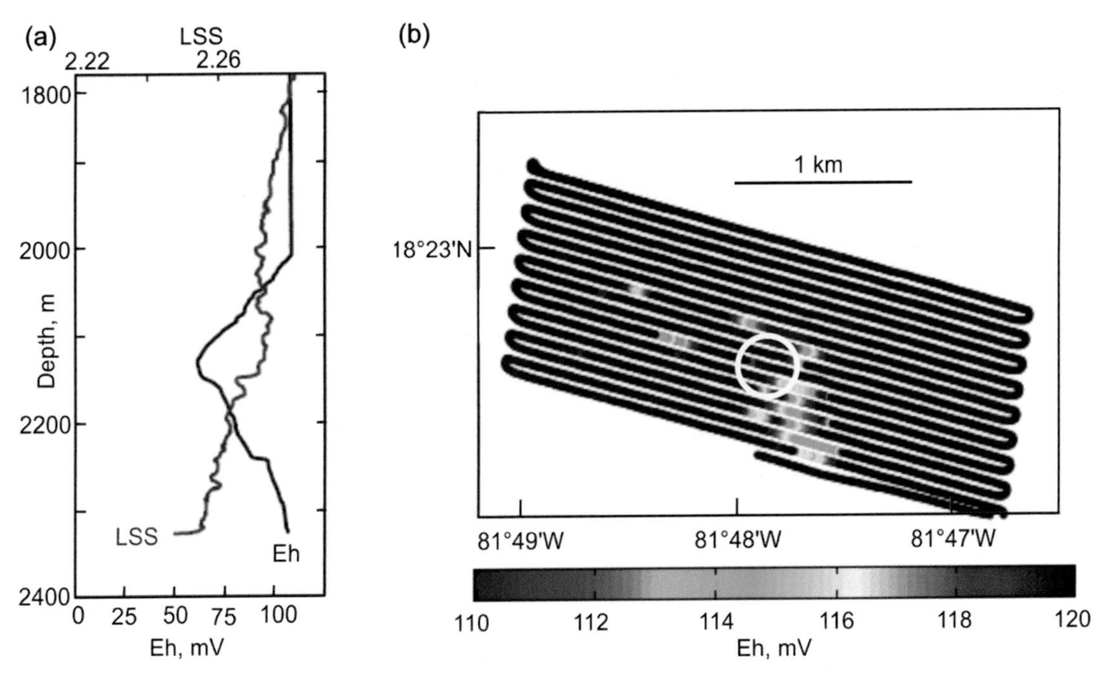


**Figure 8.5** Hydrothermal plume survey for Von Damm vent field on the Mid-Cayman Spreading Centre, after Connelly *et al.* (2012). (a) Output of light-scattering sensor (LSS) and redox potential (Eh) from a tow-yo CTD cast, showing abnormally high light scattering and low redox potential at 2000–2300 m. (b) Map of redox potential (Eh) from an AUV survey run 60 m above the sea floor. Circle: Von Damm vent field.

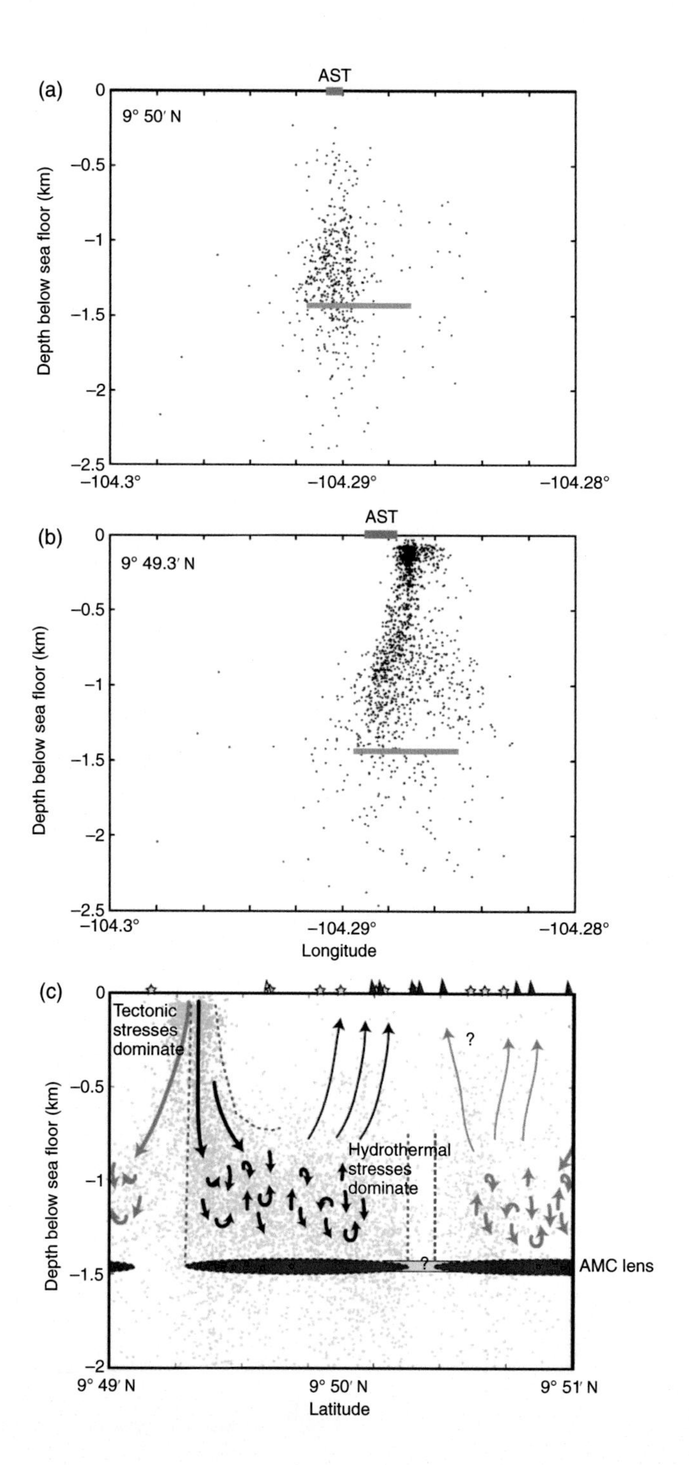


**Figure 8.11** Sections through the EPR near 9°50′ N, showing the distribution of microearthquakes and inferred hydrothermal circulation, after Tolstoy *et al.* (2008). (a) and (b) Across-axis sections with horizontal exaggeration ~1.6 showing the distribution of earthquakes near the inferred discharge and recharge zones, respectively. Grey bars at sea floor: axial summit trough (AST); grey bars at 1.4 km depth: axial magma chamber. (c) Along-axis section, with vertical exaggeration ~1.4. Blue dots: earthquakes interpreted to result from tectonic fracturing in the recharge zone at the segment end; grey dots: earthquakes inferred to result from hydrothermal stresses; red triangles and yellow stars: high- and low-temperature hydrothermal vents, respectively; black and grey arrows: possible hydrothermal flow; red: axial magma chamber, with a possible mush-filled break marked by orange band. Reprinted by permission from Macmillan Publishers Ltd: *Nature*, vol. **451**, pp. 181–184, copyright 2008.

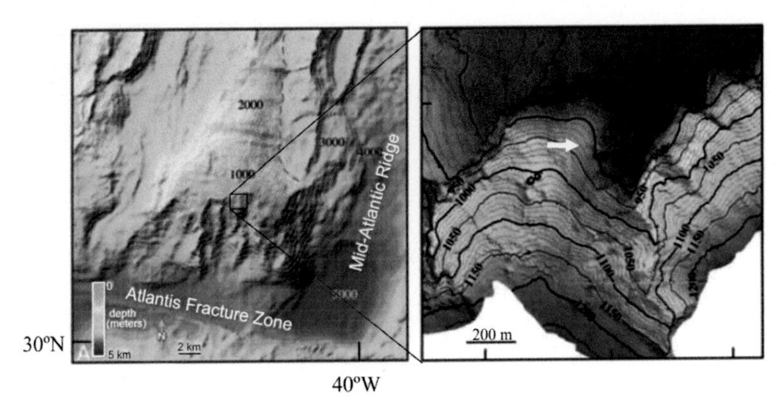


**Figure 8.12** Location (top) and cross-section (bottom) of Lost City vent field. From Kelley, D. S., *et al.*, A serpentinite-hosted ecosystem: The Lost City hydrothermal field. *Science*, vol. **307**, pp. 1428–1434, 2005. Reprinted with permission from AAAS.

**Figure 8.13** Actively venting, 10 m tall carbonate chimney at Lost City vent field. From Kelley, D. S. *et al.*, A serpentinite-hosted ecosystem: The Lost City hydrothermal field. *Science*, vol. **307**, pp. 1428–1434, 2005. Reprinted with permission from AAAS.

**Figure 8.20** Tube worms (*Riftia pachyptila* and *Tevnia jerichonana*) and brachyuran crabs (*Bythograea thermydron*) covering the sea floor at the Genesis hydrothermal vent site, EPR 13° N, after Lutz and Kennish (1993).

**Figure 8.21** Dense aggregation of Alvinocaridid shrimp on an active chimney in the Beebe vent field, Mid-Cayman Spreading Centre, after Connelly *et al.* (2012).

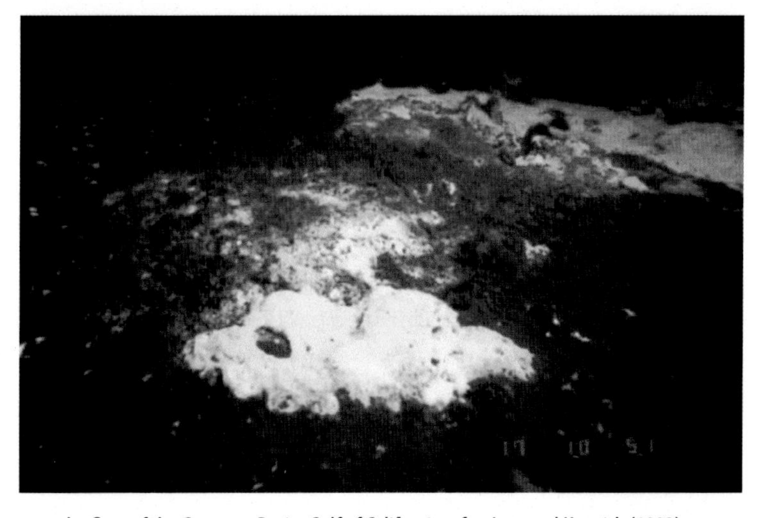

**Figure 8.22** Bacterial mats on the floor of the Guaymas Basin, Gulf of California, after Lutz and Kennish (1993).

**Figure 8.24** Alvinocaridid shrimp on the Von Damm vent field, Mid-Cayman Spreading Centre, after Connelly *et al.* (2012).

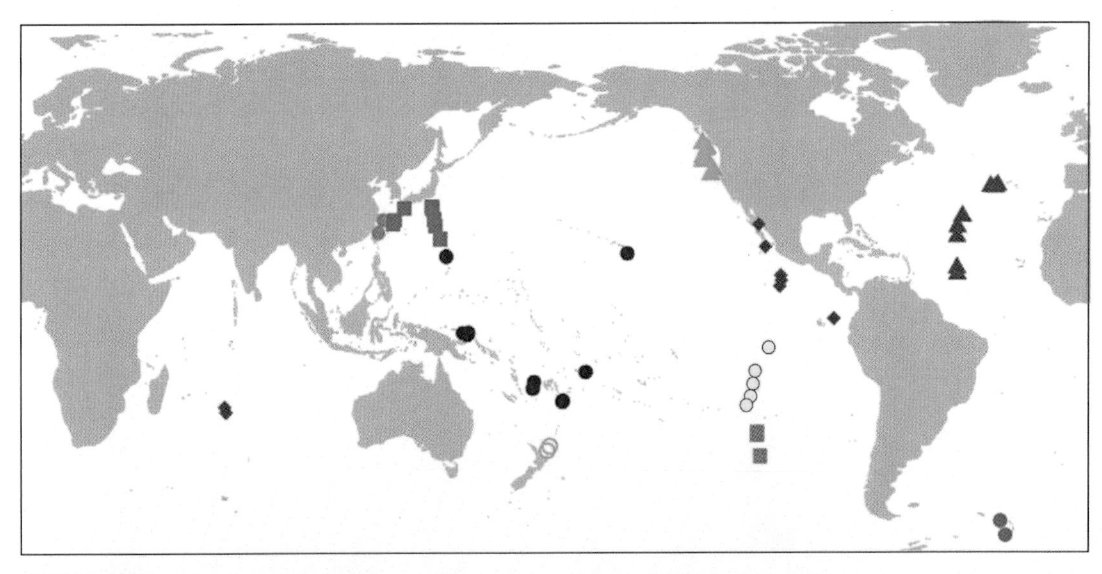

**Figure 8.25** Eleven resolved biogeographic regions identified by different symbols at the vent sites (including some non-ridge sites), after Rogers *et al.* (2012). This figure does not include recent discoveries in the Arctic (Pedersen *et al.*, 2010b), South Atlantic (German *et al.*, 2008), or Southwest Indian Ridge (Tao *et al.*, 2012).

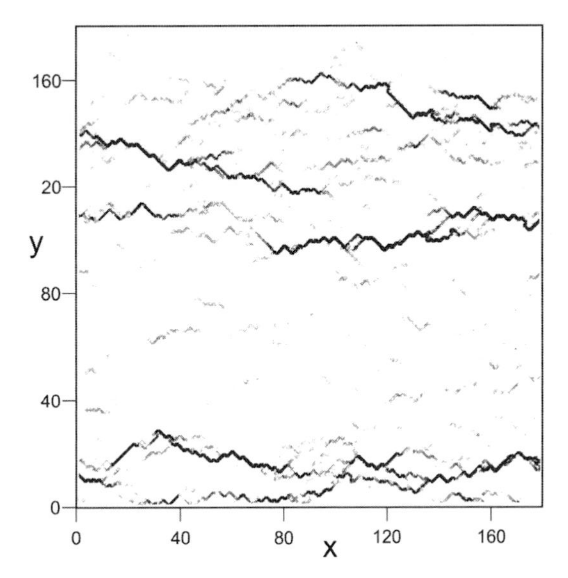


Figure 7.18 Fault pattern produced using a lattice model, after Cowie *et al.* (1993a). Model is based on an elastic plate that is progressively stretched in the *y* direction, with lattice cells failing and slipping incrementally when a threshold stress is reached. Darker shades of grey indicate greater degrees of fault displacement. As the model progresses, small individual faults grow and some link to form longer and larger-offset faults (darker grey). Those that do not link retain only small offsets (light grey). Strain is progressively concentrated on a few, large faults (black).

Figure 7.13, and has been modelled by Cowie *et al.* (1993a; Figure 7.18). Even the largest faults can hook toward and link with each other (Figures 7.14 and 7.19). Details of linkage structures on the MAR are given by McAllister and Cann (1996). Similar effects can be seen on a large scale in the thin lithosphere of fast-spreading ridges, where the ridge axis itself propagates, hooks and links, forming overlapping and 'self-decapitating' spreading centres (Macdonald and Fox, 1983; Macdonald *et al.*, 1987; Figure 4.22).

Single, isolated faults show a characteristic displacement pattern, with the largest displacement near the centre and progressively smaller displacements towards the fault tips. As faults merge, each segment of the new fault maintains this pattern, producing a complex displacement pattern along the fault (Figures 7.13 and 7.20).

There can be a complex interaction between faulting and volcanism. Some eruptions are channelled along faults (Lawson *et al.*, 1996; Standish and Sims, 2006). Also, lava may flow over faults while they continue to slip, producing so-called volcanic growth faults (Macdonald *et al.*, 1996). This latter process can produce the asymmetric horsts seen on fast-spread oceanic crust, where the inner (ridge facing) side is bounded by a steep normal fault scarp while the outer flank is a volcanic growth fault with a gentler slope. Faults in the axial zones of fast-spreading ridges are frequently buried by young lava flows, so that significant amounts of tectonic deformation may now be hidden, including 30%–100% of the several hundred metre subsidence that is needed to construct layer 2A (Escartin *et al.*, 2007).

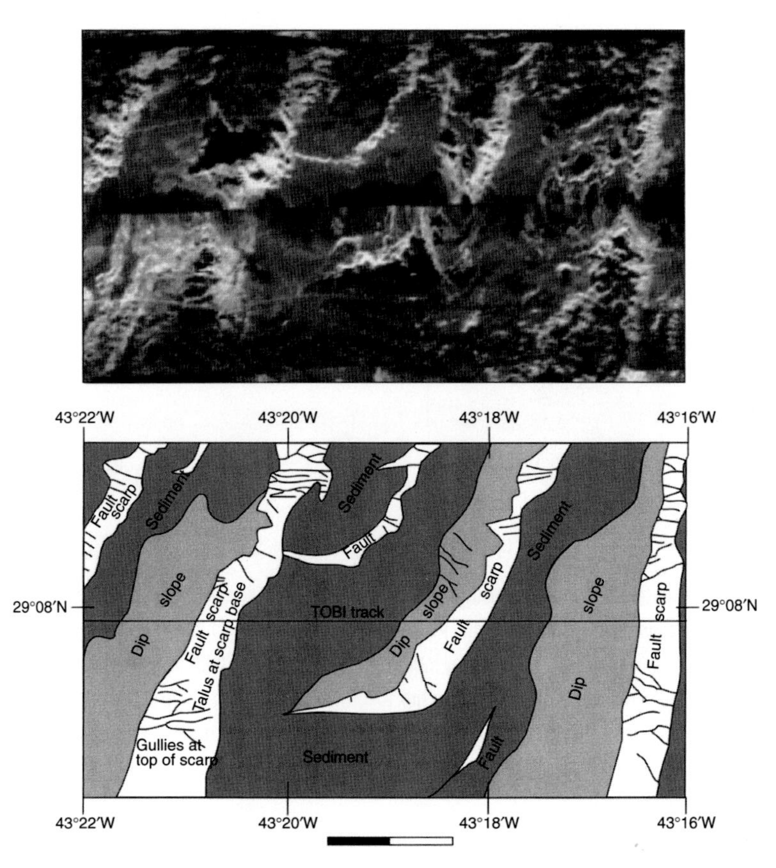


Figure 7.19 Example of large normal faults hooking towards and linking with adjacent ones on the MAR near 29° N, after Searle *et al.* (1998b). Top: single deep-towed side-scan (TOBI) swathe. Bottom: interpretation showing eroded and gullied fault scarps, volcanic dip slopes and sediment ponds between. Note insonification direction reverses at vehicle track. Reprinted with permission from Elsevier, © 1998.

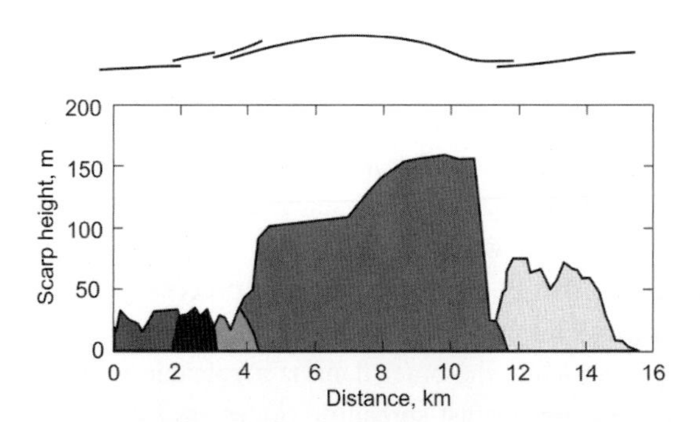


Figure 7.20 Height versus distance along a five-segment fault scarp imaged by side-scan sonar near the EPR axis near 12° N, after Cowie *et al.* (1994). Plan view of fault is shown at top. Below, the contributions of the individual segments are distinguished by different shading. Note that each segment has displacement increasing away from its ends.

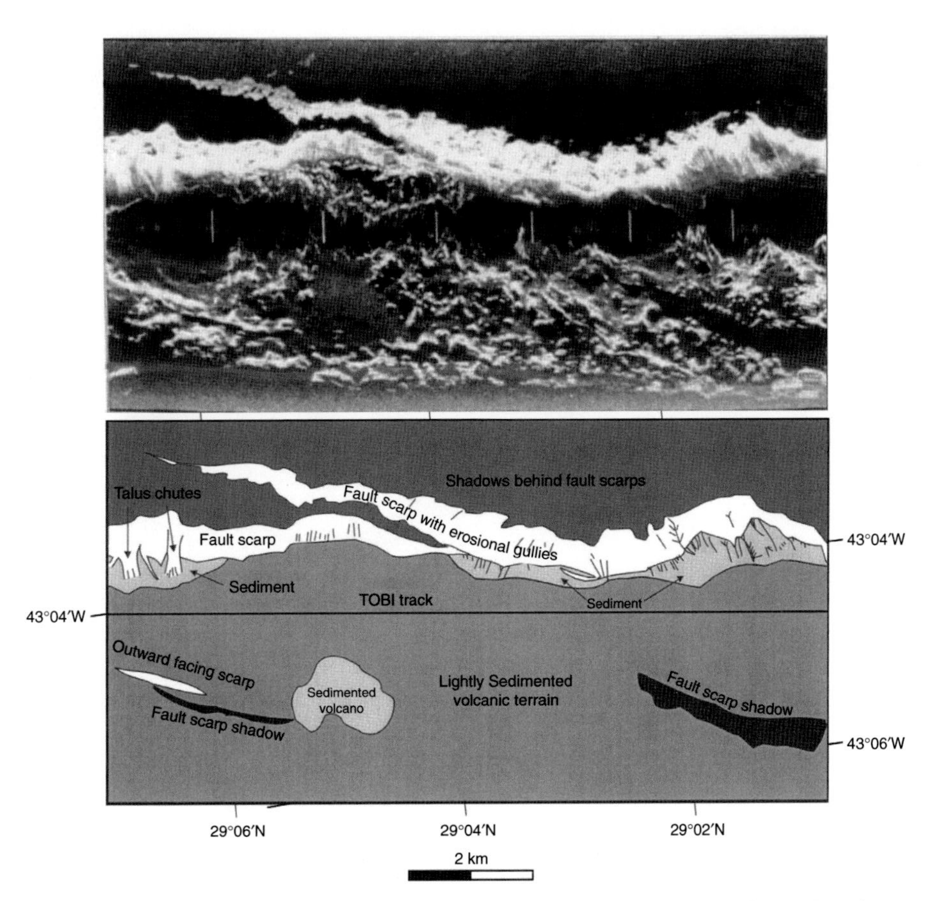


Figure 7.21 TOBI sonograph (above) and interpretation (below) of two linked faults on the MAR at 29° N, after Searle *et al.* (1998b). Note insonification direction reverses at vehicle track. Main faults are approximately 0.8 Ma old. The faults display extensive mass wasting features, including dendritic erosional gulleys, talus chutes, and sedimented talus fans. Reprinted with permission from Elsevier, © 1998.

The faulting pattern changes a few kilometres off-axis. At fast-spreading ridges, outward facing faults begin to develop, so that a horst and graben topography is produced (Carbotte and Macdonald, 1990; Bohnenstiehl and Carbotte, 2001; Figure 7.10). At slow-spreading ridges, major normal faults develop and define the edges of the median valley inner floor. These faults appear to grow to their maximum displacements of several hundred metres or more in a small 'fault generation window' only a few kilometres wide (McAllister and Cann, 1996). The inner boundary of this zone is ~2 km from the axis; McAllister and Cann propose this distance is the point at which a fault generated at the base of the lithosphere at the ridge axis breaks through the sea floor. The outer boundary of the zone is where lithosphere becomes too thick and strong for the fault to keep slipping so it is easier to create a new fault nearer the axis. Thus there may generally be only one major fault active at a time. The rate at which vertical slip accumulates on normal faults has been estimated to range from 1.2 mm $a^{-1}$ to 12 mm $a^{-1}$ (Cowie *et al.*, 1993b; McAllister and Cann, 1996).

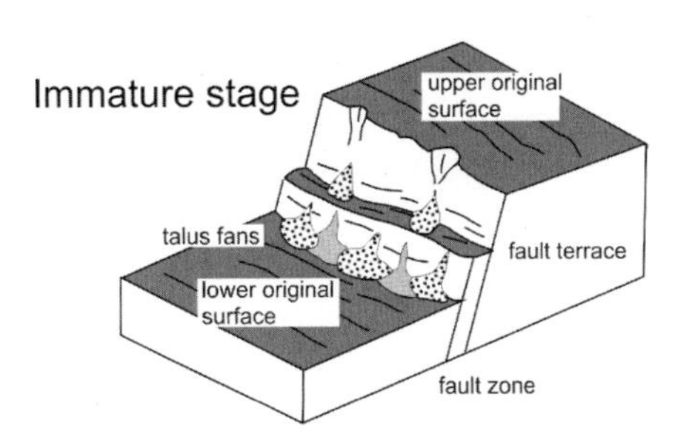


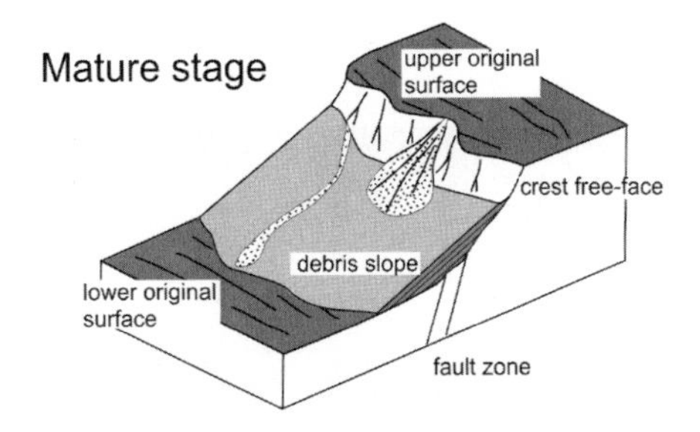


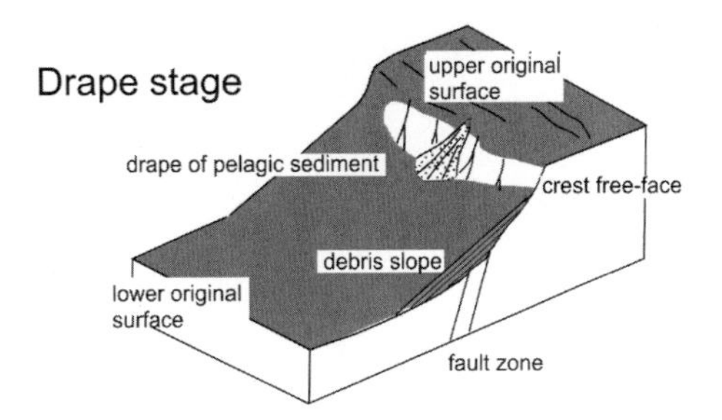


**Figure 7.22** Geomorphological development of a normal fault scarp, after Allerton *et al.* (1995), with kind permission from Springer Science and Business Media. The immature stage reflects the structure of median valley wall faults at a slow-spreading ridge shortly after they cease slipping (~100 ka to 200 ka). Erosional gullies and talus chutes are being formed on the scarp faces, with talus fans below them. At the mature stage (~1 Ma) an extensive debris slope lying at the angle of repose masks much of the lower fault face. The debris slope may itself be covered by pelagic sediment, while continuing erosion deposits talus onto it. Finally, in the drape stage (several millions of years), many of these structures are buried by pelagic sedimentation.

As major normal faults develop, the fault blocks that they bound rotate gently away from the ridge axis by up to about 10°.

Shaw and Lin (1993) modelled such development of single faults and find that the fault spacing is controlled by angle of dip, lithospheric thickness and the ratio of tectonic to magmatic extension. They suggest the McAllister and Cann model is most appropriate for the ends of slow-spreading segments, and propose an alternative for segment centres, where several faults are active simultaneously in a region of necking instability, with deformation taking place over an extended area. With this mode of extension there is a feedback mechanism whereby growth on one fault will tend to impede growth on a neighbouring

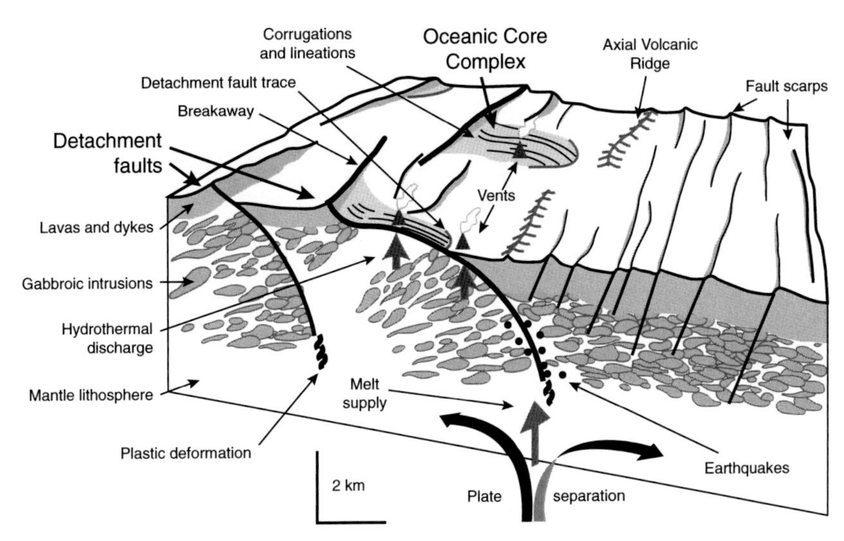


Figure 7.23 Diagram of oceanic detachment faults, oceanic core complexes and associated structures, after Escartin and Canales (2011). Black lines indicate faults: bold, curved lines are detachments, straight lines are smaller normal faults. Orange domes are oceanic core complexes. Red triangles mark hydrothermal vents; red dots, earthquakes. Proposed subsurface lithology comprises basaltic lavas and dykes (pink), variously serpentinised peridotite (white), and intruded gabbro bodies (blue), though such structures await confirmation. The 'detachment fault trace' where the fault intersects the sea floor is sometimes referred to as the 'termination' (Tucholke *et al.*, 1998) or 'emergence zone' (MacLeod *et al.*, 2009). For colour version, see plates section.

one, regulating the size of nearby faults. This mechanism is akin to the numerical model of Cowie *et al.* (1993a).

Once formed, fault scarps are modified by erosion and sedimentation. Side-scan sonar clearly shows the formation of gullies, sediment chutes and talus fans in fault scarps even as young as those bounding the median valley inner floor (Figures 7.19 and 7.21). These features evolve as the fault develops, over a time span ~1 Ma (Figure 7.22).

## 7.4 Detachment faults and oceanic core complexes

### 7.4.1 Detachment faults

Faults at the ends of slow-spreading segments tend to be larger and more widely spaced than at segment centres. Some of these faults are sufficiently well developed that they cut right through a significant thickness of the young, relatively thin lithosphere, becoming detachment faults (Dick *et al.*, 1981; Karson, 1990; Tucholke and Lin, 1994; Figure 7.23). MacLeod *et al.* (2009) suggest that lower amounts of gabbro in the lithosphere mean that normal faults have a greater chance of penetrating to peridotite; this allows ingress of water that alters the peridotite to weak minerals such as serpentine and talc, facilitating continued

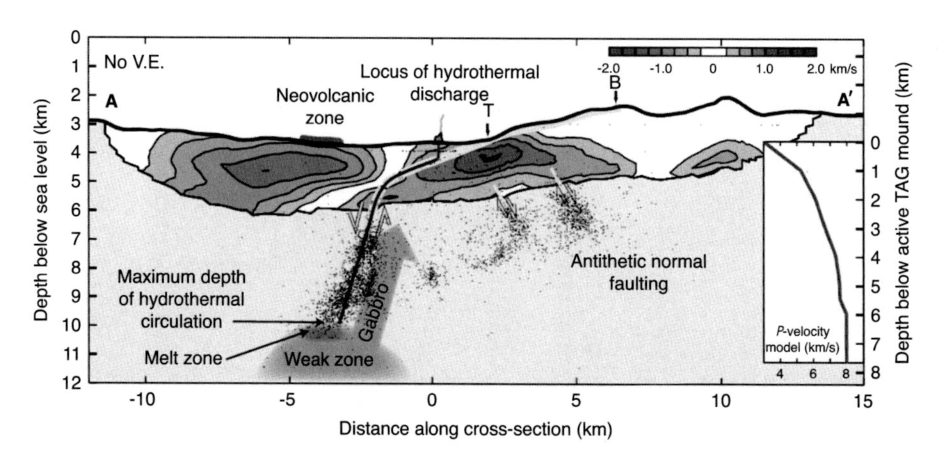


Figure 7.24 Model of the crust and detachment fault in the TAG area of the MAR at 26° N, after deMartin *et al.* (2007). Top black line is the across-axis sea floor profile, with neo-volcanic zone marked in red. Red and blue contoured areas show departures from the assumed 1D seismic velocity profile (inset) resulting from a seismic tomographic model, interpreted as volcanic crust (red) and gabbro or serpentinised peridotite (blue). Thin black line is base of tomographic model. Small dots are earthquake hypocentres projected onto plane of figure. Yellow lines indicate faults, including detachment, with arrows showing sense of offset. Detachment drawn to connect steeply dipping band of hypocentres at depth, low-angle boundary between tomographic velocity anomalies, and outcropping fault surface. No vertical exaggeration. For colour version, see plates section.

fault slip. There is debate as to where detachments root: into a sub-axial magma chamber (Dick *et al.*, 2000), at a mid-lithospheric alteration front (MacLeod *et al.*, 2002), or at the brittle–plastic transition or base of the lithosphere (Tucholke *et al.*, 1998). There is evidence from the degree of alteration, and from the style of deformation – brittle or plastic – of the fault rocks, for each of these possibilities in different detachments, and it may be that different detachments root into different structures and/or at different depths.

Where they emerge onto the sea floor, detachments have shallow dips of about 20°, but their subsurface geometry is currently uncertain. Most models envisage them as steepening downwards (Cann *et al.*, 1997; Tucholke et al., 1998), but some have suggested, at least in part, a concave up geometry (Karson and Dick, 1983; MacLeod *et al.*, 2002). The only direct evidence is from a micro-earthquake study of the MAR at 26° N, which revealed a steeply dipping zone of hypocentres in the lower crust along the supposed continuation of an outcropping detachment fault (deMartin *et al.*, 2007; Figure 7.24). However, there is a seismic gap in the upper crust there, so the question remains open.

Detachment faults may initiate as steeply dipping normal faults that then rotate to shallower dip as a result of flexure in the footwall (Smith *et al.*, 2006; Smith *et al.*, 2008; MacLeod *et al.*, 2009; Escartin and Canales, 2011). Palaeomagnetic studies of the footwalls of two detachment faults confirm that they have rotated away from the ridge axis as this model requires (Garcés and Gee, 2007; Morris *et al.*, 2009).

Detachment faults can take up large proportions, perhaps between 70% and 100%, of the relative plate motion (Searle *et al.*, 2003; Okino *et al.*, 2004; Baines *et al.*, 2008; Grimes *et al.*, 2008). Their offsets range from kilometres to tens of kilometres, and they may continue slipping for ~1 Ma or more. The largest known detachment is associated with Godzilla Mullion in the intermediate-spreading rate Parece Vela back-arc basin (Ohara *et al.*, 2003), which has an offset of ~125 km and was active for ~4 Ma (Tani *et al.*, 2011). Tucholke and Lin (1994) note that while detachments are relatively common on slow-spreading ridges, they are intermittent at intermediate-spreading rates, and rare at faster rates.

Ranero and Reston (1999) have imaged detachment faults seismically in slowly spread Cretaceous lithosphere of the eastern North Atlantic, and Reston and Ranero (2011) have discussed the 3D geometry of detachment faults, comparing the seismic images from Cretaceous crust with topographic profiles across modern oceanic detachments and numerical models. An ancient oceanic detachment fault may be exposed at the Limassol Forest in the Troodos ophiolite of Cyprus (Cann *et al.*, 2001).

## 7.4.2 Oceanic core complexes

Because oceanic detachment faults have such large offsets, they typically expose lower crustal and upper mantle rocks. At the same time, the young, thin and warm lithosphere of the detachment footwall responds to buoyancy and other forces by flexing upwards, creating a smooth dome on the sea floor (Figures 7.23, 7.25–7.26). Such domes are called oceanic core complexes (OCCs). The part of the detachment fault farthest from the ridge axis is called the 'breakaway' (Figure 7.23) and represents the earliest part of the fault to be exposed; the trace of the fault where it emerges from beneath the hanging wall is called the 'termination' and is the structurally deepest part of the fault plane to be exposed (Tucholke *et al.*, 1998). The hanging wall at the termination should have the same crustal age as the footwall at the breakaway, as they were once contiguous. Because 'termination' suggests that activity on the fault has ceased, MacLeod *et al.* (2009) preferred the term 'emergence zone' for this feature on an active OCC.

The first OCC to be recognised was Atlantis Massif (Cann *et al.*, 1997; Figure 7.25). This shows a clear pattern of topographic ridges and grooves aligned with the spreading direction. Such features, referred to by Tucholke *et al.* (1998) as 'mullion structures' but more commonly called 'corrugations', are characteristic of OCCs, and their appearance, especially in shaded relief bathymetry, is a common way of detecting core complexes. Corrugations have typical wavelengths of several hundred metres to a few kilometres and amplitudes of tens to hundreds of metres. Spencer (1999) suggested they are formed by a process similar to continuous casting, whereby the ductile footwall is pulled past the brittle hanging wall and takes up a shape following the latter's irregularities. Another possible origin is the irregular along-axis occurrence of low degrees of melting (Tucholke *et al.*, 2008); this could cause irregularities in the depth of the brittle–plastic transition, producing corrugations if the detachment roots into such an irregular layer. Side-scan sonar images reveal closer-spaced, spreading-parallel lineations usually referred to as 'striations' and presumably formed in a similar way; these have typical spacings ≤100 m, and vertical amplitudes too small (<~5 m) to be detectable (Cann *et al.*, 1997; Tucholke *et al.*, 1998; Searle et al., 2003; Mallows and Searle, 2012; Figure 7.26).

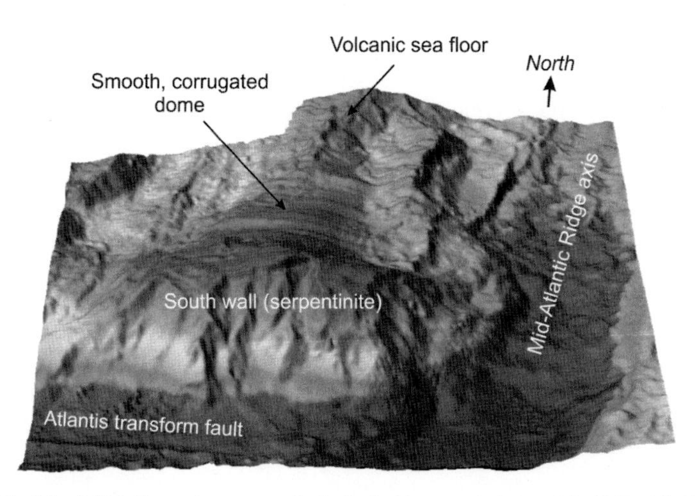


**Figure 7.25** Bathymetry of the Atlantis Massif oceanic core complex in the inside corner junction of the MAR with Atlantis transform fault, after Cann *et al.* (1997). The core complex is characterised by a smooth, corrugated dome exposing the detachment fault, which dips gently to the east and west. The corrugations are aligned along the spreading direction. The complex is surrounded by volcanic sea floor to the west, north and east, and by a wall of serpentinised peridotite flanking the transform to the south. For colour version, see plates section. Reprinted by permission from Macmillan Publishers Ltd: *Nature* © 1997.

Blackman *et al.* (2009) reviewed the geophysical signals associated with OCCs. These data reveal that OCCs usually remain active for about 1–2 Ma (Godzilla Mullion – Section 7.4.1 – is an exception). The OCCs are associated with high gravity, showing they have dense cores that reflect the relative uplift of deep crustal and mantle rocks. Although detachment faults have been imaged by seismic reflection in Cretaceous crust (Section 7.4.1), they have so far proved hard to image near MOR axes.

Seismic refraction studies reveal a heterogeneous velocity structure in the upper ~0.5–2.0 km of OCC lithosphere, which broadly correlates with both velocity gradient and detailed sea floor sampling (Canales *et al.*, 2008; Xu *et al.*, 2009; Figure 5.28). Relatively low velocities and gradients correlate with outcrops of basalt and dolerite; intermediate values (velocities of 3.4–4.2 km $s^{-1}$, gradients of 1–3 $s^{-1}$) with serpentinised peridotite, and higher velocities and gradients with gabbro.

Several detailed sampling campaigns have been carried out at OCCs, including dredging and submersible dives (e.g., MacLeod *et al.*, 1998; Tucholke *et al.*, 2001; Karson, 1999; Blackman *et al.*, 2002; MacLeod *et al.*, 2002; Karson *et al.*, 2006; Dick *et al.*, 2008; MacLeod *et al.*, 2009) and deep drilling (Dick *et al.*, 1991; Dick *et al.*, 2000; Blackman *et al.*, 2006; Blackman *et al.*, 2011). These recover mostly *in situ* gabbro and serpentinised peridotite, extensive basaltic rubble (probably spalled off the hanging wall edge as the footwall emerges), and frequently 'fault rocks' – brittlely and/or plastically deformed gabbros or peridotites and their hydrous alteration products such as serpentinite and talc. There is some indication that gabbro is more common around the terminations (MacLeod *et al.*, 2002; Xu *et al.*, 2009), although gabbro forms the majority of the uppermost 1.4 km of the footwall in the central dome of Atlantis Massif (Ildefonse *et al.*, 2007; Canales *et al.*, 2008;

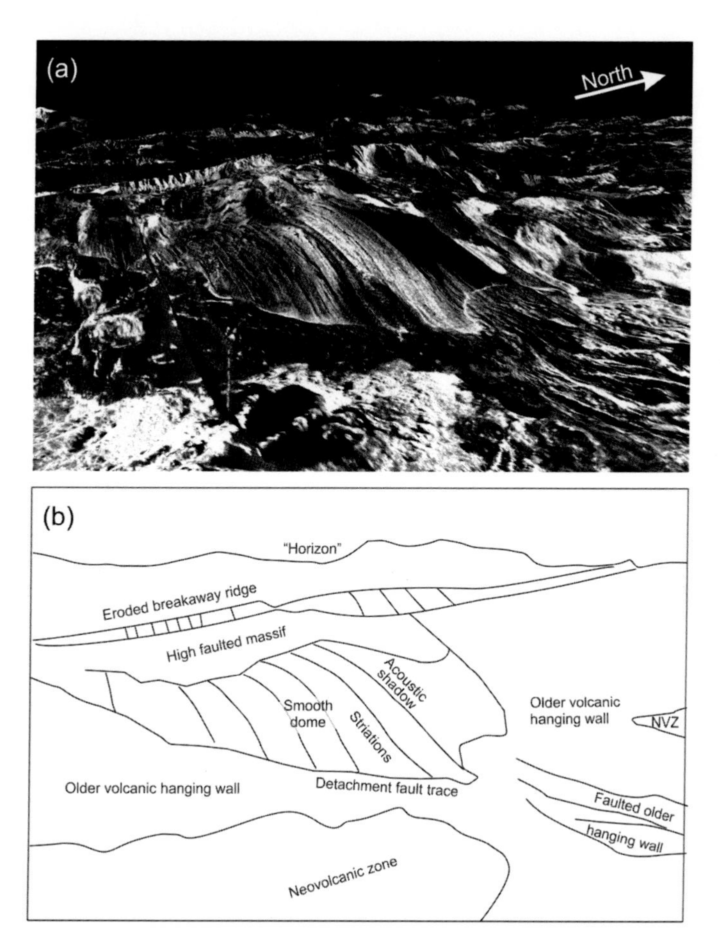


**Figure 7.26** Oblique view of an active oceanic core complex on the MAR at 13°20′ N. (a) TOBI side-scan sonar mosaic draped over bathymetry, looking NW across the ridge axis with no vertical exaggeration. Light tones represent high backscatter. (b) interpretation of main features seen in the side-scan image. Foreground shows a discontinuous neo-volcanic zone, absent opposite the OCC toe. Corrugated and striated dome of OCC is visible in middle distance. The detachment fault trace is marked by a bright fringe thought to mark talus spalled off the edge of the hanging wall. Upper, older part of detachment footwall is being broken up as footwall flexes, producing the 'high faulted massif'; this is bounded on its older (distant) side by a heavily eroded and gullied normal fault scarp, which may be the detachment 'breakaway' (MacLeod *et al.*, 2009) or a perched 'rider block', cut off from the hanging wall (Smith *et al.*, 2008).

Blackman *et al.*, 2011). Fault rocks are confined to the exposed detachment surface. MacLeod *et al.* (2009) recognise a distinction between the smooth dome of some core complexes (interpreted as the mantle section of the footwall) and a more fractured part of the footwall adjacent to the breakaway, which comprises mainly mafic rocks (basalt and dolerite) and which they interpret as the crustal section. The lateral variations in seismic velocity broadly correlate with these sampling results. The model of OCCs shown in Figure 7.23 is consistent with these findings, although many details, particularly the subsurface lithology and detachment geometry, remain to be confirmed.

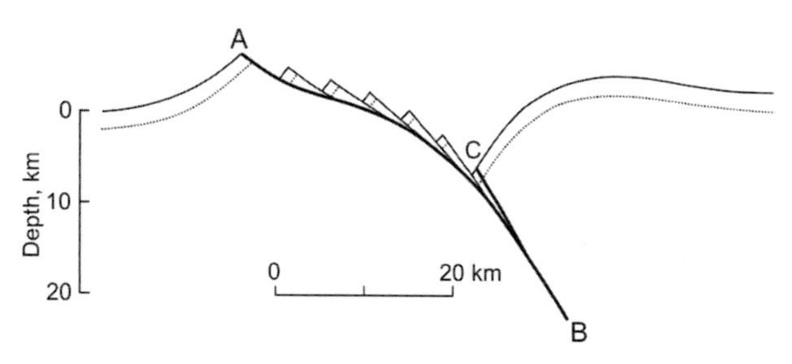


Figure 7.27 Model of the development of a flexed detachment fault and associated rider blocks, assuming an effective lithospheric elastic thickness of 1.0 km, after Buck (1988). Bold line AB is the detachment fault. Dotted line indicates depth of 2 km beneath the surface. BC is a normal fault initiated at its optimum (steep) angle of slip after the main detachment has rotated to a lower, less optimum angle; BC is cutting through the edge of the hanging wall. This cut off piece of hanging wall will be transferred to footwall as a 'rider block' and will be rafted away by slip on the detachment. Previously formed rider blocks are shown by fine lines. Flexural curvature increases as the effective elastic thickness decreases, which can happen progressively with continued plate bending.

Numerical modelling suggests that the development of OCCs requires that the 'magmatic' component of lithospheric extension fall within a critical range of ~30%–70% of the total plate separation, rather than the more normal value of ~85%–95% (Buck *et al.*, 2005; Behn and Ito, 2008; Tucholke *et al.*, 2008). At such relatively low rates of melt supply, more plate separation must take place by faulting, favouring the development of long-lived detachment faults. When the magmatic component is greater than the critical range, normal faulting and fissuring take up only a minor degree of plate separation, and the models produce a fairly symmetric pattern of accretion and normal faulting, as observed (Section 7.7). Modelling by Olive *et al.* (2010) suggests that it is the proportion of melt intruded into the brittle lithosphere, not the total proportion of melt, that is critical in determining the mode of faulting. The proportion emplaced in the underlying ductile layer determines how much gabbro is exhumed, perhaps explaining the variable lithology inferred from sampling and seismic velocities. In the hanging wall, melt is presumably injected vertically into lower crustal magma chambers, and reaches the upper crust as dykes (Chapter 6). However, in the detachment footwall, melt is thought to be injected laterally at considerable depth forming gabbro sills, and does not subsequently feed dykes or sea floor volcanism (Figure 7.23).

In some areas, such as the MAR 13° N–14° N, OCCs occur in close proximity, with apparently normal volcanic sea floor between them. It is unknown whether these are instances of isolated detachment faults that were active for relatively short times (Tucholke *et al.*, 1998; MacLeod *et al.*, 2009), or are repeated outcrops of the same detachment, linked beneath the sea floor (Smith *et al.*, 2008). In the latter case, the blocks of volcanic sea floor between OCCs could be rafted 'rider' blocks cut off from the hanging wall as the detachment rotates to lower angles (Buck, 1988; Figure 7.27).

If OCCs are episodic, their history would reflect a fluctuating melt supply, perhaps on a timescale of ~2Ma as suggested by flow line variations in gravity and crustal thickness (Tucholke and Lin, 1994; Tucholke *et al.*, 1998). OCC evolution may involve rapid jumping of the instantaneous plate boundary off-axis, followed by migration back toward the axis (MacLeod *et al.*, 2009). Detachment faulting and OCC creation would begin when the

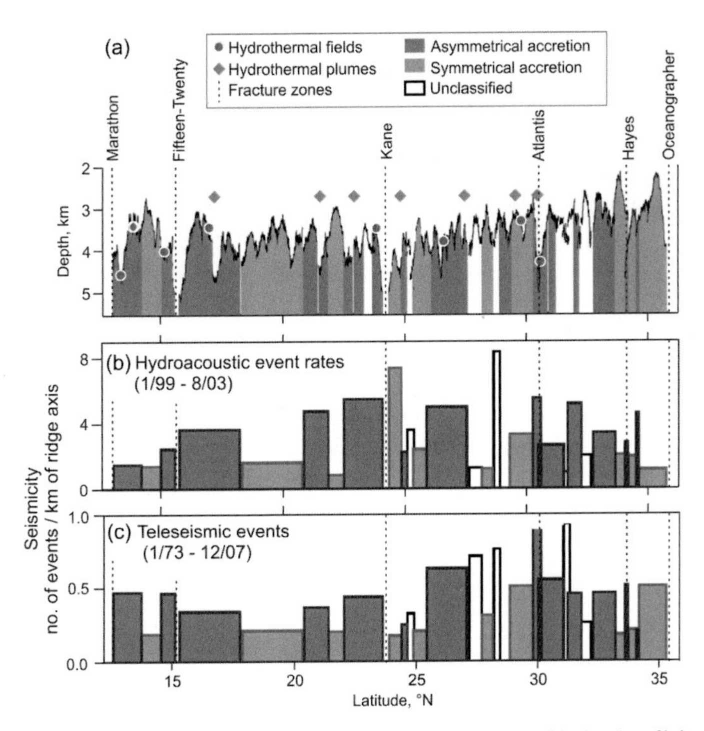


Figure 7.28 Classification of the MAR, 12.5° N–35.5° N into 'symmetric' (red) and 'asymmetric' (blue) styles of lithospheric accretion, after Escartin *et al.* (2008). (a) Axial depth profile, where segment ends are characterised by localised deeps. (b) Seismicity indicated by hydroacoustic monitoring from 1999 to 2003 using a regional array of moored acoustic detectors. (c) Seismicity from 1973 to 2007 based on global teleseismic recording stations. Note that segments undergoing asymmetric accretion generally show higher hydrothermal and seismic activity than those accreting symmetrically. For colour version, see plates section. Reprinted by permission from Macmillan Publishers Ltd: *Nature* © 2008.

melt supply falls to the critical level, and end when the melt supply resumes, either through general secular fluctuation, by propagation of high-melt zones along-axis from neighbouring melt-rich zones, or by migration of the footwall over a region of more robust melt delivery (MacLeod *et al.*, 2009).

Oceanic core complexes (OCCs) are frequently the sites of hydrothermal activity (Section 8.5). A range of vent types from basalt-hosted to ultramafic-hosted are found, with a corresponding diversity of biota. McCaig *et al.* (2007) have presented evidence for extreme focussing of hydrothermal fluids along detachment faults and a model in which the evolution of the fluids depends on their proximity to high-temperature gabbro bodies intruded into the footwall.

### 7.4.3 Asymmetric accretion

Oceanic core complexes (OCCs) constitute a profoundly asymmetric form of lithospheric accretion (Escartin *et al.*, 2008). Faulting patterns are highly asymmetric, with widely spaced, large-offset detachment faults on one flank and closely spaced, short-offset normal faults on the other; detachments have offsets an order of magnitude larger than other normal faults and take up a large proportion of the total tectonic strain; the lithosphere formed on the footwall side is largely devoid of volcanic crust and comprises an as yet uncertain mixture of variously serpentinised mantle and gabbroic intrusions; overall spreading rates tend to be highly asymmetric, with localised plate boundaries jumping and migrating across the ridge axis. In addition, sections of spreading axis associated with OCCs are more seismically active than 'normal' magmatically-spreading segments (Smith *et al.*, 2006).

An analysis of multi-beam bathymetry from the MAR by Escartin *et al.* (2008) revealed that such asymmetric accretion is much more common than previously supposed, occurring along almost half of the MAR between 12.5° N and 35° N (Figure 7.28). Hydrothermal activity is more common at asymmetric segments. Basalts erupted there show significant geochemical differences, indicating lower degrees of mantle partial melting and crystallisation at higher pressure, compared with symmetric segments, while seismicity is higher, suggesting the lithosphere there is thicker and cooler. It is therefore fair to describe lithospheric accretion at OCCs as a newly recognised mode of sea floor spreading, alongside symmetric magmatic accretion by dyke injection. See also Chapter 9.

## 7.5 Ultra-slow spreading

Ridges spreading at less than 20 km $Ma^{-1}$ full rate, such as Gakkel and the SWIR, are generally classified as ultra-slow spreading, and are characterised by a variety of tectonic processes. Symmetric, magmatic spreading is found, but is less abundant than at slow-spreading ridges; a greater proportion of segments lack or have very subdued sea floor volcanism, which tends to be focussed into widely spaced centres (Section 6.8); in addition there are segments wholly devoid of sea floor volcanism (Sauter *et al.*, 2002; Figure 6.26). In some areas, sections of ridge up to several hundred kilometres long appear to be entirely amagmatic, with mantle peridotite being emplaced directly onto the sea floor (Michael *et al.*, 2003).

The amagmatic regions are characterised by broad, segment-long, symmetric, smooth-flanked ridges up to 2000 m high, which Cannat *et al.* (2006) have called 'smooth sea floor' (Figures 7.29 and 7.30). Recent studies have shown that both the inward facing and outward facing flanks of these ridges have low slopes of 20°–35° that are composed almost entirely of serpentinised peridotite (Sauter *et al.*, 2013). There are also asymmetric ridges with inward facing peridotite slopes dipping less than 15° and steeper, outer volcanic slopes. These peridotite ridges are interpreted as upper mantle that has been directly emplaced on to the sea floor by successive low-angle normal faults of alternating polarity (Sauter *et al.*, 2013). They occur predominantly in the most oblique spreading segments, which are those with the lowest effective spreading rate (Figure 7.29).

Dick *et al.* (2003) proposed that Gakkel Ridge, spreading at 8–13 km $Ma^{-1}$, is the only truly 'ultra-slow' ridge, and that the SWIR and others spreading at 13–20 km $Ma^{-1}$ are intermediate between slow and ultra-slow. They proposed a new class of ultra-slow MOR

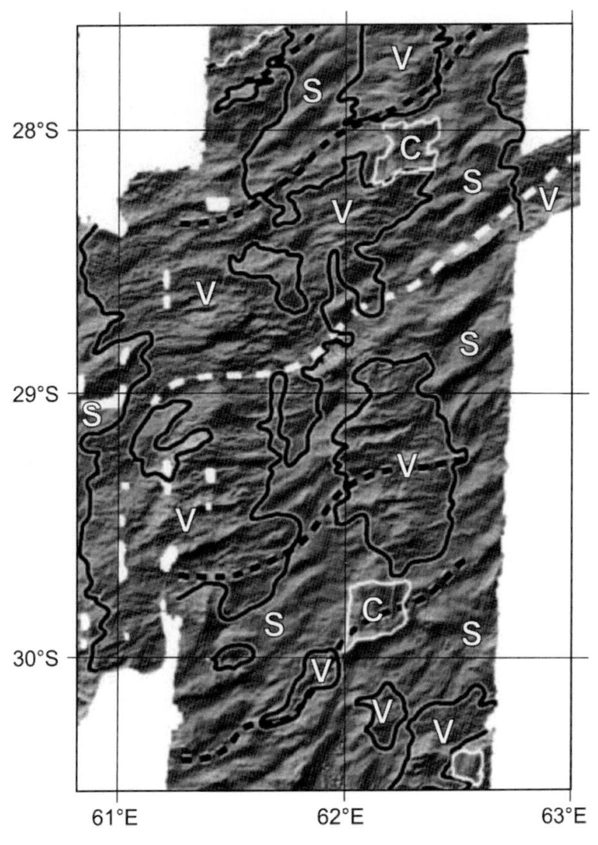


Figure 7.29 Shaded relief bathymetry of part of the Southwest Indian Ridge, showing different accretionary terranes, slightly modified after Cannat *et al.* (2006). Bathymetry is illuminated from the NW. White dashed line: ridge axis; dashed black lines: 10 Ma and 20 Ma isochrons. 'Volcanic' sea floor is outlined by black lines and labelled 'V'; 'corrugated' sea floor (oceanic core complexes) outlined in white and labelled 'C'; 'smooth' sea floor labelled 'S'. 'Smooth' sea floor is generally confined to oblique-spreading portions of the ridge, where the 'effective spreading rate for mantle upwelling' is unusually low.

characterised by linked magmatic and amagmatic segments, absence of transform faults, long oblique-spreading sections interspersed between orthogonal spreading, and discontinuous volcanism with large regions of missing or attenuated volcanic crust. A critical control on spreading processes at very low spreading rates is considered to be not the plate separation rate but the rate of mantle upwelling. Dick *et al.* (2003) define an 'effective spreading rate for mantle upwelling', which is the component of the total plate separation rate that is orthogonal to the local plate boundary. In this view the characteristics of true ultra-slow spreading occur when the effective spreading rate falls below about 12 km $Ma^{-1}$, which may happen in segments spreading orthogonally at such rates, or in highly oblique segments spreading at a full rate of 12–20 km $Ma^{-1}$.

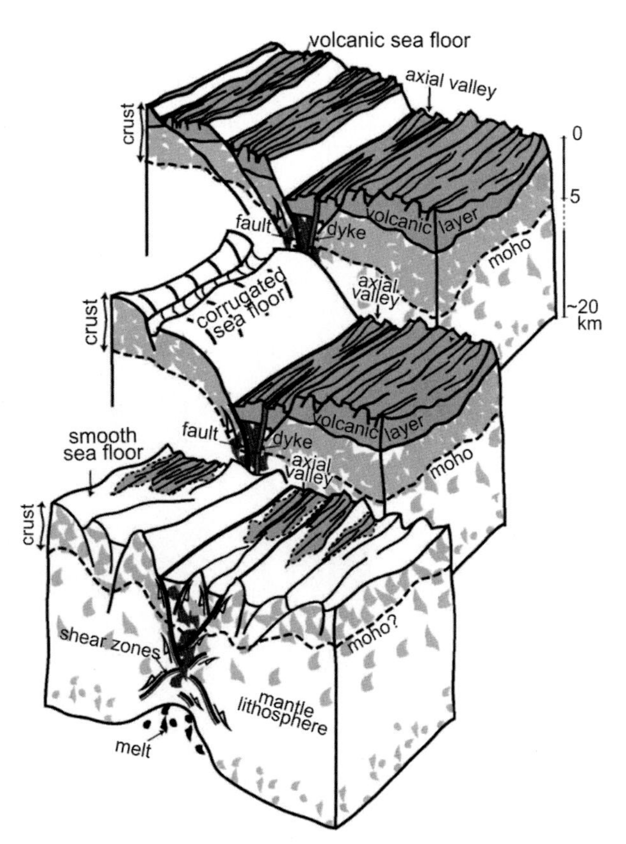


Figure 7.30 Three modes of slow sea floor spreading, producing 'volcanic', 'corrugated' and 'smooth' sea floor, after Cannat *et al.* (2006). Volcanic sea floor is prevalent at slow spreading rates, with corrugated sea floor accounting for ~25%; smooth sea floor is common only at ultra-slow spreading rates, along with the other two types. Dark grey: melt concentrations; light grey: crystallised melt.

## 7.6 Transform and strike-slip faults

A number of faulting styles are developed at transform and other high-order ridge offsets. In a fully developed transform fault, where relative plate motion is taken up on a single through-going strike-slip fault, the least compressive stress axis $\sigma_3$ is oriented at approximately 45° to the transform direction (Figures 7.6 and 7.31). Normal faults, which are aligned normal to the $\sigma_3$ direction, bend up to 45° as the transform is approached (Crane, 1976; Lonsdale, 1978; Searle, 1983; Fox and Gallo, 1984; Searle, 1986; Figure 4.16), as does the neo-volcanic zone, whose feeding fissures are also normal to $\sigma_3$.

The through-going strike slip fault represents the current plate boundary (the Principal Transform Displacement Zone, Section 4.4.1). However, local variations in geology may cause it to deviate slightly from the small circle predicted by plate tectonics theory, so

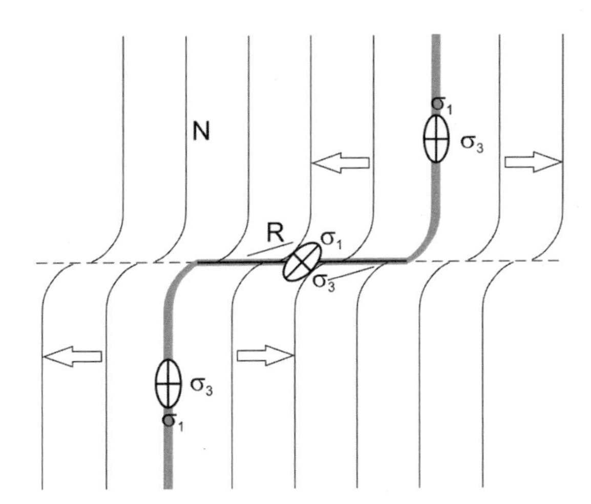


Figure 7.31 Stresses and fault patterns around a transform fault. Arrows indicate spreading direction. Stress ellipses shown with most ($\sigma_1$) and least ($\sigma_3$) compressive stress axes. Grey lines, plate boundary; heavy black line, through-going strike-slip fault; fine lines, other faults: N, normal faults, R, Riedel shears.

that its location may wander by a few kilometres either side of its mean position (Searle, 1986), producing a transform tectonised zone (Gallo *et al.*, 1986). As well as a through-going strike-slip fault aligned along the direction of relative plate movement, this zone may contain Riedel shears (short strike-slip faults oriented ~15° to the main slip direction, Figure 7.31), oblique pressure ridges and sag basins. The last two are formed by components of compression or extension across the PTDZ where it departs from being exactly parallel to the main slip direction.

A transform may contain several closely spaced strike-slip fault strands, joined by short, oblique, extensional basins (Searle, 1983; Figure 4.16), 'extensional relay zones' (Gallo *et al.*, 1986), or 'intra-transform spreading centres' (Fornari *et al.*, 1989) a few kilometres long. The inactive walls of some slow-slipping fracture zones show major scarps, several kilometres long, oriented approximately 45° to the slip direction. These have been interpreted as analogous to 'tension gashes' – essentially normal faults opened up along the direction normal to the rotated $\sigma_3$ direction (Searle, 1981).

Through-going strike slip faults do not develop at ridge offsets smaller than ~15–30 km. Instead there is an array of oblique normal and shear faults, including 45° tension gashes and Riedel shears. This faulting produces short, oblique-spreading offsets at slow-spreading ridges (Searle and Laughton, 1977; Figure 4.23) and overlapping spreading centres at fast-spreading ones (Macdonald and Fox, 1983; Figure 4.19).

A final style of faulting is so-called 'bookshelf faulting'. Here, shear occurs on a number of closely spaced, parallel faults, resulting in the whole region being sheared, similar to what happens when a row of books is tilted. Such faulting has been observed principally at propagating rifts, where the normal faults generated at the failing rift (i.e. normal to the spreading direction) are rotated up to ~60° by the passage of the propagating rift. The required shear is taken up by slip between each rotated fault block (Hey *et al.*, 1986; Figure 4.28). The interiors of oceanic microplates may also be deformed by bookshelf faulting early in their development, before they begin to undergo rigid rotation as single blocks (Larson *et al.*, 1992; Searle *et al.*, 1993).

## 7.7 Modelling faulting

Considerable progress has been made in recent years in modelling faulting at MORs. Buck (1988) developed a model for large-offset faulting including the effects of regional flexure and isostasy (Figure 7.27). The model recognises that faults form at an optimum dip angle; once the flexing plate has rotated this fault sufficiently from the optimum angle, a new, steeper fault is created. The model successfully reproduced the topography of continental metamorphic core complexes, and was subsequently applied to OCCs.

Thatcher and Hill (1995) modelled the topography produced by repeated slip and flexure on normal faults dipping 45° in a 10 km thick lithosphere, followed by infill of the flexural basin, magmatic accretion, and advection away from the spreading axis until the next fault forms. They showed that this modelled topography could closely fit that observed at the MAR by a suitable choice of faulting positions. Recently a similar model, but with faults dipping 60° in 1 km thick lithosphere, was used by Schouten *et al.* (2010) to successfully model topography produced by faults including large-offset detachments.

A series of increasingly sophisticated numerical models has been produced by Buck and co-workers. Buck and Poliakov (1998) and Lavier *et al.* (2000) developed a numerical model in which an elastic–plastic lithosphere overlies a ductile asthenosphere, with a superimposed temperature field approximating that for an MOR. The shear strength, or cohesion, of the brittle lithosphere decreases with increasing strain and strain rate, causing strain to accumulate in narrow zones that mimic faults. This model reproduces realistic topography and faulting, including a median valley and abyssal hills (Figure 4.10), and OCCs (Figure 7.32). Faults arise spontaneously, without the need to impose their positions.

Lavier *et al.* (1999) used the model to investigate detachment faulting and found that, if the reduction in strength of faults is $<\sim 10\%$ of the strength of the brittle layer, then faults lock after offsets of less than the layer thickness, and new faults form. For greater strength reduction, faults can continue to slip indefinitely, while the footwall rotates, to produce structures very like OCCs.

Poliakov and Buck (1998) extended the model by explicitly including dyke intrusion. Buck *et al.* (2005) used the model to investigate faulting in a range of environments (Figures 4.10 and 7.32). They were able to reproduce details of overall ridge morphology and fault style and spacing, controlled by two different mechanisms. At fast-spreading ridges the high heat input produces a relatively broad, shallow region of weak asthenosphere that isostatically supports an axial high; plate unbending controls the dip direction and size of fault offsets, producing small faults with alternating dip directions as observed at the EPR. At slow spreading the buoyancy effect is much reduced and lithospheric stretching dominates; a combination of stretching and variable degrees of dyke intrusion can explain the variety of faulting styles, with very large-offset normal faults (detachments) being developed only when $\sim 50\%$ of plate separation is taken up by magmatic intrusion.

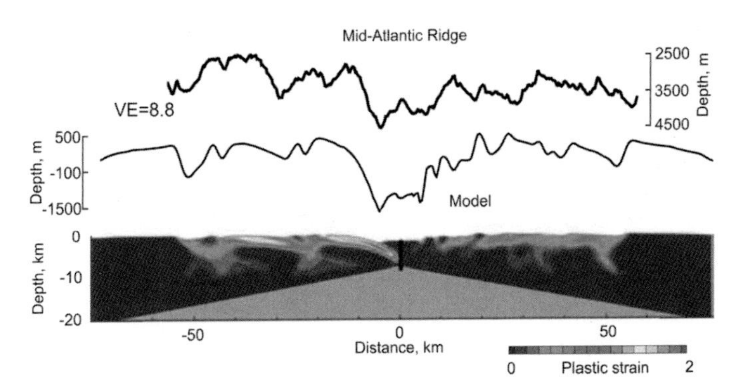


Figure 7.32 Numerical model of MOR deformation in a 'stretching-dominated' ridge with 50% of plate separation taken up by magmatic intrusion, after Buck *et al.* (2005). The lower (coloured) panel shows the degree of plastic strain developed; the curve immediately above this shows the modelled sea floor topography, and the upper curve shows a real topographic profile across the MAR and Kane OCC near 23°25′ N at the same scale. Dark blue represents brittle lithosphere and pink represents weak asthenosphere. For colour version, see plates section. Reprinted by permission from Macmillan Publishers Ltd: *Nature* © 2005.

There has been further exploration of the factors controlling faulting using similar models (e.g., Behn and Ito, 2008). Olive *et al.* (2010) investigated the effects of varying the depth at which melt is intruded (Section 7.4.2). Tucholke *et al.* (2008) used the model to investigate the effects of drastically varying the proportion, $M$, of total plate separation taken up by purely magmatic accretion. They find that long-lived detachment faults form when $M$ is ~30%–50%, with significant magmatic accretion in the footwalls of detachments. At lower values of $M$, high-angle faults form and break up the detachment hanging wall and footwall. At $M = 0$, no long-lived detachments form and all extension is taken up on relatively short-lived faults, alternating in position and dip direction, somewhat like those that bound peridotite ridges in 'smooth' sea floor. Similar results were obtained by physical analogue modelling of slow spreading by Shemenda and Grocholsky (1994).

Finally, in addition to modelling faulting in cross-section, there are models of how faults grow and interact in the horizontal plane. Rubin (1995) reviewed the theory of crack growth in the presence of magma. Other investigations include modelling the shape of overlapping spreading centres using boundary element displacement discontinuity methods (Sempéré and Macdonald, 1986), and modelling the growth and interaction of cracks by a statistical physics model (Cowie *et al.*, 1993a; Figure 7.18).

## 7.8 Summary

The lithosphere at all MOR axes is under tension, causing it to crack into fissures and faults. Dense swarms of ridge-parallel fissures form in an axial zone a few kilometres wide, some of which provide conduits for magma. As the plates separate, fissures grow and link together, the largest ones developing into normal faults several kilometres long. While most faulting takes place within a few kilometres of the axis, some active faulting may occur up to a few tens of kilometres away.

At fast-spreading ridges, the axial high is buoyed up by low-density melt, producing a concave upwards shape. As the lithosphere moves away it unbends, causing the formation of inward and outward facing ridge-parallel faults separated by 1–2 km and with vertical offsets of a few tens of metres. These faults extend the lithosphere by about 4% of the total plate separation.

As the spreading rate falls, the patterns of faulting change dramatically. The influence of tectonic stretching of the lithosphere becomes dominant, producing a median valley and larger faults that are predominantly inward dipping. At typical slow-spreading centres, over 90% of normal faults are inward facing, with lengths ~10 km (but sometimes extending the length of a spreading segment) and vertical offsets of hundreds of metres or more. The total tectonic strain may be 10%–15% of the total plate separation. Faults reflect the changing lithospheric thickness along spreading segments: segment centres have relatively thin lithosphere producing relatively small, closely spaced faults on either plate; in contrast, segment ends, especially inside corners, have thick lithosphere which supports widely spaced, large-offset faults.

Faults are typically associated with weak minerals such as serpentine and talc, produced by the hydration of olivine. This is especially the case at slow spreading rates. These weak minerals facilitate slip on the faults, which may in turn provide access of seawater to deep-seated rocks, promoting further hydration and slip.

At slow spreading rates some of the largest faults continue to slip for 1–2 Ma, accumulating many kilometres of offset and cutting right through the lithosphere as detachment faults. Such faults occur when the proportion of magmatic accretion falls to about half of the plate separation. This provides a profoundly asymmetric style of plate accretion. One plate is the detachment footwall, and grows by a combination of direct exhumation of mantle onto the sea floor, possibly coupled with intrusion of gabbroic sills into the footwall at depth, to form a domed and corrugated 'oceanic core complex'. The opposite plate, which is the detachment hanging wall, grows by relatively normal magmatic accretion and normal faulting. At certain times the whole of the plate separation may be taken up by slip on the detachment. As much as half of the oceanic crust of the central North Atlantic may have been generated by this asymmetric process.

At ultra-slow-spreading ridges the magmatic input to the spreading lithosphere is further reduced, both by the intrinsically low spreading rate and by the occurrence of long sections of oblique plate boundary which cause the component of opening normal to the plate boundary, and thus the space for magmatic input, to be further diminished. This results in yet another style of sea floor spreading, in which large, symmetric ridges of peridotite are exhumed onto the sea floor by the action of detachment faults of alternating polarity, producing so-called 'smooth' sea floor terrain, in addition to volcanic sea floor and OCCs.

MOR offsets are characterised by faulting. Transform faults contain a through-going strike-slip fault, often bordered by oblique 'tension gashes' and Riedel shears. Non-transform offsets contain similar oblique faults but no spreading-parallel fault. Near ridge offsets, the direction of the minimum compressive stress, which controls fault azimuths, can rotate up to 45° away from the general trend of the ridge axis, so that ridge-parallel normal faults tend to curve toward the direction of ridge offset, and short sections of oblique spreading may form in the offsets themselves.

Considerable progress has been made in modelling fault formation and development. Models with an elastic–plastic lithosphere in which the brittle strength reduces with strain and strain rate successfully predict patterns of ridge topography and faulting. By controlling the rate and position at which melt is introduced, fault patterns at all spreading rates can now be reproduced.

# 8 Hydrothermal processes

## 8.1 Introduction

Newly formed oceanic lithosphere is extensively fissured and faulted (Chapter 7), providing pathways for seawater to penetrate the crust. As it does so, it can become heated by hot rocks at depth, producing a highly chemically reactive fluid that undergoes significant exchange of elements with the host rocks. The heated fluid is sufficiently buoyant to return to the sea floor and emerge in 'hydrothermal vents'. Such vents may be concentrated into narrow jets, or broadly distributed as diffuse emanations. The debouching fluids range from just above ambient temperature (~2 °C) to >400 °C, and have a wide range of chemical properties. They may be acid or alkali, have salinities higher or lower than seawater, and carry a wide and variable range of dissolved minerals that they may precipitate onto the sea floor. Dissolved elements include Ca, Fe, Mn, Zn, Ni, Cu and Au and a range of gases including $H_2$, He, $CO_2$ and $CH_4$. They may produce large, potentially economically important mineral deposits, and influence the chemistry of seawater in important ways. Hydrothermal vents carry approximately one third of the global oceanic heat flux and 25% of the global crustal heat flux (Sclater *et al.*, 1981; Morton and Sleep, 1985; Stein and Stein, 1994). To dissipate this amount of heat, $\sim 5 \times 10^{14}$ kg of seawater must pass through MOR hydrothermal systems every year (Wolery and Sleep, 1976). Hydrothermal vents host unique faunal assemblages based on extremophile chemotrophic micro-organisms, which in turn support a range of specialised filter feeders and predator species. Reviews of hydrothermal systems include edited volumes by Rona *et al.* (1983), Humphris *et al.* (1995), German *et al.* (2004) and Rona *et al.* (2010).

## 8.2 Discovery and distribution of hydrothermal vents

Early in the development of sea floor spreading theory, Bostrom *et al.* (1969) noted very low aluminium and titanium and high iron and manganese contents in sediments from active MORs, and suggested that such ridges contain a source of 'volcanic emanations' containing Fe and Mn that subsequently mix with ocean bottom waters. Measured conductive heat flow in ocean basins younger than ~60 Ma is less than predicted by plate theory (Section 3.2). Based on such observations, Lister (1972) suggested the dominant heat transfer process at ridge crests is hydrothermal circulation.

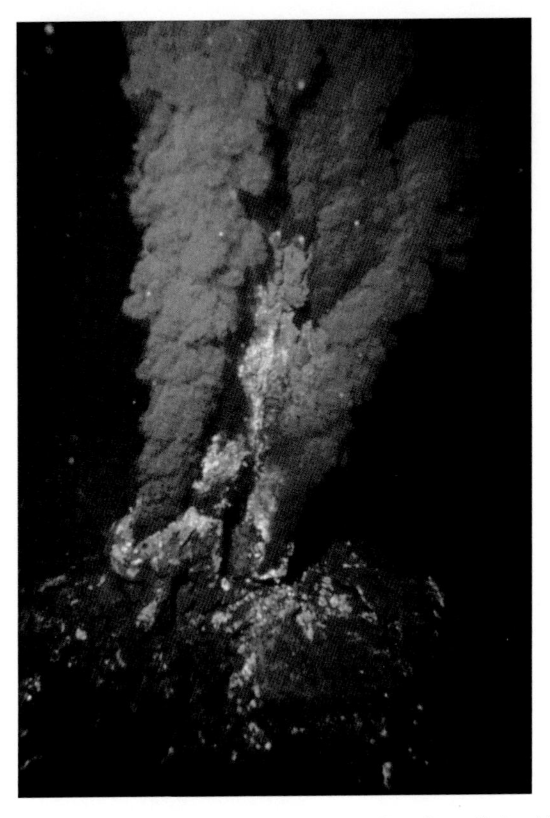

Figure 8.1 Black smoker from the MAR. Image from Wikimedia Commons, courtesy Peter Rona, National Oceanic and Atmospheric Administration. Vent chimneys are visible at foot of image. For colour version, see plates section.

One of the first reports of an active hydrothermal field was the so-called TAG (Trans-Atlantic Geotraverse) site on the MAR near 26°08′ N, based on extensive dredged samples of thick hydrothermal manganese encrustations overlying and cementing talus deposits (Scott *et al.*, 1974). However, it was over a decade later that Rona *et al.* (1986) obtained visual confirmation of active venting there. The first MOR vents to be discovered were on the Cocos–Nazca Spreading Centre in 1977 (Corliss *et al.*, 1979). These were low-temperature (≤17 °C) vents hosting animal communities inferred to be based on bacteria that obtained their entire energy supply by oxidising sulphides in the vent fluids.

High-temperature vents (380 °C ± 30 °C) were first discovered on the EPR near 22°50′ N in 1979 (Spiess *et al.*, 1980); this is below the boiling point of sea water at ambient pressure (but see Section 8.4.3). The hottest waters were blackened with sulphide precipitates, and such vents became known as 'black smokers' (Figure 8.1). Cooler 'white smokers' issued fluids at 32 °C – 330 °C with lower Fe content, while other hydrothermal springs debouched clear to milky fluids at ~20 °C. Subsequent studies revealed many hot and warm vents and often evidence of diffuse hydrothermal flow along much of the EPR (e.g., Baker *et al.*, 2002; Haymon and White, 2004), the Juan de Fuca Ridge (e.g., Baker, 1994; Lupton, 1995) and elsewhere (e.g., Pedersen *et al.*, 2010b; Rona, 2010).

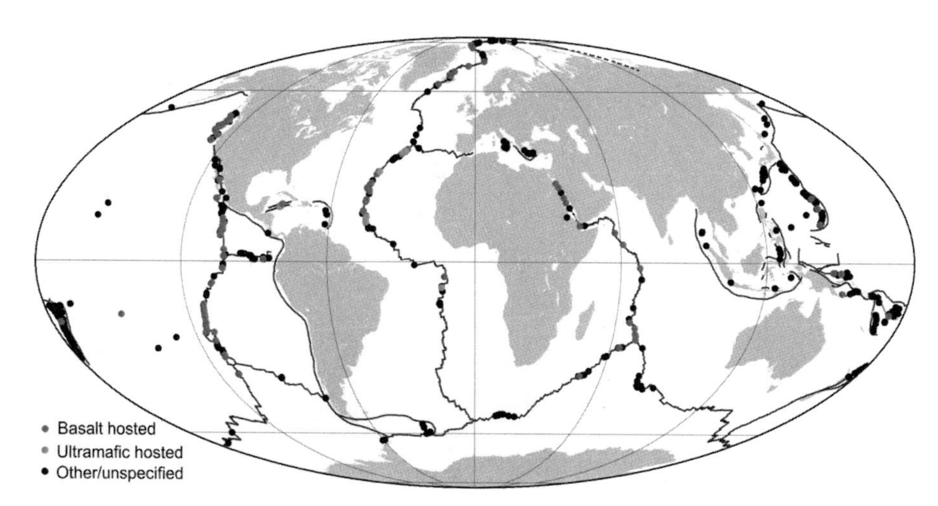


Figure 8.2 Global distribution of plate boundaries and hydrothermal vent fields, based on the data in the InterRidge vent database (http://www.interridge.org/irvents/). For colour version, see plates section.

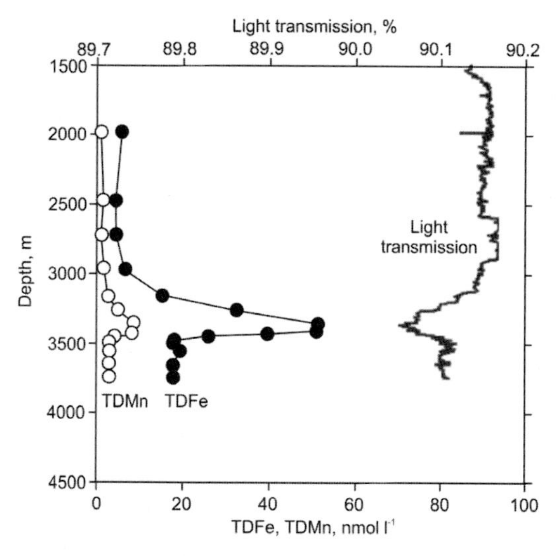


Figure 8.3 Profiles of total dissolvable Fe and Mn and light transmission in a hydrothermal plume on the MAR near 4° S, after German *et al.* (2008). The plume is clearly visible at 3400 m. Reprinted with permission from Elsevier, © 2008.

The international organisation InterRidge maintains a database of hydrothermal vent sites (http://www.interridge.org/irvents/) which at end 2012 contained 589 sites, over half of which are on MORs (Figure 8.2).

Hydrothermal vents produce buoyant plumes that rise to a level of neutral buoyancy, usually a few hundred metres above the vent (Lupton, 1995; Section 8.7). These plumes may be characterised by small anomalies in temperature, salinity, light attenuation (by sulphide precipitates), redox potential ($E_h$), dissolved manganese, $^3$He concentration, etc. (Figure 8.3). Many of these quantities can now be measured continuously *in situ*, so an

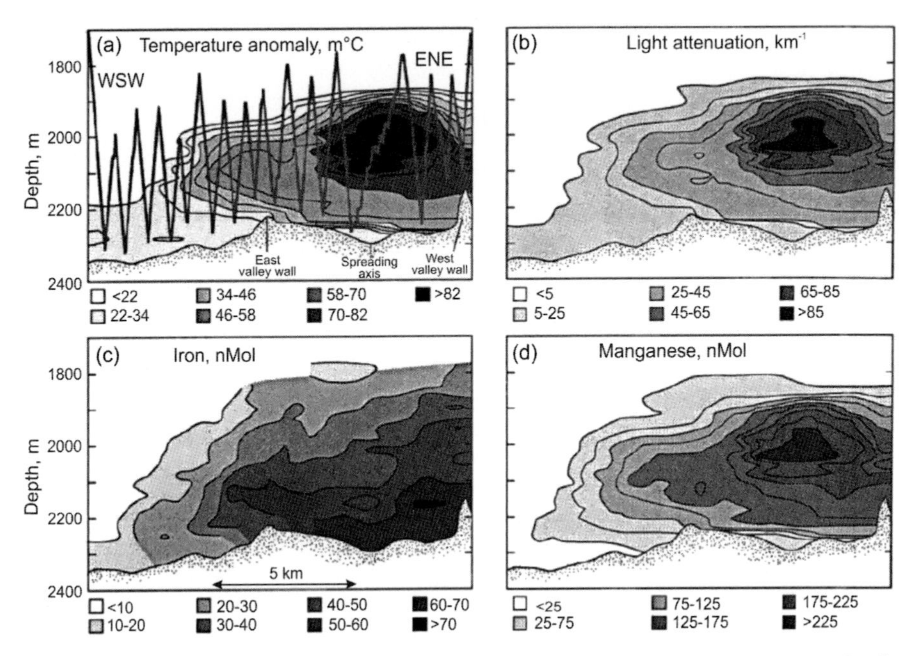


Figure 8.4 Example of a tow-yo survey across the median valley of Cleft Segment, Juan de Fuca Ridge, near 44°52′ N, after Chin *et al.* (1994). Profile is approximately 9 km long. Zigzag line in (a) shows sensor track; lower curve is sea floor. $Fe^{2+}$ has been oxidised to $Fe^{3+}$ and precipitated, so only correlates weakly with other parameters.

efficient way of surveying for hydrothermal vents is to map the distribution of plumes using these anomalies as tracers. This is done either by 'tow-yo' traverses where instruments are continuously raised and lowered through the expected depth range from a slowly moving ship (Figure 8.4) or, more recently, by using autonomous underwater vehicles (Figure 8.5).

Vents are now sought systematically by using light scattering or attenuation and the presence of excess dissolved Mn, Fe, $CH_4$, $H_2$ and $^3He$ in plumes. Optical attenuation is cheap and easy to measure, but can be ambiguous (e.g., from suspended sediment); the dissolved species are unambiguous and can verify optical predictions (Baker and German, 2004). Such techniques have enabled effective exploration for hydrothermal fields and vents on ridges where they are less frequent than the fast-spreading EPR. These include the MAR (German *et al.*, 1994; 1996b) and relatively inaccessible SWIR (German *et al.*, 1998; Tao *et al.*, 2009) and Arctic ridges (Pedersen *et al.*, 2010a).

These slower-spreading ridges have a rich diversity of hydrothermal venting, often with fluids, structures and fauna very different from those of the EPR (Fouquet *et al.*, 2010). In addition to 'black smoker' vents hosted in basaltic rocks, such as TAG, there are vents inferred to be hosted in ultramafic rocks, with very different physical and chemical properties (Section 8.5). The first high-temperature ultramafic-hosted vent field to be recognised was the Rainbow site on the MAR near 36°09′ N (German *et al.*, 1996; Donval *et al.*, 1997; Bougault *et al.*, 1998). Lost City – a low-temperature, ultramafic-hosted hydrothermal vent field with distinct properties – was discovered 15 km away from the MAR axis near 30°07′ N in 2000 (Kelley *et al.*, 2001).

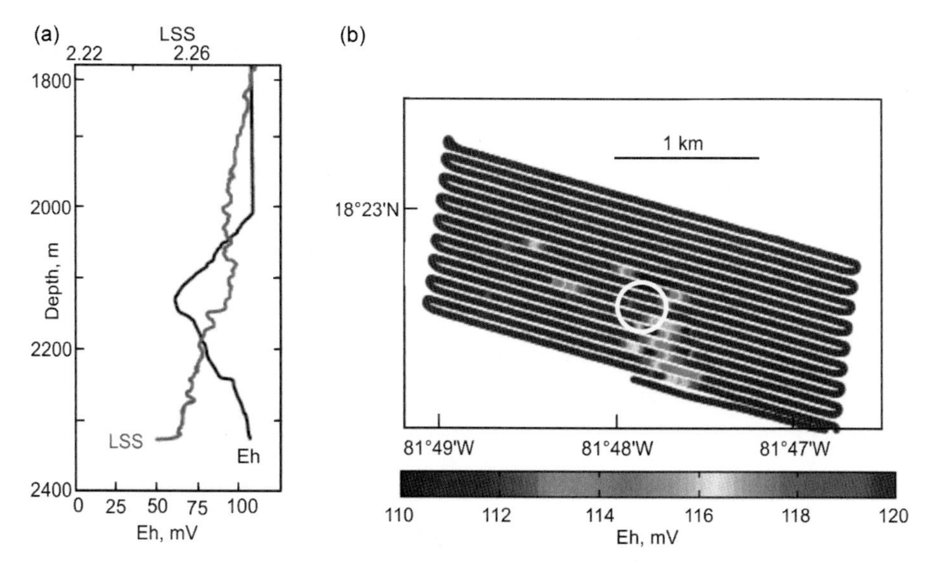


Figure 8.5 Hydrothermal plume survey for Von Damm vent field on the Mid-Cayman Spreading Centre, after Connelly *et al.* (2012). (a) Output of light-scattering sensor (LSS) and redox potential (Eh) from a tow-yo CTD cast, showing abnormally high light scattering and low redox potential at 2000–2300 m. (b) Map of redox potential (Eh) from an AUV survey run 60 m above the sea floor. Circle: Von Damm vent field. For colour version, see plates section.

Of the 319 MOR vent fields listed on the InterRidge database, 116 are reported as being hosted wholly or primarily in basalt, 19 wholly or partially in ultramafic rock, and the rest with host rocks that are unspecified (Figure 8.2). The hottest continuous venting recorded to date is 407 °C at Turtle Pits vent field near 5° S on the MAR, while the nearby Comfortless Cove field produced a brief (20 s) temperature fluctuation to 464 °C (Koschinsky *et al.*, 2008). The deepest vent field so far discovered is BeeBe at 4960 m on the Mid Cayman Spreading Centre (Connelly *et al.*, 2012), with an unconfirmed estimated exit temperature of 430 °C.

## 8.3 Basalt-hosted vent systems

Probably the commonest types of hydrothermal vent are high-temperature, basalt-hosted vents characterised by black smokers. These, and the accompanying cool vents and widespread diffuse venting, are the only types so far encountered on fast-spreading ridges, but are also found, with less frequency, on slow- and ultra-slow-spreading ridges. Those on the EPR occur in fields whose areas range from a few square metres to several hundred square metres. They are relatively short-lived, perhaps lasting a few years or decades. In contrast, those on slow-spreading ridges may be considerably larger and endure, at least intermittently, for ~100 ka.

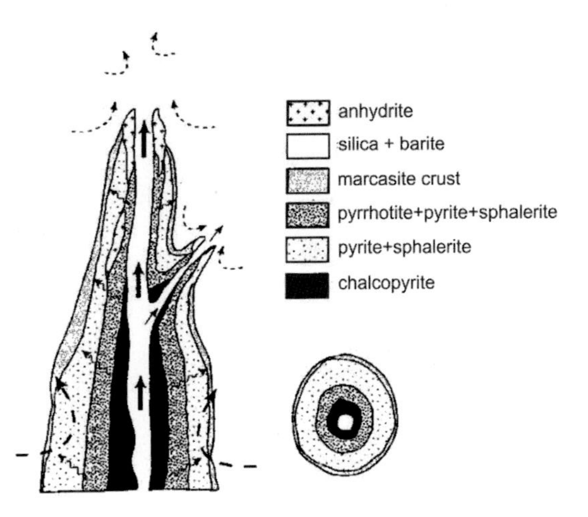


Figure 8.6 Vertical and horizontal cross-sections of a black smoker chimney, after Haymon (1983). Solid, straight arrows: direct high-temperature hydrothermal flow along open channels; wavy arrows: hydrothermal fluid diffusing through the porous chimney; broken arrows: seawater circulation. Reprinted by permission from Macmillan Publishers Ltd: *Nature* © 1983.

## 8.3.1 Black smoker vents

End-member hydrothermal fluids (i.e. before mixing with ocean-bottom water) are hot (generally about 300 °C–380 °C) and acidic (pH mostly 2.5 to 4.5). They typically have modest concentrations of $H_2$, high concentrations of $^3He$ (the latter indicating a source in the Earth's mantle), and up to 10 mmol kg$^{-1}$ of $H_2S$. The vent fluids also contain Ca, Cl, Cu, Fe, Mn, Zn, $H_2S$, and other elements (Von Damm, 1995). Magnesium is strongly depleted relative to seawater. On contact with the cold, oxidising ambient seawater, the metals combine with $H_2S$ to form sulphides that precipitate to produce the black 'smoke' of these vents.

These high-temperature vents typically build hydrothermal 'chimneys' around their orifices (Figure 8.1), although other types of structures also occur (Hannington *et al.*, 1995). The Ca and S from the hydrothermal fluid combine with oxygen to form anhydrite (anhydrous calcium sulphate, $CaSO_4$) which is precipitated, along with metal sulphides and some barite (barium sulphate, $BaSO_4$), to form the chimney. Different minerals are often arranged radially (Figure 8.6). Vent chimneys may grow to tens of metres high before collapsing to form a hydrothermal mound. The mound itself is often underlain by a mineralised 'stockwork' zone (Figures 8.7 and 8.8).

## 8.3.2 Medium- and low-temperature venting

White smokers emit cooler fluids than black smokers, and begin to form chimneys at ~100 °C–300 °C (Hannington *et al.*, 1995). Fluid flow rates tend to be lower, and chimneys smaller, than for black smokers. The white colour of the 'smoke' typically arises from

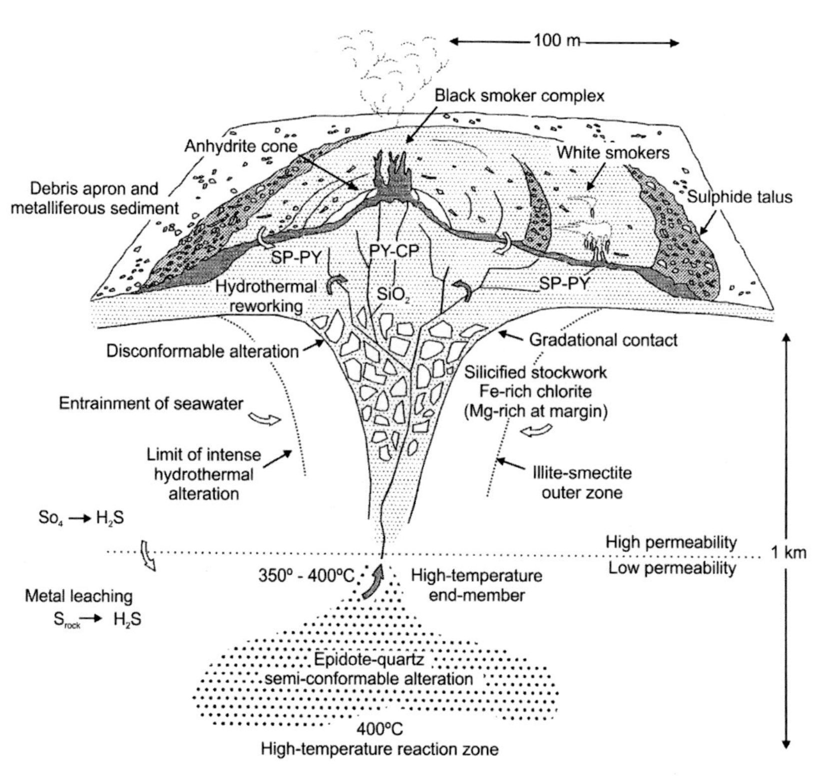


**Figure 8.7** Schematic diagram of the TAG hydrothermal mound, MAR, 26°08′ N after Hannington *et al.* (1995). Open arrows: seawater pathways; shaded arrows: hydrothermal fluid; SP: sphalerite; PY: pyrite; CP: chalcopyrite.

particles of silica ($SiO_2$), anhydrite and sometimes barite, which is common on the Juan de Fuca Ridge, but not on the EPR. White smoker chimneys contain these same compounds, and sometimes porous sulphides, indicative of higher temperatures in the past (Figure 8.9). Thus, vents may evolve from ‘white’ to ‘black’ and vice versa.

White smoker fluids may lack sulphides because their deep fluids are too cool to carry iron and sulphur to the sea floor, perhaps during early development of a hydrothermal system, or because cooling by conduction and mixing with seawater precipitated their metals in the subsurface (Hannington *et al.*, 1995; Tivey *et al.*, 1995). In the latter case, clear fluids may be vented at the sea floor. The same hydrothermal field may comprise both black and white smokers, indicating different pathways and cooling scenarios above a common heat source.

The extreme low-temperature end of hydrothermal circulation is represented by diffuse venting of cool (<100 °C), clear water. Schultz *et al.* (1992) estimate that such diffuse venting may carry five times as much heat as focussed, high-temperature venting at white and black smokers. Higher-temperature vents are often surrounded by diffuse venting. Diffuse venting is expected to be ‘ubiquitous’ and should be present along MOR crests and on ridge flanks out to 65 Ma (Schultz and Elderfield, 1997). It may result from either a modest heat source or extreme mixing of high-temperature fluids with ambient seawater and complete precipitation of minerals in the subsurface. Diffuse venting is much harder

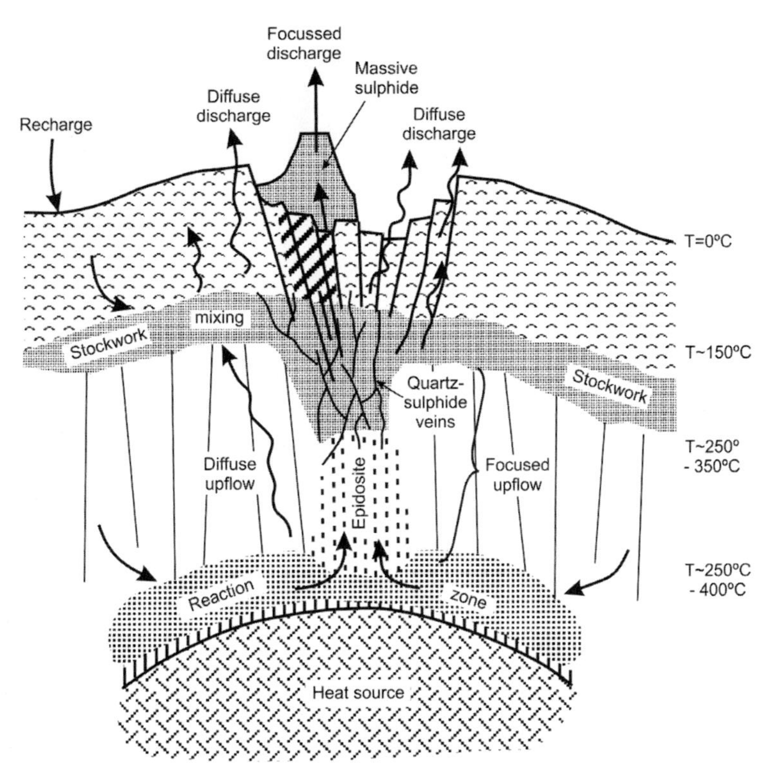


Figure 8.8 Conceptual view of a basalt-hosted hydrothermal system, after Alt (1995). The heat source is directly overlain by a reaction zone at the base of the sheeted dyke complex, and a stockwork lies at the top of the dykes and base of the lavas. Not to scale.

to detect than focussed venting, although instruments have been developed to measure its flow rates and temperatures (e.g., Schultz *et al.*, 1992, 1996; Sarrazin *et al.*, 2009).

### 8.3.3 Slow-spreading ridges

Basalt-hosted hydrothermal systems also occur on slow-spreading ridges. They are generally associated with an axial volcanic ridge or other neo-volcanic zone, and have similar chemistry to those discussed above. Examples are the Snake Pit (Karson and Brown, 1988) and Lucky Strike (Fouquet *et al.*, 1995; Langmuir *et al.*, 1997) fields on the MAR, Loki's Castle on the Mohns Ridge (Pedersen *et al.*, 2010b), and Beebe vent field on the Mid Cayman Spreading Centre (Connelly *et al.*, 2012). All are associated with neo-volcanic ridges and contain black smokers together with cooler and more diffuse venting.

Some high-temperature sites are not associated with neo-volcanic ridges, a good example being TAG. Its vent fluids are similar to those at neo-volcanic sites, but TAG is situated at the edge of the median valley ~4 km east of the neo-volcanic ridge in an area of intense cross-cutting faulting, which may control its position (Kleinrock and Humphris, 1996; but see Section 8.9). Such systems tend to be much larger and longer lived than the equivalents

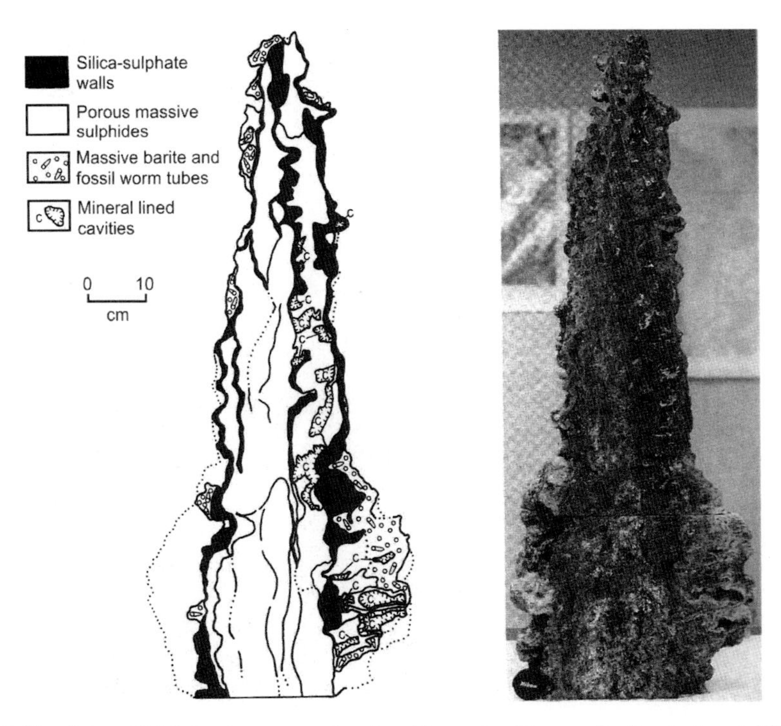


Figure 8.9 Drawing (left) and photo (right) showing cross-section of a white smoker chimney, after Hannington *et al.* (1995). Sample from CASM vent field on Axial Seamount, Juan de Fuca Ridge.

at fast-spreading ridges. For example, TAG contains several sub-fields, including a central mound ~200 m across, extending over a total area ~5 km square, and has persisted intermittently for 140 000 years (Lalou *et al.*, 1995; Humphris and Cann, 2000).

A distinctive type of vent system on slow-spreading ridges is hosted by not basaltic but ultramafic rocks. These occur typically in tectonised rather than volcanic areas, and will be considered in detail in Section 8.5.

Slow-spreading ridges have been less comprehensively surveyed than some fast-spreading ridges. Nevertheless, German *et al.* (2010) estimate that high-temperature vents occur at least every ~100 km along those ridges so far surveyed, and that half the MAR sites are long-lived, tectonically hosted systems.

## 8.3.4 Energy output

High-temperature hydrothermal systems on the EPR have power outputs of 10 MW to ~250 MW, the latter being three to six times the total theoretical heat loss for 1 km of ridge in 1 Ma (Macdonald *et al.*, 1980a; Lowell *et al.*, 1995). Ginster *et al.* (1994) estimate that individual smokers on the Juan de Fuca Ridge output from 0.1 MW to almost 100 MW, but this is only ~10% of the total advected flux estimated from measurements of the vent plumes overlying them (see Section 8.7).

German and Lin (2004) offer a conceptual model of hydrothermal activity on the EPR, in which episodic dyking occurs on an ~50 year timescale. A dyking event is followed by an event plume (Section 8.7.2) lasting a few days and releasing ~5% of the dyke's heat. Thereafter venting evolves on a timescale of 1–5 years, during which 20% of the heat is lost; finally there is a hydrothermally quiescent stage of 10–50 years with loss of the remaining 75% of heat.

Cann and Strens (1982) modelled long-lived vent systems and concluded that formation of a large sulphide deposit requires the crystallisation of ~1 $km^3$ of magma per hundred years for several hundred years. Cann *et al.* (1985) estimated that a 3 million ton sulphide deposit would need $3 \times 10^{19}$ J of heat; this could be achieved in 4000 years using 350 °C vent fluids, or 33 000 years using 250 °C fluid. These figures are comparable with the situation at TAG. The size and ages of the mineral deposits there suggest the building of the active high-temperature mound was accompanied by an average heat output of 0.5–2.5 MW, though this was intermittent and focussed into a number of short-lived (~10 year) periods of intense activity separated by gaps of thousands of years (Humphris and Cann, 2000). German and Lin (2004) consider that dyking alone may provide insufficient heat to build long-lived, tectonically hosted hydrothermal fields on slow-spreading ridges. They suggest that additional heat sources are required, such as the lower crust or exothermic serpentinisation (Section 8.5). They also suggested that along-axis focussing of melt and heat supply (Section 6.3) might help sustain the high heat flow needed at such sites.

## 8.4 Sub-sea-floor processes

Alt (1995) has given a comprehensive account of the chemical reactions occurring in basalt-hosted hydrothermal systems, largely based on the results of drilling through the upper crust at DSDP hole 504B.

### 8.4.1 Recharge zone

In the 'recharge zone', cold seawater enters the permeable volcanic layer, probably in many places, and circulates openly in the upper volcanics and to a lesser extent in the lower volcanics (Figure 8.8). There it reacts with rock at low temperature with two results. First, crustal oxidation produces minerals such as goethite (an iron oxyhydroxide); secondly, alkalis such as Na, K, Rb and Cs from seawater are fixed into crustal minerals such as celadonite (ferric mica) and the clay mineral nontronite. Off-axis, formation of zeolite minerals also contributes to alkali uptake, and formation of carbonates fixes significant amounts of $CO_2$ into the upper crust.

Circulation of seawater is more restricted in the closed-system regime of the lower volcanics and sheeted dykes, where most Mg is fixed in veins and breccias as saponite clay at $<\sim 170$ °C and in the mineral chlorite in dykes at $>200$ °C (Alt, 1995). Take-up of Mg in the crust is complemented by release of Ca into the circulating fluid. As the circulating water is heated, Ca forms anhydrite. In this regime, alkalis are leached from the crust at temperatures >150 °C, and above ~250 °C dissolved sulphate is reduced by oxidation of the crust.

Alt (1995) argues there is strong evidence for separate convection cells in the volcanic and sheeted dyke layers. The former are highly permeable and characterised by low-temperature (0° – 150 °C) reactions with high water–rock ratios (i.e. large volumes of water interacting with a given rock volume), whereas the sheeted dykes have low permeability, low water–rock ratios and reactions at temperatures of 230°–250 °C. These differing convection and reaction regimes result in a metamorphic boundary near the lava/dyke transition, which Carlson (2011) has argued may mark the boundary between seismic layers 2A and 2B (Section 5.3.2).

## 8.4.2 Reaction zone

Near the base of the sheeted dykes, the circulating fluid can come very close to the top of the magma chamber (Figure 8.8). Water cannot penetrate the magma, which is too plastic to support open fractures, so there is a 'cracking front' separating advective heat flow above from conductive heat flow below. Lister (1974) showed that such a boundary would develop at temperatures >1000 °C and depths of several kilometres. This region is referred to as the hydrothermal 'reaction zone', and is where the most intense heat and chemical exchange occurs. Temperatures and pressures here are ~340 °C–465 °C and ~35–55 MPa (Alt, 1995). At >350 °C the lower dykes and uppermost gabbros lose S and metals such as Cu and Zn to the circulating fluid. Some volatiles may pass from magma to the hydrothermal fluid, including $CO_2$ and $^3He$ (note that $^3He$, unlike $^4He$, occurs in substantial quantities only in the Earth's mantle, whereas other volatiles such as $CH_4$, Cl, $H_2$, $H_2O$ and $SO_2$ may or may not have a magmatic origin). There is some evidence for water reaching 600 °C–700 °C in the uppermost few hundred metres of gabbros, where amphibole may be formed. Hornblende and sodium plagioclase may be deposited at temperatures from ~500 °C–600 °C.

## 8.4.3 Phase separation

The composition of hydrothermal fluid is controlled not only by water–rock reactions, but often also by phase separation (Von Damm, 1995). The temperature and pressure in the reaction zone are close to the critical point of seawater, 407 °C and 29.85 MPa (Bischoff and Rosenbauer, 1988). Below this, a 'two-phase curve' separates the pressure–temperature conditions at which seawater exists as a liquid from those in which both liquid and vapour coexist (Figure 8.10). At most MOR axes the pressure at the sea floor is ≥26 MPa, so seawater below the critical temperature will normally be liquid. If seawater gets hot enough to cross the two-phase curve at less than the critical pressure it will boil into a low-salinity vapour which contains most of the dissolved gases and small quantities of metals, leaving a liquid phase with enhanced salinity but depleted in volatiles. For seawater beyond the critical point the two-phase curve continues, but when crossed in this 'supercritical' region a small amount of concentrated brine liquid condenses while the majority of the fluid and its volatiles, salts and metals go into the vapour phase. The existence of phase

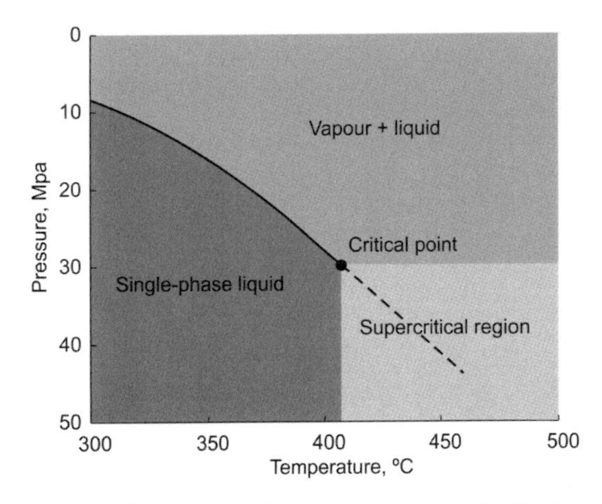


Figure 8.10 Diagram of the two-phase boundary (bold line) and critical point for seawater, after Von Damm (1995) and Koschinsky *et al.* (2008).

separation is attested by the recovery of both concentrated brines and low-salinity fluids from hydrothermal vents, as well as time-varying fluid compositions and end-member fluid temperatures at or above the critical point (e.g., Tivey *et al.*, 1990; Koschinsky, 2008).

The possibility of both sub-critical and super-critical phase separation produces a wide range of salinities and dissolved element compositions in the emerging hydrothermal fluids. The dense brines produced may be concentrated in fractures or in a closed circulation system above the heat source, communicating only episodically with the overlying hydrothermal fluids to affect their composition. Gabbros carrying hornblendes with up to 4% Cl content are attributed to reaction with such high-chlorinity brines (Alt, 1995).

## 8.4.4 Discharge zone

Near the critical point, the buoyancy of hydrothermal fluid increases dramatically, promoting focussed upflow back to the sea floor through a 'discharge zone'. Alt (1995) has suggested the return flow reaches the sea floor in a minimum of a few hours and maximum of less than a year, while the total residence time for fluid in the hydrothermal system is of the order of years.

In ophiolites, focussed upflow is recorded by ~100 m wide zones of epidosites (highly altered rocks containing epidote and quartz), occurring mainly in the lower part of the dykes and parallel to them. They could be feeder zones for the overlying massive sulphides, and are inferred to grade upwards into quartz and sulphide veins (Figure 8.8). These epidosites have been altered at 350 °C–400 °C and are enriched in Ca and Sr and depleted in Mg, Na, Zr, K, Cu, Zn and S.

On the sea floor, the exits of the upflow zones are marked by high-temperature (black smoker) vents and massive sulphide deposits, underlain by stockworks of quartz–pyrite and anhydrite veins (Figures 8.7 and 8.8), as found for example by drilling at TAG (Humphris *et al.*, 1996). The detailed structures and compositions of sea floor massive

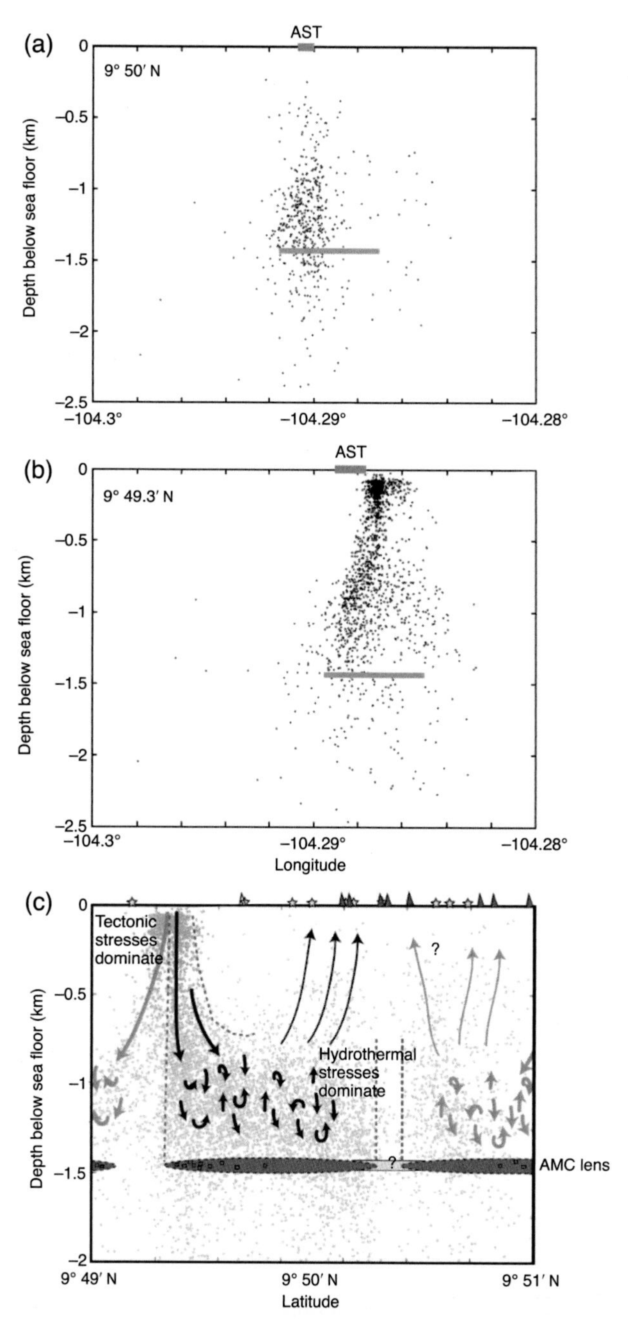


**Figure 8.11** Sections through the EPR near 9°50′ N, showing the distribution of microearthquakes and inferred hydrothermal circulation, after Tolstoy *et al.* (2008). (a) and (b) Across-axis sections with horizontal exaggeration ~1.6 showing the distribution of earthquakes near the inferred discharge and recharge zones, respectively. Grey bars at sea floor: axial summit trough (AST); grey bars at 1.4 km depth: axial magma chamber. (c) Along-axis section, with vertical exaggeration ~1.4. Blue dots: earthquakes interpreted to result from tectonic fracturing in the recharge zone at the segment end; grey dots: earthquakes inferred to result from hydrothermal stresses; red triangles and yellow stars: high- and low-temperature hydrothermal vents, respectively; black and grey arrows: possible hydrothermal flow; red: axial magma chamber, with a possible mush-filled break marked by orange band. Reprinted by permission from Macmillan Publishers Ltd: *Nature* © 2008. For colour version, see plates section.

sulphide deposits has been reviewed by Hannington *et al.* (1995). Some of the upwelling fluids may mix with seawater and deposit metals in the shallow subsurface, before exiting to the sea floor as low-temperature, metal-poor diffusive flows.

Hydrothermal cells have commonly been presented as lying in planes normal to the ridge axis, with recharge zones roughly on the same plate flow line as the vents but in older crust. However, Strens and Cann (1986) and Haymon (1996) have pointed out that the maximum permeability of the crust is likely to be parallel to the ridge-parallel faulting (Chapter 7), so that hydrothermal cells should tend to be oriented ridge-parallel. Tolstoy et al. (2008) mapped the distribution of micro-earthquakes in a well-surveyed part of the EPR, and concluded that they reflect a hydrothermal cell oriented along the ridge (Figure 8.11). They infer that a pipe-like cluster of earthquakes beneath a fourth-order discontinuity represents tectonic fracturing that generates a high permeability recharge zone (blue dots in Figure 8.11); a vertical, sheet-like band of earthquakes some 500 m wide and 2 km long (grey dots), centred under and along the spreading axis, connects with the pipe-like zone and shoals beneath mapped high-temperature hydrothermal vents; these earthquakes are interpreted as arising from hydrothermal stresses and indicate the positions of the reaction and discharge zones.

## 8.5 Ultramafic-hosted systems

### 8.5.1 Mixed basalt–ultramafic systems

A completely different class of hydrothermal vent systems is those hosted in ultramafic rock. Unlike basalt-hosted vents, to date these have only been discovered on slow- and ultra-slow-spreading ridges, where peridotite outcrops are relatively common or where seawater can readily access peridotite via major faults. Several vent fields, such as Rainbow (MAR, 36°14′N; German *et al.*, 1996a) and Nibelungen (MAR, 8°18′S; Melchert *et al.*, 2008), occur in non-transform offsets where the mafic crust is thin or patchy (Chapter 5) and large-scale normal faulting or detachment faulting is common (Chapter 7). Others, such as Logatchev (MAR, 14°45′N; Krasnov *et al.*, 1995), are in segment centres but where peridotite outcrops are common. Some, such as TAG, are sited on mafic rocks but may interact with ultramafic rocks at depth (Section 8.9).

Fluids from such ultramafic-hosted sites have high concentrations of $H_2$ and $CH_4$, which suggest a peridotite influence (Section 8.5.2). However, many sites, such as Rainbow and Logatchev, have high-temperature vents with black smokers that emit low-pH fluids moderately enriched in Si, Cu and Zn, consistent with reactions involving basalt or gabbro in addition to peridotite. A possible model to explain these differences was suggested by McCaig *et al.* (2007) and is discussed in Section 8.9.

The energy output of the Rainbow field has been estimated at ~0.5 GW from the properties of the plume lying above it – significantly larger than most basalt-hosted fields. The total fluid flux is ~450 l $s^{-1}$ and chemical fluxes are ~10 mol $s^{-1}$ for Fe and ~1 mol $s^{-1}$ for each of Mn and $CH_4$ (German *et al.*, 2010).

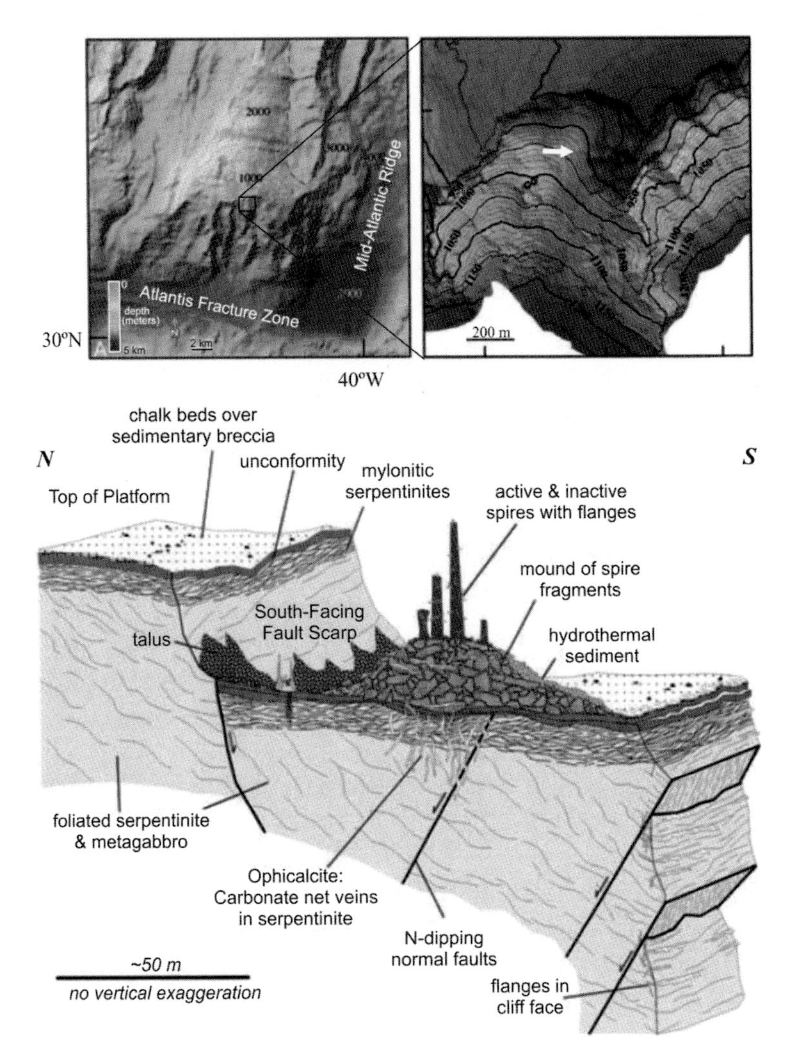


Figure 8.12 Location (top) and cross-section (bottom) of Lost City vent field. From Kelley *et al.* (2005). Reprinted with permission from AAAS. For colour version, see plates section.

## 8.5.2 'Pure' ultramafic systems – Lost City vent field

In addition to these 'mixed influence' sites, one appears to be hosted purely by ultramafic rocks and consequently displays distinct characteristics. This is Lost City vent field, on 1.5 Ma crust some ~15 km from the MAR axis at 30°07′N (Kelley *et al.*, 2001). It is located at the top of the detachment footwall associated with Atlantis Massif OCC (Chapter 7) where it intersects the north wall of Atlantis Transform, in a region of extensive serpentinite outcrop (Karson *et al.*, 2006; Figure 8.12). It is dominated by low-temperature fluids and carbonate precipitation, and is much larger and longer lived than most basalt-hosted fields (Früh-Green *et al.*, 2003; Ludwig *et al.*, 2011).

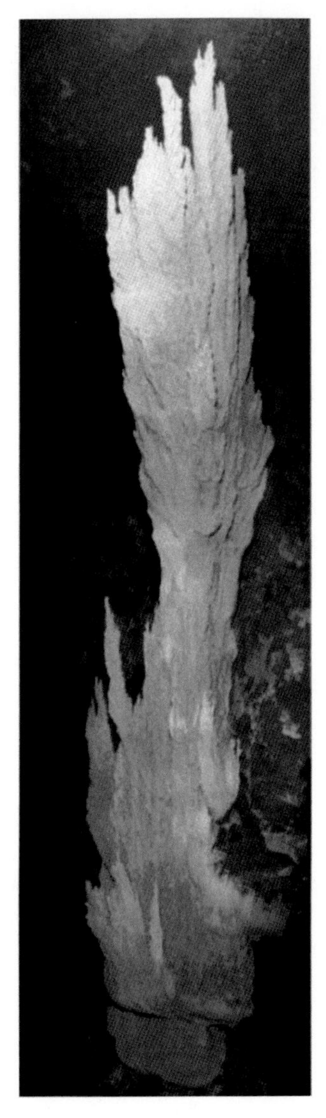

Figure 8.13 Actively venting, 10 m tall carbonate chimney at Lost City vent field. From Kelley *et al.* (2005). Reprinted with permission from AAAS. For colour version, see plates section.

Lost City hydrothermal fluids are vented at 40 °C–91 °C through a complex network of small channels that build steep-sided carbonate chimneys 30–60 m high (Figures 8.12, 8.13). Unlike basalt-hosted vents, these emit strongly alkaline fluids (high pH), rich in Ca, $H_2$ and $CH_4$ but devoid of Mg and metal sulphides (Kelley *et al.*, 2005). The fluids have near-ambient Si content and lack magmatic $CO_2$. They build chimneys not of sulphide but of calcium carbonate (Figure 8.13). Früh-Green *et al.* (2003) dated parts of the Lost City field radiometrically and documented ≥30 ka of hydrothermal activity. More recently, U/Th dating suggests activity may have been ongoing for ≥100 ka (Ludwig *et al.*, 2011).

The chemistry of ultramafic-hosted hydrothermal systems is controlled by serpentinisation reactions, whereby the minerals olivine and/or pyroxene in ultramafic rocks are hydrated, at temperatures <500 °C, to produce serpentine with the release of hydrogen and heat. For example (Schroeder *et al.*, 2002):

$$\underset{\text{olivine}}{6(\text{Mg, Fe})_2\text{SiO}_4} + 7\text{H}_2\text{O} = \underset{\text{serpentine}}{3(\text{Mg, Fe})_3\text{Si}_2\text{O}_5(\text{OH})_4} + \underset{\text{magnetite}}{\text{Fe}_3\text{O}_4} + \text{H}_2, \tag{8.1}$$

$$\underset{\text{olivine}}{4(\text{Mg, Fe})_2\text{SiO}_4} + 6\text{H}_2\text{O} = \underset{\text{serpentine}}{2(\text{Mg, Fe})_3\text{Si}_2\text{O}_5(\text{OH})_4} + \underset{\text{brucite}}{2\text{Mg(OH)}_2} + \text{H}_2, \text{ and} \tag{8.2}$$

$$\underset{\text{olivine}}{(\text{Mg, Fe})_2\text{SiO}_4} + \underset{\text{orthopyroxene}}{(\text{Mg, Fe})\text{SiO}_3} + 2\text{H}_2\text{O} = \underset{\text{serpentine}}{(\text{Mg, Fe})_3\text{Si}_2\text{O}_5(\text{OH})_4} + \text{H}_2. \tag{8.3}$$

The fluid's high pH (~10) prevents the transport of metals, so pure ultramafic systems do not precipitate metallic sulphides. Instead, the alkaline fluid promotes precipitation of carbonates and hydroxides, such as calcite, aragonite and brucite. These are deposited both on the sea floor and in the shallow subsurface (Figure 8.12). The requisite carbon is derived from ambient seawater (Kelley *et al.*, 2001; Kelley *et al.*, 2005; Ludwig *et al.*, 2006), whose dissolved $CO_2$ can be precipitated in a series of reactions, for example,

$$\underset{\text{Mg-olivine}}{4\text{Mg}_2\text{SiO}_4} + 4\text{H}_2\text{O} + 2\text{CO}_2 = \underset{\text{serpentine}}{2\text{Mg}_3\text{Si}_2\text{O}_5(\text{OH})_4} + \underset{\text{magnesite}}{2\text{MgCO}_3}, \tag{8.4}$$

followed by reactions such as

$$\underset{\text{Mg-olivine}}{4\text{Mg}_2\text{SiO}_4} + \underset{\text{diopside}}{\text{CaMgSi}_2\text{O}_6} + 6\text{H}_2\text{O} + \text{CO}_2 = \underset{\text{serpentine}}{3\text{Mg}_3\text{Si}_2\text{O}_5(\text{OH})_4} + \underset{\text{calcite}}{\text{CaCO}_3}. \tag{8.5}$$

Although it was once considered that heat from exothermic serpentinisation reactions might drive hydrothermal circulation at ultramafically hosted sites, Allen and Seyfried (2004) have shown that the heat thus produced would be insufficient at Lost City, and another heat source is required, either magmatic heat or heat deeply mined from cooling lithosphere.

## 8.6 Hydrothermal alteration of oceanic crust

The chemical exchanges described in the preceding sections produce profound changes in the chemical and mineralogical composition and possibly the physical properties of the oceanic crust. One of the most important is the hydration of the crust by the creation of such minerals as chlorite, amphibole, and serpentine. This water can be released on heating as oceanic lithosphere descends in subduction zones, promoting melting of the overlying mantle wedge. This produces island arc magmas that are more hydrous than MORB, which is derived from largely dry mantle. These hydrous magmas are more viscous (building the typical conical volcanoes of island arcs) and contain more volatiles, so that the accompanying eruptions tend to be much more explosive than those at MORs.

Other important chemical exchanges include the addition of O, Mg, Na and K to the crust, and removal of Ca, Cu, Zn, Fe, Mn, Si and S, amongst other elements, from the

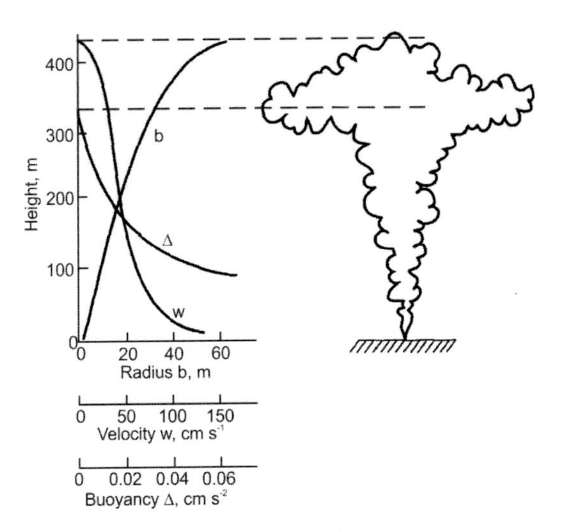


**Figure 8.14** Diagram of a hydrothermal plume, after Lupton (1995). Profiles at left show plume radius b, velocity w, and buoyancy acceleration, $\Delta = g\,(\rho - \rho_0)/\rho_0$, where $g$ is gravity, $\rho$ is density and $\rho_0$ is average local density. Calculations assume a heat flux of 60 MW and a typical density gradient for north east Pacific deep water.

reaction zone. Many of these are re-deposited on or near the sea floor to produce massive sulphide deposits, which are of potential economic importance, and are good analogues for many deposits now found in ophiolites that have been mined historically.

The serpentinisation process may be an important factor in the global carbon cycle, since $CO_2$ from the atmosphere, dissolved in seawater, can become fixed by carbonation of peridotite. It has been suggested that serpentinisation might be utilised as a means of human-induced carbon sequestration (Kelemen *et al.*, 2011).

# 8.7 Hydrothermal plumes

## 8.7.1 Buoyant plumes

Hydrothermal fluid exiting the crust is buoyant and will tend to rise, forming a hydrothermal plume whose dynamics can be calculated (Lupton, 1995). The plume may be narrow at its base (for example, if it erupts from a hydrothermal chimney), but as it rises it entrains ambient seawater by turbulent mixing and begins to widen (Figure 8.14). This portion is referred to as the 'buoyant plume'. As the plume fluid rises and entrains more seawater its buoyancy falls and so therefore does its rate of rise. When the plume reaches its level of neutral buoyancy it spreads out horizontally to form a 'non-buoyant plume' or 'effluent layer' (Lupton, 1995). At this level the plume still has some upward momentum, and tends to slightly overshoot the level of neutral buoyancy, forming a somewhat higher cap. Typical black smoker plumes rise 300–400 m above the vent and reach radii of a few tens of metres

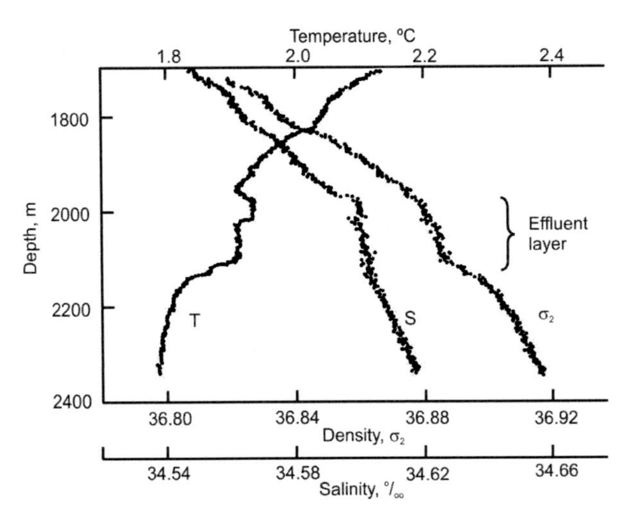


Figure 8.15 Profiles through an effluent layer 4.4 km east of Endeavour hydrothermal vent field, JdFR 47°57.5′ N, after Lupton, (1995). T: temperature; S: salinity; $\sigma_2$: potential density at 2000 m.

just below the effluent layer. The efficient detection of plumes by measuring water column properties (Section 8.2) depends on the fact that the effluent layer spreads widely, providing a reasonable-sized target. For example, the plume shown in Figure 8.5b has spread 300 m from the vent, while that in Figure 8.4 has spread several kilometres. At slow-spreading ridges with median valleys, the plume may be trapped by sea floor topography.

Temperature, salinity, density, light transmission and redox potential, together with dissolved Mn, $CH_4$, Fe, $^3He$ and other tracers, can all be used to map active plumes (Figures 8.15 and 8.16). However, many of these are non-conservative and have different plume residence times, ranging from ~days to decades (Lupton, 1995).

The EPR and Juan de Fuca Ridge have been intensively surveyed for hydrothermal plumes (Baker *et al.*, 1995a). These studies show a close correlation between plume intensity and vent location (Haymon *et al.*, 1991) and with the apparent magmatic budget as inferred from axial morphology and magma chamber characteristics (Baker *et al.*, 1994). Figure 8.16 shows mapped plumes compared with the positions of active hydrothermal vents and a recent eruption over part of the EPR axis.

## 8.7.2 Event plumes or 'megaplumes'

Some very large plumes have been mapped, and are believed to be responses to specific eruptions or other magmatic events. They are typically located some 800–1000 m above the sea floor, higher than normal hydrothermal plumes.

The first such plume was discovered over Cleft segment of the southern Juan de Fuca Ridge in 1986. It comprised a 700 m thick, 20 km wide eddy (Figure 8.17) with a mean temperature anomaly of 0.12 °C, and lay above a series of compositionally distinct plumes

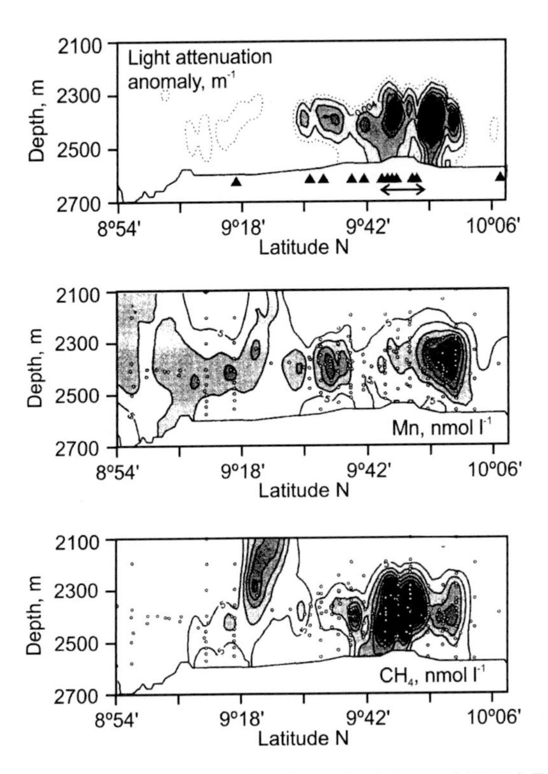


Figure 8.16 Vertical sections above the EPR axis south of Clipperton Transform, after Baker *et al.* (1995a). Top: light attenuation from tow-yo survey, contour interval 0.004 $m^{-1}$ (and additional dotted contour at 0.02 $m^{-1}$); middle and bottom: dissolved Mn and $CH_4$ from hydrocasts, contour interval 5 nmol $l^{-1}$. Triangles: mapped hydrothermal vents. Double headed arrow: extent of April 1991 eruption.

that apparently related to steady venting at the same location. This megaplume was estimated to have formed in a few days and had dissipated after 60 days. Its heat anomaly, $\geq 6.7 \times 10^{16}$ J, equalled the annual output of some 200 to 2000 high-temperature vents. It would have required 0.065 $km^3$ of hydrothermal fluid at 350 °C, representing the hydrothermal heat from some 0.4 $km^3$ of rock, to produce such an anomaly; alternatively, it could have been produced by contact cooling of ~0.1 $km^3$ of lava through 1200 °C, roughly the amount of rock accreted annually to the 55 km long spreading segment where it occurred (Baker *et al.*, 1987).

A second megaplume was observed 40 km to the north in 1987. The very high volume and heat fluxes associated with both megaplumes suggested they resulted from rift-related fracturing that abruptly increased the permeability of a hydrothermal reservoir and allowed it to empty suddenly (Baker *et al.*, 1989). Subsequent detailed surveys showed the presence of new pillow mounds and sheet flows that had been erupted between 1983 and 1987 (Figure 8.17), strongly implying a link to the megaplumes and supporting the idea that they were caused by a major volcanic rifting event (Embley and Chadwick, 1994).

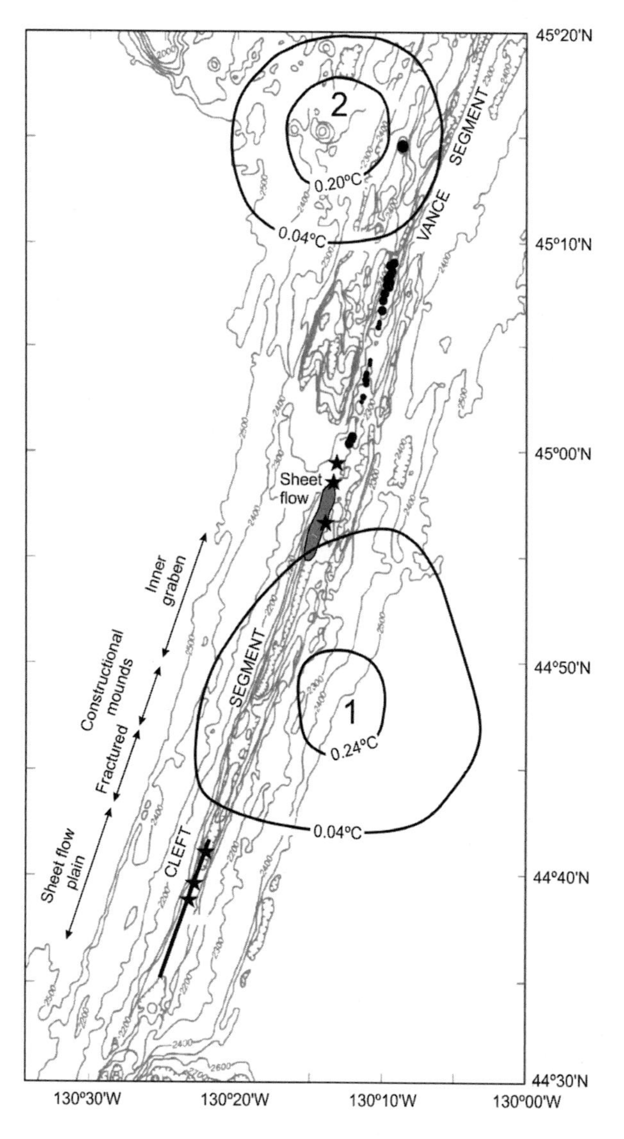


Figure 8.17 Cleft and Vance segments of the Juan de Fuca Ridge, showing locations of megaplumes and related structures, after Embley and Chadwick (1994). Megaplumes indicated by concentric contours of temperature anomaly labelled 1 and 2. Stars: high-temperature hydrothermal vents; grey: recently erupted sheet flow; black dots: pillow mounds. Grey contours: bathymetry at 100 m (and in median valley 20 m) intervals. Solid black line: the 'cleft'.

In 1993, a dyking event was observed in the Co-axial segment of the Juan de Fuca Ridge, evidenced by migrating earthquake epicentres and new eruptions, and was accompanied by the formation of two new short-lived megaplumes (Baker *et al.*, 1995b; Chadwick *et al.*, 1995; Embley *et al.*, 1995). Subsequently, other megaplumes have been observed around the world.

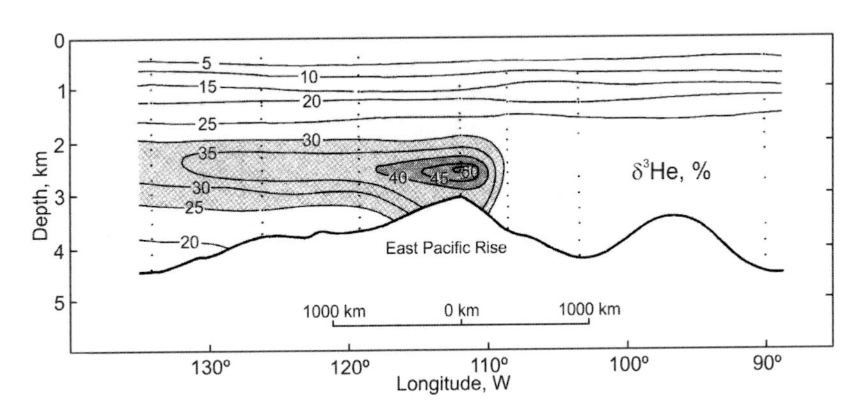


Figure 8.18 Distribution of $^3$He measured in hydrocasts across the EPR, after Lupton (1995). Heavy line is topographic profile; $\delta^3$He is the percentage departure of the observed $^3$He/$^4$He ratio from the atmospheric ratio.

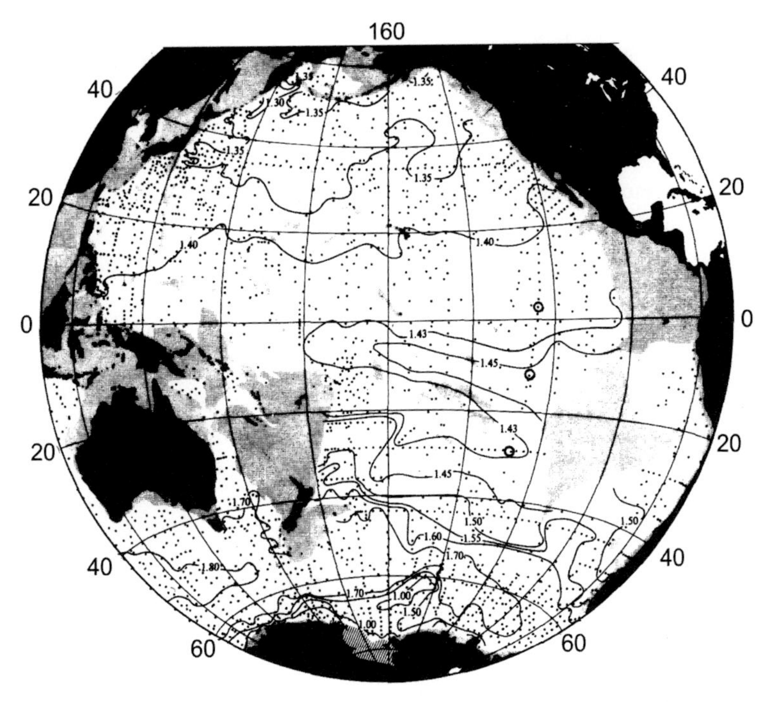


Figure 8.19 Contours of ocean potential temperature, in °C, on a constant density surface at ~2500 m depth across the Pacific, after Lupton (1995). Note the strong westward equatorial plume.

### 8.7.3 Far-field plumes

Lupton and Craig (1981) reported an ocean-scale plume of $^3$He (a conservative tracer) extending over 2000 km westward from, and at the depth of, the crest of the EPR (Figure 8.18). The same effect can be seen in temperature data across the Pacific (Reid, 1982; Figure 8.19). The asymmetric nature of this plume initially suggested a regional westward current flowing across the EPR. However, this contradicts physical oceanographic understanding, and Stommel (1982) showed that it is in fact a product of the vertical pumping of the buoyant hydrothermal flux combined with geostrophic flow. Lupton (1995) observed that the Pacific $^3He$ anomaly is at the average depth of the effluent layer from continuous venting, not at the shallower depth equivalent to event plumes, arguing that event plumes do not contribute a major fraction of the hydrothermal input to Pacific waters.

Because $^3He$ and temperature are conservative tracers, they tend to show the long-term integrated effects of hydrothermal fluxes from the rise crest. Lupton *et al.* (1985) combined observations of seawater entrainment in hydrothermal plumes with theoretical calculations of the heat flux from ridges, scaled linearly with spreading rate, to estimate that the global heat flux from hydrothermal venting at MORs is 1.98 TW. For comparison, the total global geothermal heat flux is some 40–47 TW (Pollack *et al.*, 1993; Hasterok *et al.*, 2011).

## 8.8 Hydrothermal vent biology

One of the most striking aspects of the discovery of MOR hydrothermal vents was the fact that they support rich and complex faunal ecosystems. Their food webs are based on chemoautotrophic microbes as primary producers (Lutz and Kennish, 1993); they are thus unreliant on photosynthesis as an energy source (although they do ultimately rely on it as a source of oxygen). Partly because of these novel ecosystems, including anoxic methanogenic microbes that are truly independent of photosynthesis, submarine hydrothermal vents have been suggested as sites for the origin of life on Earth (Baross and Hoffman, 1985; Corliss, 1990; Martin *et al.*, 2008). This idea continues to receive serious consideration, not least in the exobiology community (e.g., Jakosky and Shock, 1998; Russell *et al.*, 2010; Schwarz, 2011). Although the deep sea floor generally supports very little life, faunal density at hydrothermal vents is high, with biota frequently completely covering the substrate (Figures 8.20 and 8.21).

### 8.8.1 Vent fauna

Hydrothermal vent communities all rely for their base energy on endemic microbes that either live in or around vents, are symbiotes in multicellular vent organisms, or live on the bodies of such organisms. Microbes can be so plentiful they give the water a milky appearance, and may coat the sea floor in dense bacterial mats (Figure 8.22). Autotrophic microbes utilise principally the sulphide ions in vent fluids to convert $CO_2$, $H_2O$ and $NO_3$ (nitrate) into essential organic compounds via chemosynthesis (Hessler and Kaharl, 1995). In addition, heterotrophic microbes live on the by-products of other vent organisms. Vent microbes provide the base of the food chain in at least four scenarios: in the emergent and dispersing plume, in particulates used by vent-based suspension feeders, in microbial mats consumed by grazers, and as symbionts in vent macro-organisms (Tunnicliffe, 1991).

Vent microbes belong to two of the three fundamental domains of life: archaea and bacteria (Woese *et al.*, 1990). These occur near the base of the molecularly defined phylogenetic

Figure 8.20 Tube worms (*Riftia pachyptila* and *Tevnia jerichonana*) and brachyuran crabs (*Bythograea thermydron*) covering the sea floor at the Genesis hydrothermal vent site, EPR 13° N, after Lutz and Kennish (1993). For colour version, see plates section.

Figure 8.21 Dense aggregation of Alvinocaridid shrimp on an active chimney in the Beebe vent field, Mid-Cayman Spreading Centre, after Connelly *et al.* (2012). For colour version, see plates section.

tree (Figure 8.23), supporting the idea that vents may have been the origin of life on Earth (Jannasch, 1995). These creatures are extremophiles, inhabiting zones that are either too hot or too toxic for other organisms, but may have been common in the early Earth. Some bacteria can tolerate temperatures up to 121 °C (Kashefi and Lovley, 2003).

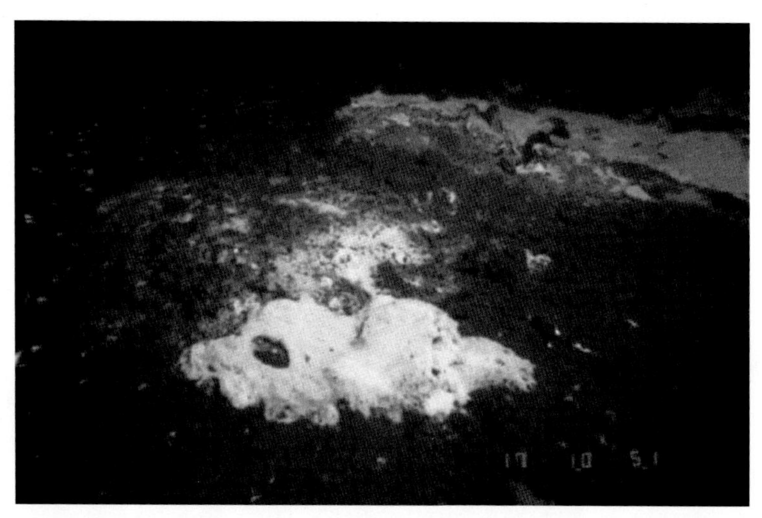

Figure 8.22 Bacterial mats on the floor of the Guaymas Basin, Gulf of California, after Lutz and Kennish (1993). For colour version, see plates section.

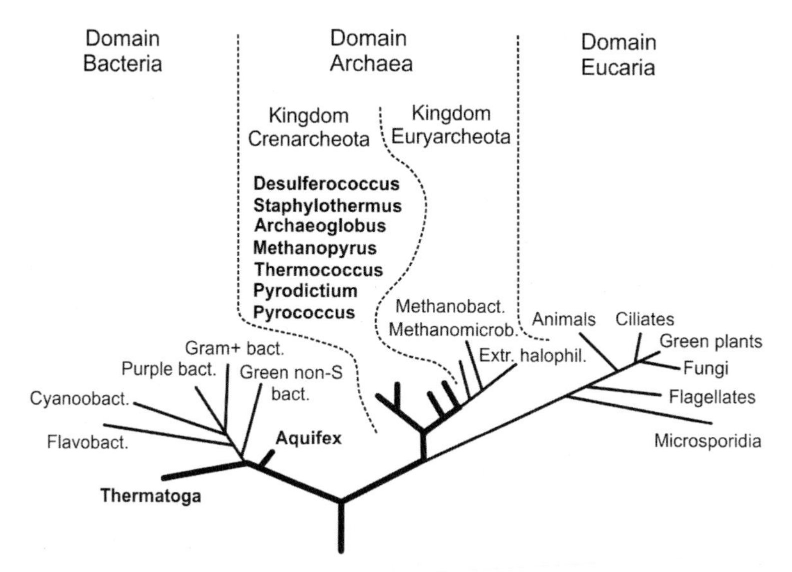


Figure 8.23 Evolutionary phylogenetic tree based on molecular analyses, after Jannasch (1995). Vent bacteria and archaea are superimposed (bold), showing their positions near the base of the tree.

In many vents, especially on the EPR and Galapagos Spreading Centre, vent openings are dominated by animals such as tubeworms (Figure 8.20), clams and mussels, all of which host endosymbiotic bacteria. In other vents, symbiotic bacteria are less important. Tubeworms are absent from Atlantic vents, which tend to be dominated by bresiliid shrimp (Figure 8.24). These feed on suspended particles produced by bacterial chemosynthesis (Williams and Rona, 1986; Vereshchaka, 1996).

Figure 8.24 Alvinocaridid shrimp on the Von Damm vent field, Mid-Cayman Spreading Centre, after Connelly *et al.* (2012). For colour version, see plates section.

Regions near but not in vents are usually dominated by sessile suspension feeders, including polychaete worms, barnacles, anemones, and bivalve molluscs, with fish, whelks and galatheid crabs living among them (Hessler and Kaharl, 1995).

Beyond the immediate vicinity of the vents can be found heterotrophic organisms that are not necessarily adapted to vent life, but live on its products and come as close as they can without being poisoned by vent fluids, which for non-vent animals can be toxic even at great dilution (Hessler and Kaharl, 1995). Such organisms include enteropneusts (acorn or spaghetti worms), dandelion siphonophores, holothurians, and some large protozoans. Rattail fish and large spider crabs may be found both here and close to vents, where they scavenge or prey on other vent creatures.

Many organisms are specially adapted for life at vents, for example those carrying symbiotic bacteria. There are steep chemical and thermal gradients close to vents, and only narrow habitable zones. The shrimp *Rimicaris exoculata*, common at Atlantic vents, is adapted to the deep sea by being devoid of eyes, but carries a high concentration of the visual pigment rhodopsin in a dorsal 'eye patch'. This allows the shrimp to detect the thermal radiation from hot vent fluids, enabling it to occupy the habitable part of the steep thermal gradient (Williams and Rona, 1986; Pelli and Chamberlain, 1989). Some adaptations allow organisms, such as bathymodiolid mussels, to move progressively from shallow to deep environments. Others, such as bresiliid shrimp, have diversified between continental margin seeps and hydrothermal vents, or entirely in vent environments (Van Dover *et al.*, 2002).

Most vent species are endemic to hydrothermal vents, though many show close evolutionary ties at a higher taxonomic level to the fauna of other deep-sea, reducing environments, such as whale drops or sea floor cold seeps (Tunnicliffe *et al.*, 1998). These species may have origins in the adjacent deep or shallow sea, but most have no extant close relatives. Vent fauna can be endemic at high taxonomic levels up to at least Order, so many taxa and species have been thought to be of considerable antiquity (Tunnicliffe, 1991; Hessler and Kaharl, 1995). This antiquity has caused some to suggest that deep-sea vents may have acted as refuges from the climatic and other primarily shallow-water effects causing faunal mass extinctions. However, Van Dover *et al.* (2002) find no strong fossil or molecular evidence for ages greater than about 100 Ma. Over 75% of vent species occur at only a single (known) site (Tunnicliffe *et al.*, 1998).

## 8.8.2 Faunal distribution and biogeography

Vent communities can change substantially over time scales of years, and some sites are re-populated by new species recruited from afar (Mullineaux *et al.*, 2010). Because vents are ephemeral, vent organisms must have efficient long-range dispersal mechanisms to preserve their lineages or rapidly colonise new sites (Van Dover *et al.*, 2002). Variations in dispersal ability may depend on such things as the feeding strategies of vent larvae (Hessler and Kaharl, 1995), and will affect species dispersal or 'gene flow'. Deep-sea currents may aid dispersion of some species (Van Dover *et al.*, 2002). Genetic studies show tubeworm dispersal is restricted to ~100 km, leading to so-called 'stepping-stone' dispersal along the ridge crest. Other organisms such as mussels and clams apparently range over thousands of kilometres. Smaller-scale barriers to dispersal may include ridge segmentation (e.g., median valleys separated by inter-segment sills, long-offset transform faults), and large variations in ridge depth (such as volcanic plateaus like the Azores or deeps such as Hess Deep) that species may not be able to bridge. The effectiveness of these different barriers at limiting dispersal will depend on the particular requirements of different species.

The biogeography and evolution of hydrothermal (and cold seep) fauna were reviewed by Van Dover *et al.* (2002). They found six distinct taxonomic groups inhabiting different geographic regions. More recently, Rogers *et al.* (2012) have suggested eleven groupings, and resolved differences along the EPR north and south of Galapagos triple junction and north and south of Easter microplate (Figure 8.25). At least some of the geographic isolation may be related to plate tectonic changes in the connectivity of spreading ridges and ocean basins (Tunnicliffe and Fowler, 1996; Hashimoto *et al.*, 2001). Some such isolations are relatively recent, for example closure of the Panama Isthmus at ~3 Ma which isolated the central Pacific from the central Atlantic oceans, and the separation of Juan de Fuca and Gorda Ridges from the northern EPR at ~28 Ma. Others are much older, such as the isolation of the northern MAR from the Indian Ocean ridges at ~90 Ma.

Recent studies have begun to fill some of the biogeographic gaps. Connelly *et al.* (2012) find assemblages in the Mid-Cayman Spreading Centre with closer affinities to the 4000 km-distant MAR than with the much nearer cold seeps in the Gulf of Mexico. German *et al.* (2008) found shrimp at Red Lion vent site on the southern MAR, but not at the nearby

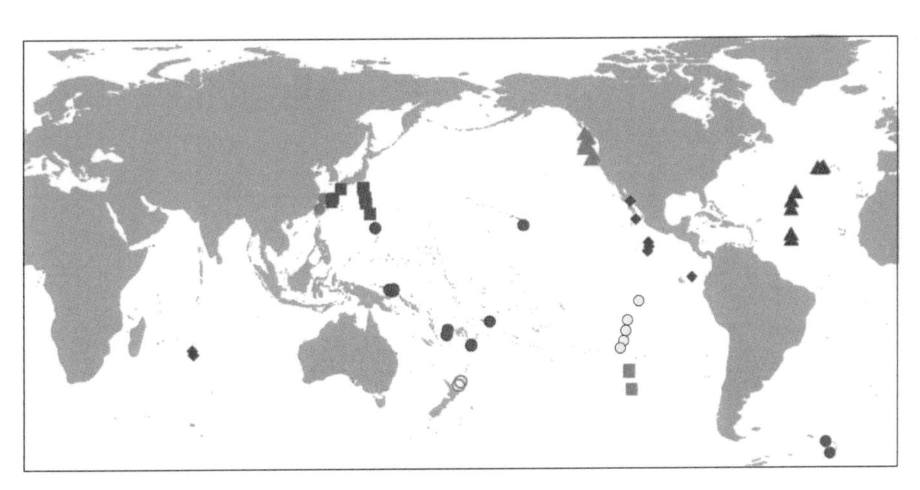

Figure 8.25 Eleven resolved biogeographic regions identified by different symbols at the vent sites (including some non-ridge sites), after Rogers *et al.* (2012). This figure does not include recent discoveries in the Arctic (Pedersen *et al.*, 2010b), South Atlantic (German *et al.*, 2008), or Southwest Indian Ridge (Tao *et al.*, 2012). For colour version, see plates section.

Turtle Pits and Wideawake sites, where bivalve shells were abundant. Vent fields at ~50° E on the SWIR have fauna that indicate a complex affinity with those on the southern MAR, Central Indian Ridge and southwest Pacific (Tao *et al.*, 2012). The Kairei field near the Rodriguez triple junction in the central Indian Ocean is dominated by shrimp and anemones, similar to vents on the MAR, but includes organisms known previously only from the Pacific; however, there is a somewhat closer affinity to the western than eastern Pacific or Atlantic (Hashimoto *et al.*, 2001). Loki's Castle vent field on the Mohns Ridge has a novel fauna distinct from that of the MAR to the south; this may have developed by migration and local specialisation from cold seeps and from the Pacific (Pedersen *et al.*, 2010b). Finally, the East Scotia Ridge appears to represent a new biogeographic province (Rogers *et al.*, 2012).

## 8.9 Controls on the distribution of hydrothermal vents

It is generally agreed that hydrothermal venting requires the combination of a suitable heat source with sufficient crustal permeability to allow seawater to access that source. Cannat *et al.* (2004) calculated that at fast-spreading ridges the combined specific heat of cooling and latent heat of crystallisation in mafic rocks can provide some 44–75 MW $km^{-1}$, compared to ~25.5 MW $km^{-1}$ at the centre of a slow-spreading ridge segment and ~16 MW $km^{-1}$ (including ~0.5 MW $km^{-1}$ from serpentinisation reactions) at magma-poor, slow-spreading segment ends.

Early studies on the EPR suggested that hydrothermal activity increases with magmatic budget (Francheteau and Ballard, 1983; Haymon, 1996). Recent work shows a good correlation between magma supply and the frequency of groups of vents, and particularly

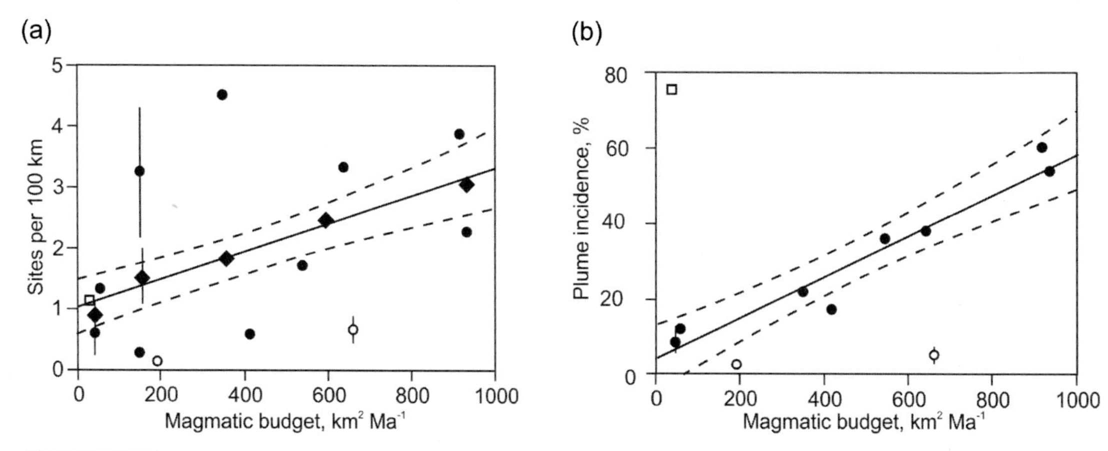


**Figure 8.26** Plots of (a) vent and (b) plume frequency versus 'magmatic budget', after Baker and German (2004). 'Magmatic budget' is full spreading rate times crustal thickness; 'plume incidence' is proportion of ridge axis overlain by hydrothermal plumes. Vertical bars: ranges of estimates, where available; open circles: hotspot-influenced ridges (Reykjanes Ridge, Southeast Indian Ridge); open square: Gakkel Ridge; black diamonds: data binned by spreading rate; solid lines: linear regression fits to un-binned data; dashed lines: 95% confidence intervals. Regression line in (a) omits hotspot ridges and in (b) omits Gakkel Ridges which has unusual hydrographic characteristics.

hydrothermal plumes, provided long sections of ridge are considered (German and Parson, 1998; Baker and German, 2004; Figure 8.26). This suggests that variability in magma (or heat) supply is the primary control on the distribution of hydrothermal sites, with permeability a secondary control. Permeability is probably increasingly important as spreading rate decreases and the larger faults developed at slow spreading rates mine deeper heat, leading to sites associated with highly fractured sea floor and OCCs.

In the case of venting associated with the neo-volcanic zone the heat source is almost certainly a body of sub-axial magma or other intruded hot basaltic rock, and this is borne out by the chemistry of vent fluids (Section 8.3). For many vents associated with ultramafic outcrops, the chemistry and/or power output suggest the involvement of both mafic and ultramafic rocks (e.g., Kelley *et al.*, 2001; Douville *et al.*, 2002; Allen and Seyfried, 2004).

Thus, in addition to slow-spreading vent fields associated with the neo-volcanic zone, there is a class that exhibit a variety of reaction temperatures and complex mixtures of features associated with both mafic and ultramafic rock. McCaig *et al.* (2007) present a model that explains such systems in terms of different stages of the development of detachment fault systems (Figure 8.27). Hot gabbroic plutons are intruded into the detachment footwall beneath the neo-volcanic zone; seawater percolates deep into the crust to mine heat from them, and is discharged via the highly permeable fault zone and overlying basalts without interacting significantly with peridotite, to form TAG-style vents (Figure 8.27a). If the seawater reaches a shallower, cooler gabbro body that has been displaced from the neo-volcanic zone by movement on the detachment fault, it can interact with both the gabbro and the serpentinised peridotite of the footwall before discharging in the median valley wall with mixed properties, like Rainbow (Figure 8.27b). Finally, off-axis, low-temperature

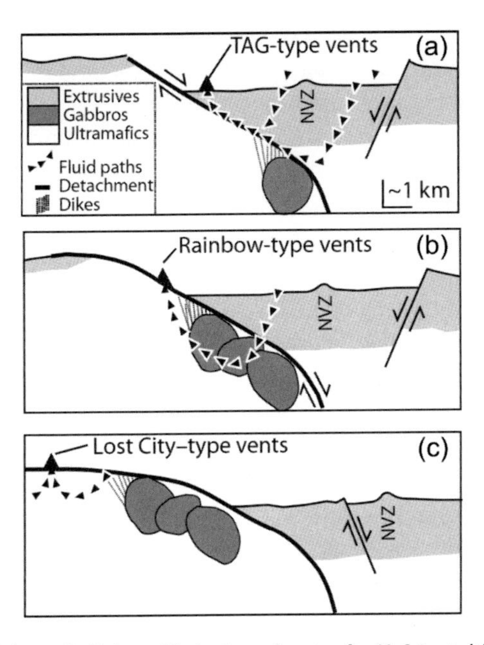


Figure 8.27 Models of different styles of ultramafically hosted hydrothermal vents, after McCaig *et al.* (2007). Bold lines: detachment faults; light lines: other normal faults; light grey: volcanic crust; dark grey: gabbro plutons; white: mantle peridotite; small triangles: water flow paths; large triangles: hydrothermal vent sites.

circulation can take place entirely within the peridotite footwall, driven mainly by residual magmatic heat, as at Lost City (Figure 8.27c).

## 8.10 Summary

The brittle, upper crust near MOR axes is permeable, allowing ingress of seawater in a so-called recharge zone. In basalt-hosted systems at fast- and parts of slow-spreading ridges, it penetrates to a reaction zone at the gabbro/dyke boundary where it reaches 350 °C to >400 °C, facilitating significant chemical exchanges with the oceanic crust. The hot water returns to the surface to debouch onto the sea floor in the discharge zone. There is a variety of venting into the ocean. Black smokers have high-temperature (up to ~400 °C), acidic fluids and carry precipitated particles of manganese and iron oxides, sulphides and other materials, including magmatic and non-magmatic volatiles. White smokers have cooler fluids (100 °C–300 °C) with less iron content and fewer dark particulates. Finally, areas of low-temperature diffuse venting surround the black and white smoker vents. The high- and medium-temperature fluids usually exit through chimneys built of their precipitates, which subsequently collapse to form broad mineral mounds. Underlying these mounds may be a rich stockwork of minerals deposited in the shallow subsurface. Small, high-temperature vent fields are common at fast-spreading ridges but may only last a few tens of years.

They appear somewhat less common at slower spreading rates, where individual vent fields may be larger and longer-lasting, up to tens of thousands of years, building large mineral deposits that may have economic value.

Vent fields such as those described in the previous paragraph are mostly sited in neo-volcanic zones, and their chemistry is indicative of interactions with basaltic rocks. At slow-spreading ridges there are also vent fields that are either hosted in ultramafic rocks or show evidence of interaction with both ultramafics and basalts. Interaction with peridotite produces alkaline fluids, and has a major effect on the nature of the lithosphere by converting strong peridotite to weak serpentine. In a purely ultramafically hosted system the hydrothermal fluids may be dominated by carbonates rather than sulphides and build carbonate-based chimneys.

Hydrothermal fluids arising from vent fields are buoyant and rise into the ocean as hydrothermal plumes, which spread out at their level of neutral buoyancy several hundred metres above the sea floor. Searching for the distinctive temperature, chemical and light-scattering signals within such plumes has proved an effective way of exploring for active vent fields. The strongly diluted signal from such plumes can spread and be detected across thousands of kilometres of ocean. Rarely, much larger plumes are observed, containing several orders of magnitude more effluent than a standard vent plume. These have been identified with individual volcanic eruptive episodes and named 'event plumes'.

Hydrothermal vents host distinct faunal ecosystems. Their food webs are based on chemotrophic microbes that rely on the direct oxidation of chemical species in the vent waters. Many of these microbes are extremophiles and amongst the most ancient phylogenies known. Other vent organisms either live symbiotically with some chemotrophes or graze on them, while higher levels predate on these or scavenge dead organisms.

Vents in different geographic areas tend to host different types and combinations of organisms, though the interrelations between these is still not fully worked out. Since vent fields are inherently scattered and ephemeral, the ways in which they are populated must depend on the dispersal strategies of vent organisms, and probably mainly on their larval stages.

# 9 Summary and synthesis

Mid-ocean ridges are one of the three types of tectonic plate boundary, where lithosphere is continually created and spread away from the ridge axis. There are profound differences across the world in the details of the processes involved, although the basic principles are clearly linked. These differences are thought mostly to reflect differing degrees of melt produced in the mantle and ultimately delivered to the crust. A convenient but approximate proxy for mantle melting is the spreading rate: generally, greater melting occurs at faster spreading rates, because heat is advected faster there. There are some important departures from this simple association, which are not yet fully understood, although variations in mantle temperature probably play a part. Important too may be variable mantle fertility: virgin mantle melts easily, but the residual mantle left over from past melting, if again brought into the melting zone, will be harder to melt. This chapter first summarises features that are common to most ridges; Sections 9.2–9.5 summarise features apparently dependent on spreading rate; and finally Section 9.6 considered the anomalies.

## 9.1 Common features

Mid-ocean ridges (MORs) are the places where plates separate and new oceanic lithosphere is formed. As that lithosphere moves away from the ridge axis it ages, cools, becomes denser and less buoyant, and slowly sinks. Sea floor depth is proportional to the square root of age, as predicted by simple thermal cooling models. The forces driving plate separation derive partly from the ridge topography, which applies a pressure gradient tending to separate the plates (the 'ridge-push' force), and partly from far-field forces such as the weight of subducting lithosphere ('slab-pull'). As plates age and cool they thicken and become stronger. The base of the lithosphere may be defined by the brittle–plastic transition, which is often approximated by the 750 °C isotherm. Thus plate thickness as well as depth increases as the square root of age, from a few kilometres at the ridge axis to tens of kilometres or more in old lithosphere.

Spreading centres, as the places where lithosphere is created, usually run roughly normal to the plate separation or 'spreading' direction, though they are offset along their lengths by a variety of structures that separate the ridge into numerous 'spreading segments'. The largest offsets, of more than 30 km, are transform faults, and are spaced hundreds of kilometres apart along MORs. Shorter, 'non-transform' offsets occur at kilometre to hundred kilometre spacing. At transform faults, plates slide past each other so these boundaries necessarily follow the spreading direction. Non-transform offsets are not so constrained, and may migrate along the ridge axis.

As plates separate, the underlying asthenosphere wells up adiabatically and begins, at a depth of $> \sim 70$ km, to undergo decompression melting. Fertile lherzolite partially melts, producing basaltic magma and leaving a dense, infertile residue of harzburgite. The degree of melting increases progressively as the mantle rises, to 40% in extreme cases. Typically the mean degree of melting is ~12% which, extended over a 60 km melting column will produce enough melt to form a crust that is 7 km thick. Melt that solidifies in the lower crust cools slowly, developing large crystals that form gabbro. At least some gabbro appears to be the solidified remains of small magma chambers that feed magma to the shallow crust and ultimately the sea floor. Magma chambers have been imaged as thin (~100 m), narrow (~1–2 km) lenses at relatively shallow crustal depths (~1–3 km), although improving techniques are beginning to image deeper melt bodies. Melt from the magma chambers passes upwards through fissures that open as the plates are pulled apart. Melt solidifying in the fissures produces vertical sheets of dolerite, called dykes, that make up the sheeted dyke layer consisting entirely of dykes intruded into each other. Any melt that reaches the sea floor erupts as lava flows, forming a variety of volcanic morphologies from flat-lying sheet flows to steep piles of pillow basalts.

The newly formed lithosphere is subjected to tension by the plate driving forces, and responds by cracking, producing fissures and faults. The horizontal component of faulting accounts for a certain proportion of plate separation, the amount of which varies from a few per cent at fast-spreading ridges to ~100% in some slow-spreading environments.

Seawater enters these fractures, and can penetrate several kilometres into the crust, exchanging chemical elements and altering the crustal mineralogy. If it reaches a substantial heat source (magma body or recently intruded rock) a very hot, corrosive fluid can result, which reacts aggressively with rock to produce hydrothermal fluids rich in dissolved compounds. On their buoyant return to the sea floor they emerge at hydrothermal vents, where the dissolved compounds may be precipitated to form vent chimneys and hydrothermal mounds, or rise into the water column as hydrothermal plumes. Such hydrothermal vent fields typically host ecosystems containing organisms that ultimately utilise vent chemicals for energy.

## 9.2 Fast-spreading ridges

At fast-spreading ridges the melt supply is high (~95% of the amount needed to fill the space created by separating plates). Consequently the heat supply is also high, producing a thin, weak lithosphere only a few kilometres thick. Unable to support large loads, it is warped by the buoyancy of underlying magma and hot rock, forming a small axial high. This weak lithosphere is relatively unstable and makes the plate boundaries sensitive to small changes in driving forces. The ridge axis is mostly normal to the spreading direction, except close to ridge offsets where it curves into the offset in response to shearing forces there. Overlapping spreading centres are common and can migrate along the ridge and evolve rapidly, e.g. by 'self-decapitation'; the lithosphere between the overlapping limbs is subject to intense deformation. Transform faults may be unstable at the fastest spreading rates, where ridge offsets are accomplished by overlapping spreading centres of various sizes, culminating in oceanic microplates for offsets $>\sim100$ km. Faulting is small-scale, with throws of a few tens of metres, and total tectonic strain is ~5%. Both inward and outward dipping faults occur, the latter created by unbending of the lithosphere as it moves off the axial high. The resultant horsts define low-amplitude abyssal hills.

The ridge axis is underlain by a more-or-less continuous axial magma chamber at about 1 km below the sea floor. This is underlain by a mainly gabbroic lower crust, and overlain by a sheeted dyke layer that feeds magma into the axial neo-volcanic zone, which includes a small axial summit trough. Lava may flow over the sides of the trough and down the ridge flanks for several kilometres, often in channels, and small eruptions may occur on these flanks. A large proportion of lavas are sheet flows, with pillow basalts occurring along channel margins and in small pillow mounds. There may be small melt lenses in the lower crust and perhaps uppermost mantle, in addition to the high-level sub-axial magma chamber. This volcanic construction leads to the classic three-layer, ~6 km thick oceanic crust: gabbroic layer 3, volcanic and dyke-intrusive layer 2, and surficial sedimentary layer 1, confirmed by sampling in 'tectonic windows' and deep boreholes. These layers appear relatively uniform along and across axis.

Hydrothermal activity is common at fast-spreading ridges, with an average of 2–3 sites per 100 km of ridge length. Hydrothermal sites include high-temperature black smokers and lower-temperature vents, but the sites are relatively small and low powered compared with some at slower spreading rates.

## 9.3 Intermediate spreading

As the spreading rate, mantle temperature or fertility falls, less melt and heat are delivered to the lithosphere, which becomes cooler and stronger. It is less capable of flexing over axial buoyancy, and more capable of supporting brittle deformation. This results in a diminishing axial high and increasing tendency to rift and neck, forming a small median valley. At many intermediate-spreading ridges the axial topography is almost flat, and may alternate along-axis between small axial high and embryonic median valley. The balance is often tipped by factors other than spreading rate as discussed in Section 9.6.

As the lithosphere becomes stronger and thicker, faulting intially becomes more prominent; however, as the axial high diminishes and its unbending lessens, the generation of outward facing faults declines. Faulting is predominantly inward facing and of somewhat larger throw than at fast-spreading ridges. However, fault spacing remains similar, probably controlled by the thickness of the axial lithosphere, which remains only a few kilometres at the point of its creation.

The main change in plate boundary geometry is that propagating rifts become an important mode of offsetting the spreading axis. The lithosphere is now strong enough to support coherent traces of the migrating rift tip, producing a characteristic V-shaped pair of 'pseudo-faults', whose angle to the ridge axis depends on the spreading and propagation rates. Propagating rifting is an important mechanism for realigning plate boundaries.

The igneous construction of the crust is similar to that at faster spreading rates and is confirmed by deep sampling. The axial magma chamber is somewhat deeper, reflecting the changed balance between heat advected into the crust mainly by emplaced melt and carried out of it by conduction and hydrothermal advection. Analogy with slower-spreading ridges suggests the thickness of the crustal layers may become somewhat more variable. The average crustal thickness remains remarkably constant at ~6–7 km.

There is a decrease in the incidence of sheet flows and an increase in pillow lavas. Pillow mounds may be aggregated into small axial volcanic ridges a few kilometres across and tens of kilometres long. Larger axial volcanoes occur from time to time, supported by the stronger lithosphere.

Hydrothermal venting follows similar patterns to that at fast-spreading ridges, though with a slightly reduced frequency along the ridge axis, as expected as a result of reduced magmatic heat input.

## 9.4 Slow spreading

As the full spreading rate falls below 60–70 km $Ma^{-1}$, an abrupt change occurs. The melt supply falls to perhaps 80%–90% of that required by plate separation, and a sub-axial magma chamber is episodic and rare. The effect of a buoyant axial zone becomes negligible, and the cooler, stronger lithosphere becomes subject to strong rifting and necking. The ridge axis is characterised by a well-defined median valley, some 1–2 km deep and 15–30 km wide. The amplitude of this valley is variable along-axis: it is generally greatest at the ends of spreading segments, reflecting a lower amount of magmatic heat input and greater tendency to rifting there. The strong lithosphere can support larger loads, so normal fault offsets increase to typically hundreds and sometimes thousands of metres. Almost all faults are now inward dipping and provide the brittle expression of lithospheric necking. Large, mostly fault-bounded abyssal hills are common.

The stronger lithosphere also affects the plate boundary geometry. Spreading centres remain generally normal to plate relative motion vectors, although a small (~5°–10°) obliquity is common. Where this occurs, features that form close to the axis (such as fissures and linear volcanoes erupted from them) are more nearly normal to the spreading direction, being less constrained by the thinner, weaker lithosphere; by contrast, the major axial-valley-bounding faults that develop several kilometres off-axis follow the oblique regional trend of the ridge, being controlled by the large-scale architecture of the lithosphere. Transform faults occur more frequently along-axis than at faster spreading rates, with an average spacing ~500 km. The commonest form of non-transform offset is a short (~20–30 km) section of 45° oblique spreading, or a slightly shorter en echelon offset of the median valley. Overlapping spreading centres are rare, although AVRs may overlap within the median valley offset. Propagating rifts with well-defined pseudofaults are absent, but non-transform offsets may migrate up and down the ridge axis, leaving trails of poorly defined shallow V-shaped valleys or lines of basins.

As at faster spreading rates, the average crustal thickness remains about 6–7 km, but is more variable along-axis. It may be up to 8 km thick at segment centres, but thins to 3–4 km at segment ends, particularly at transform faults. This variation probably reflects the focussing of melt supply toward segment centres. Sea floor sampling, particularly along the scarps of transform valleys, is consistent with the layered model of ocean crust found at faster spreading rates, but as yet there is little direct evidence of deep subsurface lithological structure. Along-axis, basaltic rocks are commonly recovered from segment centres, but peridotite becomes common toward segment ends, where a layered crust may give way to a 'crust' in which isolated gabbroic plutons are embedded in a peridotite matrix. Because peridotite reacts easily with seawater to produce serpentine, most recovered peridotites are serpentinised.

At slow spreading, volcanism is dominated by AVRs which extend the length of most spreading segments. They are built almost entirely of small, monogenetic cones usually called 'hummocks', which consist mainly of pillow lavas, and are probably erupted from fissures where linear eruptions are unstable and rapidly break down into lines of point sources. They appear similar to the fast-spreading pillow mounds. AVRs are often surrounded by flat-lying lavas consisting of sheet flows and/or flattened, lobate pillows. Rarely, segments display no AVR and just flat-lying lavas. In addition to the hummocks associated with AVRs, small numbers of larger, flat-topped seamounts are scattered throughout segments, though they tend to be more common at segment ends. They range in diameter from about one to a few kilometres in diameter, and have a characteristic diameter : height ratio of ~10. There is some evidence that flat-topped seamounts and flat-lying lavas have slightly different magmatic sources than AVRs. It has been suggested that AVRs and flat lavas mark different stages of a volcano–tectonic cycle; however, this is still moot. The deep lithological structure of slowly spread crust is poorly known, with only limited deep sampling at unrepresentative sites.

Tectonic extension begins at the ridge axis with the production of small fissures, which are found within AVRs and become numerous in the surrounding sea floor. Small faults and fissures grow in both length and offset, linking to form larger faults. Eventually, a few kilometres from the axis, some grow to have offsets of hundreds of metres and form the first axial-valley wall faults, spaced a few kilometres apart. The pattern of faulting varies along axis: at segment centres, faults have moderate offsets and are roughly symmetrical across-axis; however, at segment ends there are much larger-offset and wider-spaced faults on the inside corner and smaller-offset, closer-spaced faults on the outside corner. These differences reflect the varying lithospheric thickness and strength along the segment, with thinner lithosphere at the segment centre resulting from the increased melt and heat input there. Despite the variation in faulting patterns, total tectonic strain is roughly constant at about 10%–15% of total plate separation.

Numerical modelling suggests that a dramatic change in the style of extension should occur if the melt supply falls as low as ~50% of that required to accommodate plate separation. This appears to happen in places along slow-spreading ridges, particularly at the lower end of the slow-spreading range, at segment ends, or where the mantle is cold or infertile. Then, instead of new faults being continually initiated at the edge of the median valley, one fault may continue slipping for millions of years, achieving an offset of many kilometres and producing an oceanic 'detachment fault'. The thin lithosphere in the footwall of such faults flexes easily under the bending stresses present, and forms a domed structure called an 'oceanic core complex' (or OCC). The dome characteristically displays small corrugations parallel to the spreading direction, and sea floor terrain containing such structures may be termed 'corrugated'. The emerging footwall of the detachment fault forms the newly created lithosphere. Whether there is truly a 'crust' here is debatable; current evidence suggests that the footwall may comprise strongly serpentinised peridotite possibly heavily intruded by gabbros emplaced deep below the ridge axis before the surrounding material is exhumed. Certainly there is little or no sheeted dyke or volcanic layer (though small, isolated slivers of volcanic crust may ride on the footwall). The details of this process are the subject of current research. The hanging wall of the detachment forms the conjugate plate, and appears to be formed by volcanic processes as elsewhere at slow spreading ridges, though details are scarce at present. Once thought rare, it now appears that OCCs and their conjugate hanging walls may make up some 50% of North Atlantic slow-spread crust.

High-temperature, basalt-hosted hydrothermal systems occur on slow-spreading ridges, but appear less common than at faster spreading rates. Another type, the ultramafic-hosted system, becomes relatively common, resulting from the interaction of hydrothermal fluids with the readily available peridotite in OCCs and elsewhere. Unlike basalt-hosted systems, their effluent fluids are alkaline and have high concentrations of $H_2$ and $CH_4$ resulting from the reaction of hot seawater with olivine during serpentinisation. One purely ultramafic-hosted system is known, but many others have mixed basalt- and ultramafic-hosted characteristics. Many of these hydrothermal systems are located on or very near OCCs, though they probably need a nearby magmatic heat source to drive them.

## 9.5 Ultra-slow ridges

At ultra-slow spreading rates the component of opening normal to the local ridge axis is the important control on melt supply. The lithosphere is thick and strong, and can support a deep median valley and large axial volcanic constructs. Melt appears to be focussed, by an as yet unknown mechanism, into a few segments separated by ~150 km, which may have large volcanic massifs at their centres, sometimes almost filling the median valley. Between these highly volcanic segments are segments dominated by intense tectonism with much less, and often no, volcanism. Large areas of regional oblique spreading, oblique-spreading segments and large, oblique, non-transform offsets are common, tending to reduce the effective spreading rate.

Modelling suggests that highly tectonised spreading may be associated with a further fall in melt supply to <30%–50% of that required to match plate separation. Increasing obliquity can bring a ridge section into this critical regime. When the melt supply falls below this threshold, sea floor spreading appears to be largely devoid of magmatism and is accomplished mainly by faulting. Large detachment faults form continually, but dipping in alternating directions, on a timescale of ~1 Ma. These form large, segment-long ridges of serpentinised peridotite. Volcanism is limited to small areas of hummocky volcanoes, mostly around the ends of the serpentinite ridges. Possibly because of the absence of a hard volcanic hanging wall, their flanks are not corrugated, so this style of sea floor terrain has been named 'smooth'.

A few basalt-hosted hydrothermal vents have been discovered on ultra-slow-spreading ridges, but no ultramafic-hosted ones to date. However, exploration of such ridges is in its infancy, and it is too soon to draw firm conclusions on the distribution of hydrothermal vents there.

## 9.6 'Anomalous' ridges

The foregoing description sets out a unified view of the variation of sea floor spreading processes as a function of melt supply, which was assumed to be itself a function of spreading rate. However, other factors can control the melt supply, particularly the mantle potential temperature and the amount of fertile (i.e. capable of melting) material in the upper mantle. Changes in these properties, particularly when the spreading rate is near a critical value, can tip the balance from one spreading style to another.

Perhaps the most obvious examples are ridges which approach or cross mantle 'hot-spots' – places that exhibit anomalously high degrees of volcanism. These include the Iceland hotspot, flanked by the Reykjanes Ridge and Kolbeinsey Ridge, the Galapagos hotspot close to the Cocos–Nazca Spreading Centre, the Marion hotspot near the SWIR, and the Kerguelan hotspot on the flank of the Southeast Indian Ridge. In each case, the proximity of the hotspot causes the ridge to locally take on some of the characteristics of a faster-spreading ridge. On the Reykjanes Ridge, for example, the hotspot causes the ridge topography to change from the median valley typical of its slow spreading rate to an axial high more typical of a fast-spreading ridge, presumably by increasing the amount of melt delivered to the ridge axis.

The opposite change can occur where the mantle is anomalously cool. This has been suggested, for example, at the Australian Antarctic Discordance and towards the eastern end of the SWIR. It can also occur near ridge–transform intersections, where the opposing plate is old and cold, lowering the melt supply to the adjacent spreading centre. An example is the intersection of the northern EPR with Tamayo Transform, where an axial high characteristic of fast spreading changes to a median valley near the transform. Another possible example of poor mantle fertility occurs on the MAR around 14° N, where frequent OCCs occur on a ridge that is spreading well within the slow-spreading range.

## 9.7 Summary

In summary, mid-ocean ridges are tectonic plate boundaries where new lithosphere is created and spread apart. The details of the processes involved depend strongly on the melt supplied to the new lithosphere, which is closely dependent on spreading rate but also influenced by mantle temperature and fertility and plate boundary geometry. At the fastest spreading rates almost all plate separation is taken up by magmatic accretion; lithosphere is thin and weak, the ridge axis is dominated by buoyancy and an axial high, and hydrothermal vents hosted in hot basaltic rocks are common. At progressively slower spreading rates the melt supply diminishes, lithosphere becomes thicker and stronger, and axial processes become dominated by rifting and large-scale faulting producing a large axial valley. Basalt-hosted hydrothermal vents are less common but larger and longer-lived, and are supplemented by peridotite-hosted vents as ultramafic outcrops increase and faults more easily access subsurface peridotite. As the melt supply falls to ~50% the spreading style changes again to a highly asymmetric mode with the development of widely spaced, long-lived detachment faults that form oceanic core complexes and directly exhume ductile mantle onto the sea floor. Volcanism is restricted to the conjugate plate, and ultramafic-hosted hydrothermal venting is common. Finally, at even lower degrees of melt input, volcanism may virtually cease, and most spreading is accomplished by exhumation of ductile mantle along series of detachment faults of alternating polarity.

# Appendix A Glossary of terms

**ABE** Autonomous Benthic Explorer. An AUV developed by Woods Hole Oceanographic Institution.

**Actinolite** Type of amphibole.

**Admittance function** Mathematical function relating the spectrum of gravitational variation to that of the underlying topography. Can be used to distinguish different isostatic models of oceanic crust.

**Air gun** Source of seismic energy in which highly pressured air is suddenly released into the sea.

**Airy isostasy** Form of isostatic compensation, named after the astronomer Airy, in which topographic features are balanced by variations in crustal thickness.

**Albite** Sodium end-member of plagioclase feldspar: $NaAlSi_3O_8$.

**Alvin** Manned research submersible operated by Woods Hole Oceanographic Institution.

**Alvinocarid shrimp** Family of shrimp species that inhabit hydrothermal vents.

**Amphibole** Type of iron magnesium silicate; sometimes with Ca or Na. Includes actinolite, tremolite, and hornblende.

**ANGUS** Acoustically-Navigated Geological Undersea Surveyor. A deep-towed photographic system developed by Woods Hole Oceanographic Institution.

**Anhydrite** Anhydrous calcium sulphate, $CaSO_4$. A common mineral associated with hydrothermal vents.

**Aragonite** One of the minerals formed by calcium carbonate, $CaCO_3$.

**Aseismic ridge** Oceanic ridge that is not an active plate boundary. Generally characterised by thick crust and believed to result from voluminous volcanism above a mantle plume.

**Asthenosphere** Ductile, convecting part of the Earth's upper mantle, lying directly beneath the lithosphere.

**Autotrophe** Organism that produces its own complex organic molecules without reliance on a living food source.

**AUV** Autonomous Underwater Vehicle. An instrument platform that can carry out detailed near-bottom and other surveys following a pre-programmed plan.

**AVR** Axial Volcanic Ridge.

**Axial high** Small topographic rise a few kilometres wide and a few hundred metres high, superimposed on the axial regions of fast-spreading ridges.

**Axial magma chamber** Shallow sub-axial magma chamber found beneath the axes of fast-spreading and some other ridges.

**Axial summit caldera** Another term for axial summit trough.

**Axial summit graben** Another term for axial summit trough.

**Axial summit trough** Narrow trough lying along the crest of the axial high on fast-spreading ridges and the site of most volcanic eruptions.

**Axial valley** Another name for median valley.

**Axial Volcanic Ridge** A ridge, usually tens of kilometres long, a few kilometres wide and a few hundred metres high, formed entirely of volcanic products and commonly found at the axis of slow-spreading ridges.

**Back-arc spreading centre** Spreading centre formed behind a volcanic arc and above a subduction zone. Has many similarities to a mid-ocean ridge spreading centre.

**Barite** Common mineral associated with hydrothermal vents, composed of barium sulphate, $BaSO_4$.

**Basalt** Fine-grained volcanic rock produced by the partial melting of mantle peridotite and characteristic of ocean floor lavas.

**Bathymetry** Strictly, the technique of measuring depth. Loosely used to mean ocean-floor topography.

**Blocking temperature** In rock magnetism, the temperature below which a natural remanent magnetisation is effectively permanent. Usually slightly below the Curie temperature. Above the blocking temperature, the NRM gradually decays as a result of thermal perturbation.

**Body wave** Seismic wave (P-wave or S-wave) that has propagated through the body of the Earth.

**Body-wave magnitude** Measure of the size of an earthquake, based on the recorded amplitude of P-waves, corrected for epicentral distance. Symbol $m_b$.

**Bouguer anomaly** Gravity anomaly produced by correcting raw observations for the effects of varying height, latitude and topography. In marine work, assumes that seawater is replaced by rock of a given density.

**Breakaway** In an oceanic core complex, the locus of the intersection of the sea floor with the shallowest part of the detachment fault; the oldest and most distant boundary of the OCC from the active plate margin.

**Bresiliid shrimp** Family of shrimp species that inhabit hydrothermal vents.

**Brittle regime** Rheological regime in which materials behave elastically up to their yield strength and then fracture; typical of the lithosphere.

**Brucite** Mineral composed of magnesium hydroxide, $Mg(OH)_2$.

**Brunhes** Most recent 'normal' magnetic polarity chron, lasting from 0.78 Ma to present, named after French physicist Bernard Brunhes.

**Brittle–plastic transition** Point at which deformation changes from brittle to plastic. Dependent on temperature, strain-rate, composition and other factors.

**b-value** The slope of a plot of cumulative earthquake numbers against magnitude. Different b-values characterise tectonic and volcanic earthquakes.

**Calcite** One of the minerals formed by calcium carbonate, $CaCO_3$.

**Carbonate compensation depth** Depth at which the rate of solution of calcium carbonate exceeds the rate of supply, so that no solid carbonate is precipitated. Varies between about 4200 m and 5000 m depth.

**Cataclastic** Of rock, composed of small broken fragments.

**Celadonite** A mica group mineral commonly formed by low-temperature hydrothermal alteration of basalt. Contains Al, Fe, H, K, Mg, O and Si.

**Centroid depth** A weighted estimate of the average depth of rupture in an earthquake.

**Chalcopyrite** Common mineral associated with hydrothermal vents, composed of copper iron sulphide, $CuFeS_2$.

**Chemical Remanent Magnetisation (CRM)** Any permanent magnetisation acquired by a rock through chemical processes. A common type of CRM in sea floor rocks is acquired when magnetite is formed during the alteration of olivine to serpentine.

**Chemoautotrophe** An organism that derives its energy by direct oxidation of chemicals and synthesises its organic compounds directly from $CO_2$.

**Chlorite** Silicate mineral containing Mg, Fe, Ni, or Mn. Formed during hydrothermal alteration of the lower volcanics and sheeted dyke complex of the oceanic crust.

**Clinopyroxene** Group of minerals that are major constituents of peridotite. Usually contains two or more of Al, Ca, Fe, Mg, Na, with O and Si.

**Cohesion** That part of a rock's shear strength that is independent of internal friction.

**Compliance** Degree to which a body deforms under a given force. The compliance technique measures small-scale variations in sea floor depth induced by the varying pressure caused by the passage of sea-surface waves.

**Conductivity** The intrinsic ability of materials to conduct electricity. Inverse of resistivity.

**Crestal high** Another term for 'axial high'.

**CRM** Chemical Remanent Magnetisation.

**Crust** Outermost compositional layer of the Earth, characterised by low-density rocks and minerals containing high proportions of light elements such as aluminium, sodium and potassium, silicon and oxygen.

**Crystal mush zone** Broad region containing a mix of crystals and surrounding melt; has very high viscosity and cannot convect or erupt.

**CTD** An instrument that continuously records Conductivity, Temperature and Depth, which characterise different water masses.

**Cumulate** An igneous rock formed from an accumulation of precipitated crystals.

**Curie temperature** Temperature below which ferromagnetic or ferrimagnetic minerals maintain a permanent magnetic field. About 580 °C for magnetite.

**Deep Sea Drilling Project** Multi-institution US programme operating a drilling ship for scientific sampling of the oceans from 1968 to 1985.

**Deep Tow** The original deep-towed geophysical vehicle developed by Scripps Institution of Oceanography.

**Deflated** Of a magma chamber, containing a relatively small volume of magma at relatively low pressure.

**Detachment fault** Large-offset normal fault that cuts entirely through the lithosphere.

**DEVAL** DEViation from Axial Linearity. Type of fourth-order non-transform offset found on fast-spreading ridges, characterised by a small change in the orientation of the spreading axis.

**Diabase** US term for fresh dolerite.

**Diatoms** Group of algae whose siliceous skeletons contribute to siliceous ooze.

**Diopside** A pyroxene mineral with composition $MgCaSi_2O_6$.

**Dip** The angle that a fault or other plane makes with the horizontal.

**Dolerite** Medium- to fine-grained intrusive igneous rock with the composition of basalt, produced by the partial melting of mantle peridotite and cooled and crystallised at relatively shallow levels in the crust. Prime constituent of oceanic crustal dykes.

**Doomed rift** In a propagating rift system, the spreading centre that is shortening at the expense of the propagating rift.

**Downward continuation** Geophysical technique that allows the equivalent gravity or magnetic field to be calculated on a horizontal surface at a lower level than that it was originally measured on.

**Dredge** Strong bucket with attached bag for collecting rock samples from the sea floor.

**Ductile** Capable of being deformed plastically, with no tendency to revert to its previous shape.

**Dunite** Type of ultramafic rock consisting of $>90\%$ olivine.

**Dyke** ('Dike' in USA.) A narrow, usually vertical, sheet of rock that has been emplaced into a crack while molten.

**Dyking** ('Diking' in USA.) The process of dyke emplacement.

**Effective elastic thickness** The thickness of a purely elastic plate that can reproduce observed sea floor topography by elastic flexure. Ranges from ~2–50 km in oceanic lithosphere.

**Effective spreading rate** The component of the full spreading rate that is normal to the local plate boundary.

**$E_h$** Redox potential – a measure of the tendency of a chemical species to be reduced.

**Elastic** Rheological regime in which applied stress produces a proportional strain, which is recoverable when the stress is removed.

**Electromagnetic sounding** Geophysical technique for determining subsurface electrical conductivity using electromagnetically induced currents.

**Emergence zone** In an active oceanic core complex, the locus of the places where the detachment fault footwall emerges from under the hanging wall.

**E-MORB** Enriched Mid-Ocean Ridge Basalt. Basalt of a composition intermediate between N-MORB and OIB; thought to reflect mixing with sub-hotspot mantle.

**Endemic** In ecology, unique to a particular habitat or location.

**Enteropneust** Class of invertebrate marine worms sometimes found around hydrothermal vents.

**Epicentre** Point on the Earth's surface immediately above an earthquake location.

**Epidosite** Highly altered rock containing abundant epidote, common in ophiolites where it is formed by hydrothermal alteration of the sheeted dykes.

**Epidote** A calcium–aluminium–iron silicate mineral, commonly formed by hydrothermal alteration of crustal rocks.

**EPR** East Pacific Rise.

**Euler pole** In plate kinematics, one of a pair of points where a rotation axis describing the relative motion of two plates cuts the Earth's surface.

**Euler's theorem** States that any motion on a sphere can be represented as a single rotation about an axis passing through the centre of the sphere.

**Evolved** In petrology, an igneous rock that differs in composition from its mantle source, having experienced crystal fractionation in a magma chamber.

**Exponential distribution** Statistical distribution in which the probability of an object being larger than size $x$ is proportional to $e^{-x}$.

**Extremophile** An organism that thrives under extreme conditions.

**Extrusive** Of igneous rocks, synonymous with 'lava'.

**Farallon plate** An ancient tectonic plate of the eastern Pacific, which split into the present-day Juan de Fuca and Cocos plates at ~30 Ma.

**Fast Fourier transform** Algorithm for rapidly taking the Fourier transform of a function. Widely used in processing and interpretation of geophysical data.

**Failed rift** In a propagating rift system, the trace of the erstwhile doomed rift.

**Fault** In geology, a plane offsetting two bodies of rock.

**Fault plane solution** Result of analysing seismograms to determine the direction of fault slip during an earthquake.

**Feldspar** Group of calcium-, potassium-, sodium-minerals that are a major component of crustal rocks.

**Ferrimagnetic** Material consisting of two sets of oppositely aligned atomic magnetic dipoles of unequal strengths, producing a net external magnetic field.

**FFT** Fast Fourier Transform.

**Fissure** Tensional crack with small amount of horizontal but no vertical offset.

**Focal mechanism** The rock movements giving rise to an earthquake, deduced from the first motions or whole waveforms of several seismograms.

**Footwall** The block of rock below a dipping fault plane.

**Foraminifera** Class of small planktonic organisms whose calcium carbonate shells compose calcareous sea floor sediment.

**Forward modelling** Geophysical interpretation technique in which a model is assumed and its effect calculated, compared to observations and adjusted manually to obtain a good fit.

**Fourier transform** A mathematical technique to decompose a profile or time series into its frequency components and vice versa. Utilised in many powerful geophysical interpretation techniques.

**Fracture zone** Strictly, the inactive trace of a transform fault. Loosely, includes the active transform as well.

**Free-air anomaly** A method of reducing gravity data in which observations are corrected for height and latitude. Gives a rough indication of degree of isostatic equilibrium.

**Gabbro** A coarse-grained intrusive igneous rock of basaltic composition, produced by partial melting of mantle peridotite and subsequent cooling and crystallisation deep in the crust.

**Galatheid crab** Type of squat lobster; common scavenger around hydrothermal vents.

**Garnet** Minor constituent of some lherzolites. Composition $Mg_3Al_2Si_3O_{12}$.

**Geostrophic flow** Fluid motion in which the pressure gradient is balanced by the Coriolis force.

**Geotherm** Temperature–depth profile in the Earth.

**GLORIA** Geological Long-Range Inclined Asdic. Surface-towed side-scan sonar with range ≤30 km, extensively used to study mid-ocean ridges in the 1970s and 1980s.

**Goethite** Iron oxyhydroxide mineral with chemical composition FeO(OH), typically formed by low-temperature oxidation of basalt.

**GPS** Global Positioning System. A satellite navigation system in use since 1983 that provides continuous estimates of position based on estimation of the ranges of three or more satellite-based radio transmitters.

**Graben** Valley formed by a pair of inward facing normal faults.

**Great circle** A circle that marks a diameter of a sphere.

**Half-space model** A model of lithospheric temperature, with a lower boundary condition of constant mantle temperature at infinite depth. Successfully predicts depth and heat flow for sea floor <50 Ma.

**Hangingwall** The block of rock above a dipping fault plane.

**Harzburgite** Type of peridotite; an ultramafic rock forming the residual upper mantle after the extraction of basaltic melt.

**Haystack** Small, steep-sided pile of pillow lavas, typically several metres wide and high.

**Heat flow** The rate at which heat is transferred from Earth's interior to the surface.

**Heave** The horizontal displacement on a normal fault.

**Heterotroph** An organism that cannot fix carbon but utilises pre-existing organic carbon for growth.

**Holothurian** Sea cucumber. Type of marine invertebrate found around hydrothermal vents.

**Hotspot** In geodynamics, a region of the Earth exhibiting abnormally high volcanic activity; often inferred to overlie a rising mantle plume.

**Hummock** Small (<~500 m diameter) monogenetic volcano, cone- or dome-shaped, formed principally of pillow and lobate lavas.

**Hummocky mound** Another term for hummocky volcano.

**Hummocky shield** Another term for hummocky volcano.

**Hummocky volcano** A small (<~1 km diameter) volcano built of an agglomeration of hummocks.

**Hydro-acoustic monitoring** The use of moored hydrophone arrays to monitor acoustic signals in the ocean and detect earthquakes.

**Hydrophone** Device for detecting underwater acoustic signals.

**Hydrothermal vent** Sea floor vent of warm or hot water carrying dissolved minerals and produced by interaction of circulating seawater with hot subsurface rock.

**Hypocentre** The position of an earthquake in 3D.

**Igneous** In geology, rocks produced from the solidification of magma.

**IGRF** International Geomagnetic Reference Field.

**Incompatible element** An element not fitting the structure of a crystallising mineral, which tends to remain in the melt.

**Inflated** Of a magma chamber, containing a relatively large volume of magma at relatively high pressure.

**Inside corner high** A topographic high typically developed in the inside corner of the plate boundary between spreading centres and transform faults.

**Integrated Ocean Drilling Program** International programme operating two drilling ships and other platforms for scientific sampling of the oceans from 2004 to 2013. Succeeded ODP.

**International Geomagnetic Reference Field** Internationally agreed model of the average geomagnetic field, updated every five years and used for reducing marine magnetic observations.

**International Gravity Formula** Theoretical formula giving the approximate variation of gravity with latitude, assuming the Earth is a rotating ellipsoid of revolution.

**Inverse modelling** Geophysical technique in which an approximate starting model is automatically modified to achieve an optimum fit. Has the advantages of independence of modeller bias and automatic estimation of goodness of fit.

**IODP** Integrated Ocean Drilling Program.

**Isostasy** Tendency of the lithosphere to attain hydrostatic equilibrium by rising or sinking through the viscous asthenosphere as loads vary.

**Isostatic compensation** Compensation of topographic loads by subsurface density variations.

**Isostatic equilibrium** State of the lithosphere in which it is in hydrostatic equilibrium with respect to the asthenosphere.

**Isotherm** Contour of equal temperature.

**JdFR** Juan de Fuca Ridge.

**Kilobar** Unit of pressure equal to 100 kPa.

**Königsberger ratio** Ratio of remanent to induced magnetisation.

**Lava lake** Extensive area of molten lava. On mid-ocean ridges, often used to describe the solidified remains of such a body, including extensive collapse areas and lava pillars.

**Lava pillar** Pillar of lava that has supported the collapsed crust of a lava lake or sheet flow. Probably formed by freezing of lava around a rising column of cooler water.

**LBL** Long Base Line acoustic navigation.

**Lead line** Early depth-sounding device, employing a lead weight on the end of a thin line.

**Lherzolite** Type of peridotite; an ultramafic rock that makes up much of the upper mantle and is the primary source for oceanic crustal rocks.

**Liquidus** Temperature above which a rock will be completely molten.

**Lithosphere** Outermost mechanical layer of the Earth, characterised by material that is strong and brittle over timescales $>\sim 1$ ka and forming the tectonic plates.

**Lobate lava flow** Type of bulbous lava flow wider and flatter than pillow lava, typical of intermediate effusion rates or flow on moderate slopes.

**Long Base Line navigation** Acoustic navigation system in which a platform is navigated near the sea floor by measuring ranges from moored acoustic transponders.

**LORAN** (LOng RAnge Navigation). Navigation system based on measuring ranges from two or more land-based radio transmitters.

**Low-velocity zone** Region of relatively low seismic velocity, particularly in upper mantle where it marks the asthenosphere.

**LVZ** Low-velocity zone.

**Mafic** A mineral or rock rich in magnesium and iron; in ocean crust refers principally to basalt, dolerite and gabbro.

**Magma** Underground molten rock.

**Magma chamber** A subsurface body of connected melt.

**Magnesite** Mineral composed of magnesium carbonate, $MgCO_3$.

**Magnetite** Iron oxide mineral, composition $Fe^{2+}Fe^{3+}{}_2O_4$. A minor component of erupted basalt, also formed during serpentinisation of olivine. Important as the main carrier of magnetisation in the oceanic lithosphere.

**Magnitude** In seismology, a measure of the power of an earthquake. Approximately proportional to the logarithm of released energy, one unit of magnitude representing approximately thirtyfold increase in energy.

**Major elements** The elements that comprise the bulk of a rock, typically Si, O, Fe, Mg, Mn, Ti, Al, Ca, K, Na.

**Mantle** The thick layer of the Earth between crust and core. Contains high proportions of relatively dense elements such as iron and manganese.

**Mantle Bouguer anomaly** A way of reducing marine gravity data that corrects for variations in latitude and the effects of observed water depth and an assumed constant thickness and density crust.

**Mantle convection** The slow motion of ductile mantle driven by Earth's internal heat.

**MAR** Mid-Atlantic Ridge.

**Marcasite** A mineral associated with hydrothermal vents. Form of iron sulphide, $FeS_2$.

**Massive sulphide deposit** A sea floor deposit of mostly metal sulphides produced by hydrothermal activity.

**Matuyama** Most recent reversed magnetic polarity chron, 0.99–0.78 Ma, named after Japanese geophysicist Motonori Matuyama.

**Median valley** Broad valley marking the axis of a slow-spreading mid-ocean ridge.

**Megatumulus** Large volcanic mound formed by inflation of a distal lava pond by the influx of new lava.

**Metamorphism** The alteration of rocks by chemical interaction or changes in temperature or pressure.

**Methanogen** A microbe that lives in anoxic conditions and produces methane.

**Microplate** A small tectonic plate of a few hundred kilometres diameter lying between and rotated by two or more adjacent major plates.

**Mid-ocean ridge basalt** Basalt of the composition typically found at mid-ocean ridges.

**Mid-ocean rise** Sometimes applied to fast-spreading mid-ocean ridges to indicate their gentler relief compared to slow-spreading ridges.

**Moho** Mohorovičić discontinuity.

**Mohorovičić discontinuity** The depth in the Earth where the seismic P-wave velocity first excedes $\sim 8$ km $s^{-1}$, roughly at the base of the crust.

**MOR** Mid-ocean ridge.

**MORB** Mid-ocean ridge basalt.

**Multi-beam echosounder** An echosounder with a large number of narrow beams deployed in a fan normal to the direction of travel. May have over 100 beams in a fan $>90°$ wide.

**Nanotesla** A unit of magnetic field strength.

**Natural remanent magnetisation** Any permanent magnetisation acquired by a rock through natural processes. Common types of NRM are Thermal Remanent Magnetisation and Chemical Remanent Magnetisation.

**Neo-volcanic zone** The central region of a ridge axis where new oceanic crust is forming by volcanic activity.

**N-MORB** 'Normal' mid-ocean ridge basalt, of a composition found along most of the mid-ocean ridge system.

**Nodal plane** Plane along which an earthquake radiates no energy. For simple faulting events, the nodal planes are parallel and normal to the active fault.

**Non-transform discontinuity** Another term for non-transform offset.

**Non-transform offset** Short (<30 km) offset of a spreading centre with no through-going transform fault.

**Nontronite** Iron-rich clay mineral, also containing Ca or Na, Si and Al. Formed by low-temperature hydrothermal alteration of basalt.

**Normal fault** A fault whose plane dips into the Earth, with the hanging wall (overlying mass) moving or having moved downwards relative to the underlying mass or footwall.

**Normal magnetisation** Natural remanent magnetisation parallel to the present geomagnetic field.

**NRM** Natural Remanent Magnetisation.

**nT** Nanotesla.

**Nusselt number** The dimensionless ratio of convective to conductive heat flow. Used in thermal models of mid-ocean ridges to parameterise the effect of hydrothermal circulation.

**NVZ** Neo-volcanic zone.

**Obduction** The process whereby a sliver of oceanic crust and upper mantle is detached from its parent tectonic plate and emplaced onto a continent, forming an ophiolite.

**OCC** Oceanic core complex.

**Ocean Drilling Program** An international programme operating a drilling ship for scientific sampling of the oceans from 1985 to 2004.

**Ocean island basalt** Basalt highly enriched in incompatible elements characteristic of ocean islands and hotspots.

**Oceanic core complex** Large-scale structural feature comprising the domed and corrugated footwall of an oceanic detachment fault.

**ODP** Ocean Drilling Program.

**OIB** Ocean island basalt.

**Olivine** Mineral of chemical composition $(Mg, Fe)_2SiO_4$, that is the main constituent of dunite, peridotite and the upper mantle.

**Ooze** Fine-grained pelagic sediment typically composed of foraminiferal or radiolarian shells.

**Ophiolite** Suite of rocks found on land, typically comprising layers of peridotite, gabbro, sheeted dyke unit, pillow basalt and pelagic sediment, interpreted to be an example of ancient oceanic crust.

**Orthopyroxene** Iron–magnesium silicate mineral that is a major component of peridotite. Composition $(Mg,Fe)SiO_3$.

**Overlapping spreading centre** A form of non-transform ridge offset common on fast-spreading ridges where the offset spreading centres overlap each other for several kilometres to tens of kilometres.

**Pahoehoe** Type of sheet flow lava characterised by folded, ropey or whorled patterns.

**Palaeomagnetism** The measurement of naturally acquired rock magnetisations to deduce the original orientation of the rock.

**Peridotite** Family of ultramafic rocks composed of olivine and pyroxene, which are the main constituents of the upper mantle.

**Petrological Moho** Point where mantle rocks are first encountered, marking the base of the crust.

**Pillow lava** Bulbous lava flow typical of eruptions with low effusion rate or on steep surfaces. Shows characteristic radial jointing when fractured.

**Pillow mound** Similar to volcanic hummock.

**Pinger** Acoustic device that emits a short pulse of sound at regular intervals. Can be attached to near-bottom instruments to show their height off-bottom, or combined with acoustic transducers to form a navigation network.

**Plagioclase** Feldspar mineral of composition $NaAlSi_3O_8$ to $CaAl_2Si_2O_8$. A major constituent of basalt.

**Plagiogranite** Another name for trondhjemite.

**Plankton** Free-floating marine organisms.

**Plastic** Rheological regime in which materials deform continuously and without recovery while strain is applied. Typical of rocks at high temperature and pressure and low strain rate.

**Plate** In plate tectonics, one of the thin, rigid, spherical caps that form the outer layer of the Earth, the lithosphere.

**Plate kinematics** The motions of lithospheric plates.

**Plate model** Model of the lithosphere with a lower boundary condition of constant mantle temperature at a constant finite depth $\sim$100 km.

**Plate tectonics** Study of the geodynamic causes and consequences of the movements, creation and destruction of tectonic plates.

**Plume** A narrow jet of fluid rising under its own buoyancy. Mantle plumes are hypothesised to be hot, ductile mantle rising under crustal 'hot spots' such as Iceland. Hydrothermal plumes are jets of warm water rising above hydrothermal vents.

**Plutonic** In geology, generally any large, deep-seated intrusive rock. At mid-ocean ridges, usually means the gabbroic layer of the crust.

**Poisson's ratio** The ratio of axial extension to radial compression when an elastic body is stretched. Relates P-wave velocity to S-wave velocity.

**Pole** In plate kinematics, the point where a rotation axis intersects the Earth's surface.

**Polychaete worm** Type of marine worm characterised by numerous bristles, sometimes called 'bristle worms'. May be found around hydrothermal vents.

**Potential density** The density a fluid would have if brought adiabatically to a given depth.

**Potential temperature** The temperature a material would have if brought adiabatically to a standard pressure. Of the mantle, the temperature it would have if it rose adiabatically to Earth's surface.

**Primitive** In petrology, an igneous rock that is close in composition to its mantle source, having experienced little or no crystal fractionation.

**Principal transform displacement zone** The currently active strike-slip fault or narrow band of faults and other elements that marks the instantaneous plate boundary in a transform fault.

**Propagating rift** *Sensu lato*: a type of ridge axis discontinuity, common at intermediate spreading rates, where the offset migrates along the plate boundary leaving two 'pseudofaults' in a V-shaped wake. *Sensu stricto*: the spreading centre that is actively propagating.

**Proton precession magnetometer** A common form of marine magnetometer.

**Pseudofault** One of the two oblique traces of the past position of a propagating rift tip.

**P-wave** 'Primary wave'. First seismic wave to arrive at a seismometer, whose ground motion is compression/dilatation along the direction of propagation.

**Pyrite** A mineral commonly associated with hydrothermal vents, composed of iron sulphide, $FeS_2$.

**Pyroclastic** Volcanic deposit consisting of small shards of solidified lava produced in explosive eruptions.

**Pyroxene** Large group of silicate minerals comprising mainly Fe or Mg and often Ca and other elements, common in ultramafic rocks.

**Pyroxenite** An ultramafic rock composed essentially of pyroxene, commonly found in the basal units of ophiolites.

**Pyrrhotite** A mineral associated with hydrothermal vents, composed of iron sulphide, with formula $Fe_{1-x}S$, where $x$ can vary from 0 to 0.2.

**Radiolaria** Group of small marine plankton whose siliceous shells are a major constituent of siliceous ooze.

**Rayleigh–Taylor instability** A gravitational instability driven by density differences, possibly controlling mantle and melt upwelling beneath mid-ocean ridges and leading to ridge segmentation.

**Ray parameter** In a spherical Earth, the seismic ray parameter $p = \frac{r \sin i}{v}$ is a constant, where $v$ is the velocity at radius $r$ and $i$ is the angle of incidence at that point.

**Ray-tracing** In seismology, the process of tracing the trajectory of a seismic ray given a model of the velocity structure.

**Redox potential** Short for 'reduction/oxidation potential'. Symbol $E_h$. A measure of a chemical species' ability to accept electrons and thus be reduced.

**Reduction to the pole** Means of transforming magnetic anomalies to remove the inherent asymmetry that arises when a magnetic boundary is aligned oblique to the Earth's magnetic field.

**Reflection seismology** Use of low-frequency sound at near-vertical incidence to elucidate crustal structure by imaging subsurface reflections.

**Refraction seismology** Use of artificial acoustic sources recorded at large horizontal offsets following subsurface refraction to determine Earth structure.

**Remotely Operated Vehicle** Near-bottom instrument platform, tethered to a ship but with limited independent manoeuverability.

**Residual mantle Bouguer anomaly** Result of correcting the mantle Bouguer anomaly for the predicted effects of lithospheric cooling.

**Resistivity** The intrinsic ability of materials to resist electrical conduction. Inverse of conductivity.

**Reverse magnetisation** Natural remanent magnetisation that is anti-parallel to the present geomagnetic field.

**Rheology** Science of how materials flow and deform, or the style of that deformation.

**Rider block** In an oceanic core complex, a small ridge of shallow crust cut off the hanging wall and transferred to the footwall, on which it now 'rides'.

**Ridge-push force** A plate driving force originating at mid-ocean ridges, derived from the tendency of the ridge flanks to slide apart under gravity.

**Riedel shear** A secondary strike-slip fault lying at an angle to the main direction of shearing.

**Rift mountains** The relatively shallow areas flanking the median valley at slow-spreading ridges.

**Rift valley** Any large, fault-bounded valley formed by tectonic extension. Often used specifically for the broad valley marking the axes of slow-spreading mid-ocean ridges. The latter is also called the median valley or axial valley.

**RMBA** Residual mantle Bouguer anomaly.

**Rock magnetism** The study of magnetisation of rocks, including the magnetic properties of individual minerals, yielding information on the mineralogical makeup of a rock.

**Rotation pole** In plate kinematics, one of a pair of points where a rotation axis describing the relative motion of two plates cuts the Earth's surface.

**ROV** Remotely Operated Vehicle.

**Salinity** A measure of the salt content (mostly NaCl) of water. Usually expressed as parts per thousand (‰); typically ~35‰ for seawater.

**Saponite** A clay mineral formed by hydrothermal alteration in the lower basalts and sheeted dykes of oceanic crust. Composition $Ca_{0.25}(Mg,Fe)_3((Si,Al)_4O_{10})(OH)_2.n(H_2O)$.

**SBL** Short Base Line acoustic navigation.

**SeaBeam** The world's first non-military multi-beam echosounder, comprising 16 beams in a 45° fan.

**Segment** See 'spreading segment'.

**Segmentation** The division of spreading centres into short segments, each being a few tens of kilometres long and bounded at each end by transform or non-transform offsets.

**Seismic Moho** The crustal depth where the seismic P-wave velocity first exceeds 8 km $s^{-1}$. May be associated with ultramafic cumulates derived from crustal magma chambers, not necessarily the top of the mantle (see 'petrological Moho').

**Seismicity** Earthquake activity; distribution of earthquakes in time or space.

**Seismology** The study of earthquakes, or use of earthquakes or artificial explosions to study Earth structure.

**Self-decapitation** The process whereby one limb of an overlapping spreading centre propagates towards and links to the opposite limb, thus cutting off the latter's tip.

**Self-potential** An electrical polarisation of rocks arising from their minerals' different electrochemical potentials. Significant self-potential anomalies exist around massive sulphide deposits.

**Serpentine** Group of weak minerals formed by hydration of olivine at $<500$ °C, with composition $(Mg, Fe)_3Si_2O_5(OH)_4$.

**Serpentinite** A rock composed mainly of serpentine minerals. Produced by hydration of peridotite.

**Sheared zone** In a propagating rift system, the region between the inner pseudofault and the failed rift which has been sheared by the passage of the propagating rift tip.

**Sheet flow** Type of flat-lying lava flow typical of eruptions at high effusion rate or onto low slopes. May be decorated with ropy, folded, whorled or jumbled patterns.

**Sheeted dyke complex** A component of ophiolites formed completely of doleritic dykes, thought to be roughly equivalent to seismic layer 2C.

**Short Base Line acoustic navigation** Method of acoustic navigation in which a remote instrument is navigated by measuring ranges from acoustic beacons mounted on a ship's hull.

**Side-scan sonar** A means of imaging the sea floor by plotting acoustic backscatter against travel time recorded from a fan-shaped beam, with successive 'pings' building up a 2D image.

**Silica** Silicon dioxide, $SiO_2$.

**Siphonophore** Type of colonial marine invertebrate found around some hydrothermal vents.

**Slab pull force** A plate driving force derived from the gravitational drag of dense, descending plates in subduction zones.

**Small circle** Any circle on the surface of a sphere other than a great circle.

**SNOO** Small, Non-Overlapping Offset. Type of fourth-order non-transform offset found on fast-spreading ridges, characterised by a very small offset of the spreading axis.

**Solidus** Temperature below which a rock is completely solid.

**Sparker** Type of marine seismic source in which a high-energy electric spark instantaneously vaporises a volume of water.

**Sphalerite** A mineral comprising crystalline zinc sulphide with a variable component of iron, associated with hydrothermal vents.

**Spinel** A minor mineral component of some lherzolites; composition $MgAl_2O_4$.

**Spreading axis** Common term for an extensional plate boundary. The locus of youngest sea floor at such a boundary.

**Spreading centre** Another term for spreading axis.

**Spreading rate** Strictly, the rate at which a plate spreads away from the ridge axis (sometimes referred to as the half spreading rate or plate accretion rate). Also used (usually as 'full spreading rate') to indicate the rate of separation of two plates about a common ridge axis.

**Spreading segment** A length of mid-ocean ridge, typically a few tens of kilometres long, bounded at each end by transform or non-transform offsets.

**Strain** The amount of deformation undergone by a rock.

**Strain rate** The rate at which strain accumulates: an important control on rheology.

**Strike** In geology, the azimuth of a feature.

**Strike-slip fault** Type of fault in which the two blocks slide horizontally with respect to each other.

**Surface wave** A seismic wave that propagates along the surface of the Earth.

**Surface wave dispersion** The variation of surface wave velocity as a function of wavelength. Since longer waves sample deeper velocities, can be used to estimate seismic velocity at various depths.

**Susceptibility (magnetic)** Ratio of induced to inducing magnetic field.

**S-wave** 'Secondary wave'. Second seismic wave to arrive at a seismograph, whose ground motion is normal to the propagation direction.

**SWIR** Southwest Indian Ridge.

**TAG** Trans-Atlantic Geotraverse. Also used for the hydrothermal vent field at 26°08′N where the geotraverse crosses the MAR axis.

**Talc** A very weak mineral, composition $Mg_3Si_4O_{10}(OH)_2$, produced by the hydration of olivine.

**Talus** Coarsely broken rock usually resulting from faulting; scree.

**Tau-p transformation** A means of transforming seismic records from time vs. distance to intercept (tau) vs. travel-time slope (p). Provides considerable advantages in computation and interpretation and can yield a continuous velocity–depth function with uncertainty limits.

**Teleseismic** Earthquakes recorded by distant, usually global, networks.

**Tension gash** Extensional opening (fissure or basin) formed in response to tension at ~45° to a shearing direction.

**Termination** In an oceanic core complex, the intersection of the sea floor with the structurally deepest part of the detachment fault; the nearest OCC boundary to the spreading axis.

**Thermal conductivity** Measure of the ability of material to conduct heat.

**Thermal Remanent Magnetisation** A permanent magnetisation acquired by a rock through natural thermal processes; TRM in sea floor rocks is commonly acquired when newly erupted lavas cool through their Curie and magnetic blocking temperatures.

**Throw** The vertical displacement on a normal fault.

**Titanomagnetite** Naturally occurring ferrimagnetic mineral that carries a component of NRM in sea floor rocks. Composition is a solid solution between magnetite ($Fe_3 SiO_4$), ulvöspinel ($Fe_2TiO_4$) and ilmenite ($FeTiO_3$).

**TOBI** Towed Ocean Bottom Instrument. Deep-towed geophysical vehicle developed by the British Institute of Oceanographic Sciences in the late 1980s carrying a 30 kHz, 3 km range side-scan sonar.

**Tomography** In seismology, the process of determining 3D velocity structure by analysing data from a large number of source–receiver paths. Similar in principle to medical CAT scanning.

**Trace elements** Elements present in rocks in only trace amounts. Typically Rb, Sr, Zr, Ba, Th, Nb and various rare-Earth elements: Ce, Gd, La, Lu, Nd, Sm and Y. May characterise the source-rock of MORB.

**Transcurrent fault** Another term for strike-slip fault.

**Transform domain** The region that contains all transform-related tectonic elements such as the transform valley, flanking ridge(s) and transform-related faults.

**Transform fault** Tectonic plate boundary in which the plates move past each other with pure slip parallel to the spreading direction.

**Transform tectonised zone** The zone containing all faults parallel to a transform fault axis that take up, singly or jointly, the strike-slip motion of the transform.

**Transit** A form of satellite navigation system used in the 1960s to 1980s, in which position lines were inferred from the Doppler shift in signals broadcast by satellites.

**Transverse ridge** A large ridge parallel to and on one (or occasionally both) flanks of a transform fault or fracture zone.

**Trench** Strictly, a deep-sea trench formed at a subduction zone. In plate kinematics, sometimes used as shorthand for a subduction plate boundary.

**Triple junction** Junction between three tectonic plates.

**TRM** Thermal Remanent Magnetisation.

**Trondhjemite** Light-coloured plutonic rock with higher silica content than gabbro. Also known as plagiogranite.

**Tumulus** In volcanology, a low dome or blister on a lava flow caused by local inflation and drain-back.

**Turbidites** Flat-lying sediments deposited by underwater sediment avalanches.

**T-wave** Acoustic wave travelling in the water column, produced by conversion of seismic P-wave energy at the sea floor and detected by hydrophones.

**Ultramafic** Types of rock or mineral that have very low silica content and high magnesium and iron, characteristic of the Earth's mantle.

**Upward continuation** Geophysical technique that allows the equivalent gravity or magnetic field to be calculated on a horizontal surface at a higher level than it was originally measured on.

**U-series dating** A method of radiometric dating using the Uranium-series decay chain, some of which have half-lives of only a few thousand years so can resolve ages within the Brunhes magnetic chron.

**Velocity diagram** Another term for velocity triangle.

**Velocity triangle** Plot of the relative velocities between three tectonic plates at a triple junction.

**Water gun** Source of seismic energy similar to the air gun, in which compressed air is used to create a void in the sea water, which subsequently collapses forming the seismic pulse.

**Wehrlite** An ultramafic rock consisting of olivine and clinopyroxene. A constituent of lower crustal cumulate rocks.

**WWSSN** World-Wide Standardised Seismograph Network. The first global network of standardised seismograph stations, established by the US Geological Survey in the early 1960s.

**Zeolite** A class of aluminosilicate mineral formed by low-temperature hydrothermal alteration of basalt.

# Appendix B Directory of named features

The following table gives the locations of features mentioned in the text. Where the feature covers a significant extent, the approximate location of its centre is given. Locations are given to either the nearest degree, minute, or tenth of minute, as appropriate.

| Feature name | Feature type | General location | Latitude | Longitude |
|---|---|---|---|---|
| **Albatross Plateau (obsolete)** | Mid-ocean ridge | Eastern equatorial Pacific | 0° | 100° W |
| **American–Antarctic Ridge** | Mid-ocean ridge | Southern Atlantic | 57° S | 6° W |
| **Ashadze vent field** | Hydrothermal vent field | Mid-Atlantic Ridge | 12°58′ N | 44°52′ W |
| **Atlantis Bank** | Transform ridge | Southwest Indian Ridge | 32°40′ S | 57°17′ E |
| **Atlantis Massif** | Oceanic core complex | Northern Mid-Atlantic Ridge | 30°08′ N | 42°08′ W |
| **Atlantis transform fault** | Transform fault | Northern Mid-Atlantic Ridge | 30°02′ N | 42°22′ W |
| **Australian–Antarctic Discordance** | Mid-ocean ridge | Southeast Indian Ridge | 50° S | 117° E |
| **Axial Seamount** | Seamount | Juan de Fuca Ridge | 46° N | 130° W |
| **Azores spreading centre** | Spreading centre | Central North Atlantic | 38° N | 27° W |
| **Azores triple junction** | Plate triple junction | Mid-Atlantic Ridge | 39° N | 30° W |
| **Bauer microplate** | Ancient oceanic microplate | Eastern central Pacific | 12° S | 99° W |
| **Beebe vent field** | Hydrothermal vent field | Mid-Cayman Spreading Centre | 18°33′ N | 81°43′ W |
| **Blanco transform** | Transform fault | Northeast Pacific | 43°45′ N | 128°30′ W |
| **Carlsberg Ridge** | Mid-ocean ridge | Northwest Indian Ocean | 6° N | 61° E |
| **Central Indian Ridge** | Mid-ocean ridge | Central Indian Ocean | 7° S | 68° E |
| **Chain transform fault** | Transform fault | Equatorial Atlantic | 1° S | 14°30′ W |
| **Challenger Ridge (obsolete)** | Mid-ocean ridge | Southern Mid-Atlantic Ridge | 20° S | 13° W |
| **Charlie–Gibbs transform** | Transform fault | Northern Mid-Atlantic Ridge | 52°30′ N | 32°30′ W |
| **Chile Rise** | Mid-ocean ridge | South-eastern Pacific | 40° S | 91° W |
| **Chile triple junction** | Plate triple junction | Offshore Chile | 46°20′ S | 75°45′ S |
| **Clarion fracture zone** | Fracture zone | Eastern Pacific | 16° N | 141° W |
| **Cleft segment** | Spreading segment | Juan de Fuca Ridge | 44°40′ N | 130°20′ W |
| **Clipperton fracture zone** | Fracture zone | Eastern Pacific | 6° N | 136° W |

| Feature name | Feature type | General location | Latitude | Longitude |
|---|---|---|---|---|
| **Co-axial segment** | Spreading segment | Juan de Fuca Ridge | 46°24′N | 129°40′W |
| **Cocos–Nazca Spreading Centre** | Mid-ocean ridge | Eastern central Pacific | 2°N | 94°W |
| **Dante's Domes** | Oceanic core complex | Mid-Atlantic Ridge | 26°40′N | 44°20′W |
| **DSDP Hole 504B** | Deep borehole | Eastern Pacific | 01°13.6′N | 83°43.8′W |
| **Dolphin Ridge or Rise (obsolete)** | Mid-ocean ridge | Mid-Atlantic Ridge | 35°N | 32°W |
| **East Pacific Rise (north)** | Mid-ocean ridge | Eastern Pacific Ocean | 12°N | 104°W |
| **East Pacific Rise (south)** | Mid-ocean ridge | Eastern Pacific Ocean | 12°S | 111°W |
| **Easter microplate** | Oceanic microplate | Eastern Pacific Rise | 25°S | 114°W |
| **East Scotia Ridge** | Back-arc spreading centre | Scotia Sea | 58°S | 30°W |
| **Explorer Ridge** | Mid-ocean ridge | Northeast Pacific | 49°40′N | 130°20′W |
| **FAMOUS** | Study area | Mid-Atlantic Ridge | 36°48′N | 33°16′W |
| **Farallon Plate (now defunct)** | Ancient plate, now partly subducted | Eastern Pacific | | |
| **Fifteen-Twenty transform** | Transform fault | Mid-Atlantic Ridge | 15°20′N | 45°45′W |
| **Gakkel Ridge** | Mid-ocean ridge | Arctic Ocean | 86°N | 75°E |
| **Galapagos Rift** | Mid-ocean ridge | Eastern central Pacific | 2°N | 94°W |
| **Galapagos triple junction** | Plate triple junctiom | Eastern equatorial Pacific | 2°N | 102°W |
| **Godzilla Mullion** | Oceanic core complex | Parece Vela Basin, West Pacific | 16°N | 139°E |
| **Gorda Ridge** | Mid-ocean ridge | Northeast Pacific | 42°N | 127°W |
| **Gulf of Aden spreading centre** | Mid-ocean ridge | Gulf of Aden, NW Indian Ocean | 13°N | 49°E |
| **Hess Deep** | Spreading segment | Cocos–Nazca Spreading Centre | 2°10′N | 101°30′W |
| **IODP hole 1256D** | Deep borehole | Eastern Pacific Ocean | 6°44.2′N | 91°56.1′W |
| **Juan de Fuca Ridge** | Mid-ocean ridge | Northeast Pacific | 47°N | 129°W |
| **Juan Fernandez microplate** | Oceanic microplate | Eastern South Pacific | 33°S | 111°W |
| **Kairei vent field** | Hydrothermal vent field | Central Indian Ridge | 25°19′S | 70°02′E |
| **Kane transform fault** | Transform fault | Mid-Atlantic Ridge | 23°45′N | 45°40′W |
| **Kane oceanic core complex** | Oceanic core complex | Mid-Atlantic Ridge | 23°30′N | 45°20′W |
| **Kolbeinsey Ridge** | Mid-ocean ridge | Norwegian Basin | 69°N | 17°W |
| **Kurchatov fracture zone** | Fracture zone | Mid-Atlantic Ridge | 40°30′N | 29°30′W |
| **Laptev Sea Rift** | Rift system | Arctic Ocean | 75°N | 127°E |
| **Lau Spreading Centre** | Back-arc spreading centre | Lau Basin, SW Pacific Ocean | 21°12′S | 176°20′W |
| **Logatchev vent field** | Hydrothermal vent field | Mid-Atlantic Ridge | 14°45′N | 44°59′W |

(*continued*)

| Feature name | Feature type | General location | Latitude | Longitude |
|---|---|---|---|---|
| **Loki's Castle vent field** | Hydrothermal vent field | Mohns Ridge | 73°33′N | 8°09′E |
| **Lost City vent field** | Hydrothermal vent field | Mid-Atlantic Ridge | 30°07′N | 42°07′W |
| **Lucky Strike vent field** | Hydrothermal vent field | Mid-Atlantic Ridge | 37°18′N | 32°16′W |
| **Madeira Abyssal Plain** | Abyssal plain | Mid-Atlantic Ridge flank | 30° N | 22° W |
| **MARK** | Study area | Mid-Atlantic Ridge | 23° N | 45° W |
| **Marquesas fracture zone** | Fracture zone | Pacific ocean | 10°50′S | 135°45′W |
| **Mathematician microplate** | Ancient oceanic microplate | Central Eastern Pacific | 17° N | 109° W |
| **Melville fracture zone** | Fracture zone | Southwest Indian Ridge | 29°40′S | 60°40′E |
| **Mendocino fracture zone** | Fracture zone | Northern Pacific | 40° N | 139° W |
| **Mid-Atlantic Ridge (north)** | Mid-ocean ridge | Atlantic Ocean | 35° N | 36° W |
| **Mid-Atlantic Ridge (south)** | Mid-ocean ridge | Atlantic Ocean | 26° S | 14° W |
| **Mid-Cayman Spreading Centre** | Spreading centre | Caribbean Sea | 18°20′N | 81°40′W |
| **Mohns Ridge** | Mid-ocean ridge | Norwegian Sea | 74° N | 8° E |
| **Molokai fracture zone** | Fracture zone | Eastern Pacific | 24° N | 136° W |
| **Murray fracture zone** | Fracture zone | Eastern Pacific | 31° N | 145° W |
| **Nibelungen vent field** | Hydrothermal vent field | Mid-Atlantic Ridge | 8°18′S | 13°31′W |
| **ODP hole 735B** | Deep borehole | Southwest Indian Ridge | 32°43.4′S | 57°16.0′E |
| **Oriente transform fault** | Transform fault | Caribbean Sea | 19° N | 80° W |
| **Orozco transform fault** | Transform fault | Eastern Pacific | 15°25′N | 105°05′W |
| **Pacific-Antarctic Rise** | Mid-ocean ridge | Southern Pacific | 55° S | 130° W |
| **Pito Deep** | Spreading segment | Easter microplate | 23° S | 112° W |
| **Rainbow vent field** | Hydrothermal vent field | Mid-Atlantic Ridge | 36°14′N | 33°54′W |
| **Red Lion vent site** | Hydrothermal vent field | Mid-Atlantic Ridge | 4°48′S | 12°22′W |
| **Reykjanes Ridge** | Mid-ocean ridge | North Atlantic | 61° N | 28° W |
| **Rodrigues triple junction** | Plate triple junction | Central Indian Ocean | 25°33′S | 70°00′E |
| **Romanche transform fault** | Transform fault | Equatorial Atlantic | 0°25′S | 20°00′W |
| **Saint Paul transform fault** | Transform fault | Equatorial Atlantic | 1°00′N | 28°00′W |
| **San Andreas fault** | Transform fault | California | 36°27′N | 121°03′W |
| **Serocki volcano** | Volcano | Mid-Atlantic Ridge, MARK area | 22°55′N | 44°57′W |
| **Snake Pit vent field** | Hydrothermal vent field | Mid-Atlantic Ridge, MARK area | 23°22′N | 44°57′W |
| **Southeast Indian Ridge** | Mid-ocean ridge | Southern Ocean | 45° S | 95° E |
| **Southwest Indian Ridge** | Mid-ocean ridge | Southern Ocean | 44° S | 39° E |
| **Swan Islands transform fault** | Transform fault | Caribbean Sea | 17°27′N | 83°00′W |
| **TAG** | Hydrothermal vent field | Central North Atlantic | 26°08′N | 44°50′W |
| **Tamayo transform fault** | Transform fault | Eastern Pacific | 23°07′N | 108°20′N |

| Feature name | Feature type | General location | Latitude | Longitude |
|---|---|---|---|---|
| **Telegraphic Plateau (obsolete)** | Mid-ocean ridge | North Atlantic | 50° N | 30° W |
| **Turtle Pits vent field** | Hydrothermal vent field | Mid-Atlantic Ridge | 4°48′ S | 12°22′ W |
| **Vema transform** | Transform fault | Southern North Atlantic | 10°48′ N | 42°15′ W |
| **Valu Fa Ridge** | Back-arc spreading centre | Lau Basin, SW Pacific Ocean | 22°17′ S | 176°38′ W |
| **Von Damm vent site** | Hydrothermal vent field | Mid-Cayman Spreading Centre | 18°23′ N | 81°48′ W |
| **Walvis Ridge** | Aseismic ridge | Eastern South Atlantic | 27° S | 4° E |
| **Wideawake vent field** | Hydrothermal vent field | Mid-Atlantic Ridge | 4°49′ S | 12°22′ W |

# References

Abrams, L. J., Detrick, R. S. and Fox, P. J. (1988). Morphology and crustal structure of the Kane fracture zone transverse ridge. *Journal of Geophysical Research*, **93**, 3195–3210.

Alexander, R. T. and Macdonald, K. C. (1996). Sea Beam, SeaMARC II and ALVIN-based studies of faulting on the East Pacific Rise 9°20′N–9°50′N. *Marine Geophysical Researches*, **18**, (5), 557–587.

Allen, D. E. and Seyfried, W. E. (2004). Serpentinization and heat generation: constraints from Lost City and Rainbow hydrothermal systems. *Geochimica et Cosmochimica Acta*, **68**, (6), 1347–1354.

Allerton, S. and MacLeod, C. (1999). New wireline seafloor drill augers well. *EOS, Transactions of the American Geophysical Union*, **80**, (33), 367.

Allerton, S. and Tivey, M. A. (2001). Magnetic polarity structure of the lower oceanic crust. *Geophysical Research Letters*, **28**, (3), 423–426.

Allerton, S., Murton, B. J., Searle, R. C. and Jones, M. (1995). Extensional faulting and segmentation of the Mid-Atlantic Ridge north of the Kane fracture zone (24°00′N to 24°40′N). *Marine Geophysical Researches*, **17**, 37–61.

Allerton, S., Escartin, J. and Searle, R. C. (2000). Extremely asymmetric magmatic accretion of oceanic crust at the ends of slow-spreading ridge-segments. *Geology*, **28**, 179–182.

Alt, J. C. (1995). Subseafloor processes in mid-ocean ridge hydrothermal systems. In *Seafloor Hydrothermal Systems: Physical, Chemical, Biological, and Geological Interactions, Geophysical Monograph no. 91*, ed. S. E. Humphris, R. A. Zierenberg, L. S. Mullineaux and R. E. Thomson. Washington, D.C.: American Geophysical Union, pp. 85–114.

Anderson, E. M. (1951). *The Dynamics of Faulting and Dyke Formation with Implications to Britain*, 2nd edition. Edinburgh & London: Oliver & Boyd.

Anderson, R. N., Honnorez, J., Becker, K. *et al.* (1982). DSDP hole-504B, the 1st reference section over 1km through layer-2 of the oceanic-crust. *Nature*, **300**, (5893), 589–594.

Anderson-Fontana, S., Engeln, J. F., Lundgren, P., Larson, R. L. and Stein, S. (1986). Tectonics and evolution of the Juan Fernandez microplate at the Pacific–Nazca–Antarctic triple junction. *Journal of Geophysical Research*, **91**, 2005–2018.

Anonymous (1993). News from national ridge research programs. *InterRidge News*, **2**, (1), 20–25.

Atwater, T. (1970). Implications of plate tectonics for the Cenozoic tectonic evolution of western North America. *Geological Society of America Bulletin*, **81**, 3513–3536.

Auzende, J.-M., Bideau, D., Bonatti, E. *et al.* (1989). Direct observation of a section through slow-spreading oceanic crust. *Nature*, **337**, 726–729.

Baines, A. G., Cheadle, M. J., John, B. E. and Schwartz, J. J. (2008). The rate of detachment faulting at Atlantis Bank, SW Indian Ridge. *Earth and Planetary Science Letters*, **273**, 105–114.

Baker, E. T. (1994). A 6-year time-series of hydrothermal plumes over the Cleft segment of the Juan-de-Fuca Ridge. *Journal of Geophysical Research-Solid Earth*, **99**, (B3), 4889–4904.

Baker, E. T. and German, C. (2004). On the global distribution of hydrothermal vent fields. In *Mid-Ocean Ridges: Hydrothermal Interactions Between the Lithosphere and Oceans, Geophysical Monograph 148*, ed. C. German, J. Lin and L. M. Parson. Washington, D.C.: American Geophysical Union, pp. 1–18.

Baker, E. T., Massoth, G. J. and Feely, R. A. (1987). Cataclysmic hydrothermal venting on the Juan-de-Fuca ridge. *Nature*, **329**, (6135), 149–151.

Baker, E. T., Lavelle, J. W., Feely, R. A., Massoth, G. J. and Walker, S. L. (1989). Episodic venting of hydrothermal fluids from the Juan de Fuca ridge. *Journal of Geophysical Research-Solid Earth and Planets*, **94**, (B7), 9237–9250.

Baker, E. T., Feely, R. A., Mottl, M. J. *et al.* (1994). Hydrothermal plumes along the East Pacific Rise, 8°40′ to 11°50′N – plume distribution and relationship to the apparent magmatic budget. *Earth and Planetary Science Letters*, **128**, (1–2), 1–17.

Baker, E. T., German, C. R. and Elderfield, H. (1995a). Hydrothermal plumes: global distributions and geological inferences. In *Seafloor Hydrothermal Systems: Physical, Chemical, Biological and Geological Interactions, Geophysical Monograph 91*, ed. S. E. Humphris, R. A. Zierenberg, L. S. Mullineaux and R. E. Thomson. Washington, D.C.: American Geophysical Union, pp. 47–71.

Baker, E. T., Massoth, G. J., Feely, R. A. *et al.* (1995b). Hydrothermal event plumes from the Coaxial sea-floor eruption site, Juan de Fuca ridge. *Geophysical Research Letters*, **22**, (2), 147–150.

Baker, E. T., Hey, R. N., Lupton, J. E. *et al.* (2002). Hydrothermal venting along Earth's fastest spreading center: East Pacific Rise, 27.5°–32.3° S. *Journal of Geophysical Research*, **107**, (B7), EPM2, doi:10.1029/2001JB000651.

Ballard, R. D. (1993). The Medea-Jason Remotely Operated Vehicle system. *Deep-Sea Research Part I-Oceanographic Research Papers*, **40**, (8), 1673.

Ballard, R. D. and Moore, J. G. (1977). *Photographic Atlas of the Mid-Atlantic Ridge Rift Valley*. Heidelberg: Springer-Verlag, 114 pp.

Ballard, R. D. and Van Andel, T. H. (1977). Morphology and tectonics of the inner rift valley at lat. 36°50′N on the Mid-Atlantic Ridge. *Geological Society of America Bulletin*, **88**, 507–530.

Ballard, R. D., Bryan, W. B., Heirtzler, J. R. *et al.* (1975). Manned submersible observations in FAMOUS area – Mid-Atlantic Ridge. *Science*, **190**, (4210), 103–108.

Ballard, R. D., Holcomb, R. T. and Van Andel, T. H. (1979). The Galapagos Rift at 86° W: 3. Sheet flows, collapse pits, and lava lakes of the rift valley. *Journal of Geophysical Research*, **84**, (B10), 5407–5422.

Ballard, R. D., Van Andel, T. H. and Holcomb, R. T. (1982). The Galapagos Rift at 86° W: 5. Variations in volcanism, structure and hydrothermal activity along a 30-kilometer segment of the rift valley. *Journal of Geophysical Research*, **87**, (B2), 1149–1161.

Ballu, V., Dubois, J., Deplus, C., Diament, M. and Bonvalor, S. (1998). Crustal structure of the Mid-Atlantic Ridge south of Kane Fracture Zone from seafloor and sea surface gravity data. *Journal of Geophysical Research, B*, **103**, (2), 2615–2631.

Baragar, W., Lambert, M., Baglow, N. and Gibson, I. (1990). The sheeted dyke zone in the Troodos ophiolite. In *Proceedings of the Symposium on Ophiolites and Oceanic Lithosphere – TROODOS 87*, ed. J. Malpas, E. M. Moores, A. Panayiotou and C. Xenophontos. Nicosia, Cyprus: Geological Survey Department, Ministry of Agriculture and Natural Resources, pp. 37–51.

Baran, J. M., Cochran, J. R., Carbotte, S. M. and Nedimovic, M. R. (2005). Variations in upper crustal structure due to variable mantle temperature along the Southeast Indian Ridge. *Geochemistry Geophysics Geosystems*, **6**, Q11002, doi:10.1029/2005GC000943.

Barclay, A. H., Toomey, D. R. and Solomon, S. C. (1998). Seismic structure and crustal magmatism at the Mid-Atlantic Ridge, 35° N. *Journal of Geophysical Research*, **103**, (B8), 17 827–17 844.

Barclay, A. H., Toomey, D. R. and Solomon, S. C. (2001). Microearthquake characteristics and crustal Vp/Vs structure at the Mid-Atlantic Ridge, 35° N. *Journal of Geophysical Research-Solid Earth*, **106**, (B2), 2017–2034.

Baross, J. A. and Hoffman, S. E. (1985). Submarine hydrothermal vents and associated gradient environments as sites for the origin and evolution of life. *Origins of Life and Evolution of the Biosphere*, **15**, (4), 327–345.

Batiza, R., Fox, P. J., Vogt, P. R. *et al.* (1989). Morphology, abundance, and chemistry of near-ridge seamounts in the vicinity of the Mid-Atlantic Ridge ~26° S. *Journal of Geology*, **97**, 209–220.

Becker, K., Sakai, H., Adamson, A. C. *et al.* (1989). Drilling deep into young oceanic-crust, Hole-504B, Costa-Rica Rift. *Reviews of Geophysics*, **27**, (1), 79–102.

Behn, M. D. and Ito, G. (2008). Magmatic and tectonic extension at mid-ocean ridges: 1. Controls on fault characteristics. *Geochemistry Geophysics Geosystems*, **9**, doi:10.1029/2008gc001965.

Bell, D. R. and Rossman, G. R. (1992). Water in Earth's mantle – the role of nominally anhydrous minerals. *Science*, **255**, (5050), 1391–1397.

Bell, R. E. and Buck, W. R. (1992). Crustal control of ridge segmentation inferred from observations of the Reykjanes Ridge. *Nature*, **357**, 583–586.

Beltenev, V., Ivanov, V., Rozhdestvenskaya, I. *et al.* (2007). New data about hydrothermal fields on the Mid-Atlantic Ridge between 11°–14° N: 32nd cruise of R/V Professor Logatchev. *InterRidge News*, **18**, 14–18.

Bergman, E. A. and Solomon, S. C. (1988). Transform fault earthquakes in the North Atlantic: source mechanisms and depth of faulting. *Journal of Geophysical Research, B*, **93**, (8), 9027–9057.

Bessonova, E. N., Fishman, V. M., Ryaboyi, V. Z. and Sitnikova, G. A. (1974). The tau method for inversion of travel times – I. Deep seismic sounding data. *Geophysical Journal of the Royal Astronomical Society*, **36**, (2), 377–398.

Best, M. G. and Christiansen, E. H. (2001). *Igneous Petrology*. Oxford: Blackwell Science, 458 pp.

Bicknell, J. D., Sempere, J.-C. and Macdonald, K. C. (1988). Tectonics of a fast spreading center: a Deep-Tow and Sea Beam survey on the East Pacific Rise at 19°30′ S. *Marine Geophysical Researches*, **9**, (1), 25–45.

Bird, R. T., Naar, D. F., Larson, R. L., Searle, R. C. and Scotese, C. R. (1998). Plate tectonic reconstructions of the Juan Fernandez microplate: transformation from internal shear to rigid rotation. *Journal of Geophysical Research*, **103**, (B4), 7049–7067.

Bird, R. T., Tebbens, S. F. and Kleinrock, M. C. (1999). Episodic triple-junction migration by rift propagation and microplates. *Geology*, **27**, 911–914.

Bischoff, J. L. and Rosenbauer, R. J. (1988). Liquid–vapor relations in the critical region of the system NaCl–$H_2O$ from 380 °C to 415 °C – a refined determination of the critical-point and 2-phase boundary of seawater. *Geochimica et Cosmochimica Acta*, **52**, (8), 2121–2126.

Blackinton, J. G., Hussong, D. M. and Kosslos, J. (1983). First results from a combination side scan sonar and seafloor mapping system (SeaMARC II). In *OffshoreTechnology Conference, OTC 4478*, pp. 307–311.

Blackman, D. K. and Forsyth, D. W. (1991). Isostatic compensation of tectonic features of the Mid-Atlantic Ridge: 25–27°30′ S. *Journal of Geophysical Research*, **96**, 11 741–11 758.

Blackman, D. K., Karson, J. A., Kelley, D. S. *et al.* (2002). Geology of the Atlantis Massif (Mid-Atlantic Ridge, 30° N): implications for the evolution of an ultramafic oceanic core complex. *Marine Geophysical Researches*, **23**, (5–6), 443–469.

Blackman, D. K., Ildefonse, B., John, B. E. *et al.* (2006). Oceanic core complex formation, Atlantis Massif: Expeditions 304 and 305 of the riserless drilling platform from and to Ponta Delgada, Azores (Portugal), Sites U1309–U1311, 17 November 2004–7 January 2005, and from and to Ponta Delgada, Azores (Portugal), Site U1309, 7 January–2 March 2005. *Proceedings of the Integrated Ocean Drilling Program*, **304/305**, doi:10.2204/iodp.proc.304305.2006.

Blackman, D. K., Karner, G. D. and Searle, R. C. (2008). Three-dimensional structure of oceanic core complexes: effects on gravity signature and ridge flank morphology, Mid-Atlantic Ridge 30° N. *Geochemistry Geophysics Geosystems*, **9**, (6), Q06007, doi:10.1029/2008GC001951.

Blackman, D. K., Canales, J. P. and Harding, A. (2009). Geophysical signatures of oceanic core complexes. *Geophysical Journal International*, **178** (2), 593–613.

Blackman, D. K., Ildefonse, B., John, B. E. *et al.* (2011). Drilling constraints on lithospheric accretion and evolution at Atlantis Massif, Mid-Atlantic Ridge 30° N. *Journal of Geophysical Research*, **116**, doi:10.1029/2010JB007931.

Blakely, R. and Cox, A. (1972). Identification of short polarity events by transforming marine magnetic profiles to the pole. *Journal of Geophysical Research*, **77**, 4339–4349.

Bohnenstiehl, D. R. and Carbotte, S. M. (2001). Faulting patterns near 19°30′ S on the East Pacific Rise: fault formation and growth at a superfast spreading center. *Geochemistry Geophysics Geosystems*, **2**, art. no. 2001GC000156.

Bonatti, E. (1976). Serpentine protrusions in the oceanic crust. *Earth and Planetary Science Letters*, **32**, 107–113.

Bonatti, E. (1978). Vertical tectonism in oceanic fracture zones. *Earth and Planetary Science Letters*, **37**, 369–379.

Bonatti, E. and Harrison, C. G. A. (1988). Eruption styles of basalt in oceanic spreading ridges and seamounts – effect of magma temperature and viscosity. *Journal of Geophysical Research-Solid Earth and Planets*, **93**, (B4), 2967–2980.

Bostrom, K., Peterson, M., Joensuu, O. and Fisher, D. (1969). Aluminum-poor ferromanganoan sediments on active oceanic ridges. *Journal of Geophysical Research*, **74** (12), 3261–3270.

Bott, M. H. P. (1982). *The Interior of the Earth, its Structure, Constitution and Evolution*, 2nd edition. London: Edward Arnold.

Boudier, F. and Nicolas, A. (1988). The ophiolites of Oman – preface. *Tectonophysics*, **151**, (1–4), R7–R8.

Bougault, H., Aballea, M., Radford-Knoery, J. *et al.* (1998). FAMOUS and AMAR segments on the Mid-Atlantic Ridge: ubiquitous hydrothermal Mn, $CH_4$, delta He-3 signals along the rift valley walls and rift offsets. *Earth and Planetary Science Letters*, **161**, (1–4), 1–17.

Bowen, A. N. and White, R. S. (1986). Deep-tow seismic profiles from the Vema transform and ridge-transform intersection. *Journal of the Geological Society of London*, **143**, 807–817.

Bown, J. and White, R. (1994). Variation with spreading rate of oceanic crustal thickness and geochemistry. *Earth and Planetary Science Letters*, **121**, (3–4), 435–449.

Bratt, S. R. and Purdy, G. M. (1984). Structure and variability of oceanic crust on the flanks of the East Pacific Rise between 11° and 13° N. *Journal of Geophysical Research*, **89**, 6111–6125.

Briais, A., Sloan, H., Parson, L. and Murton, B. (2000). Accretionary processes in the axial valley of the Mid-Atlantic Ridge 27°– 30° N from TOBI side-scan sonar images. *Marine Geophysical Researches*, **21**, 87–119.

Bryan, W. B., Humphris, S. E., Thompson, G. and Casey, J. F. (1994). Comparative volcanology of small axial eruptive centers in the MARK area. *Journal of Geophysical Research*, **99**, 2973–2984.

Buck, W. R. (1988). Flexural rotation of normal faults. *Tectonics*, **7**, (5), 959–973.

Buck, W. R. and Poliakov, A. N. B. (1998). Abyssal hills formed by stretching oceanic lithosphere. *Nature*, **392**, 272–275.

Buck, W. R. and Su, W. (1989). Focused mantle upwelling below mid-ocean ridges due to feedback between viscosity and melting. *Geophysical Research Letters*, **16**, 641–644.

Buck, W. R., Lavier, L. L. and Poliakov, A. N. B. (2005). Modes of faulting at mid-ocean ridges. *Nature*, **434**, 719–723.

Bullard, E. C. (1952). Heat flow through the floor of the eastern North Pacific ocean. *Nature*, **170**, (4318), 199–200.

Bullard, E. C. (1963). The flow of heat through the floor of the ocean. In *The Sea, volume 3*, ed. M. N. Hill. New York: Interscience Publishers, pp. 218–232.

Bullard, E. C. and Mason, R. G. (1961). The magnetic field astern of a ship. *Deep-Sea Research*, **8**, 20–27.

Bullard, E. C. and Mason, R. G. (1963). The magnetic field over the oceans. In *The Sea, volume 3*, ed. M. N. Hill. New York: Interscience Publishers, pp. 175–217.

Bullard, E. C., Maxwell, A. E. and Revelle, R. (1956). Heat flow through the deep sea floor. In *Advances in Geophysics*, ed. H. E. Landsberg. New York: Academic Press, pp. 153–181.

Byerlee, J. D. (1978). Friction of rocks. *Pure and Applied Geophysics*, **116**, 615–626.

Canales, J. P., Detrick, R. S., Toomey, D. R. and Wilcock, W. S. D. (2003). Segment-scale variations in the crustal structure of 150–300 kyr old fast spreading oceanic crust (East Pacific Rise, 8°15′N–10°5′N) from wide-angle seismic refraction profiles. *Geophysical Journal International*, **152**, (3), 766–794.

Canales, J. P., Singh, S. C., Detrick, R. S. *et al.* (2006). Seismic evidence for variations in axial magma chamber properties along the southern Juan de Fuca Ridge. *Earth and Planetary Science Letters*, **246**, (3–4), 353–366.

Canales, J. P., Tucholke, B. E., Xu, M., Collins, J. A. and DuBois, D. L. (2008). Seismic evidence for large-scale compositional heterogeneity of oceanic core complexes. *Geochemistry Geophysics Geosystems*, **9**, (8), doi:10.1029/2008GC002009.

Canales, J. P., Nedimovic, M. R., Kent, G. M., Carbotte, S. M. and Detrick, R. S. (2009). Seismic reflection images of a near-axis melt sill within the lower crust at the Juan de Fuca ridge. *Nature*, **460**, (7251), 89–99.

Cande, S. and Kent, D. (1995). Revised calibration of the geomagnetic polarity timescale for the late Cretaceous and Cenozoic. *Journal of Geophysical Research*, **100**, (B4), 6093–6095.

Cann, J. R. (1970). New model for structure of ocean crust. *Nature*, **226**, (5249), 928–930.

Cann, J. R. (1974). Model for oceanic crustal structure developed. *Geophysical Journal of the Royal Astronomical Society*, **39**, (1), 169–187.

Cann, J. R. and Smith, D. K. (2005). Evolution of volcanism and faulting in a segment of the Mid-Atlantic Ridge at 25° N. *Geochemistry Geophysics Geosystems*, **6**, doi:10.1029/2005gc000954.

Cann, J. R. and Strens, M. R. (1982). Black smokers fuelled by freezing magma. *Nature*, **298**, 147–149.

Cann, J. R., Strens, M. R. and Rice, A. (1985). A simple magma-driven thermal balance model for the formation of volcanogenic massive sulphides. *Earth and Planetary Science Letters*, **76**, 123–134.

Cann, J. R., Blackman, D. K., Smith, D. K. *et al.* (1997). Corrugated slip surfaces formed at ridge-transform intersections on the Mid Atlantic Ridge. *Nature*, **385**, 329–332.

Cann, J. R., Elderfield, H. and Laughton, A. S. (1999). *Mid-Ocean Ridges; Dynamics of Processes Associated with the Creation of New Oceanic Crust*. Cambridge: Cambridge University Press.

Cann, J. R., Prichard, H. M., Malpas, J. G. and Xenophontos, C. (2001). Oceanic inside corner detachments of the Limassol Forest area, Troodos ophiolite, Cyprus. *Journal of the Geological Society*, **158**, 757–767.

Cannat, M., Mevel, C., Maia, M. *et al.* (1995). Thin crust, ultramafic exposures, and rugged faulting patterns at Mid-Atlantic Ridge (22°–24° N). *Geology*, **23**, (1), 49–52.

Cannat, M., Rommevaux-Jestin, C., Sauter, D., Deplus, C. and Mendel, V. (1999). Formation of the axial relief at the very slow spreading Southwest Indian Ridge (49° to 69° E). *Journal of Geophysical Research*, **104**, (B10), 22 825–22 843.

Cannat, M., Cann, J. and Maclennan, J. (2004). Some hard rock constraints on the supply of heat to mid-ocean ridges. In *Mid-Ocean Ridges: Hydrothermal Interactions between the Lithosphere and Oceans, Geophysical Monograph*, **148**, ed. German, C. R., Lin, J. and Parson, L. M., Washington, D.C.: AGU, pp. 111–149.

Cannat, M., Sauter, D., Mendel, V. *et al.* (2006). Modes of seafloor generation at a melt-poor ultraslow-spreading ridge. *Geology*, **34**, (7), 605–608; doi: 10.1130/G22486.1.

Carbotte, S. M. and Macdonald, K. C. (1990). Causes of variation in fault-facing direction on the ocean floor. *Geology*, **18**, 749–752.

Carbotte, S. and Macdonald, K. (1992). East Pacific Rise 8°–10°30′ N: evolution of ridge segments and discontinuities from SeaMARC II and three-dimensional magnetic studies. *Journal of Geophysical Research*, **97**, (B5), 6959–6982.

Carbotte, S. M. and Macdonald, K. C. (1994a). The axial topographic high at intermediate and fast spreading ridges. *Earth and Planetary Science Letters*, **128**, (3–4), 85–97.

Carbotte, S. M. and Macdonald, K. C. (1994b). Comparison of seafloor tectonic fabric at intermediate, fast, and super fast spreading ridges: influence of spreading rate, plate motions, and ridge segmentation on fault patterns. *Journal of Geophysical Research*, **99**, (B7), 13 609–13 631.

Carbotte, S. M. and Scheirer, D. S. (2004). Variability of ocean crustal structure created along the global mid-ocean ridge. In *Hydrogeology of the Oceanic Lithosphere*, ed. E. E. Davis and H. Elderfield. Cambridge: Cambridge University Press, pp. 59–107.

Carbotte, S. M., Detrick, R. S., Harding, A. *et al.* (2006). Rift topography linked to magmatism at the intermediate spreading Juan de Fuca Ridge. *Geology*, **34**, (3), 209–212.

Carlson, R. L. (1998). Seismic velocities in the uppermost oceanic crust: age dependence and the fate of layer 2A. *Journal of Geophysical Research-Solid Earth*, **103**, (B4), 7069–7077.

Carlson, R. L. (2011). The effect of hydrothermal alteration on the seismic structure of the upper oceanic crust: evidence from Holes 504B and 1256D. *Geochemistry Geophysics Geosystems*, **12**, doi:10.1029/2011gc003624.

Carpine-Lancre, J. (2001). Oceanographic Sovereigns: Prince Albert I of Monaco and King Carlos I of Portugal. In *Understanding the Oceans: A Century of Ocean Exploration*, ed. M. Deacon, T. Rice and C. Summerhayes. London: UCL Press, pp. 56–68.

Casey, J. F. and Karson, J. A. (1981). Magma chamber profiles from the Bay-of-Islands ophiolite complex. *Nature*, **292**, (5821), 295–301.

Chadwick, W. W. and Embley, R. W. (1994). Lava flows from a mid-1980s submarine eruption on the cleft segment, Juan-de-Fuca Ridge. *Journal of Geophysical Research-Solid Earth*, **99**, (B3), 4761–4776.

Chadwick, W. W., Embley, R. W. and Fox, C. G. (1995). Seabeam depth changes associated with recent lava flows, Coaxial segment, Juan-de-Fuca Ridge – evidence for multiple eruptions between 1981–1993. *Geophysical Research Letters*, **22**, (2), 167–170.

Chadwick, W. W., Embley, R. W. and Shank, T. M. (1998). The 1996 Gorda Ridge eruption: geologic mapping, sidescan sonar, and SeaBeam comparison results. *Deep-Sea Research Part II-Topical Studies in Oceanography*, **45**, (12), 2547–2569.

Chadwick, W. W., Embley, R. W., Milburn, H. B., Meinig, C. and Stapp, M. (1999). Evidence for deformation associated with the 1998 eruption of Axial Volcano, Juan de

Fuca Ridge, from acoustic extensometer measurements. *Geophysical Research Letters*, **26**, (23), 3441–3444.

Chadwick, W. W., Nooner, S. L., Zumberge, M. A., Embley, R. W. and Fox, C. G. (2006). Vertical deformation monitoring at Axial Seamount since its 1998 eruption using deep-sea pressure sensors. *Journal of Volcanology and Geothermal Research*, **150**, (1–3), 313–327.

Chapman, C. H. and Drummond, R. (1982). Body-wave seismograms in inhomogeneous media using Maslov asymptotic theory. *Bulletin of the Seismological Society of America*, **72**, S277–S317.

Chave, A. D. and Cox, C. S. (1982). Controlled electromagnetic sources for measuring electrical-conductivity beneath the oceans .1. Forward problem and model study. *Journal of Geophysical Research*, **87**, (B7), 5327–5338.

Chayes, D. N. (1983). Evolution of Sea MARC I. In *IEEE Proceedings of the 3rd Working Symposium on Oceanographic Data Systems*. New York: IEEE Computer Society Press, pp. 103–108.

Chen, Y. J. (1992). Oceanic crustal thickness versus spreading rate. *Geophysical Research Letters*, **19**, (8), 753–756.

Chen, Y. (2004). Modelling the thermal structure of the oceanic crust. In *Mid-Ocean Ridges: Hydrothermal Interactions between the Lithosphere and Oceans, Geophysical Monograph 148*, ed. C. German, J. Lin and L. M. Parson. Washington, D.C.: American Geophysical Union, pp. 95–110.

Chen, Y. S. J. and Lin, J. (2004). High sensitivity of ocean ridge thermal structure to changes in magma supply: the Galapagos Spreading Center. *Earth and Planetary Science Letters*, **221**, (1–4), 263–273.

Chen, Y. and Morgan, W. J. (1990a). A nonlinear rheology model for mid-ocean ridge axis topography. *Journal of Geophysical Research*, **95**, 17 583–17 604.

Chen, Y. and Morgan, W. J. (1990b). Rift valley/no rift valley transition at mid-ocean ridges. *Journal of Geophysical Research*, **95**, 17 571–17 581.

Chen, Y. J. and Phipps Morgan, J. (1996). The effects of spreading rate, the magma budget, and the geometry of magma emplacement on the axial heat flux at mid-ocean ridges. *Journal of Geophysical Research*, **101**, (B5), 11 475–11 482.

Chesterman, W. D., Clynick, P. R. and Stride, A. H. (1958). An acoustic aid to sea bed survey. *Acustica*, **8**, 285–290.

Chin, C. S., Coale, K. H., Elrod, V. A. *et al.* (1994). In-situ observations of dissolved iron and manganese in hydrothermal vent plumes, Juan-de-Fuca Ridge. *Journal of Geophysical Research-Solid Earth*, **99**, (B3), 4969–4984.

Christeson, G. L., Purdy, G. M. and Fryer, G. J. (1994). Seismic constraints on shallow crustal emplacement processes at the fast spreading East Pacific Rise. *Journal of Geophysical Research*, **99**, (B9), 17 957–17 973.

Christeson, G. L., McIntosh, K. D. and Karson, J. A. (2010). Inconsistent correlation of seismic layer 2a and lava layer thickness in oceanic crust. *Nature*, **445**, 418–421.

Clague, D. A., Moore, J. G. and Reynolds, J. R. (2000a). Formation of submarine flat-topped volcanic cones in Hawai'i. *Bulletin of Volcanology*, **62**, (3), 214–233.

Clague, D. A., Reynolds, J. R. and Davis, A. S. (2000b). Near-ridge seamount chains in the northeastern Pacific Ocean. *Journal of Geophysical Research-Solid Earth*, **105**, (B7), 16 541–16 561.

Clague, D. A., Paduan, J. B. and Davis, A. S. (2009). Widespread strombolian eruptions of mid-ocean ridge basalt. *Journal of Volcanology and Geothermal Research*, **180**, (2–4), 171–188.

Cochran, J. R. (1979). An analysis of isostasy in the worlds oceans: 2. Mid-ocean ridge crests. *Journal of Geophysical Research*, **84**, 4713–4729.

Cochran, J. R. (1986). Variations in subsidence rates along intermediate and fast spreading midocean ridges. *Geophysical Journal of the Royal Astronomical Society*, **87**, (2), 421–454.

Cochran, J. R. (2008). Seamount volcanism along the Gakkel Ridge, Arctic Ocean. *Geophysical Journal International*, **174**, 1153–1173.

Cochran, J. R., Kurras, G. J., Edwards, M. H. and Coakley, B. J. (2003). The Gakkel Ridge: bathymetry, gravity anomalies, and crustal accretion at extremely slow spreading rates. *Journal of Geophysical Research*, **108**, (B2), doi:10.1029/2002JB001830.

Coffin, M. F. and 30 others (2001). Earth, Oceans and Life: Scientific Investigation of the Earth System Using Multiple Drilling Platforms and New Technologies. Integrated Ocean Drilling Program Initial Science Plan, 2003–2013. Washington, D.C., 20036–2102: IODP International Working Group Support Office.

Cogné, J. P., Francheteau, J., Coutillot, V. *et al.* (1995). Large rotation of the Easter microplate as evidenced by oriented paleomagnetic samples from the ocean floor. *Earth and Planetary Science Letters*, **136**, 213–222.

Collette, B. J. (1974). Thermal contraction joints in a spreading seafloor as origin of fracture zones. *Nature*, **251**, 299–300.

Collier, J. S. and Singh, S. C. (1998). Poisson's ratio structure of young oceanic crust. *Journal of Geophysical Research, B*, **103**, 20 981–20 996.

Collier, J. and Sinha, M. (1990). Seismic images of a magma chamber beneath the Lau Basin back-arc spreading center. *Nature*, **346**, (6285), 646–648.

Collins, J. A., Blackman, D. K., Harris, A. and Carlson, R. L. (2009). Seismic and drilling constraints on velocity structure and reflectivity near IODP Hole U1309D on the central dome of Atlantis Massif, Mid-Atlantic Ridge 30° N. *Geochemistry Geophysics Geosystems*, **10**, doi:10.1029/2008GC002121.

Connelly, D., Copley, J., Murton, B. *et al.* (2012). Hydrothermal vent fields and chemosynthetic biota on the world's deepest seafloor spreading centre. *Nature Communications*, **3**, doi:10.1038/ncomms1636.

Cooper, K. M., Goldstein, S. J., Sims, K. W. W. and Murrell, M. T. (2003). Uranium-series chronology of Gorda Ridge volcanism: new evidence from the 1996 eruption. *Earth and Planetary Science Letters*, **206**, (3–4), 459–475.

Corliss, J. B. (1990). Hot-springs and the origin of life. *Nature*, **347**, (6294), 624.

Corliss, J. B., Dymond, J., Gordon, L. I. *et al.* (1979). Submarine thermal springs on the Galapagos Rift. *Science*, **203**, 1073–1083.

Cormier, M.-H. (1997). The ultra-fast East Pacific Rise: instability of the plate boundary and implications for accretionary processes. *Philosophical Transactions of the Royal Society of London, Series A*, **355**, 341–367.

Cormier, M.-H. and Macdonald, K. C. (1994). East Pacific Rise at 18°–19° S: asymmetric spreading and ridge reorientation by ultrafast migration of axial discontinuities. *Journal of Geophysical Research*, **99**, (B1), 543–564.

Cormier, M.-H., Scheirer, D. S. and Macdonald, K. C. (1996). Evolution of the East Pacific Rise at 16°–19° S since 5 Ma: bisection of overlapping spreading centers by new, rapidly propagating ridge segments. *Marine Geophysical Researches*, **18**, 53–84.

Cormier, M. H., Ryan, W. B. F., Shah, A. K. *et al.* (2003). Waxing and waning volcanism along the East Pacific Rise on a millennium time scale. *Geology*, **31**, (7), 633–636.

Courtillot, V. (1982). Propagating rifts and continental breakup. *Tectonics*, **1**, 239–250.

Cowie, P. A., Vanneste, C. and Sornette, D. (1993a). Statistical physics model for the spatio-temporal evolution of faults. *Journal of Geophysical Research*, **98**, 21 809–21 821.

Cowie, P. A., Scholz, C. H., Edwards, M. and Malinverno, A. (1993b). Fault strain and seismic coupling on mid-ocean ridges. *Journal of Geophysical Research*, **98**, 17 911–17 920.

Cowie, P. A., Malinverno, A., Ryan, W. B. F. and Edwards, M. H. (1994). Quantitative fault studies on the East Pacific Rise: a comparison of sonar imaging techniques. *Journal of Geophysical Research*, **99**, (B8), 15 205–15 218.

Cox, A. and Hart, R. B. (1986). *Plate Tectonics – How It Works*. Blackwell Scientific Publications, Inc., 392 pp.

Cox, A., Dalrymple, G. B. and Doell, R. R. (1963). Geomagnetic polarity epochs and pleistocene geochronometry. *Nature*, **198**, (488), 1049–1051.

Crane, K. (1976). The intersection of the Siqueiros transform fault and the East Pacific Rise. *Marine Geology*, **21**, 25–46.

Crane, K. (1987). Structural evolution of the East Pacific Rise axis from 13°10′N to 10°35′N – interpretations from SeaMARC-I data. *Tectonophysics*, **136**, (1–2), 65–124.

Crawford, W. C. and Webb, S. C. (2002). Variations in the distribution of magma in the lower crust and at the Moho beneath the East Pacific Rise at 9°–10° N. *Earth and Planetary Science Letters*, **203**, (1), 117–130.

Crawford, W. C., Webb, S. C. and Hildebrand, J. A. (1991). Sea-floor compliance observed by long-period pressure and displacement measurements. *Journal of Geophysical Research-Solid Earth*, **96**, (B10), 16 151–16 160.

Crawford, W. C., Webb, S. C. and Hildebrand, J. A. (1998). Estimating shear velocities in the oceanic crust from compliance measurements by two-dimensional finite difference modeling. *Journal of Geophysical Research, B*, **103**, (5), 9895–9916.

Crawford, W. C., Webb, S. C. and Hildebrand, J. A. (1999). Constraints on melt in the lower crust and Moho at the East Pacific Rise, 9°48′N, using seafloor compliance measurements. *Journal of Geophysical Research-Solid Earth*, **104**, (B2), 2923–2939.

Crawford, W. C., Singh, S. C., Seher, T. *et al.* (2010). Crustal structure, magma chamber, and faulting beneath the Lucky Strike hydrothermal vent field. In *Diversity of Hydrothermal Systems on Slow Spreading Ocean Ridges, Geophysical Monograph 188*, ed. P. A. Rona, C. W. Devey, J. Dyment and B. J. Murton. Washington, D.C.: American Geophysical Union, pp. 113–152.

Crowder, L. K. and Macdonald, K. C. (2000). New constraints on the width of the zone of active faulting on the East Pacific Rise 8°30′N–10°00′N from Sea Beam Bathymetry and SeaMARC II Side-scan Sonar. *Marine Geophysical Researches* **21**, (6), 513–527.

CYAMEX (1981). First manned submersible dives on the East Pacific Rise at 21° N (Project RITA): general results. *Marine Geophysical Researches*, **4**, 345–379.

Danchik, R. J. (1984). The navy navigation satellite system (Transit). *Johns Hopkins APL Technical Digest*, **5**, (4), 323–329.

Davis, E. E. (1982). Evidence for extensive basalt flows on the sea-floor. *Geological Society of America Bulletin*, **93**, (10), 1023–1029.

Davis, E. E. and Lister, C. R. B. (1974). Fundamentals of ridge crest topography. *Earth and Planetary Science Letters*, **21**, 405–413.

Delaney, J. R. (1989). RIDGE Initial Science Plan February 1989. Seattle: School of Oceanography, University of Washington, 90 pp.

deMartin, B. J., Sohn, R. A., Canales, J. P. and Humphris, S. E. (2007). Kinematics and geometry of active detachment faulting beneath the Trans-Atlantic Geotraverse (TAG) hydrothermal field on the Mid-Atlantic Ridge. *Geology*, **35**, (8), 711–714; doi:10.1130/G23718A.1.

DeMets, C., Gordon, R. G., Argus, D. F. and Stein, S. (1990). Current plate motions. *Geophysical Journal International*, **101**, 425–478.

DeMets, C., Gordon, R. G., Argus, D. F. and Stein, S. (1994). Effect of recent revisions to the geomagnetic reversal timescale on estimates of current plate motions. *Geophysical Research Letters*, **21**, (20), 2191–2194.

Deschamps, A., Tivey, M., Embley, R. W. and Chadwick, W. W. (2007). Quantitative study of the deformation at Southern Explorer Ridge using high-resolution bathymetric data. *Earth and Planetary Science Letters*, **259**, (1–2), 1–17.

Detrick, R. S., Cormier, M. H., Prince, R. A., Forsyth, D. W. and Ambos, E. L. (1982). Seismic constraints on the crustal structure of the Vema fracture zone. *Journal of Geophysical Research*, **87**, 10 599–10 612.

Detrick, R. S., Buhl, P., Vera, E. *et al.* (1987). Multi-channel seismic imaging of a crustal magma chamber along the East Pacific Rise. *Nature*, **326**, 35–41.

Detrick, R., White, R. and Purdy, G. (1993a). Crustal structure of north-Atlantic fracture-zones. *Reviews of Geophysics*, **31**, (4), 439–458.

Detrick, R. S., Harding, A. J., Kent, G. M. *et al.* (1993b). Seismic structure of the southern East Pacific Rise. *Science*, **259**, (5094), 499–503.

Detrick, R., Collins, J. and Swift, S. (1994). In-situ evidence for the nature of the seismic layer 2/3 boundary in oceanic crust. *Nature*, **370**, 288–290.

Detrick, R. S., Needham, H. D. and Renard, V. (1995). Gravity anomalies and crustal thickness variations along the Mid-Atlantic Ridge between 33° N and 40° N. *Journal of Geophysical Research*, **100**, (B3), 3767–3787.

Detrick, R. S., Sinton, J. M., Ito, G. *et al.* (2002). Correlated geophysical, geochemical, and volcanological manifestations of plume-ridge interaction along the Galapagos Spreading Center. *Geochemistry Geophysics Geosystems*, **3**, doi:10.1029/2002gc000350.

Dick, H. J. B., Thompson, W. B. and Bryan, W. B. (1981). Low angle faulting and steady-state emplacement of plutonic rocks at ridge-transform intersections. *EOS, Transactions of the American Geophysical Union*, **62**, 406.

Dick, H. J. B., Fisher, R. L. and Bryan, W. B. (1984). Mineralogic variability of the uppermost mantle along mid-ocean ridges. *Earth and Planetary Science Letters*, **69**, (1), 88–106.

Dick, H. J. B., Meyer, P. S., Bloomer, S., Stakes, D. and Mawer, C. (1991). Lithostratigraphic evolution of an in situ section of oceanic layer 3. *Proceedings of the Ocean Drilling Program, Scientific Results*, **118**, 439–538.

Dick, H. J. B., Natland, J. H., Alt, J. C. *et al.* (2000). A long in situ section of the lower ocean crust: results of ODP Leg 176 drilling at the Southwest Indian Ridge. *Earth and Planetary Science Letters*, **179**, (1), 31–51.

Dick, H. J. B., Lin, J. and Schouten, H. (2003). An ultraslow-spreading class of ocean ridge. *Nature*, **426**, (6965), 405–412.

Dick, H. J. B., Tivey, M. A. and Tucholke, B. E. (2008). Plutonic foundation of a slow-spreading ridge segment: oceanic core complex at Kane Megamullion, 23°30′ N, 45°20′ W. *Geochemistry Geophysics Geosystems*, **9**, (9), Q05014.

Dietz, R. S. (1961). Continental and ocean basin evolution by spreading of the sea floor. *Nature*, **190**, 854–7.

Dilek, Y., Moores, E., Elthon, D. and Nicolas, A. (2000). *Ophiolites and Oceanic Crust: New Insights from Field Studies and the Ocean Drilling Program*. Boulder: Geological Society of America, 552 pp.

Donovan, D. T. and Stride, A. H. (1961). An acoustic survey of the sea floor south of Dorset and its geological interpretation. *Philosophical Transactions of the Royal Society of London*, **B244**, 299–330.

Donval, J. P. *et al.* (1997). High $H_2$ and $CH_4$ content in hydrothermal fluids from Rainbow site newly sampled at 36°14′ N on the AMAR segment, Mid-Atlantic Ridge (diving FLORES cruise, July 1997). Comparison with other MAR sites. *EOS, Transactions of the American Geophysical Union*, **78**, 832.

Douville, E., Charlou, J. L., Oelkers, E. H. *et al.* (2002). The rainbow vent fluids (36°14′ N, MAR): the influence of ultramafic rocks and phase separation on trace metal content in Mid-Atlantic Ridge hydrothermal fluids. *Chemical Geology*, **184**, (1–2), 37–48.

Dunlop, D. J. and Prévot, M. (1982). Magnetic properties and opaque mineralogy of drilled submarine intrusive rocks. *Geophysical Journal of the Royal Astronomical Society*, **69**, (3), 763–802.

Dunn, R. A., Toomey, D. R. and Solomon, S. C. (2000). Three-dimensional seismic structure and physical properties of the crust and shallow mantle beneath the East Pacific Rise at 9°30′ N. *Journal of Geophysical Research-Solid Earth*, **105**, (B10), 23 537–23 555.

Dunn, R. A., Lekic, V., Detrick, R. S. and Toomey, D. R. (2005). Three-dimensional seismic structure of the Mid-Atlantic Ridge (35° N): evidence for focused melt supply and lower crustal dike injection. *Journal of Geophysical Research-Solid Earth*, **110**, (B9), doi:10.1029/2004JB003473.

Duven, D. J. and Artis, D. A. (1985). Global Positioning System surface navigation accuracy study. *Marine Geodesy*, **9**, (2), 145–173.

Dyment, J., Arkani-Hamed, J. and Ghods, A. (1997). Contribution of serpentinized ultramafics to marine magnetic anomalies at slow and intermediate spreading centres: insights from the shape of the anomalies. *Geophysical Journal International*, **129**, (3), 691–701.

Dziak, R. P. and Fox, C. G. (1999). Long-term seismicity and ground deformation at Axial Volcano, Juan de Fuca Ridge. *Geophysical Research Letters*, **26**, (24), 3641–3644.

Dziak, R. P., Hammond, S. R. and Fox, C. G. (2011). A 20-year hydroacoustic time series of seismic and volcanic events in the northeast Pacific Ocean. *Oceanography*, **24**, (3), 280–293.

Eakins, B. W. and Lonsdale, P. F. (2003). Structural patterns and tectonic history of the Bauer microplate, Eastern Tropical Pacific. *Marine Geophysical Researches*, **24**, (3–4), 171–205.

Edwards, R. N., Law, L. K. and Delaurier, J. M. (1981). On measuring the electrical-conductivity of the oceanic-crust by a modified magnetometric resistivity method. *Journal of Geophysical Research*, **86**, (B12), 1609–1615.

Eittreim, S. and Ewing, J. (1975). Vema Fracture Zone transform fault. *Geology*, **3**, (10), 555–558.

Elthon, D. (1991). Geochemical evidence for formation of the Bay-of-Islands ophiolite above a subduction zone. *Nature*, **354**, (6349), 140–143.

Embley, R. W. and Chadwick, W. W. (1994). Volcanic and hydrothermal processes associated with a recent phase of sea-floor spreading at the northern Cleft segment – Juan-de-Fuca Ridge. *Journal of Geophysical Research-Solid Earth*, **99**, (B3), 4741–4760.

Embley, R. W. and Wilson, D. S. (1992). Morphology of the Blanco Transform Fault Zone – NE Pacific: implications for its tectonic evolution. *Marine Geophysical Researches*, **14**, 25–45.

Embley, R. W., Chadwick, W. W., Jonasson, I. R., Butterfield, D. A. and Baker, E. T. (1995). Initial results of the rapid response to the 1993 Coaxial event – relationships between hydrothermal and volcanic processes. *Geophysical Research Letters*, **22**, (2), 143–146.

Embley, R. W., Chadwick, W. W., Clague, D. and Stakes, D. (1999). 1998 Eruption of Axial Volcano: multibeam anomalies and sea-floor observations. *Geophysical Research Letters*, **26**, (23), 3425–3428.

Engeln, J. F., Wiens, D. A. and Stein, S. (1986). Mechanisms and depths of Atlantic transform earthquakes. *Journal of Geophysical Research-Solid Earth and Planets*, **91**, (B1), 548–577.

Escartin, J. and Canales, J. P. (2011). Detachments in Oceanic Lithosphere: Deformation, Magmatism, Fluid Flow, and Ecosystems, AGU Chapman Conference on Oceanic Detachments; Agros, Cyprus, 8–15 May 2010. *EOS, Transactions of the American Geophysical Union*, **92**, (4), 31–32.

Escartin, J. and Lin, J. (1995). Ridge offsets, normal faulting, and gravity anomalies of slow spreading ridges. *Journal of Geophysical Research*, **100**, (B4), 6163–6177.

Escartin, J. and Lin, J. (1998). Tectonic modification of axial structure: evidence from spectral analyses of gravity and bathymetry of the Mid-Atlantic Ridge flanks (25.5°–17.5° N). *Earth and Planetary Science Letters*, **154**, (1–4), 279–293.

Escartin, J., Hirth, G. and Evans, B. (1997). Effects of serpentinization on the lithospheric strength and the style of normal faulting at slow-spreading ridges. *Earth and Planetary Science Letters*, **151**, 181–189.

Escartin, J., Cowie, P. A., Searle, R. C. *et al.* (1999). Quantifying tectonic strain and magmatic accretion at a slow-spreading ridge segment, Mid-Atlantic Ridge, 29° N. *Journal of Geophysical Research, B*, **104**, 10 421–10 437.

Escartin, J., Soule, S. A., Fornari, D. J. *et al.* (2007). Interplay between faults and lava flows in construction of the upper oceanic crust: the East Pacific Rise crest 9°25′–9°58′ N. *Geochemistry Geophysics Geosystems*, **8**, doi:10.1029/2006gc001399.

Escartin, J., Smith, D. K., Cann, J. *et al.* (2008). Central role of detachment faults in accretion of slow-spreading oceanic lithosphere. *Nature*, **455**, 790–795.

Evans, R. L., Constable, S. C., Sinha, M. C., Cox, C. S. and Unsworth, M. J. (1991). Upper crustal resistivity structure of the East Pacific Rise near 13° N. *Geophysical Research Letters*, **18**, (10), 1917–1920.

Ewing, J. and Ewing, M. (1959). Seismic refraction measurements in the Atlantic Ocean basins, in the Mediterranean, on the Mid-Atlantic Ridge, and in the Norwegian Sea. *Bulletin of the Geological Society of America*, **70**, 291–318.

Ewing, J. I. and Tirey, G. B. (1961). Seismic profiler. *Journal of Geophysical Research*, **66**, (9), 2917.

Ewing, J. I. and Zaunere, R. (1964). Seismic profiling with a pneumatic sound source. *Journal of Geophysical Research*, **69**, 4913–4915.

Ewing, M., Crary, A. P. and Rutherford, H. M. (1937). Geophysical investigations in the emerged and submerged Atlantic coastal plain, part I: methods and results. *Bulletin of the Geological Society of America*, **48**, 753–802.

Ewing, M., Worzel, J. L. and Vine, A. C. (1967). Early development of ocean-bottom photography at Woods Hole Oceanographic Institution and Lamont Geological Observatory. In *Deep-Sea Photography*, ed. J. B. Hersey. Baltimore: The Johns Hopkins Press, pp. 13–41.

Fleming, J. A. (1937). Magnetic survey of the oceans. In *International Aspects of Oceanography*, ed. T. W. Vaughan. Washington, D.C.: National Academy of Sciences.

Flewellen, C., Millard, N. and Rouse, I. (1993). TOBI, a vehicle for deep ocean survey. *Electronics and Communication Engineering Journal*, (April 1993), 85–93.

Fornari, D. J., Gallo, D. G., Edwards, M. H. *et al.* (1989). Structure and topography of the Siqueiros Transform Fault System: evidence for the development of intra-transform spreading centers. *Marine Geophysical Researches*, **11**, 263–299.

Fornari, D. J., Haymon, R. M., Perfit, M. R., Gregg, T. K. P. and Edwards, M. H. (1998). Axial summit trough of the East Pacific Rise 9°–10° N: geological characteristics and evolution of the axial zone on fast spreading mid-ocean ridges. *Journal of Geophysical Research, B*, **103**, (5), 9827–9855.

Fornari, D., Tivey, M., Schouten, H. *et al.* (2004). Submarine lava flow emplacement at the East Pacific Rise 9°50′ N: implications for uppermost ocean crust stratigraphy and hydrothermal fluid circulation. In *Mid-Ocean Ridges: Hydrothermal Interactions Between the Lithosphere and Oceans: Geophysical Monograph 148*, ed. C. R. German, J. Lin and L. M. Parson. Washington, D.C.: American Geophysical Union, pp. 187–217.

Forsyth, D. W. (1975). Early structural evolution and anisotropy of oceanic upper mantle. *Geophysical Journal of the Royal Astronomical Society*, **43**, (1), 103–162.

Forsyth, D. and Uyeda, S. (1975). Relative importance of driving forces of plate motion. *Geophysical Journal of the Royal Astronomical Society*, **43**, (1), 163–200.

Fouquet, Y., Ondreas, H., Charlou, J. L. *et al.* (1995). Atlantic lava lakes and hot vents. *Nature*, **377**, 201.

Fouquet, Y., Cambon, P., Etoubleau, J. *et al.* (2010). Geodiversity of hydrothermal processes along the Mid-Atlantic Ridge and ultramafic-hosted mineralization: a new type of oceanic Cu-Zn-Co-Au volcanogenic massive sulfide deposit. In *Diversity of Hydrothermal Systems on Slow Spreading Ocean Ridges, Geophysical Monograph 188*, ed. P. A. Rona, C. W. Devey, J. Dyment and B. J. Murton. Washington, D.C.: American Geophysical Union, pp. 321–367.

Fowler, C. M. R. (2005). *The Solid Earth: An Introduction to Global Geophysics*, 2nd edition. Cambridge: Cambridge University Press, 685 pp.

Fox, C. G., Murphy, K. M. and Embley, R. W. (1988). Automated display and statistical-analysis of interpreted deep-sea bottom photographs. *Marine Geology*, **78**, (3–4), 199–216.

Fox, C. G., Matsumoto, H. and Lau, T. K. A. (2001). Monitoring Pacific Ocean seismicity from an autonomous hydrophone array. *Journal of Geophysical Research-Solid Earth*, **106**, (B3), 4183–4206.

Fox, P. J. and Gallo, D. G. (1984). A tectonic model for ridge–transform–ridge plate boundaries: implications for the structure of oceanic lithosphere. *Tectonophysics*, **104**, (3–4), 205–242.

Fox, P. J. and Stroup, J. B. (1981). The plutonic foundation of the oceanic crust. In *The Sea,* volume **7**, ed. C. Emiliani. New York: Wiley, pp. 119–218.

Fox, P. J., Schreiber, E and Peterson, J. J. (1973). Geology of oceanic crust – compressional wave velocities of oceanic rocks. *Journal of Geophysical Research*, **78**, (23), 5155–5172.

Fox, P. J., Grindlay, N. R. and Macdonald, K. C. (1991). The Mid-Atlantic Ridge (31° S–34°30′ S): temporal and spatial variations of accretionary processes. *Marine Geophysical Researches*, **13**, 1–20.

Francheteau, J. and Ballard, R. D. (1983). The East Pacific Rise near 21° N, 13° N, and 20° S: inferences for along-strike variability of axial processes of the mid-ocean ridge. *Earth and Planetary Science Letters*, **64**, 93–116.

Francheteau, J., Choukroune, P., Hekinian, R., Le Pichon, X. and Needham, H. D. (1976). Oceanic fracture zones do not provide deep sections in the crust. *Canadian Journal of Earth Sciences*, **13**, 1223–1235.

Francheteau, J., Juteau, T. and Rangan, C. (1979). Basaltic pillars in collapsed lava-pools on the deep ocean-floor. *Nature*, **281**, (5728), 209–211.

Francis, T. J. G. (1985). Resistivity measurements of an ocean floor sulphide mineral deposit from the submersible Cyana. *Marine Geophysical Researches*, **7**, 419–438.

Freed, A. M., Lin, J. and Melosh, H. J. (1995). Long-term survival of the axial valley morphology at abandoned slow-spreading centers. *Geology*, **23**, (11), 971–974.

Freudenthal, T. and Wefer, G. (2007). Scientific drilling with the sea floor drill rig MeBo. *Scientific Drilling*, **5**, 63–66.

Früh-Green, G. L., Kelley, D. S., Bernasconi, S. M. *et al.* (2003). 30,000 years of hydrothermal activity at the Lost City vent field. *Science*, **301**, 495–498.

Fuchs, K. (1968). The reflection of spherical waves from transition zones with arbitrary depth-dependent elastic moduli and density. *Journal of the Physics of the Earth*, **16** (Special Issue), 27.

Fuchs, K. and Müller, G. (1971). Computation of synthetic seismograms with the reflectivity method and comparison with observations. *Geophysical Journal of the Royal Astronomical Society*, **23**, 417–433.

Fujiwara, T., Lin, J., Matsumoto, T. *et al.* (2003). Crustal evolution of the Mid-Atlantic Ridge near Fifteen-Twenty Fracture Zone in the last 5 Ma. *Geochemistry, Geophysics, Geosystems*, **4**, article 1024, doi: 10.1029/2002GC000364.

Fundis, A. T., Soule, S. A., Fornari, D. J. and Perfit, M. R. (2010). Paving the seafloor: volcanic emplacement processes during the 2005–2006 eruptions at the fast spreading East Pacific Rise, 9°50′N. *Geochemistry Geophysics Geosystems*, **11**, doi:10.1029/2010GC003058.

Gac, S., Dyment, J., Tisseau, C. and Goslin, J. (2003). Axial magnetic anomalies over slow-spreading ridge segments: insights from numerical 3-D thermal and physical modelling. *Geophysical Journal International*, **154**, (3), 618–632.

Gallo, D. G., Kidd, W. S. F., Fox, P. J. *et al.* (1984). Tectonics at the intersection of the East Pacific Rise with the Tamayo Transform Fault. *Marine Geophysical Researches*, **6**, (2), 159–185.

Gallo, D. G., Fox, P. J. and Macdonald, K. C. (1986). A Sea Beam investigation of the Clipperton Transform Fault: the morphotectonic expression of a fast-slipping transform boundary. *Journal of Geophysical Research*, **91**, 3455–3467.

Galperin, E. I. and Kosminskaya, I. P. (1958). Characteristics of the methods of deep seismic sounding on the sea. *Bulletin of the Academy of Sciences USSR, Geophysical series*, (7), 475–483.

Gaherty, J. B., Kato, M. and Jordan, T. H. (1999). Seismological structure of the upper mantle: a regional comparison of seismic layering. *Physics of the Earth and Planetary Interiors*, **110**, (1–2), 21–41.

Garcés, M. and Gee, J. S. (2007). Paleomagnetic evidence of large footwall rotations associated with low-angle faults at the Mid-Atlantic Ridge. *Geology*, **35**, (3), 279–282; doi: 10.1130/G23165A.1.

Gass, I. G. (1968). Is Troodos Massif of Cyprus a fragment of Mesozoic ocean floor? *Nature*, **220**, (5162), 39.

Gass, I. G. (1990). Ophiolites and oceanic lithosphere. In *Proceedings of the Symposium on Ophiolites and Oceanic Lithosphere – TROODOS 87*, ed. J. Malpas, E. M. Moores, A. Panayiotou and C. Xenophontos. Nicosia, Cyprus: Geological Survey Department, pp. 1–10.

Gass, I. G. and Masson Smith, D. C. (1963). The geology and gravity anomalies of the Troodos Massif, Cyprus. *Philosophical Transactions of the Royal Society of London, Series A*, **255**, 417–467.

Gass, I. G., Lippard, S. J. and Shelton, A. W. (1984). *Ophiolites and Oceanic Lithosphere, Special Publication 13*. London: Geological Society, 413 pp.

Gee, J. S. and Kent, D. V. (2007). Source of oceanic magnetic anomalies and the geomagnetic polarity timescale. In *Treatise on Geophysics*, Volume **5**, ed. G. Schubert. Amsterdam: Elsevier, pp. 455–507.

Gee, J. and Meurer, W. P. (2002). Slow cooling of middle and lower oceanic crust inferred from multicomponent magnetizations of gabbroic rocks from the Mid-Atlantic Ridge south of the Kane fracture zone. *Journal of Geophysical Research*, **107**, (B7), EPM3, doi10.1029/2001JB000062.

Gente, P., Pockalny, R. A., Durand, C. *et al.* (1995). Characteristics and evolution of the segmentation of the Mid-Atlantic Ridge between 20° N and 24° N during the last 10 million years. *Earth and Planetary Science Letters*, **129**, (1–4), 55–71.

Gee, J., Schneider, D. A. and Kent., D. V. (1996). Marine magnetic anomalies as recorders of geomagnetic intensity variations. *Earth and Planetary Science Letters*, **144**, 327–335.

Gerard, R., Ewing, M. and Langseth, M. G. (1962). Thermal gradient measurements in water and bottom sediment of western Atlantic. *Journal of Geophysical Research*, **67**, (2), 785–803.

German, C. R. and Lin, J. (2004). The thermal structure of the oceanic crust, ridge-spreading and hydrothermal circulation: how well do we understand their inter-connections? *Mid-Ocean Ridges: Hydrothermal Interactions between the Lithosphere and Oceans*, **148**, 1–18.

German, C. R. and Parson, L. M. (1998). Hydrothermal activity along the Mid-Atlantic Ridge: an interplay between magmatic and tectonic processes. *Earth and Planetary Science Letters*, **160**, 327–341.

German, C. R., Briem, J., Chin, C. *et al.* (1994). Hydrothermal activity on the Reykjanes Ridge: the Steinahóll vent field at 63°06′ N. *Earth and Planetary Science Letters*, **121**, (3–4), 647–654.

German, C. R., Klinkhammer, G. P. and Rudnicki, M. D. (1996a). The Rainbow hydrothermal plume, 36.15 N, MAR. *Geophysical Research Letters*, **23**, (21), 2979–2982.

German, C. R., Parson, L. M., Bougault, H. *et al.* (1996b). Hydrothermal exploration near the Azores Triple-Junction: tectonic control of venting at slow-spreading ridges? *Earth and Planetary Science Letters*, **138**, (1–4), 93–104.

German, C. R., Baker, E. T., Mével, C. A., Tamaki, K. and FUJI Scientific Team (1998). Hydrothermal activity along the South West Indian Ridge. *Nature*, **395**, 490–492.

German, C., Tyler, P. and Griffiths, G. (2003). The maiden voyage of UK ROV "Isis". *Ocean Challenge*, **12**, (3), 16–18.

German, C., Lin, J. and Parson, L. M. (2004). *Mid-Ocean Ridges: Hydrothermal Interactions Between the Lithosphere and Oceans*. Washington, D.C.: American Geophysical Union, 311 pp.

German, C. R., Bennett, S. A., Connelly, D. P. *et al.* (2008). Hydrothermal activity on the southern Mid-Atlantic Ridge: tectonically- and volcanically-controlled venting at 4–5° S. *Earth and Planetary Science Letters*, **273**, (3–4), 332–344.

German, C., Thurnherr, A., Knoery, J. *et al.* (2010). Heat, volume and chemical fluxes from submarine venting: a synthesis of results from the Rainbow hydrothermal field,

36° N MAR. *Deep-Sea Research Part I-Oceanographic Research Papers*, **57** (4), 518–527.

Gill, R. (2010). *Igneous Rocks and Processes*. Chichester: Wiley-Blackwell, 428 pp.

Ginster, U., Mottl, M. J. and Von Herzen, R. P. (1994). Heat-flux from black smokers on the Endeavor and Cleft segments, Juan de Fuca Ridge. *Journal of Geophysical Research-Solid Earth*, **99**, (B3), 4937–4950.

Girardeau, J. and Nicolas, A. (1981). The structures of two ophiolite massifs, Bay-of-Islands, Newfoundland: a model for the oceanic-crust and upper mantle. *Tectonophysics*, **77**, (1–2), 1–34.

Glen, W. (1982). *The Road to Jaramillo: Critical Years of the Revolution in Earth Science*. Stanford: Stanford University Press, 480 pp.

Glenn, M. F. (1970). Introducing an operational multi-beam array sonar. *International Hydrographic Review*, **47**, 35–39.

Goetze, C. (1978). The mechanism of solid state creep. *Philosophical Transactions of the Royal Society of London, Series A*, **288**, 99–119.

Goldstein, S. J., Perfit, M. R., Batiza, R., Fornari, D. J. and Murrell, M. T. (1994). Off-axis volcanism at the East Pacific Rise detected by uranium-series dating of basalts. *Nature*, **367**, (6459), 157–159.

Goss, A. R., Perfit, M. R., Ridley, W. I. *et al.* (2010). Geochemistry of lavas from the 2005–2006 eruption at the East Pacific Rise, 9°46′N–9°56′N: implications for ridge crest plumbing and decadal changes in magma chamber compositions. *Geochemistry Geophysics Geosystems*, **11**, doi:10.1029/2009gc002977.

Gregg, T. K. P. and Fink, J. H. (1995). Quantification of submarine lava-flow morphology through analog experiments. *Geology*, **23**, 73–76.

Gregg, T. K. P. and Smith, D. K. (2003). Volcanic investigations of the Puna Ridge, Hawai'i: relations of lava flow morphologies and underlying slopes. *Journal of Volcanology and Geothermal Research*, **126**, (1–2), 63–77.

Gregg, T. K. P., Fornari, D. J., Perfit, M. R., Haymon, R. M. and Fink, J. H. (1996). Rapid emplacement of a mid-ocean ridge lava flow on the East Pacific Rise at 9°46′–51′N. *Earth and Planetary Science Letters*, **144**, (3–4), E1–E7.

Griffiths, R. W. and Fink, J. H. (1992). Solidification and morphology of submarine lavas: a dependence on extrusion rate. *Journal of Geophysical Research*, **97**, 19 729–19 737.

Grimes, C. B., John, B. E., Cheadle, M. J. and Wooden, J. L. (2008). Protracted construction of gabbroic crust at a slow spreading ridge: constraints from $^{206}Pb/^{238}U$ zircon ages from Atlantis Massif and IODP Hole U1309D (30° N, MAR). *Geochemistry Geophysics Geosystems*, **9**, doi10.1029/2008GC002063.

Grindlay, N. R., Fox, P. J. and Macdonald, K. C. (1991). Second-order ridge axis discontinuities in the South Atlantic: morphology, structure, and evolution. *Marine Geophysical Researches*, **13**, 21–49.

Grindlay, N. R., Fox, P. J. and Vogt, P. R. (1992). Morphology and tectonics of the Mid-Atlantic Ridge (25°–27°30′ S) from Sea Beam and magnetic data. *Journal of Geophysical Research*, **97**, (B5), 6983–7010.

Grindlay, N. R., Madsen, J. A., Rommevaux-Jestin, C. and Sclater, J. (1998). A different pattern of ridge segmentation and mantle Bouguer gravity anomalies along the ultra-slow spreading Southwest Indian Ridge (15°30′E–25°E). *Earth and Planetary Science Letters*, **161**, 243–253.

Guspi, F. (1987). Frequency-domain reduction of potential field measurements to a horizontal plane. *Geoexploration*, **24**, 87–98.

Gutenberg, B. and Richter, C. F. (1954). *Seismicity of the Earth and Associated Phenomena*. Princeton, NJ: Princeton University Press.

Hall, J. M., Fisher, B. E., Walls, C. C. *et al.* (1987). Vertical distribution and alteration of dikes in a profile through the Troodos ophiolite. *Nature*, **326**, (6115), 780–782.

Hannington, M. D., Jonasson, I. R., Herzig, P. M. and Petersen, S. (1995). Physical and chemical processes of seafloor mineralization at mid-ocean ridges. In *Seafloor Hydrothermal Systems: Physical, Chemical, Biological and Geological Interactions, Geophysical Monograph 91*, ed. S. E. Humphris, R. A. Zierenberg, L. S. Mullineaux and R. E. Thomson. Washington, D.C.: American Geophysical Union, pp. 115–157.

Harding, A. J., Kent, G. M. and Orcutt, J. A. (1993). A multichannel seismic investigation of upper crustal structure at 9° N on the East Pacific Rise: implications for crustal accretion. *Journal of Geophysical Research-Solid Earth*, **98**, (B8), 13 925–13 944.

Harrison, C. G. A. (1976). Magnetization of oceanic-crust. *Geophysical Journal of the Royal Astronomical Society*, **47**, (2), 257–283.

Hashimoto, J., Ohta, S., Gamo, T. *et al.* (2001). First hydrothermal vent communities from the Indian Ocean discovered. *Zoological Science*, **18**, (5), 717–721.

Hasterok, D., Chapman, D. S. and Davis, E. E. (2011). Oceanic heat flow: implications for global heat loss. *Earth and Planetary Science Letters*, **311**, (3–4), 386–395.

Haxby, W. F. and Weissel, J. K. (1986). Evidence for small-scale mantle convection from Seasat altimeter data. *Journal of Geophysical Research*, **91**, 3507–3520.

Hayes, D. E. (1988). Age–depth relationships and depth anomalies in the southeast Indian Ocean and south Atlantic Ocean. *Journal of Geophysical Research*, **93**, (B4), 2937–2954.

Haymon, R. M. (1983). Growth history of hydrothermal black smoker chimneys. *Nature*, **301**, (5902), 695–698.

Haymon, R. M. (1996). The response of ridge crest hydrothermal systems to segmented, episodic magma supply. In *Tectonic, Hydrothermal and Biological Segmentation at Mid-Ocean Ridges*, ed. C. J. MacLeod, P. A. Tyler and C. L. Walker. London: Geological Society, pp. 157–168.

Haymon, R. and White, S. (2004). Fine-scale segmentation of volcanic/hydrothermal systems along fast-spreading ridge crests. *Earth and Planetary Science Letters*, **226**, (3–4), 367–382.

Haymon, R. M., Fornari, D. J., Edwards, M. H. *et al.* (1991). Hydrothermal vent distribution along the East Pacific Rise crest (9°09′–54′N) and its relationship to magmnatic and tectonic processes on fast-spreading mid-ocean ridges. *Earth and Planetary Science Letters*, **104**, 513–534.

Haymon, R. M., Fornari, D. J., Von Damm, K. L. *et al.* (1993). Volcanic eruption of the mid-ocean ridge along the East Pacific Rise crest at 9°45–52′N: direct submersible observations of seafloor phenomena associated with an eruption. *Earth and Planetary Science Letters*, **119**, 85–101.

Head, J. W., Wilson, L. and Smith, D. K. (1996). Mid-ocean ridge eruptive vent morphology and structure: evidence for dike widths, eruption rates, and evolution of eruptions and axial volcanic ridges. *Journal of Geophysical Research*, **101**, (B12), 28 265–28 280.

Heezen, B. C. (1960). The rift in the ocean floor. *Scientific American*, **203**, (4), 98–110.

Heezen, B. C. (1962). The deep-sea floor. In *Continental Drift*, ed. S. K. Runcorn. New York: Academic Press, pp. 235–268.

Heezen, B. C. (1969). World rift system – an introduction to the symposium. *Tectonophysics*, **8**, (4–6), 269–279.

Heezen, B. C. and Menard, H. W. (1963). Chapter 12: Topography of the deep-sea floor. In *The Sea*, ed. M. N. Hill. New York: John Wiley, pp. 233–280.

Heezen, B. C. and Rawson, M. (1977). Visual observations of sea-floor subduction line in Middle-America Trench. *Science*, **196**, (4288), 423–426.

Heezen, B. C. and Tharp, M. (1954). Physiographic diagram of the western North Atlantic. *Geological Society of America Bulletin*, **65**, (12), 1260–1261.

Heezen, B. C. and Tharp, M. (1957). *Physiographic Diagram of the North Atlantic*. New York: Geological Society of America.

Heezen, B. C. and Tharp, M. (1961). *Physiographic Diagram of the South Atlantic, the Caribbean, the Scotia Sea, and the Eastern Margin of the South Pacific Ocean*. New York: Geological Society of America.

Heezen, B. C. and Tharp, M. (1964). *Physiographic Diagram of the Indian Ocean*. New York: Geological Society of America.

Heezen, B. C. and Tharp, M. (1971). *Physiographic Diagram of the Western Pacific Ocean*. Boulder, Colorado: Geological Society of America.

Heezen, B. C. and Tharp, M. (1977). World Ocean Floor Panorama, Marie Tharp Maps, NY 10976.

Heezen, B. C., Tharp, M. and Gerard, R. D. (1964a). Vema fracture zone in equatorial Atlantic. *Journal of Geophysical Research*, **69**, (4), 733–739.

Heezen, B. C., Bunce, E. T., Hersey, J. B. and Tharp, M. (1964b). Chain and Romanche fracture zones. *Deep-Sea Research*, **11**, 11–33.

Heirtzler, J. R., Le Pichon, X. and Baron, J. G. (1966). Magnetic anomalies over the Reykjanes Ridge. *Deep-Sea Research*, **13**, 427–443.

Hekinian, R., Thompson, G. and Bideau, D. (1989). Axial and off-axial heterogeneity of basaltic rocks from the East Pacific Rise at 12°35′N–12°51′N and 11°26′N–11°30′N. *Journal of Geophysical Research-Solid Earth and Planets*, **94**, (B12), 17 437–17 463.

Helmberger, D. V. and Moms, G. B. (1969). A travel time and amplitude interpretation of a marine refraction profile: primary waves. *Journal of Geophysical Research*, **74**, 483–494.

Helo, C., Longpre, M.-A., Shimizu, N., Clague, D. A. and Stix, J. (2011). Explosive eruptions at mid-ocean ridges driven by $CO_2$-rich magmas. *Nature Geoscience*, **4**, (4), 260–263.

Hersey, J. B. (1963). 4. Continuous reflection profiling. In *The Sea, volume 3*, ed. M. N. Hill. New York: Interscience Publishers, pp. 47–72.

Hess, H. H. (1962). History of ocean basins. In *Petrologic Studies: A Volume in Honour of A. F. Buddington*, ed. A. E. J. Engel, H. L. James and B. F. Leonard. New York: Geological Society of America, pp. 599–620.

Hessler, R. R. and Kaharl, V. A. (1995). Tectonic and volcanic controls on hydrothermal processes at the mid-ocean ridge: an overview based on near-bottom and submersible studies. In *Seafloor Hydrothermal Systems: Physical, Chemical, Biological, and Geological Interactions, Geophysical Monograph 91*, ed. S. E. Humphris, R. A. Zierenberg, L. S. Mullineaux and R. E. Thomson. Washington, D.C.: American Geophysical Union, pp. 72–84.

Hey, R. N. (1977). A new class of pseudofaults and their bearing on plate tectonics: a propagating rift model. *Earth and Planetary Science Letters*, **37**, 321–325.

Hey, R. N. and Wilson, D. S. (1982). Propagating rift explanation for the tectonic evolution of the NE Pacific – the pseudomovie. *Earth and Planetary Science Letters*, **58**, 147–188.

Hey, R. N., Duennebier, F. K. and Morgan, W. J. (1980). Propagating rifts on mid-ocean ridges. *Journal of Geophysical Research*, **85**, 2647–2658.

Hey, R. N., Naar, D. F., Kleinrock, M. C. *et al.* (1985). Microplate tectonics along a superfast seafloor spreading system near Easter Island. *Nature*, **317**, 320–325.

Hey, R. N., Kleinrock, M. C., Miller, S. P., Atwater, T. M. and Searle, R. C. (1986). Sea Beam/Deep-tow investigation of an active oceanic propagating rift system, Galapagos 95.5° W. *Journal of Geophysical Research*, **91**, 3369–3393.

Hill, M. N. (1952). Seismic refraction shooting in an area of the Eastern Atlantic. *Philosophical Transactions of the Royal Society of London*, **244**, 561–596.

Hill, M. (1959). A ship-borne nuclear-spin magnetometer. *Deep-Sea Research*, **5**, 309–311.

Hill, M. N. (1957). Recent geophysical exploration of the ocean floor. *Physics and Chemistry of the Earth*, **2**, 129–163.

Hill, M. N. (1963). 3. Single-ship seismic refraction shooting. In *The Sea, volume 3, The Earth Beneath the Sea*, ed. M. N. Hill. New York: John Wiley & Sons, pp. 39–46.

Hooft, E. E. E., Schouten, H. and Detrick, R. S. (1996). Constraining crustal emplacement processes from the variation in seismic layer 2A thickness at the East Pacific Rise. *Earth and Planetary Science Letters*, **142**, (3–4), 289–309.

Hooft, E. E. E., Detrick, R. S., Toomey, D. R., Collins, J. A. and Lin, J. (2000). Crustal thickness and structure along three contrasting spreading segments of the Mid-Atlantic Ridge, 33.5°–35° N. *Journal of Geophysical Research, B*, **105**, 8205–8226.

Horton, E. E. (1961). Preliminary drilling phase of Mohole project I. Summary of drilling operations (La Jolla and Guadalupe sites). *Bulletin of the American Association of Petroleum Geologists*, **45**, (11), 1789–1792.

Hosford, A., Lin, J. and Detrick, R. S. (2001). Crustal evolution over the last 2 m.y. at the Mid-Atlantic Ridge OH-1 segment, 35° N. *Journal of Geophysical Research*, **106**, (B7), 13 269–13 285.

Houtz, R. E. and Ewing, J. I. (1976). Upper crustal structure as a function of plate age. *Journal of Geophysical Research*, **81**, 2490–2498.

Huang, P. Y. and Solomon, S. C. (1988). Centroid depths of mid-ocean ridge earthquakes: dependence on spreading rate. *Journal of Geophysical Research*, **93**, 13 445–13 477.

Huang, P. Y., Solomon, S. C., Bergman, E. A. and Nabelek, J. L. (1986). Focal depths and mechanisms of Mid-Atlantic Ridge earthquakes from body wave-form inversion. *Journal of Geophysical Research-Solid Earth and Planets*, **91**, (B1), 579–598.

Huggett, Q. (1990). Long-range underwater photography in the deep ocean. *Marine Geophysical Researches*, **12**, (1–2), 69–81.

Humphris, S. E. and Cann, J. R. (2000). Constraints on the energy and chemical balances of the modern TAG and ancient Cyprus seafloor sulfide deposits. *Journal of Geophysical Research-Solid Earth*, **105**, (B12), 28 477–28 488.

Humphris, S. E., Bryan, W. B., Thompson, G. and Autio, L. K. (1990). Morphology, geochemistry, and evolution of Serocki Volcano. In *Proceedings of the Ocean Drilling Program, Scientific Results*, ed. R. Detrick, J. Honnorez, W. B. Bryan, T. Juteau *et al.* College Station, TX, Ocean Drilling Program, pp. 67–84.

Humphris, S. E., Zierenberg, R. A., Mullineaux, L. S. and Thomson, R. E. (1995). *Seafloor Hydrothermal Systems: Physical, Chemical, Biological, and Geological Interactions, Geophysical Monograph no. 91*. Washington, D.C.: American Geophysical Union.

Humphris, S. E., Herzig, P. M., Miller, D. J. and others (1996). 1. Introduction and principal results. In *Proceedings of the Ocean Drilling Program, Initial Reports, 158*, ed. S. E. Humphris, P. M. Herzig and D. J. Miller. College Station, Texas, pp. 5–14.

Hussenoeder, S., Tivey, M. A. and Schouten, H. (1995). Direct inversion of potential field data from an irregular observation surface with application to the Mid-Atlantic Ridge. *Geophysical Research Letters*, **22**, (23), 3131–3134.

Hussenoeder, S. A., Tivey, M. A., Schouten, H. and Searle, R. C. (1996). Near-bottom magnetic survey of the Mid-Atlantic Ridge axis, 24°–24°40′ N: implications for crustal accretion at slow spreading ridges. *Journal of Geophysical Research*, **101**, (B10), 22 051–22 069.

IAGA Working Group V-MOD (2010). International Geomagnetic Reference Field: the eleventh generation. *Geophysical Journal International*, **183**, (3), 1216–1230.

Ildefonse, B., Blackman, D., John, B. E. *et al.* (2007). Oceanic core complexes and crustal accretion at slow-spreading ridges. *Geology*, **35**, (7), 623–626; doi: 10.1130/G23531A.1.

Irving, E., Robertson, W. A. and Aumento, F. (1970). The Mid-Atlantic Ridge near 45° N, VI: remanent intensity, susceptibility and iron content of dredge sample. *Canadian Journal of Earth Sciences*, **7**, 226–238.

Isacks, B. L., Oliver, J. and Sykes, L. R. (1968). Seismology and the new global tectonics. *Journal of Geophysical Research*, **73**, 5855–5900.

Isezaki, N. (1986). A new shipboard three-component magnetometer. *Geophysics*, **51**, (10), 1992–1998.

Ito, G. and Behn, M. D. (2008). Magmatic and tectonic extension at mid-ocean ridges: 2. Origin of axial morphology. *Geochemistry Geophysics Geosystems*, **9**, doi:10.1029/2008gc001970.

Ito, G. and Martel, S. J. (2002). Focusing of magma in the upper mantle through dike interaction. *Journal of Geophysical Research-Solid Earth*, **107**, (B10), art. no. 2223, pp. ECV 6-1–ECV 6-17.

Jackson, H. R., Reid, I. and Falconer, R. K. H. (1982). Crustal structure near the Arctic mid-ocean ridge. *Journal of Geophysical Research*, **87**, (B3), 1773–1783.

Jakosky, B. M. and Shock, E. L. (1998). The biological potential of Mars, the early Earth, and Europa. *Journal of Geophysical Research-Planets*, **103**, (E8), 19 359–19 364.

Jannasch, H. W. (1995). Microbial interactions with hydrothermal fluids. In *Seafloor Hydrothermal Systems: Physical, Chemical, Biological, and Geological Interactions, Geophysical Monograph no. 91*, ed. S. E. Humphris, R. A. Zierenberg, S. Mullineaux and R. E. Thomson. Washington D.C.: American Geophysical Union, pp. 273–296.

Johnson, H. P. and Atwater, T. (1977). Magnetic study of basalts from the Mid-Atlantic Ridge, lat 37° N. *Geological Society of America Bulletin*, **88**, 637–647.

Johnson, H. P. and Merrill, R. (1973). Low temperature oxidation of a titanomagnetite and the implications for paleomagnetism. *Journal of Geophysical Research*, **78**, 4938–4949.

Johnson, H. P. and Tivey, M. A. (1995). Magnetic-properties of zero-age oceanic-crust – a new submarine lava flow on the Juan-De-Fuca Ridge. *Geophysical Research Letters*, **22**, (2), 175–178.

Johnson, H. P., Becker, K. and Von Herzen, R. (1993). Near-axis heat-flow measurements on the northern Juan-de-Fuca ridge – implications for fluid circulation in oceanic-crust. *Geophysical Research Letters*, **20**, (17), 1875–1878.

Jones, E. J. W. (1967). *Seismic reflection profiling at sea with a pneumatic sound source.* PhD thesis, Cambridge University.

Jones, E. J. W. (1999). *Marine Geophysics*. Chichester: John Wiley & Sons Ltd., 466 pp.

Kappel, E. S. and Ryan, W. B. F. (1986). Volcanic episodicity and a non-steady state rift valley along northeast Pacific spreading centers: evidence from SeaMARC I. *Journal of Geophysical Research*, **91**, 13 925–13 940.

Karson, J. A. (1990). Seafloor spreading on the Mid-Atlantic Ridge: implications for the structure of ophiolites and oceanic lithosphere produced in slow-spreading environments. In *Proceedings of the Symposium on Ophiolites and Oceanic Lithosphere – TROODOS 87*, ed. J. Malpas, E. M. Moores, A. Panayiotou and C. Xenophontos. Nicosia, Cyprus: Geological Survey Department, Ministry of Agriculture and Natural Resources, pp. 547–555.

Karson, J. A. (1998). Internal structure of oceanic lithosphere: a perspective from tectonic windows. In *Faulting and Magmatism at Mid-Ocean Ridges: Geophysical Monograph 106*, ed. W. R. Buck, P. T. Delaney, J. A. Karson and Y. Lagabrielle. Washington, D.C.: American Geophysical Union, pp. 177–218.

Karson, J. A. (1999). Geological investigation of a lineated massif at the Kane Transform Fault: implications for oceanic core complexes. *Philosophical Transactions of the Royal Society of London Series A-Mathematical Physical and Engineering Sciences*, **357**, (1753), 713–736.

Karson, J. A. and Brown, J. R. (1988). Geologic setting of the Snake Pit hydrothermal site: an active vent field on the Mid-Atlantic Ridge. *Marine Geophysical Researches*, **10**, (1/2), 91–107.

Karson, J. A. and Dick, H. J. B. (1983). Tectonics of ridge-transform intersections at the Kane fracture-zone. *Marine Geophysical Researches*, **6**, (1), 51–98.

Karson, J. A., Thompson, G., Humphris, S. E. *et al.* (1987). Along-axis variations in seafloor spreading in the MARK area. *Nature*, **328**, 681–685.

Karson, J. A., Tivey, M. A. and Delaney, J. R. (2002a). Internal structure of uppermost oceanic crust along the Western Blanco Transform Scarp: implications for subaxial accretion and deformation at the Juan de Fuca Ridge. *Journal of Geophysical Research-Solid Earth*, **107**, (B9), doi:10.1029/2000JB000051.

Karson, J. A., Klein, E. M., Hurst, S. D. *et al.* (2002b). Structure of uppermost fast-spread oceanic crust exposed at the Hess Deep Rift: implications for subaxial processes at the East Pacific Rise. *Geochemistry Geophysics Geosystems*, **3**, doi10.1029/2001GC000155.

Karson, J. A., Frueh-Green, G. L., Kelley, D. S. *et al.* (2006). Detachment shear zone of Atlantic Massif core complex, Mid-Atlantic Ridge, 30° N. *Geochemistry Geophysics Geosystems*, **7**, doi:10.1029/2005GC001109.

Kashefi, K. and Lovley, D. (2003). Extending the upper temperature limit for life. *Science*, **301**, 934.

Kastens, K. A., Macdonald, K. C., Becker, K. and Crane, K. (1979). The Tamayo transform fault in the mouth of the Gulf of California. *Marine Geophysical Researches*, **4**, 129–151.

Katsumata, K., Sato, T., Kasahara, J. *et al.* (2001). Microearthquake seismicity and focal mechanisms at the Rodriguez Triple Junction in the Indian Ocean using ocean bottom seismometers. *Journal of Geophysical Research-Solid Earth*, **106**, (B12), 30 689–30 699.

Kearey, P. and Brooks, M. (1991). *An Introduction to Geophysical Exploration*, 2nd edition, Oxford: Blackwell Scientific Publications, 254 pp.

Kearey, P., Klepeis, K. A. and Vine, F. J. (2009). *Global Tectonics*, 3rd edition. Wiley-Blackwell 482 pp.

Keen, C. and Tramontini, C. (1970). A seismic refraction survey on the Mid-Atlantic Ridge. *Geophysical Journal of the Royal Astronomical Society*, **20**, (5), 473–491.

Kelemen, P. B. and Aharonov, E. (1998). Periodic formation of magma fractures and generation of layered gabbros in the lower crust beneath oceanic spreading ridges. In *Faulting and Magmatism at Mid-Ocean Ridges – Geophysical Monograph 106*, ed. W. R. Buck, P. T. Delaney, J. A. Karson and Y. Lagabrielle. Washington, D.C.: American Geophysical Union, pp. 267–289.

Kelemen, P. B., Shimizu, N. and Salters, V. J. M. (1995). Extraction of mid-ocean-ridge basalt from the upwelling mantle by focused flow of melt in dunite channels. *Nature*, **375**, (6534), 747–753.

Kelemen, P. B., Braun, M. and Hirth, G. (2000). Spatial distribution of melt conduits in the mantle beneath oceanic spreading centers: observations from the Ingalls and Oman ophiolites. *Geochemistry Geophysics Geosystems*, **1**, doi:1999GC000012.

Kelemen, P. B., Matter, J., Streit, E. E. *et al.* (2011). Rates and mechanisms of mineral carbonation in peridotite: natural processes and recipes for enhanced, in situ $CO_2$ capture and storage. In *Annual Review of Earth and Planetary Sciences, Vol 39*, ed. R. Jeanloz and K. H. Freeman, pp. 545–576.

Kelley, D. S., Karson, J. A., Blackman, D. K. *et al.* (2001). An off-axis hydrothermal vent field near the Mid-Atlantic Ridge at 30° N. *Nature*, **412**, (6843), 145–149.

Kelley, D. S., Karson, J. A., Fruh-Green, G. L. *et al.* (2005). A serpentinite-hosted ecosystem: the Lost City hydrothermal field. *Science*, **307**, (5714), 1428–1434.

Kennett, B. L. N. and Orcutt, J. A. (1976). Comparison of travel time inversions for marine refraction profiles. *Journal of Geophysical Research*, **81**, (23), 4061–4070.

Kennett, B. L. N., Bunch, A. W. H., Orcutt, J. A. and Raitt, R. W. (1977). Variations in crustal structure on East Pacific Rise crest – travel time inversion approach. *Earth and Planetary Science Letters*, **34**, (3), 439–444.

Kennish, M. J. and Lutz, R. A. (1998). Morphology and distribution of lava flows on mid-ocean ridges: a review. *Earth-Science Reviews*, **43**, (3–4), 63–90.

Kent, G. M., Harding, A. J. and Orcutt, J. A. (1990). Evidence for a smaller magma chamber beneath the East Pacific Rise at 9°30′ N. *Nature*, **344**, 650–653.

Kent, G. M., Harding, A. J. and Orcutt, J. A. (1993). Distribution of magma beneath the East Pacific Rise between the Clipperton Transform and the 9°17′ N deval from forward modeling of common depth point data. *Journal of Geophysical Research*, **98**, (B8), 13 971–13 995.

Kent, G. M., Harding, A. J., Orcutt, J. A. *et al.* (1994). Uniform accretion of oceanic-crust south of the Garrett Transform at 14°15′ S on the East Pacific Rise. *Journal of Geophysical Research-Solid Earth*, **99**, (B5), 9097–9116.

Kirk, R. E., Whitmarsh, R. B. and Langford, J. J. (1982). A 3-component ocean bottom seismograph for controlled source and earthquake seismology. *Marine Geophysical Researches*, **5**, (3), 327–341.

Kitahara, A., Seama, N. and Isezaki, N. (1994). A subduction angle of the Pacific Plate beneath the Japan Trench inferred from three component geomagnetic anomalies. *Journal of Geomagnetism and Geolectricity*, **46**, 455–462.

Kleinrock, M. C. and Hey, R. N. (1989a). Detailed tectonics near the tip of the Galapagos 95.5° W propagator: how the lithosphere tears and a spreading center develops. *Journal of Geophysical Research*, **94**, 13 801–13 838.

Kleinrock, M. C. and Hey, R. N. (1989b). Migrating transform zone and lithosphere transfer at the Galapagos 95.5° W propagator. *Journal of Geophysical Research*, **94**, 13 859–13 878.

Kleinrock, M. C. and Humphris, S. E. (1996). Structural control on sea-floor hydrothermal activity at the TAG active mound. *Nature*, **382**, (6587), 149–153.

Koelsch, D. E., Witzell, W. E. S., Broda, J. E., Wooding, F. B. and Purdy, G. M. (1986). A deep towed explosive source for seismic experiments on the ocean floor. *Marine Geophysical Researches*, **8**, 345–361.

Kohnen, W. (2009). Human exploration of the deep seas: fifty years and the inspiration continues. *Marine Technology Society Journal*, **43**, (5), 42–62.

Kong, L. S. L., Solomon, S. C. and Purdy, G. M. (1986). Microearthquakes near the TAG hydrothermal field, Mid-Atlantic Ridge, 26° N. *EOS, Transactions of the American Geophysical Union*, **67**, 1021.

Kong, L. S., Detrick, R. S., Fox, P. J., Mayer, L. A. and Ryan, W. B. F. (1988/89). The morphology and tectonics of the MARK area from Sea Beam and Sea MARC I observations (Mid-Atlantic Ridge 23° N). *Marine Geophysical Researches*, **10**, 59–90.

Kong, L. S. L., Solomon, S. C. and Purdy, G. M. (1992). Microearthquake characteristics of a midocean ridge along-axis high. *Journal of Geophysical Research*, **97**, (B2), 1659–1685.

Koschinsky, A. *et al.* (2008). Hydrothermal venting at pressure–temperature conditions above the critical point of seawater, 5° S on the Mid-Atlantic Ridge. *Geology*, **36**, 615–618.

Krasnov, S. G., Cherkashev, G. A., Sepanova, T. V. *et al.* (1995). Detailed geological studies of hydrothermal fields in the North Atlantic. In *Hydrothermal Vents and Processes, Special Publication 87*, ed. L. M. Parson, C. L. Walker and D. R. Dixon. London: Geological Society, pp. 43–64.

Kuo, B. Y. and Forsyth, D. W. (1988). Gravity anomalies of the ridge-transform system in the South Atlantic between 31 and 34.5° S: upwelling centers and variations in crustal thickness. *Marine Geophysical Researches*, **10**, (3–4), 205–232.

Kuo, B. Y., Forsyth, D. W. and Parmentier, E. M. (1986). Flexure and thickening of the lithosphere at the East Pacific Rise. *Geophysical Research Letters*, **13**, 681–684.

Lalou, C., Reyss, J. L., Brichet, E., Rona, P. A. and Thompson, G. (1995). Hydrothermal activity on a 105 year scale at a slow-spreading ridge, TAG hydrothermal field, Mid-Atlantic Ridge 26° N. *Journal of Geophysical Research-Solid Earth*, **100**, (B9), 17 855–17 862.

Langmuir, C. H., Klein, E. M. and Plank, T. (1992). Petrological systematics of mid-ocean ridge basalts: constraints on melt generation beneath ocean ridges. In *Mantle Flow and Melt Generation at Mid-Ocean Ridges, Geophysical Monograph 71*, ed. J. P. Morgan, D. K. Blackman and J. M. Sinton. Washington, D.C.: American Geophysical Union, pp. 183–280.

Langmuir, C., Humphris, S., Fornari, D. *et al.* (1997). Hydrothermal vents near a mantle hot spot: the Lucky Strike vent field at 37° N on the Mid-Atlantic Ridge. *Earth and Planetary Science Letters*, **148**, (1–2), 69–91.

Langston, C. A. (1981). Source inversion of seismic waveforms – the Koyna, India, earthquakes of 13 September 1967. *Bulletin of the Seismological Society of America*, **71**, (1), 1–24.

Larson, R. L. and Spiess, F. N. (1969). East Pacific Rise crest – a near-bottom geophysical profile. *Science*, **163**, (3862), 68–71.

Larson, R. L., Searle, R. C., Kleinrock, M. C. *et al.* (1992). Roller-bearing tectonic evolution of the Juan Fernandez microplate. *Nature*, **356**, 571–576.

Laughton, A. S. (1981). The first decade of GLORIA. *Journal of Geophysical Research*, **86**, 11 511–11 534.

Laughton, A. S., Searle, R. C. and Roberts, D. G. (1979). The Reykjanes Ridge crest and the transition between its rifted and non-rifted regions. *Tectonophysics*, **55**, 173–177.

Lavier, L. L., Buck, W. R. and Poliakov, A. N. B. (1999). Self-consistent rolling-hinge model for the evolution of large-offset low-angle normal faults. *Geology*, **27**, (12), 1127–1130.

Lavier, L. L., Buck, W. R. and Poliakov, A. N. B. (2000). Factors controlling normal fault offset in an ideal brittle layer. *Journal of Geophysical Research-Solid Earth*, **105**, (B10), 23 431–23 442.

Lawson, K., Searle, R. C., Pearce, J. A., Browning, P. and Kempton, P. (1996). Detailed volcanic geology of the MARNOK area, Mid-Atlantic Ridge north of Kane transform. In *Tectonic, Magmatic, Hydrothermal and Biological Segmentation of Mid-Ocean Ridges, Geological Society of London, Special Publication*, **118**, ed. C. J. MacLeod, P. A. Tyler and C. L. Walker. London: Geological Society, pp. 61–102.

Le Pichon, X. (1968). Sea-floor spreading and continental drift. *Journal of Geophysical Research*, **73**, 3661–3697.

Le Pichon, X. and Hayes, D. E. (1971). Marginal offsets, fracture zones and the early opening of the South Atlantic. *Journal of Geophysical Research*, **76**, 6283–6293.

Lee, S.-M. and Searle, R. C. (2000). Crustal magnetization of the Reykjanes Ridge and implications for its along-axis variability and the formation of axial volcanic ridges. *Journal of Geophysical Research*, **105**, 5907–5930.

Leeds, A. R., Kausel, E. and Knopoff, L. (1974). Variations of upper mantle structure under the Pacific Ocean. *Science*, **186**, 141–143.

Lilwall, R. C., Francis, T. J. G. and Porter, I. T. (1977). Ocean bottom seismograph observations on the Mid-Atlantic Ridge near 45° N. *Geophysical Journal of the Royal Astronomical Society*, **51**, 357–370.

Lilwall, R. C., Francis, T. J. G. and Porter, I. T. (1978). Ocean bottom seismograph observations on the Mid-Atlantic Ridge near 45° N – further results. *Geophysical Journal of the Royal Astronomical Society*, **55**, 255–262.

Lilwall, R. C., Francis, T. J. G. and Porter, I. T. (1981). A microearthquake survey at the junction of the East Pacific Rise and the Wilkes (9° S) fracture zone. *Geophysical Journal of the Royal Astronomical Society*, **66**, 407–416.

Lin, J. and Phipps Morgan, J. (1992). The spreading rate dependence of three-dimensional mid-ocean ridge gravity structure. *Geophysical Research Letters*, **19**, (1), 13–16.

Lin, J., Purdy, G. M., Schouten, H., Sempéré, J.-C. and Zervas, C. (1990). Evidence from gravity data for focused magmatic accretion along the Mid-Atlantic Ridge. *Nature*, **344**, 627–632.

Lippard, S. J., Shelton, A. W. and Gass, I. G. (1986). *The Ophiolite of Northern Oman, Memoir 11*. London: Geological Society.

Lister, C. R. B. (1970). Measurement of in-situ sediment conductivity by means of a Bullard-type probe. *Geophysical Journal of the Royal Astronomical Society*, **19**, (5), 521–532.

Lister, C. R. B. (1972). Thermal balance of a mid-ocean ridge. *Geophysical Journal of the Royal Astronomical Society*, **26**, (5), 515.

Lister, C. R. B. (1974). Penetration of water into hot rock. *Geophysical Journal of the Royal Astronomical Society*, **39**, (3), 465–509.

Lonsdale, P. (1977). Structural geomorphology of a fast-spreading rise crest: the East Pacific Rise near 3°25′ S. *Marine Geophysical Researches*, **3**, 251–293.

Lonsdale, P. (1978). Near-bottom reconnaissance of a fast-slipping transform fault zone at the Pacific–Nazca plate boundary. *Journal of Geology*, **86**, 451–472.

Lonsdale, P. (1988). Structural pattern of the Galapagos microplate and evolution of the Galapagos triple junctions. *Journal of Geophysical Research*, **93**, 13 551–13 574.

Lonsdale, P. (1994). Geomorphology and structural segmentation of the crest of the southern (Pacific–Antarctic) East Pacific Rise. *Journal of Geophysical Research-Solid Earth*, **99**, (B3), 4683–4702.

Lonsdale, P. (1995). Segmentation and disruption of the East Pacific Rise in the mouth of the Gulf of California. *Marine Geophysical Researches*, **17**, (4), 323–359.

Lonsdale, P. (2005). Creation of the Cocos and Nazca plates by fission of the Farallon plate. *Tectonophysics*, **404**, (3–4), 237–264.

Louden, K. E., White, R. S., Potts, C. G. and Forsyth, D. W. (1986). Structure and seismotectonics of the Vema Fracture Zone, Atlantic Ocean. *Journal of the Geological Society of London*, **143**, 795–805.

Lowell, R. P., Rona, P. A. and Von Herzen, R. P. (1995). Sea-floor hydrothermal systems. *Journal of Geophysical Research-Solid Earth*, **100**, (B1), 327–352.

Lowrie, W. (1974). Oceanic basalt magnetic-properties and Vine and Matthews hypothesis. *Journal of Geophysics-Zeitschrift Fur Geophysik*, **40**, (4), 513–536.

Lowrie, W. (1997). *Fundamentals of Geophysics*. Cambridge: Cambridge University Press, 354 pp.

Ludwig, K. A., Kelley, D. S., Butterfield, D. A., Nelson, B. K. and Früh-Green, G. L. (2006). Formation and evolution of carbonate chimneys at the Lost City Hydrothermal Field. *Geochimica et Cosmochimica Acta*, **70**, 3625–3645.

Ludwig, K. A., Shen, C.-C., Kelley, D. S., Cheng, H. and Edwards, R. L. (2011). U–Th systematic and $^{230}$Th ages of carbonate chimneys at the Lost City Hydrothermal Field. *Geochimica et Cosmochimica Acta*, **75**, 1869–1888.

Lupton, J. E. (1995). Hydrothermal plumes: near and far field. In *Seafloor Hydrothermal Systems: Physical, Chemical, Biological, and Geological Interactions, Geophysical Monograph no. 91*, ed. S. E. Humphris, R. A. Zierenberg, S. Mullineaux and R. E. Thomson. Washington, D.C.: American Geophysical Union, pp. 317–346.

Lupton, J. E. and Craig, H. (1981). A major helium-3 source at 15° S on the East Pacific Rise. *Science*, **214**, 13–18.

Lupton, J., Delaney, J., Johnson, H. and Tivey, M. (1985). Entrainment and vertical transport of deep-ocean water by buoyant hydrothermal plumes. *Nature*, **316**, 621–623.

Lutz, R. A. and Kennish, M. J. (1993). Ecology of deep-sea hydrothermal vent communities: a review. *Reviews of Geophysics*, **31**, (3), 211–242.

Macdonald, K. C. (1977). Near-bottom magnetic anomalies, asymmetric spreading, oblique spreading, and tectonics of the Mid-Atlantic Ridge near lat 37° N. *Geological Society of America Bulletin*, **88**, 541–555.

Macdonald, K. C. (1982). Mid-ocean ridges: fine sale tectonic, volcanic and hydrothermal processes within the plate boundary zone. *Annual Reviews of Earth and Planetary Science*, **10**, 155–189.

Macdonald, K. C. (1998). Linkages between faulting, volcanism, hydrothermal activity and segmentation on fast-spreading centers. In *Faulting and Magmatism at Mid-Ocean Ridges – Geophysical Monograph 106*, ed. W. R. Buck, P. T. Delaney, J. A. Karson and Y. Lagabrielle. Washington, D.C.: American Geophysical Union, pp. 27–59.

Macdonald, K. C. and Fox, P. J. (1983). Overlapping spreading centres: new accretion geometry on the East Pacific Rise. *Nature*, **302**, 55–57.

Macdonald, K. C. and Fox, P. J. (1988). The axial summit graben and cross-sectional shape of the East Pacific Rise as indicators of axial magma chambers and recent volcanic eruptions. *Earth and Planetary Science Letters*, **88**, 119–131.

Macdonald, K. C., Kastens, K. A., Spiess, F. N. and Miller, S. P. (1979). Deep tow studies of the Tamayo Transform Fault. *Marine Geophysical Researches*, **4**, 37–70.

Macdonald, K. C., Becker, K., Spiess, F. N. and Ballard, R. D. (1980a). Hydrothermal heat flux of the "black smoker" vents on the East Pacific Rise. *Earth and Planetary Science Letters*, **48**, 1–7.

Macdonald, K. C., Miller, S. P., Huestis, S. P. and Spiess, F. N. (1980b). Three-dimensional modelling of a magnetic reversal boundary from inversion of deep-tow measurements. *Journal of Geophysical Research*, **85**, 3670–3680.

Macdonald, K. C., Sempere, J.-C. and Fox, P. J. (1984). East Pacific Rise from Siqueiros to Orozco fracture zones: along-strike continuity of the axial neovolcanic zone and structure and evolution of overlapping spreading centers. *Journal of Geophysical Research*, **89**, 6049–6069.

Macdonald, K. C., Sempere, J.-C. and Fox, P. J. (1986a). Reply: the debate concerning overlapping spreading centers and mid-ocean ridge processes. *Journal of Geophysical Research*, **91**, 10 501–10 510.

Macdonald, K. C., Castillo, D., Miller, S. *et al.* (1986b). Deep-tow studies of the Vema Fracture Zone: 1 – The tectonics of a major slow-slipping transform fault and its intersection with the Mid-Atlantic Ridge. *Journal of Geophysical Research*, **91**, 3334–3354.

Macdonald, K. C., Sempere, J.-C., Fox, P. J. and Tyce, R. (1987). Tectonic evolution of ridge-axis discontinuities by the meeting, linking, or self-decapitation of neighbouring ridge segments. *Geology*, **15**, 993–997.

Macdonald, K. C., Haymon, R. M., Miller, S. P., Sempere, J.-C. and Fox, P. J. (1988a). Deep-tow and Sea Beam studies of duelling propagating ridges on the East Pacific Rise near 20°40′ S. *Journal of Geophysical Research*, **93**, 2875–2898.

Macdonald, K. C., Fox, P. J., Perram, L. J. *et al.* (1988b). A new view of the mid-ocean ridge from the behaviour of ridge-axis discontinuities. *Nature*, **335**, 217–225.

Macdonald, K. C., Haymon, R. and Shor, A. (1989). A 220 km$^2$ recently erupted lava field on the East Pacific Rise near lat. 8° S. *Geology*, **17**, 212–216.

Macdonald, K. C., Scheirer, D. S. and Carbotte, S. M. (1991). Mid-ocean ridges: discontinuities, segments and giant cracks. *Science*, **253**, 986–994.

Macdonald, K. C., Fox, P. J., Miller, S. *et al.* (1992). The East Pacific Rise and its flanks 8–18° N – history of segmentation, propagation and spreading direction based on Seamarc-II and Sea Beam studies. *Marine Geophysical Researches*, **14**, (4), 299.

Macdonald, K. C., Fox, P. J., Alexander, R. T., Pockalny, R. and Gente, P. (1996). Volcanic growth faults and the origin of Pacific abyssal hills. *Nature*, **380**, (6570), 125–129.

MacGregor, L. M., Constable, S. and Sinha, M. C. (1998). The RAMESSES experiment–III. Controlled-source electromagnetic sounding of the Reykjanes Ridge at 57°45′ N. *Geophysical Journal International*, **135**, (3), 773–789.

Macleod, C. and Rothery, D. (1992). Ridge axial segmentation in the Oman ophiolite: evidence from along-strike variations in the sheeted dyke complex. In *Ophiolites and their Modern Oceanic Analogues*, ed. L. M. Parson, B. Murton and P. Browning. London: Geological Society, pp. 39–63.

MacLeod, C. J. and Yaouancq, G. (2000). A fossil melt lens in the Oman ophiolite: implications for magma chamber processes at fast spreading ridges. *Earth and Planetary Science Letters*, **176**, (3–4), 357–373.

MacLeod, C. J., Dick, H. J. B., Allerton, S. *et al.* (1998). Geological mapping of slow-spread lower ocean crust: a deep-towed video and wireline rock drilling survey of Atlantis Bank (ODP Site 735, Southwest Indian Ridge). *InterRidge News*, **7**, 39–43.

MacLeod, C. J., Escartin, J., Banerji, D. *et al.* (2002). Direct geological evidence for oceanic detachment faulting: the Mid-Atlantic Ridge, 15°45′N. *Geology*, **30**, 879–882.

MacLeod, C. J., Searle, R. C., Murton, B. J. *et al.* (2009). Life cycle and internal structure of oceanic core complexes. *Earth and Planetary Science Letters*, **287**, (3–4), 333–344.

Madsen, J. A., Forsyth, D. W. and Detrick, R. S. (1984). A new isostatic model for the East Pacific Rise Crest. *Journal of Geophysical Research*, **89**, 9997–10 015.

Madsen, J. A., Fox, P. J. and Macdonald, K. C. (1986). Morphotectonic fabric of the Orozco Transform Fault: results from a Sea Beam investigation. *Journal of Geophysical Research*, **91**, 3439–3454.

Magde, L. S. and Smith, D. K. (1995). Seamount volcanism at the Reykjanes Ridge: relationship to the Iceland hot spot. *Journal of Geophysical Research*, **100**, (B5), 8449–8468.

Magde, L. S. and Sparks, D. W. (1997). Three-dimensional mantle upwelling, melt generation, and melt migration beneath segment slow spreading ridges. *Journal of Geophysical Research, B*, **102**, (9), 20 571–20 583.

Magde, L. S., Detrick, R. S., Kent, G. M. *et al.* (1995). Crustal and upper-mantle contribution to the axial gravity-anomaly at the southern East Pacific Rise. *Journal of Geophysical Research-Solid Earth*, **100**, (B3), 3747–3766.

Magde, L. S., Sparks, D. W. and Detrick, R. S. (1997). The relationship between buoyant mantle flow, melt migration, and gravity bull's eyes at the Mid-Atlantic Ridge between 33° N and 35° N. *Earth and Planetary Science Letters*, **148**, (1–2), 59–67.

Magde, L. S., Barclay, A. H., Toomey, D. R., Detrick, R. S. and Collins, J. A. (2000). Crustal magma plumbing within a segment of the Mid-Atlantic Ridge, 35° N. *Earth and Planetary Science Letters*, **175**, 55–67.

Maia, M. and Gente, P. (1998). Three-dimensional gravity and bathymetric analysis of the Mid-Atlantic Ridge between 20° N and 24° N: flow geometry and temporal evolution of segmentation. *Journal of Geophysical Research*, **103B**, (1), 951–974.

Mallows, C. and Searle, R. C. (2012). A geophysical study of oceanic core complexes and surrounding terrain, Mid-Atlantic Ridge at 13°–14° N. *Geochemistry Geophysics Geosystems*, **13**, doi:10.1029/2012GC004075.

Malpas, J. (1978). Magma generation in upper mantle, field evidence from ophiolite suites, and application to generation of oceanic lithosphere. *Philosophical Transactions of the Royal Society of London Series A-Mathematical Physical and Engineering Sciences*, **288**, (1355), 527–545.

Malpas, J. (1990). Crustal accretionary processes in the Troodos ophiolite, Cyprus: evidence from field mapping and deep crustal drilling. In *Proceedings of the Symposium on*

*Ophiolites and Oceanic Lithosphere – TROODOS 87*, ed. J. Malpas, E. M. Moores, A. Panayiotou and C. Xenophontas. Nicosia, Cyprus: Geological Survey Department, pp. 65–74.

Malpas, J., Moores, E. M., Panayiotou, A. and Xenophontos, C. (1990). *Proceedings of the Symposium on Ophiolites and Oceanic Lithosphere – TROODOS 87*. Nicosia, Cyprus: Geological Survey Department, Ministry of Agriculture and Natural Resources, 733 pp.

Mammerickx, J., Naar, D. F. and Tyce, R. L. (1988). The Mathematician paleoplate. *Journal of Geophysical Research*, **93**, 3025–3040.

Marks, K. M., Vogt, P. R. and Hall, S. A. (1990). Residual depth anomalies and the origin of the Australian–Antarctic Discordance zone. *Journal of Geophysical Research*, **95**, (B11), 17 325–17 337.

Marmer, H. A. (1933). Recent major oceanographic expeditions: a review of the work of the Meteor, Carnegie, Dana, and Snellius expeditions. *Geographical Review*, **23**, (2), 299–305.

Martin, W., Baross, J., Kelley, D. and Russell, M. J. (2008). Hydrothermal vents and the origins of life. *Nature Reviews Microbiology*, **6**, 805–814.

Mason, R. (1985). Ophiolites. *Geology Today*, **1**, 136–140.

Mason, R. G. (1958). A magnetic survey off the west coast of the United States between latitudes 32° and 36° N, longitudes 121° and 128° W. *Geophysical Journal of the Royal Astronomical Society*, **1**, 320–329.

Mason, R. G. and Raff, A. D. (1961). Magnetic survey off the west coast of North America, 32° N latitude to 42° N latitude. *Geological Society of America Bulletin*, **72**, (8), 1259–1265.

Mastin, L. G. and Pollard, D. D. (1988). Surface deformation and shallow dike intrusion processes at Inyo Craters, Long Valley, California. *Journal of Geophysical Research-Solid Earth and Planets*, **93**, (B11), 13 221–13 235.

Maury, M. F. (1860). *The Physical Geography of the Sea*. London: T. Nelson and Sons.

Maxwell, A. E., von Herzen, R. P., Hsu, K. J. *et al.* (1970). Deep sea drilling in the South Atlantic. *Science*, **168**, 1047–1059.

McAllister, E. and Cann, J. (1996). Initiation and evolution of boundary wall faults along the Mid-Atlantic Ridge, 25–29° N. In *Tectonic, Magmatic, Hydrothermal and Biological Segmentation of Mid-Ocean Ridges*, ed. C. J. MacLeod, P. A. Tyler and C. L. Walker. London: Geological Society Special Publication 118, pp. 29–48.

McCaig, A. M., Cliff, R. A., Escartin, J., Fallick, A. E. and MacLeod, C. J. (2007). Oceanic detachment faults focus very large volumes of black smoker fluids. *Geology*, **35**, (10), 935–938; doi: 10.1130/G23657A.1.

McKenzie, D. P. (1967). Some remarks on heat flow and gravity anomalies. *Journal of Geophysical Research*, **72**, (24), 6261–6273.

McKenzie, D. (1986). The geometry of propagating rifts. *Earth and Planetary Science Letters*, **77**, 176–186.

McKenzie, D. P. and Bickle, M. J. (1988). The volume and composition of melt generated by extension of the lithosphere. *Journal of Petrology*, **29**, 625–679.

McKenzie, D. P. and Bowin, C. O. (1976). The relationship between bathymetry and gravity in the Atlantic Ocean. *Journal of Geophysical Research*, **81**, 1903–1915.

McKenzie, D. P. and Morgan, W. J. (1969). Evolution of triple junctions. *Nature*, **224**, (5215), 125–133.

McKenzie, D. P. and Parker, R. L. (1967). The North Pacific: an example of tectonics on a sphere. *Nature*, **216**, 1276–1280.

Melchert, B., Devey, C. W., German, C. R. *et al.* (2008). First evidence for high-temperature off-axis venting of deep crustal/mantle heat: the Nibelungen hydrothermal vent field, southern Mid-Atlantic Ridge. *Earth and Planetary Science Letters*, **275**, 61–69.

Menard, H. W. (1960). East Pacific Rise. *Science*, **132**, (3441), 1737–1746.

Menard, H. W. (1967). Extension of Northeastern–Pacific fracture zones. *Science*, **155**, 72–74.

Menard, H. W. and Atwater, T. (1968). Changes in direction of seafloor spreading. *Nature*, **219**, 463–467.

Menard, H. W. and Atwater, T. (1969). Origin of fracture-zone topography. *Nature*, **222**, 1037–1040.

Mendel, V. and Sauter, D. (1997). Seamount volcanism at the super slow-spreading southwest Indian ridge between 57° E and 70° E. *Geology*, **25**, (2), 99–102.

Mendel, V., Sauter, D., Parson, L., Vanney, J.-R. and Munschy, M. (1997). Segmentation and morphotectonic variations along a super slow-spreading center: the Southwest Indian Ridge (57° E–70° E). *Marine Geophysical Researches*, **19**, (6), 505–533.

Meurer, W. P. and Gee, J. (2002). Evidence for the protracted construction of slow-spread oceanic crust by small magmatic injections. *Earth and Planetary Science Letters*, **201**, 45–55.

Michael, P. J., Langmuir, C. H., Dick, H. J. B. *et al.* (2003). Magmatic and amagmatic seafloor generation at the ultraslow-spreading Gakkel ridge, Arctic Ocean. *Nature*, **423**, (6943), 956–961.

Millard, N. W., Griffiths, G., Finegan, G. *et al.* (1998). Versatile autonomous submersibles – the realising and testing of a practical vehicle. *Underwater Technology*, **23**, (1), 7–17.

Miller, E. and Ewing, M. (1956). Geomagnetic measurements in the Gulf of Mexico and in the vicinity of Caryn Peak. *Geophysics*, **21**, 406–432.

Minshull, T. A., Muller, M. R. and White, R. S. (2006). Crustal structure of the Southwest Indian Ridge at 66° E: seismic constraints. *Geophysical Journal International*, **166**, (1), 135–147.

Minster, J. B., Jordan, T. H., Molnar, P. and Haines, E. (1974). Numerical modelling of instantaneous plate tectonics. *Geophysical Journal of the Royal Astronomical Society*, **36**, 541–576.

Mitchell, N. C. and Searle, R. C. (1999). Fault scarp statistics at the Galapagos Spreading Centre from Deep Tow data. *Marine Geophysical Researches*, **20**, (3), 183–193.

Moore, D. E. and Lockner, D. A. (2007). Comparative deformation behavior of minerals in serpentinized ultramafic rock: application to the slab–mantle interface in subduction zones. *International Geology Review*, **49**, (5), 401–415.

Moore, D. E. and Rymer, M. J. (2007). Talc-bearing serpentinite and the creeping section of the San Andreas fault. *Nature*, **448**, doi:10.1038/nature06064.

Moores, E. M. (1982). Origin and emplacement of ophiolites. *Reviews of Geophysics*, **20**, (4), 735–760.

Morgan, W. J. (1968). Rises, trenches, great faults and crustal blocks. *Journal of Geophysical Research*, **73**, 1959–1982.

Moritz, H. (1980). Geodetic Reference System 1980: International Union of Geodesy and Geophysics Resolution no. 7, pp. 128–133.

Morley, L. W. and Larochelle, A. (1964). Paleomagnetism as a means of dating geological events. *Royal Society of Canada, Special Publication*, **9**, 40–51.

Morris, A., Gee, J. S., Pressling, N. *et al.* (2009). Footwall rotation in an oceanic core complex quantified using reoriented Integrated Ocean Drilling Program core samples. *Earth and Planetary Science Letters*, **28**, 217–228.

Morton, J. L. and Sleep, N. H. (1985). Seismic reflections from a Lau Basin magma chamber. In *Geology and Offshore Resources of Pacific Island Arcs – Tonga Region*, ed. D. W. Scholl and T. L. Vallier. Houston: Circum-Pacific Council for Energy and Mineral Resources, pp. 441–453.

Muller, M. R., Robinson, C. J., T. A. Minshull, White, R. S. and Bickle, M. J. (1997). Thin crust beneath Ocean Drilling Program borehole 735B at the Southwest Indian Ridge? *Earth and Planetary Science Letters*, **148**, 93–107.

Muller, M. R., Minshull, T. A. and White, R. S. (1999). Segmentation and melt supply at the Southwest Indian Ridge. *Geology*, **27**, (10), 867–870.

Muller, M. R., Minshull, T. A. and White, R. S. (2000). Crustal structure of the Southwest Indian Ridge at the Atlantis II Fracture Zone. *Journal of Geophysical Research, B*, **105**, 25 809–25 828.

Müller, R. D., Sdrolias, M., Gaina, C. and Roest, W. R. (2008). Age, spreading rates, and spreading asymmetry of the world's ocean crust. *Geochemistry Geophysics Geosystems*, **9**, doi:10.1029/2007GC001743.

Mullineaux, L. S., Adams, D. K., Mills, S. W. and Beaulieu, S. E. (2010). Larvae from afar colonize deep-sea hydrothermal vents after a catastrophic eruption. *Proceedings of the National Academy of Sciences*, **107** (17), 7829–7834.

Murray, J. and Hjort, J. (1912). *The Depths of the Ocean*. London: Macmillan & Co.

Murton, B. J., Schroth, N., LeBas, T. *et al.* (2013). Formation of volcanic crust at slow spreading mid-ocean ridges by steady state processes. *Nature Geoscience*, under review.

Mutter, C. Z. and Mutter, J. C. (1993). Variations in thickness of layer-3 dominate oceanic crustal structure. *Earth and Planetary Science Letters*, **117**, (1–2), 295–317.

Naar, D. F. and Hey, R. N. (1989). Speed limit for oceanic transform faults. *Geology*, **17**, 420–422.

Naar, D. F. and Hey, R. N. (1991). Tectonic evolution of the Easter microplate. *Journal of Geophysical Research*, **96**, 7961–7973.

Navin, D. A., Peirce, C. and Sinha, M. C. (1998). The RAMESSES experiment – II. Evidence for accumulated melt beneath a slow spreading ridge from wide-angle refraction and multichannel reflection seismic profiles. *Geophysical Journal International*, **135**, (3), 746–772.

Nazarova, K. (1994). Serpentinized peridotites as a possible source for oceanic magnetic anomalies. *Marine Geophysical Researches*, **16**, 455–462.

Neumann, G. A. and Forsyth, D. W. (1993). The paradox of the axial profile: isostatic compensation along the axis of the Mid-Atlantic Ridge? *Journal of Geophysical Research*, **98**, (B10), 17 891–17 910.

Neves, M. C., Searle, R. C. and Bott, M. H. P. (2003). Easter microplate dynamics. *Journal of Geophysical Research*, **108**, (B4), ETG14, doi:10.1029/2001JB000908.

Nicolas, A. (1989). *Structure of Ophiolites and Dynamics of Oceanic Lithosphere*. Dordrecht: Kluwer Academic Publishers.

Nicolas, A., Boudier, E., Ildefonse, B. and Ball, E. (2000). Accretion of Oman and United Arab Emirates ophiolite – discussion of a new structural map. *Marine Geophysical Research*, **21**, (3–4), 147–179.

Niu, Y. and O'Hara, M. J. (2007). Global correlations of ocean ridge basalt chemistry with axial depth: a new perspective. *Journal of Petrology*, **49**, (4), 633–664.

Nooner, S. L. and Chadwick, W. W. (2009). Volcanic inflation measured in the caldera of Axial Seamount: implications for magma supply and future eruptions. *Geochemistry Geophysics Geosystems*, **10**, doi:10.1029/2008gc002315.

Normark, W. R., Morton, J. L. and Ross, S. L. (1987). Submersible observations along the southern Juan-de-Fuca Ridge – 1984 Alvin program. *Journal of Geophysical Research-Solid Earth and Planets*, **92**, (B11), 11 283–11 290.

Officer, C. B., Ewing, J. I., Hennion, J. F., Harkrider, D. G. and Miller, D. E. (1959). Geophysical investigations in the eastern Caribbean: summary of 1955 and 1956 cruises. In *Physics and Chemistry of the Earth, volume 3*. London: Pergamon, pp. 17–26.

Ohara, Y., Fujioka, K., Ishii, T. and Yurimoto, H. (2003). Peridotites and gabbros from the Parece Vela backarc basin: unique tectonic window in an extinct backarc spreading ridge. *Geochemistry Geophysics Geosystems*, **4**, 8611, doi:10.1029/2002GC000469.

Okino, K., Matsuda, K., Christie, D. M., Nogi, Y. and Koizumi, K.-I. (2004). Development of oceanic detachment and asymmetric spreading at the Australian–Antarctic Discordance. *Geochemistry Geophysics Geosystems*, **5**, (12), Q12012, doi:10.1029/2004GC000793.

Oldenburg, D. W. (1975). A physical model for the creation of the lithosphere. *Geophysical Journal of the Royal Astronomical Society*, **43**, (2), 425–451.

Oldenburg, D. W. and Brune, J. N. (1972). Ridge transform fault spreading pattern in freezing wax. *Science*, **178**, 301–304.

Oldenburg, D. W. and Brune, J. N. (1975). An explanation for the orthogonality of ocean ridges and transform faults. *Journal of Geophysical Research*, **80**, 2575–2585.

Olive, J.-A., Behn, M. D. and Tucholke, B. E. (2010). The structure of oceanic core complexes controlled by the depth distribution of magma emplacement. *Nature Geoscience*, **3**, 491–495.

O'Neill, H. S. C. and Jenner, F. E. (2012). The global pattern of trace-element distributions in ocean floor basalts. *Nature*, **491**, (7426), 698–705.

Oufi, O., Cannat, M. and Horen, H. (2002). Magnetic properties of variably serpentinized abyssal peridotites. *Journal of Geophysical Research-Solid Earth*, **107**, (B5), EPM 3–1–EPM 3–19, doi:10.1029/2001jb000549.

Paduan, J. B., Caress, D. W., Clague, D. A., Paull, C. K. and Thomas, H. (2009). High-resolution mapping of mass wasting, tectonic, and volcanic hazards using the MBARI Mapping AUV. In *Extended Abstracts of the International Conference on Seafloor Mapping for Geohazard Assessment, May 11–13, Forio d'Ischia*, ed. F. L. Chiocci, D. Ridente, D. Casalbore and A. Bosman, Forio d'Ischia, Italy: Società Geologica Italiana.

Pariso, J. E. and Johnson, H. P. (1991). Alteration processes at Deep-Sea Drilling Project Ocean Drilling Program Hole 504B at the Costa-Rica Rift – implications for magnetization of oceanic-crust. *Journal of Geophysical Research-Solid Earth and Planets*, **96**, (B7), 11 703–11 722.

Pariso, J. E. and Johnson, H. P. (1993a). Do lower crustal rocks record reversals of the earth's magnetic-field? Magnetic petrology of oceanic gabbros from Ocean Drilling Program hole-735b. *Journal of Geophysical Research*, **98**, (B9), 16 013–16 032.

Pariso, J. E. and Johnson, H. P. (1993b). Do layer-3 rocks make a significant contribution to marine magnetic-anomalies – in-situ magnetization of gabbros at Ocean Drilling Program Hole-735b. *Journal of Geophysical Research-Solid Earth*, **98**, (B9), 16 033–16 052.

Parker, R. L. (1972). The rapid calculation of potential anomalies. *Geophysical Journal of the Royal Astronomical Society*, **31**, 447–455.

Parker, R. L. and Huestis, S. P. (1974). The inversion of magnetic anomalies in the presence of topography. *Journal of Geophysical Research*, **79**, 1587–1593.

Parker, R. L. and Klitgord, K. D. (1972). Magnetic upward continuation from an uneven track. *Geophysics*, **37**, (4), 662–668.

Parker, R. L. and Oldenburg, D. W. (1973). Thermal model of ocean ridges. *Nature Physical Science*, **242**, 137–139.

Parson, L. M., Murton, B. J. and Browning, P. (1992). *Ophiolites and their Modern Oceanic Analogues, Special Publication, 60*. London: Geological Society, 330 pp.

Parson, L. M., Murton, B. J., Searle, R. C. *et al.* (1993). En echelon volcanic ridges at the Reykjanes Ridge: a life cycle of volcanism and tectonics. *Earth and Planetary Science Letters*, **117**, 73–87.

Parsons, B. and Sclater, J. G. (1977). An analysis of the variation of ocean floor bathymetry and heat flow with age. *Journal of Geophysical Research*, **82**, 803–827.

Patriat, P. and Courtillot, V. (1984). On the stability of triple junctions and its relation to episodicity in spreading. *Tectonics*, **3**, 317–332.

Patterson, R. B. (1972). Increased Ranges for Conventional Underwater Cameras. *Proc. S.P.I.E.* (*Proceedings of the International Society for Optics and Photonics*), **24**, 153–161.

Pearce, J. A., Lippard, S. J. and Roberts, S. (1984). Characteristics and tectonic significance of supra-subduction zone ophiolites. In *Marginal Basin Geology, Special Publication, 16*, ed. B. P. Kokelaar and M. F. Howells. London: Geological Society, pp. 77–94.

Pedersen, R. B., Thorseth, I. H., Nygård, T. E., Lilley, M. D. and Kelley, D. S. (2010a). Hydrothermal activity at the Arctic mid-ocean ridges. In *Diversity of Hydrothermal Systems on Slow Spreading Ocean Ridges, Geophysical Monograph 188*, ed. P. A. Rona, C. W. Devey, J. Dyment and B. J. Murton. Washington, D.C.: American Geophysical Union, pp. 67–89.

Pedersen, R. B., Rapp, H. T., Thorseth, I. H. *et al.* (2010b). Discovery of a black smoker vent field and vent fauna at the Arctic Mid-Ocean Ridge. *Nature Communications*, **1**, 126, doi:10.1038/ncomms1124.

Pelli, D. G. and Chamberlain, S. C. (1989). The visibility of 350 °C black-body radiation by the shrimp *Rimicaris exoculata* and man. *Nature*, **337**, (6206), 460–461.

Penrose (1972). Penrose field conference on ophiolites. *Geotimes*, **17**, 24–25.

Perfit, M. R. and Chadwick, W. W. (1998). Magmatism at mid-ocean ridges: constraints from volcanological and geochemical investigations. In *Faulting and Magmatism at Mid-Ocean Ridges, Geophysical Monograph 106*, ed. W. R. Buck, P. T. Delaney, J. A. Karson and Y. Lagabrielle. Washington, D.C.: American Geophysical Union, pp. 59–115.

Perfit, M. R., Fornari, D. J., Smith, M. C. *et al.* (1994). Small-scale spatial and temporal variations in midocean ridge crest magmatic processes. *Geology*, **22**, (4), 375–379.

Perk, N. W., Coogan, L. A., Karson, J. A., Klein, E. M. and Hanna, H. D. (2007). Petrology and geochemistry of primitive lower oceanic crust from Pito Deep: implications for the accretion of the lower crust at the Southern East Pacific Rise. *Contributions to Mineralogy and Petrology*, **154**, (5), 575–590.

Phillips, J. D., Driscoll, A. H., Peal, K. R., Marquet, W. M. and Owen, D. M. (1979). A new undersea geological survey tool: ANGUS. *Deep Sea Research Part A*, **26**, (2), 211–225.

Phipps Morgan, J. and Chen, Y. J. (1993). Dependence of ridge-axis morphology on magma supply and spreading rate. *Nature*, **364**, (19 August), 706–708.

Phipps Morgan, J. and Forsyth, D. W. (1988). 3-D flow and temperature perturbations due to a transform offset: effects on oceanic crustal and upper mantle structure. *Journal of Geophysical Research*, **93**, 2955–2966.

Phipps Morgan, J. and Parmentier, E. M. (1985). Causes and rate-limiting mechanism of ridge propagations: a fracture mechanics model. *Journal of Geophysical Research*, **90**, 8603–8612.

Pockalny, R. A., Detrick, R. S. and Fox, P. J. (1988). Morphology and tectonics of the Kane Transform from Sea Beam bathymetry data. *Journal of Geophysical Research*, **93**, 3179–3193.

Pockalny, R. A., Smith, A. and Gente, P. (1995). Spatial and temporal variability of crustal magnetization of a slowly spreading ridge: Mid-Atlantic Ridge (20°–24° N). *Marine Geophysical Researches*, **17**, (3), 301–320.

Pockalny, R. A., Gente, P. and Buck, R. (1996). Oceanic transverse ridges: a flexural response to fracture-zone-normal extension. *Geology*, **24**, (1), 71–74.

Poliakov, A. N. B. and Buck, W. R. (1998). Mechanics of stretching elastic-plastic-viscous layers: applications to slow-spreading mid-ocean ridges. In *Faulting and Magmatism at Mid-Ocean Ridges*, ed. W. R. Buck, P. T. Delaney, J. A. Karson and Y. Lagabrielle. Washington, D.C.: American Geophysical Union, pp. 305–323.

Pollack, H. N., Hurter, S. J. and Johnson, J. R. (1993). Heat-flow from the Earth's interior – analysis of the global data set. *Reviews of Geophysics*, **31**, (3), 267–280.

Pollard, D. D. and Aydin, A. (1984). Propagation and linkage of oceanic ridge segments. *Journal of Geophysical Research*, **89**, 10 017–10 028.

Pollard, D. D., Delaney, P. T., Duffield, W. A., Endo, E. T. and Okamura, A. T. (1983). Surface deformation in volcanic rift zones. *Tectonophysics*, **94**, (1–4), 541–584.

Powell, C. (1971). Decca, LORAN and Omega – amphibious aids to navigation. *Radio and Electronic Engineer*, **41**, (12), S184.

Prévot, M. and Grommé, S. (1975). Intensity of magnetization of sub-aerial and submarine basalts and its possible change with time. *Geophysical Journal of the Royal Astronomical Society*, **40**, 207–224.

Prince, R. A. and Forsyth, D. W. (1988). Horizontal extent of anomalously thin crust near the Vema fracture-zone from the 3-dimensional analysis of gravity-anomalies. *Journal of Geophysical Research-Solid Earth and Planets*, **93**, (B7), 8051–8063.

Pruis, M. J. and Johnson, H. P. (2004). Tapping into the sub-seafloor: examining diffuse flow and temperature from an active seamount on the Juan de Fuca Ridge. *Earth and Planetary Science Letters*, **217**, (3–4), 379–388.

Purdy, G. M. and Detrick, R. S. (1986). Crustal structure of the Mid-Atlantic Ridge at 23° N from seismic refraction studies. *Journal of Geophysical Research-Solid Earth and Planets*, **91**, (B3), 3739–3762.

Purdy, G. M., Sempere, J.-C., Schouten, H., DuBois, D. L. and Goldsmith, R. (1990). Bathymetry of the Mid-Atlantic Ridge, 24°–31° N: a map series. *Marine Geophysical Researches*, **12**, 247–252.

Quick, J. E. and Delinger, R. P. (1993). Ductile deformation and the origin of layered gabbro in ophiolites. *Journal of Geophysical Research*, **98**, (B8), 14 015–14 027.

Raff, A. D. and Mason, R. G. (1961). Magnetic survey off the west coast of North-America, 40° N latitude to 52° N latitude. *Geological Society of America Bulletin*, **72**, (8), 1267–1270.

Raitt, R. W. (1956). Seismic-refraction studies of the Pacific ocean basin: part I: crustal thickness of the central equatorial Pacific. *Bulletin of the Geological Society of America*, **67**, 1623–1640.

Raitt, R. W. (1963). Chapter 6. The crustal rocks. In *The Sea, volume 3*, ed. M. N. Hill. New York: John Wiley, pp. 85–102.

Ranero, C. R. and Reston, T. J. (1999). Detachment faulting at ocean core complexes. *Geology*, **27**, (11), 983–986.

Reid, J. L. (1982). Evidence of an effect of heat-flux from the East Pacific Rise upon the characteristics of the mid-depth waters. *Geophysical Research Letters*, **9**, (4), 381–384.

Reinke-Kunze, C. (1994). *Welt der Forschungsschiffe*. Hamburg: DSV-Verlag GmbH, 192 pp.

Renard, V. and Allenou, J. P. (1979). Sea Beam, multi-beam echo-sounding in Jean Charcot – description, evaluation and 1st results. *International Hydrographic Review*, **56**, (1), 35–67.

Reston, T. J. and Ranero, C. R. (2011). The 3-D geometry of detachment faulting at mid-ocean ridges. *Geochemistry Geophysics Geosystems*, **12**, doi:10.1029/2011gc003666.

Reynolds, J. R., Langmuir, C. H., Bender, J. F., Kastens, K. A. and Ryan, W. B. F. (1992). Spatial and temporal variability in the geochemistry of basalts from the East Pacific Rise. *Nature*, **359**, (6395), 493–499.

Riedel, W. R., Ladd, H. S., Tracey, J. I. Jr. and Bramlette, M. N. (1961). Preliminary drilling phase of Mohole project. *Bulletin of the American Association of Petroleum Geologists*, **45**, 1793–1798.

Ritsema, J., Deuss, A., van Heijst, H. J. and Woodhouse, J. H. (2011). S40RTS: a degree-40 shear-velocity model for the mantle from new Rayleigh wave dispersion, teleseismic traveltime and normal-mode splitting function measurements. *Geophysical Journal International*, **184**, (3), 1223–1236.

Robertson, A. and Xenophontos, C. (1993). Development of concepts concerning the Troodos ophiolite and adjacent units in Cyprus. In *Magmatic Processes and Plate Tectonics*, ed. H. M. Prichard, T. Alabaster, N. B. W. Harris and C. R. Neary. London: Geological Society, pp. 85–119.

Robinson, P. T., Melson, W. G., Ohearn, T. and Schmincke, H. U. (1983). Volcanic glass compositions of the Troodos ophiolite, Cyprus. *Geology*, **11**, (7), 400–404.

Rogers, A. D., Tyler, P. A., Connelly, D. P. *et al.* (2012). The discovery of new deep-sea hydrothermal vent communities in the Southern Ocean and implications for biogeography. *Plos Biology*, **10**, (1), doi:10.1371/journal.pbio.1001234.

Rona, P. A. (2010). Emerging diversity of hydrothermal systems on slow spreading ocean ridges. In *Diversity of Hydrothermal Systems on Slow Spreading Ocean Ridges, Geophysical Monograph 188*, ed. P. A. Rona, C. W. Devey, J. Dyment and B. J. Murton. Washington, D.C.: American Geophysical Union, pp. 5–10.

Rona, P. A., Boström, K., Laubier, L. and Smith, K. L. (1983). *Hydrothermal Processes at Seafloor Spreading Centers*. New York: Plenum, 796 pp.

Rona, P. A., Klinkhammer, G., Nelsen, T. A., Trefry, J. H. and Elderfield, H. (1986). Black smoker, massive sulfides and vent biota at the Mid-Atlantic Ridge. *Nature*, **321**, 33–37.

Rona, P. A., Jackson, D. R., Wen, T. *et al.* (1997). Acoustic mapping of diffuse flow at a seafloor hydrothermal site: Monolith Vent, Juan de Fuca Ridge. *Geophysical Research Letters*, **24**, (19), 2351–2354.

Rona, P. A., Devey, C. W., Dyment, J. and Murton, B. J. (editors) (2010). *Diversity of Hydrothermal Systems on Slow Spreading Ocean Ridges, Geophysical Monograph 188*. Washington, D.C.: American Geophysical Union, 431 pp.

Rubin, A. M. (1995). Propagation of magma-filled cracks. *Annual Review of Earth and Planetary Sciences*, **23**, 287–336.

Rusby, J. S. M., Dobson, R., Edge, R. H., Pierce, F. E. and Somers, M. L. (1969). Records obtained from the trials of a long range side-scan sonar (GLORIA Project). *Nature*, **223**, (5212), 125–126.

Rusby, R. I. and Searle, R. C. (1993). Intraplate thrusting near the Easter microplate. *Geology*, **21**, 311–314.

Rusby, R. I. and Searle, R. C. (1995). A history of the Easter microplate, 5.25 Ma to present. *Journal of Geophysical Research*, **100**, (B7), 12 617–12 640.

Russell, M. J., Hall, A. J. and Martin, W. (2010). Serpentinization as a source of energy at the origin of life. *Geobiology*, **8**, (5), 355–371.

Ryan, W. B. F., Carbotte, S. M., Coplan, J. O. *et al.* (2009). Global multi-resolution topography synthesis. *Geochemistry Geophysics Geosystems*, **10**, doi:10.1029/2008gc002332.

Rychert, C. A. and Shearer, P. M. (2009). A global view of the lithosphere–asthenosphere boundary. *Science* **324**, 495–498.

Rychert, C. A. and Shearer, P. M. (2011). Imaging the lithosphere–asthenosphere boundary beneath the Pacific using SS waveform modeling. *Journal of Geophysical Research*, **116**, doi:10.1029/2010JB008070.

Salisbury, M. H. and Christensen, N. I. (1978). Seismic velocity structure of a traverse through Bay of Islands ophiolite complex, Newfoundland, an exposure of oceanic-crust and upper mantle. *Journal of Geophysical Research*, **83**, (B2), 805–817.

Sandwell, D. T. (1991). Geophysical applications of satellite altimetry. *Reviews of Geophysics*, **29**, part 1, supplement S, 132–137.

Sandwell, D. T. and Smith, W. H. F. (2009). Global marine gravity from retracked Geosat and ERS-1 altimetry: Ridge segmentation versus spreading rate. *Journal of Geophysical Research*, **114**, B0141, doi:10.1029/2008JB006008.

Sarrazin, J., Rodier, P., Tivey, M. K. *et al.* (2009). A dual sensor device to estimate fluid flow velocity at diffuse hydrothermal vents. *Deep-Sea Research Part I-Oceanographic Research Papers*, **56**, (11), 2065–2074.

Sauter, D., Parson, L., Mendel, V. *et al.* (2002). TOBI sidescan sonar imagery of the very slow-spreading Southwest Indian Ridge: evidence for along-axis magma distribution. *Earth and Planetary Science Letters*, **199**, (1–2), 81–95.

Sauter, D., Mendel, V., Rommevau-Jestin, C. *et al.* (2004). Focused magmatism versus amagmatic spreading along the ultra-slow spreading Southwest Indian Ridge: evidence from TOBI side scan imagery. *Geochemistry Geophysics Geosystems*, **5**, (10), doi: 10.1029/2004GC000738.

Sauter, D., Cannat, M., Rouméjon, S. *et al.* (2013). 11 Myr-continuous exhumation of mantle-derived rocks at the Southwest Indian Ridge. *Nature Geoscience*, doi:10.1038/NGE01771.

Scheirer, D. S. and Macdonald, K. C. (1995). Near-axis seamounts on the flanks of the East Pacific Rise, 8° N to 17° N. *Journal of Geophysical Research*, **100**, (B2), 2239–2259.

Schmerr, N. (2012). The Gutenberg discontinuity: melt at the lithosphere–asthenosphere boundary. *Science*, **335**, (6075), 1480–1483.

Schouten, H. and Denham, C. (1979). Modelling the oceanic magnetic source layer. In *Deep Drilling Results in the Atlantic Ocean*, ed. M. Talwani. Washington, D.C.: American Geophysical Union, pp. 151–159.

Schouten, H. and McCamy, K. (1972). Filtering marine magnetic anomalies. *Journal of Geophysical Research*, **77**, 7089–7099.

Schouten, H. and White, R. S. (1980). Zero-offset fracture zones. *Geology*, **8**, 175–179.

Schouten, H., Klitgord, K. D. and Whitehead, J. A. (1985). Segmentation of mid-ocean ridges. *Nature*, **317**, 225–229.

Schouten, H., Klitgord, K. D. and Gallo, D. G. (1993). Edge-driven microplate kinematics. *Journal of Geophysical Research*, **98**, 6689–6701.

Schouten, H., Tivey, M. A., Fornari, D. J. and Cochran, J. R. (1999). Central anomaly magnetization high: constraints on the volcanic construction and architecture of seismic layer 2A at a fast-spreading mid-ocean ridge, the East Pacific Rise at 9°30′–50′ N. *Earth and Planetary Science Letters*, **169**, 37–50.

Schouten, H., Smith, D. K., Cann, J. R. and Escartín, J. (2010). Tectonic versus magmatic extension in the presence of core complexes at slow-spreading ridges from a visualization of faulted seafloor topography. *Geology*, **38**, (7), 615–618.

Schroeder, T., John, B. and Frost, B. R. (2002). Geologic implications of seawater circulation through peridotite exposed at slow-spreading mid-ocean ridges. *Geology*, **30**, (4), 367–370.

Schultz, A. and Elderfield, H. (1997). Controls on the physics and chemistry of seafloor hydrothermal circulation. *Philosophical Transactions of the Royal Society of London, Series A*, **355**, (1723), 387–425.

Schultz, A., Delaney, J. R. and McDuff, R. E. (1992). On the partitioning of heat-flux between diffuse and point-source sea-floor venting. *Journal of Geophysical Research-Solid Earth*, **97**, (B9), 12 299–12 314.

Schultz, A., Dickson, P. and Elderfield, H. (1996). Temporal variations in diffuse hydrothermal flow at TAG. *Geophysical Research Letters*, **23**, (23), 3471–3474.

Schwarz, A. (2011). Papers from Origins 2011, Origins of Life and Evolution of Biospheres, Special Issue: **41**, 495–632.

Sclater, J. G., Anderson, R. N. and Bell, M. L. (1971). Elevation of ridges and evolution of central eastern Pacific. *Journal of Geophysical Research*, **76**, (32), 7888–7915.

Sclater, J. G., Parsons, B. and Jaupart, C. (1981). Oceans and continents: similarities and differences in the mechanisms of heat loss. *Journal of Geophysical Research*, **86**, 11 535–11 552.

Scott, R. B., Rona, P. A., McGregor, B. A. and Scott, M. R. (1974). TAG hydrothermal field. *Nature*, **251**, (5473), 301–302.

Seama, N., Nogi, Y. and Isezaki, N. (1993). A new method for precise determination of the position and strike of magnetic boundaries using vector data of the geomagnetic anomaly field. *Geophysical Journal International*, **113**, 155–164.

Searle, R. C. (1980). Tectonic pattern of the Azores spreading centre and triple junction. *Earth and Planetary Science Letters*, **51**, 415–434.

Searle, R. C. (1981). The active part of Charlie–Gibbs Fracture Zone: a study using sonar and other geophysical techniques. *Journal of Geophysical Research*, **86**, 243–262.

Searle, R. C. (1983). Multiple, closely spaced transform faults in fast-slipping fracture zones. *Geology*, **11**, 607–610.

Searle, R. C. (1984). GLORIA survey of the East Pacific Rise near 3.5° S: tectonic and volcanic characteristics of a fast-spreading mid-ocean rise. *Tectonophysics*, **101**, 319–344.

Searle, R. C. (1986). GLORIA investigations of oceanic fracture zones: comparative studies of the transform fault zone. *Journal of the Geological Society of London*, **143**, 743–756.

Searle, R. C. (2012). Are axial volcanic ridges where all the (volcanic) action is? Paper OS11E-06, American Geophysical Union Fall Meeting, San Francisco.

Searle, R. C. and Escartín, J. (2004). The rheology and morphology of oceanic lithosphere and mid-ocean ridges. In *Mid-Ocean Ridges: Hydrothermal Interactions between the Lithosphere and Oceans, Geophysical Monograph 148*, ed. C. German, J. Lin and L. M. Parson. Washington, D.C.: American Geophysical Union, pp. 63–94.

Searle, R. C. and Hey, R. N. (1983). GLORIA observations of the propagating rift at 95.5° W on the Cocos–Nazca spreading center. *Journal of Geophysical Research*, **88**, 6433–6447.

Searle, R. C. and Laughton, A. S. (1977). Sonar studies of the Mid-Atlantic Ridge crest near Kurchatov Fracture Zone. *Journal of Geophysical Research*, **82**, 5313–5328.

Searle, R. C., Rusby, R. I., Engeln, J. *et al.* (1989). Comprehensive sonar imaging of the Easter microplate. *Nature*, **341**, 701–705.

Searle, R. C., Le Bas, T. P., Mitchell, N. C. *et al.* (1990). GLORIA image processing: the state of the art. *Marine Geophysical Researches*, **12**, 21–39.

Searle, R. C., Bird, R. T., Rusby, R. I. and Naar, D. F. (1993). The development of two oceanic microplates: Easter and Juan Fernandez microplates, East Pacific Rise. *Journal of the Geological Society of London*, **150**, 965–976.

Searle, R. C., Keeton, J. A., Lee, S. M. *et al.* (1998a). The Reykjanes Ridge: structure and tectonics of a hot-spot influenced, slow-spreading ridge, from multibeam bathymetric, gravity and magnetic investigations. *Earth and Planetary Science Letters*, **160**, 463–478.

Searle, R. C., Cowie, P. A., Mitchell, N. C. *et al.* (1998b). Fault structure and detailed evolution of a slow spreading ridge segment: the Mid-Atlantic Ridge at 29° N. *Earth and Planetary Science Letters*, **154**, (1–4), 167–183.

Searle, R. C., Cannat, M., Fujioka, K. *et al.* (2003). The FUJI Dome: A large detachment fault near 64° E on the very slow-spreading southwest Indian Ridge. *Geochemistry Geophysics Geosystems*, **4**, (8), 9105, doi:10.1029/2003GC000519.

Searle, R. C., Francheteau, J. and Armijo, R. (2006). Compressional deformation north of the Easter microplate: a manned submersible and seafloor gravity investigation. *Geophysical Journal International*, **164**, 359–369, doi:10.1111/j.1365–246X.2005.02812.x.

Searle, R. C., Murton, B. J., Achenbach, K. *et al.* (2010). Structure and development of an axial volcanic ridge: Mid-Atlantic Ridge, 45° N. *Earth and Planetary Science Letters*, **299**, 228–241.

Sempéré, J.-C. and Macdonald, K. C. (1986). Overlapping spreading centers: implications from crack growth by the displacement discontinuity method. *Tectonics*, **5**, 151–163.

Sempéré, J.-C., Lin, J., Brown, H. S., Schouten, H. and Purdy, G. M. (1993). Segmentation and morphotectonic variations along a slow spreading center: the Mid-Atlantic Ridge (24°00′ N–30°40′ N). *Marine Geophysical Researches*, **15**, (3), 153–200.

Sempéré, J.-C., Cochran, J. R. and SEIR Scientific Team (1997). The Southeast Indian Ridge between 88° E and 118° E: variations in crustal accretion at constant spreading rate. *Journal of Geophysical Research*, **102**, (B7), 15 489–15 505.

Severinghaus, J. P. and Macdonald, K. C. (1988). High inside corners at ridge-transform intersections. *Marine Geophysical Researches*, **9**, 353–367.

Shah, A. K. and Buck, W. R. (2001). Causes for axial high topography at mid-ocean ridges and the role of crustal thermal structure. *Journal of Geophysical Research-Solid Earth*, **106**, (B12), 30 865–30 879.

Shah, A. K. and Buck, W. R. (2003). Plate bending stresses at axial highs, and implications for faulting behavior. *Earth and Planetary Science Letters*, **211**, (3–4), 343–356.

Shaw, P. (1992). Ridge segmentation, faulting and crustal thickness in the Atlantic Ocean. *Nature*, **358**, 490–493.

Shaw, P. R. and Lin, J. (1993). Causes and consequences of variations in faulting style at the Mid-Atlantic Ridge. *Journal of Geophysical Research*, **98**, (B12), 21 839–21 851.

Shemenda, A. I. and Grocholsky, A. L. (1994). Physical modelling of slow seafloor spreading. *Journal of Geophysical Research*, **99**, (B5), 9137–9153.

Shepard, F. P. (1948). *Submarine Geology*. New York: Harper and Brothers, 348 pp.

Shepard, F. P. (1959). *The Earth Beneath the Sea*. London: Oxford University Press, 275 pp.

Sichler, B. and Hekinian, R. (2002). Three-dimensional inversion of marine magnetic anomalies on the equatorial Atlantic Ridge (St. Paul Fracture Zone): delayed magnetization in a magmatically starved spreading center? *Journal of Geophysical Research-Solid Earth*, **107**, (B12), doi:10.1029/2001JB000401.

Sims, K. W. W., Blichert-Toft, J., Fornari, D. J. *et al.* (2003). Aberrant youth: chemical and isotopic constraints on the origin of off-axis lavas from the East Pacific Rise, 9°–10° N. *Geochemistry Geophysics Geosystems*, **4**, (10), 8621, doi:10.1029/2002GC000443.

Singh, S. C., Crawford, W. C., Carton, H. *et al.* (2006a). Discovery of a magma chamber and faults beneath a Mid-Atlantic Ridge hydrothermal field. *Nature*, **442**, (7106), 1029–1032.

Singh, S. C., Harding, A. J., Kent, G. M. *et al.* (2006b). Seismic reflection images of the Moho underlying melt sills at the East Pacific Rise. *Nature*, **442**, (7100), 287–290.

Sinha, M. and Evans, R. L. (2004). Geophysical constraints on the thermal regime of the oceanic crust. In *Mid-Ocean Ridges: Hydrothermal Interactions between Lithosphere and Oceans, Geophysical Monograph 148*, ed. C. German, J. Lin, R. Tribuzio *et al.* Washington, D.C.: American Geophysical Union, pp. 19–62.

Sinha, M. C., Patel, P. D., Unsworth, M. J., Owen, T. R. E. and MacCormack, M. R. G. (1990). An active source electromagnetic sounding system for marine use. *Marine Geophysical Researches*, **12**, (1–2), 59–68.

Sinha, M. C., Constable, S. C., Peirce, C. *et al.* (1998). Magmatic processes at slow-spreading ridges: implications of the RAMESSES experiment at 57°45′N on the Mid-Atlantic Ridge. *Geophysical Journal International*, **135**, (3), 731–745.

Sinton, J. M. and Detrick, R. S. (1992). Mid-ocean ridge magma chambers. *Journal of Geophysical Research*, **97**, 197–216.

Sinton, J. M., Smaglik, S. M., Mahoney, J. J. and Macdonald, K. C. (1991). Magmatic processes at superfast spreading midocean ridges – glass compositional variations along the East Pacific Rise 13°–23° S. *Journal of Geophysical Research-Solid Earth and Planets*, **96**, (B4), 6133–6155.

Sinton, J. M., Bergmanis, E., Rubin, K. *et al.* (2002). Volcanic eruptions on mid-ocean ridges: new evidence from the superfast spreading East Pacific Rise, 17°–19° S. *Journal of Geophysical Research*, **107**, (B6), ECV3 1–20, doi:10.1029/2001JB000090.

Sleep, N. H. and Biehler, S. (1970). Topography and tectonics at the intersection of fracture zones with central rifts. *Journal of Geophysical Research*, **75**, 2748–2752.

Small, C. and Sandwell, D. T. (1989). An abrupt change in ridge-axis gravity with spreading rate. *Journal of Geophysical Research*, **94**, (B12), 17 388–17 392.

Smallwood, J. R. and White, R. S. (1998). Crustal accretion at the Reykjanes-Ridge, 61–62° N. *Journal of Geophysical Research*, **103**, 5185–5201.

Smith, D. K. and Cann, J. R. (1990). Hundreds of small volcanoes on the median valley floor of the Mid-Atlantic Ridge at 24–30° N. *Nature*, **348**, 152–155.

Smith, D. K. and Cann, J. R. (1992). The role of seamount volcanism in crustal construction at the Mid-Atlantic Ridge (24°–30° N). *Journal of Geophysical Research*, **97**, 1645–1658.

Smith, D. K. and Cann, J. R. (1999). Constructing the upper crust of the Mid-Atlantic Ridge: a reinterpretation based on Puna Ridge, Kilauea Volcano. *Journal of Geophysical Research, B*, **104**, 25 379–25 399.

Smith, D. K., Humphris, S. E. and Bryan, W. B. (1995a). A comparison of volcanic edifices at the Reykjanes Ridge and the Mid-Atlantic Ridge at 24°–30° N. *Journal of Geophysical Research*, **100**, (B11), 22 485–22 498.

Smith, D. K., Cann, J. R., Dougherty, M. E. *et al.* (1995b). Mid-Atlantic Ridge volcanism from deep-towed side-scan sonar images, 25°–29° N. *Journal of Volcanology and Geothermal Research*, **67**, (4), 233–262.

Smith, D. K., Tolstoy, M., Fox, C. G. *et al.* (2002). Hydroacoustic monitoring of seismicity at the slow-spreading Mid-Atlantic Ridge. *Geophysical Research Letters*, **29**, (11), 13-1–13-4.

Smith, D. K., Escartin, J., Cannat, M. *et al.* (2003). Spatial and temporal distribution of seismicity along the northern Mid-Atlantic Ridge (15°–35° N). *Journal of Geophysical Research*, **108**, (B3), 2167, doi:10.1029/2002JB001964.

Smith, D. K., Cann, J. R. and Escartin, J. (2006). Widespread active detachment faulting and core complex formation near 13° N on the Mid-Atlantic Ridge. *Nature*, **442**, 440–443, doi:10.1038/nature04950.

Smith, D. K., Escartin, J., Schouten, H. and Cann, J. R. (2008). Fault rotation and core complex formation: significant processes in seafloor formation at slow-spreading midocean ridges (Mid-Atlantic Ridge, 13°–15° N). *Geochemistry Geophysics Geosystems*, **9**, (3), doi:10.1029/2007GC001699.

Smith, M. C., Perfit, M. R. and Jonasson, I. R. (1994). Petrology and geochemistry of basalts from the southern Juan-de-Fuca Ridge – controls on the spatial and temporal evolution of midocean ridge basalt. *Journal of Geophysical Research-Solid Earth*, **99**, (B3), 4787–4812.

Sohn, R. A. and Sims, K. W. W. (2005). Bending as a mechanism for triggering off-axis volcanism on the East Pacific Rise. *Geology*, **33**, 93–96.

Sohn, R. A., Webb, S. C., Hildebrand, J. A. and Cornuelle, B. D. (1997). Three-dimensional tomographic velocity structure of upper crust, co-axial segment, Juan de Fuca Ridge: implications for on-axis evolution and hydrothermal circulation. *Journal of Geophysical Research*, **102**, (B8), 17 679–17 695.

Sohn, R. A., Hildebrand, J. A. and Webb, S. C. (1998). Postrifting seismicity and a model for the 1993 diking event on the CoAxial segment, Juan de Fuca Ridge. *Journal of Geophysical Research, B*, **103**, (5), 9867–9877.

Sohn, R. A., Hildebrand, J. A. and Webb, S. C. (1999). A microearthquake survey of the high-temperature vent fields on the volcanically active East Pacific Rise (9°50′ N). *Journal of Geophysical Research-Solid Earth*, **104**, (B11), 25 367–25 377.

Sohn, R. and 21 others (2008). Explosive volcanism on the ultraslow-spreading Gakkel ridge, Arctic Ocean. *Nature*, **453**, (7199), 1236–1238.

Soule, S. A., Fornari, D. J., Perfit, M. R. *et al.* (2005). Channelized lava flows at the East Pacific Rise crest 9°–10° N: the importance of off-axis lava transport in developing the architecture of young oceanic crust. *Geochemistry Geophysics Geosystems*, **6**, doi: 10.1029/2005gc000912.

Soule, S. A., Fornari, D. J., Perfit, M. R. and Rubin, K. H. (2007). New insights into mid-ocean ridge volcanic processes from the 2005–2006 eruption of the East Pacific Rise, 9°46′ N–9°56′ N. *Geology*, **35**, (12), 1079–1082.

Soule, S. A., Escartin, J. and Fornari, D. J. (2009). A record of eruption and intrusion at a fast spreading ridge axis: axial summit trough of the East Pacific Rise at 9–10° N. *Geochemistry Geophysics Geosystems*, **10**, doi: 10.1029/2008gc002354.

Sparks, D. W. and Parmentier, E. M. (1991). Melt extraction from the mantle beneath spreading centers. *Earth and Planetary Science Letters*, **105**, 368–377.

Sparks, D. W., Parmentier, E. M. and Morgan, J. P. (1993). Three-dimensional mantle convection beneath a segmented spreading center: implications for along-axis variations in crustal thickness and gravity. *Journal of Geophysical Research*, **98**, 21 977–21 995.

Spencer, J. E. (1999). Geologic continuous casting below continental and deep-sea detachment faults and at the striated extrusion of Sacsayhuaman, Peru. *Geology*, **27**, (4), 327–330.

Spiess, F. N. and Maxwell, A. E. (1964). Search for Thresher. *Science*, **145**, (3630), 349–355.

Spiess, F. N. and Mudie, J. D. (1970). Small scale topographic and magnetic features. In *The Sea, volume 4 (Part I)*, ed. A. E. Maxwell, E. Bullard and J. L. Worzel. New York: Wiley-Interscience, pp. 205–250.

Spiess, F. N., Macdonald, K. C., Atwater, T. *et al.* (1980). East Pacific Rise – hot springs and geophysical experiments. *Science*, **207**, (4438), 1421–1433.

Spudich, P. and Orcutt, J. (1980). A new look at the seismic velocity structure of the oceanic-crust. *Reviews of Geophysics*, **18**, (3), 627–645.

Stakes, D. S., Holloway, G. L., Tucker, P. *et al.* (1997). Diamond rotary coring from an ROV or submersible for hardrock sample recovery and instrument deployment: the MBARI Multiple-Barrel Rock Coring System. *Marine Technology Society Journal*, **31**, (3), 11–20.

Stakes, D. S., Perfit, M. R., Tivey, M. A. *et al.* (2006). The Cleft revealed: geologic, magnetic, and morphologic evidence for construction of upper oceanic crust along the southern Juan de Fuca Ridge. *Geochemistry Geophysics Geosystems*, **7**, doi: 10.1029/2005gc001038.

Standish, J. J. and Sims, K. W. W. (2006). Lava emplacement and crustal architecture within an ultraslow-spreading rift valley. *Geochimica et Cosmochimica Acta*, **70**, (18), A611.

Standish, J. J. and Sims, K. W. W. (2010). Young off-axis volcanism along the ultraslow-spreading Southwest Indian Ridge. *Nature Geoscience*, **3**, (4), 286–292.

Stein, C. A. and Stein, S. (1992). A model for the global variation in oceanic depth and heat flow with lithospheric age. *Nature*, **359**, 123–129.

Stein, C. A. and Stein, S. (1994). Constraints on hydrothermal heat flux through the oceanic lithosphere from global heat flow. *Journal of Geophysical Research*, **99**, 3081–3096.

Stein, C. A. and Stein, S. (1996). Thermo-mechanical evolution of oceanic lithosphere: implications for the subduction process and deep earthquakes (overview). In *Subduction: Top to Bottom. Geophysical Monograph 96*, ed. G. E. Bebout, *et al.* Washington: American Geophysical Union, pp. 1–17.

Stommel, H. (1982). Is the South Pacific He-3 plume dynamically active? *Earth and Planetary Science Letters*, **61**, (1), 63–67.

Strens, M. R. and Cann, J. R. (1986). A fracture-loop thermal balance model of black smoker circulation. *Tectonophysics*, **122**, (3–4), 307–324.

Stride, A. (2010). 14. Side-scan sonar – a tool for seafloor geology. In *Of Seas and Ships and Scientists*, ed. A. Laughton, J. Gould, T. Tucker and H. Roe. Cambridge: Lutterworth Press, pp. 193–207.

Sturm, M. E., Goldstein, S. J., Klein, E. M., Karson, J. A. and Murrell, M. T. (2000). Uranium-series age constraints on lavas from the median valley of the Mid-Atlantic Ridge, MARK area. *Earth and Planetary Science Letters*, **181**, (1–2), 61–70.

Swift, S. A. and Stephen, R. A. (1992). How much gabbro is in ocean seismic layer-3? *Geophysical Research Letters*, **19**, (18), 1871–1874.

Swift, S. A., Lizarralde, D., Stephen, R. A. and Hoskins, H. (1998). Velocity structure in upper ocean crust at Hole 504B from vertical seismic profiles. *Journal of Geophysical Research-Solid Earth*, **103**, (B7), 15 361–15 376.

Sykes, L. R. (1967). Mechanism of earthquakes and nature of faulting on the mid-oceanic ridges. *Journal of Geophysical Research*, **72**, 5–27.

Talwani, M. (1965). Computation with the help of a digital computer of magnetic anomalies caused by bodies of arbitrary shape. *Geophysics*, **30**, 797–817.

Talwani, M. and Heirtzler, J. R. (1964). Computation of magnetic anomalies caused by two-dimensional structures of arbitrary shape. *Computers in the Mineral Industries, part 1, Stanford University Publications in the Geological Sciences*, **9**, 464–480.

Talwani, M., Heezen, B. C. and Worzel, J. L. (1961). Gravity anomalies, physiography, and crustal structure of Mid-Atlantic Ridge. *Journal of Geophysical Research*, **66**, (8), 2565.

Talwani, M., Le Pichon, X. and Ewing, M. (1965). Crustal structure of the mid-ocean ridges 2. Computed model from gravity and seismic refraction data. *Journal of Geophysical Research*, **70**, 341–352.

Talwani, M., Windisch, C. C. and Langseth, M. G. (1971). Reykjanes Ridge crest: a detailed geophysical study. *Journal of Geophysical Research*, **76**, 473–577.

Tan, Y. and Helmberger, D. V. (2007). Trans-Pacific upper mantle shear velocity structure. *Journal of Geophysical Research-Solid Earth*, **112**, (B8), doi: 10.1029/2006jb004853.

Tani, K., Dunkley, D. J. and Ohara, Y. (2011). Termination of backarc spreading: zircon dating of a giant oceanic core complex. *Geology*, **39**, (1), 47–50.

Tao, C. *et al.* (2009). New hydrothermal fields found along the SWIR during the Legs 5–7 of the Chinese DY115–20 Expedition. In *American Geophysical Union, Fall Meeting*, paper OS21A-1150.

Tao, C., Lin, J., Guo, S. *et al.* (2012). First active hydrothermal vents on an ultraslow-spreading center: Southwest Indian Ridge. *Geology*, **40**, (1), 47–50.

Tapponnier, P. and Francheteau, J. (1978). Necking of the lithosphere and the mechanics of slowly accreting plate boundaries. *Journal of Geophysical Research*, **83**, 3955–3970.

Tauxe, L. (1998). *Paleomagnetic Principles and Practice*. Dordrecht: Kluwer Academic Publishers, 299 pp.

Taylor, B., Crook, K. and Sinton, J. (1994). Extensional transform zones and oblique spreading centers. *Journal of Geophysical Research, B*, **99**, (10), 19 707–19 718.

Teagle, D. A. H., Ildefonse, B., Blum, P. and IODP Expedition 335 Scientists (2012). IODP Expedition 335: Deep Sampling in ODP Hole 1256D. *Scientific Drilling*, (13), 28–34.

Telford, W. M., Geldart, L. P. and Sheriff, R. E. (1990). *Applied Geophysics*. 2nd edition. Cambridge: Cambridge University Press, 792 pp.

Tharp, M. (1982). Mapping the ocean floor – 1947 to 1977. In *The Ocean Floor*, ed. R. A. Scrutton and M. Talwani. Chichester: John Wiley & Sons Ltd., pp. 19–31.

Thatcher, W. and Hill, D. P. (1995). A simple model for the fault-generated morphology of slow-spreading mid-oceanic ridges. *Journal of Geophysical Research*, **100**, (B1), 561–570.

Thomson, C. W. (1877). *The Atlantic, A Preliminary Account of the General Results of the Exploring Voyage of H. M. S. Challenger*. London: Macmillan.

Thurber, C. H. (1983). Earthquake locations and three-dimensional crustal structure in the Coyote Lake area, central California. *Journal of Geophysical Research*, **88**, 8226–8236.

Tivey, M. A. (1996). Vertical magnetic structure of ocean crust determined from near-bottom magnetic field measurements. *Journal of Geophysical Research-Solid Earth*, **101**, (B9), 20 275–20 296.

Tivey, M. A. and Johnson, H. P. (1987). The central anomaly magnetic high: implications for ocean crust construction and evolution. *Journal of Geophysical Research*, **92**, 12 685–12 694.

Tivey, M. A. and Johnson, H. P. (1993). Variations in oceanic crustal structure and implications for the fine-scale magnetic anomaly signal. *Geophysical Research Letters*, **20**, (17), 1879–1882.

Tivey, M. K., Olson, L. O., Miller, V. W. and Light, R. D. (1990). Temperature measurements during initiation and growth of a black smoker chimney. *Nature* **346**, 51–54.

Tivey, M. K., Humphris, S. E., Thompson, G., Hannington, M. D. and Rona, P. A. (1995). Deducing patterns of fluid-flow and mixing within the TAG active hydrothermal mound using mineralogical and geochemical data. *Journal of Geophysical Research-Solid Earth*, **100**, (B7), 12 527–12 555.

Tivey, M. A., Schouten, H. and Kleinrock, M. C. (2003). A near-bottom magnetic survey of the Mid-Atlantic Ridge axis at 26° N: implications for the tectonic evolution of the TAG segment. *Journal of Geophysical Research-Solid Earth*, **108**, (B5), doi: 10.1029/2002JB001967.

Tizard, T. H. (1876). 'Report on Temperatures' and 'General Summary of Atlantic Ocean temperatures'. In *HMS Challenger, no. 7. Report on ocean soundings and temperatures.* London: Admiralty.

Tolstoy, M., Harding, A. J. and Orcutt, J. A. (1993). Crustal thickness on the Mid-Atlantic Ridge – Bull's-eye gravity-anomalies and focused accretion. *Science*, **262**, (5134), 726–729.

Tolstoy, M., Harding, A. J., Orcutt, J. A. and the TERA Group (1997). Deepening of the axial magma chamber on the southern East Pacific Rise toward the Garrett Fracture Zone. *Journal of Geophysical Research*, **102**, (B2), 3097–3108.

Tolstoy, M., Cowen, J. P., Baker, E. T. *et al.* (2006). A sea-floor spreading event captured by seismometers. *Science* **314**, 1920–1922.

Tolstoy, M., Waldhauser, F., Bohnenstiehl, D. R., Weekly, R. T. and Kim, W.-Y. (2008). Seismic identification of along-axis hydrothermal flow on the East Pacific Rise. *Nature*, **451**, (7175), doi: 10.1038/nature06424.

Toomey, D. R., Solomon, S. C., Purdy, G. M. and Murray, M. H. (1985). Microearthquakes beneath the median valley of the Mid-Atlantic Ridge near 23° N: hypocenters and focal mechanisms. *Journal of Geophysical Research*, **90**, 5443–5458.

Toomey, D. R., Solomon, S. C. and Purdy, G. M. (1988). Microearthquakes beneath the median valley of the Mid-Atlantic Ridge near 23° N: tomography and tectonics. *Journal of Geophysical Research*, **93**, (B8), 9093–9112.

Toomey, D. R., Purdy, G. M., Solomon, S. C. and Wilcock, W. S. D. (1990). The three-dimensional seismic velocity structure of the East Pacific Rise near latitude 9°30′N. *Nature*, **347**, 639–645.

Toomey, D. R., Purdy, G. M., Barclay, A. H., Wolfe, C. J. and Solomon, S. C. (1993). FARA microearthquake experiments IV: implications of Mid-Atlantic Ridge seismicity for models of young oceanic lithosphere. *EOS, Transactions of the American Geophysical Union*, **74**, (43), 601.

Toomey, D. R., Solomon, S. C. and Purdy, G. M. (1994). Tomographic imaging of the shallow crustal structure of the East Pacific Rise at 9°30′N. *Journal of Geophysical Research*, **99**, (B12), 24 135–24 157.

Toomey, D. R., Wilcock, W. S. D., Conder, J. A. *et al.* (2002). Asymmetric mantle dynamics in the MELT region of the East Pacific Rise. *Earth and Planetary Science Letters*, **200**, (3–4), 287–295.

Tréhu, A. M. (1975). Depth versus (age)$^{1/2}$: a perspective on mid-ocean rises. *Earth and Planetary Science Letters*, **27**, 287–304.

Triantafyllou, M. and Hoover, F. (1990). Cable dynamics for tethered underwater vehicles. In *MIT SeaGrant College Program Report 90–4*. Cambridge, Massachussetts: Massachussetts Institute of Technology.

Tucholke, B. E. and Lin, J. (1994). A geological model for the structure of ridge segments in slow spreading ocean crust. *Journal of Geophysical Research*, **99**, 11 937–11 958.

Tucholke, B. E. and Schouten, H. (1988/89). Kane Fracture Zone. *Marine Geophysical Researches*, **10**, 1–39.

Tucholke, B., Lin, J., Kleinrock, M. *et al.* (1997). Segmentation and crustal structure of the western Mid-Atlantic Ridge flank, 25°25′–27°10′ N and 0–29 m.y. *Journal of Geophysical Research*, **102**, (B5), 10 203–10 223.

Tucholke, B., Lin, J. and Kleinrock, M. (1998). Megamullions and mullion structure defining oceanic metamorphic core complexes on the Mid-Atlantic Ridge. *Journal of Geophysical Research, B*, **103**, (5), 9857–9866.

Tucholke, B. E., Fujioka, K., Ishihara, T., Hirth, G. and Kinoshita, M. (2001). Submersible study of an oceanic megamullion in the central North Atlantic. *Journal of Geophysical Research, B*, **106**, 16 145–16 161.

Tucholke, B. E., Behn, M. D., Buck, W. R. and Lin, J. (2008). Role of melt supply in oceanic detachment faulting and formation of megamullions. *Geology*, **36**, 455–458.

Tucker, T. and McCartney, B. (2010). 16: Engineering and applied physics. In *Of Seas and Ships and Scientists*, ed. A. Laughton, J. Gould, T. Tucker and H. Roe. Cambridge: Lutterworth Press, pp. 233–267.

Tunnicliffe, V. (1991). The biology of hydrothermal vents – ecology and evolution. *Oceanography and Marine Biology*, **29**, 319–407.

Tunnicliffe, V. and Fowler, C. M. R. (1996). Influence of sea-floor spreading on the global hydrothermal vent fauna. *Nature*, **379**, (6565), 531–533.

Tunnicliffe, V., McArthur, A. G. and McHugh, D. (1998). A biogeographical perspective of the deep-sea hydrothermal vent fauna. *Advances in Marine Biology*, **34**, 353–442.

Turcotte, D. L. (1974). Are transform faults thermal contraction cracks? *Journal of Geophysical Research*, **79**, 2573–2577.

Turcotte, D. L. and Schubert, G. (1982). *Geodynamics: Applications of Continuum Physics to Geological Problems*. New York: John Wiley & Sons, 450 pp.

Turner, S., Beier, C., Niu, Y. and Cook, C. (2011). U–Th–Ra disequilibria and the extent of off-axis volcanism across the East Pacific Rise at 9°30′ N, 10°30′ N, and 11°20′ N. *Geochemistry Geophysics Geosystems*, **12**, doi: 10.1029/2010gc003403.

Unsworth, M. (1994). Exploration of midocean ridges with a frequency-domain electromagnetic system. *Geophysical Journal International*, **116**, (2), 447–467.

Van Andel, T. H. and Ballard, R. D. (1979). The Galapagos Rift at 86° W: 2. Volcanism, structure and evolution of the rift valley. *Journal of Geophysical Research*, **84**, (B10), 5390–5406.

Van Avendonk, H. J. A., Harding, A. J., Orcutt, J. A. and McClain, J. S. (2001). Contrast in crustal structure across the Clipperton transform fault from travel time tomography. *Journal of Geophysical Research-Solid Earth*, **106**, (B6), 10 961–10 981.

Van Dover, C. L., German, C. R., Speer, K. G., Parson, L. M. and Vrijenhoek, R. C. (2002). Evolution and biogeography of deep-sea vent and seep invertebrates. *Science*, **295**, (5558), 1253–1257.

Vening Meinesz, F. A. (1929). *Theory and Practice of Pendulum Observations at Sea*. Delft: Waltman.

Vera, E. E. and Diebold, J. B. (1994). Seismic imaging of oceanic layer 2a between 9°30′ N and 10° N on the East Pacific Rise from 2-ship wide-aperture profiles. *Journal of Geophysical Research-Solid Earth*, **99**, (B2), 3031–3041.

Vereshchaka, A. L. (1996). A new genus and species of caridean shrimp (crustacea: decapoda: alvinocarididae) from North Atlantic hydrothermal vents. *Journal of the Marine Biological Association, U.K.*, **76**, 951–961.

Vine, F. J. (1966). Spreading of the ocean floor: new evidence. *Science*, **154**, 1405–1415.

Vine, F. J. and Matthews, D. H. (1963). Magnetic anomalies over ocean ridges. *Nature*, **199**, 947–949.

Vine, F. J. and Smith, G. C. (1990). Structure and physical properties of the Troodos crustal section at ICRDG drillholes CY1, 1A and 4. In *Proceedings of the Symposium on Ophiolites and Oceanic Lithosphere – TROODOS 87*, ed. J. Malpas, E. M. Moores, A. Panayiotou and C. Xenophontas. Nicosia, Cyprus: Geological Survey Department, pp. 113–130.

Vogt, P. R. and deBoer, J. (1976). Morphology, magnetic anomalies, and basalt magnetization at the ends of the Galapagos high-amplitude zone. *Earth and Planetary Science Letters*, **33**, 145–163.

Von Damm, K. L. (1995). Controls on the chemistry and temporal variability of seafloor hydrothermal fluids. In *Seafloor Hydrothermal Systems: Physical, Chemical, Biological, and Geological Interactions, Geophysical Monograph, 91*, ed. S. E. Humphris, R. A. Zierenberg, S. Mullineaux and R. E. Thomson. Washington D.C.: American Geophysical Union, pp. 222–247.

Von Herzen, R. P., Cordery, M. J., Detrick, R. S. and Fang, C. (1989). Heat-flow and the thermal origin of hot spot swells – the Hawaiian swell revisited. *Journal of Geophysical Research-Solid Earth and Planets*, **94**, (B10), 13 783–13 799.

Von Herzen, R. P., Kirklin, J. and Becker, K. (1996). Geoelectrical measurements at the TAG hydrothermal mound. *Geophysical Research Letters*, **23**, 3451–3454.

Wang, X. and Cochran, J. R. (1993). Gravity anomalies, isostasy, and mantle flow at the East Pacific Rise crest. *Journal of Geophysical Research*, **98**, 19 505–19 532.

Watts, A. B. (1978). An analysis of isostasy in the worlds oceans: 1. Hawaiian–Emperor seamount chain. *Journal of Geophysical Research*, **83**, 5989–6004.

Watts, A. B., Bodine, J. H. and Steckler, M. S. (1980). Observations of flexure and the state of stress in the oceanic lithosphere. *Journal of Geophysical Research*, **85**, 6369–6376.

Wegener, A. (1912). Die Entstehung der Kontinente. *Geologische Rundschau*, **3**, (4), 276–292.

Wegener, A. (1966). *The Origin of the Continents and Oceans* (translated from 4th German edition by J. Biram). London: Methuen.

White, R. S. (1984). Atlantic oceanic crust: seismic structure of a slow-spreading ridge. In *Ophiolites and Oceanic Lithosphere, Special Publication of the Geological Society, London, 13*, ed. I. G. Gass, S. J. Lippard and A. W. Shelton. Oxford: Blackwell Scientific Publications, pp. 101–111.

White, R. S., Detrick, R. S., Sinha, M. C. and Cormier, M. H. (1984). Anomalous seismic crustal structure of oceanic fracture zones. *Geophysical Journal of the Royal Astronomical Society*, **79**, 779–798.

White, R., Mckenzie, D. and O'Nions, R. (1992). Oceanic crustal thickness from seismic measurements and rare-earth element inversions. *Journal of Geophysical Research*, **97**, (B13), 19 683–19 715.

White, S. M., Haymon, R. M., Fornari, D. J., Perfit, M. R. and Macdonald, K. C. (2002). Correlation between tectonic and volcanic segmentation of fast-spreading ridges: evidence from volcanic structures and lava flow morphology on the East Pacific Rise at 9°–10° N. *Journal of Geophysical Research, B*, **107**, (8), EPM7, 1–20, doi:10.1029/2001JB000571.

White, S. M., McClinton, J. T., Sinton, J. M. *et al.* (2010). Resolving volcanic eruptions: new fine-scale mapping by AUV Sentry of Galápagos Spreading Center 92° W and 95° W. In *American Geophysical Union Fall Meeting*. San Francisco: American Geophysical Union, pp. V52A-07.

Whitehead, J. A., Dick, H. J. B. and Schouten, H. (1984). A mechanism for magmatic accretion under spreading centres. *Nature*, **312**, 146–148.

Whitmarsh, R. B. (1970). An ocean bottom pop-up seismic recorder. *Marine Geophysical Researches*, **1**, 91–98.

Whitmarsh, R. B. and Laughton, A. S. (1975). The fault pattern of a slow-spreading ridge near a fracture zone. *Nature*, **258**, 509–510.

Wiens, D. A. and Stein, S. (1983). Age dependence of oceanic intraplate seismicity and implications for lithospheric evolution. *Journal of Geophysical Research*, **88**, 6455–6468.

Wilcock, W. S. D., Purdy, G. M. and Solomon, S. C. (1990). Microearthquake evidence for extension across the Kane transform fault. *Journal of Geophysical Research, B*, **95**, (10), 15 439–15 462.

Wilcock, W. S. D., Purdy, G. M., Solomon, S. C., Dubois, D. L. and Toomey, D. R. (1992). Microearthquakes on and near the East Pacific Rise, 9° N–10° N. *Geophysical Research Letters*, **19**, (21), 2131–2134.

Wilcock, W. S. D., Archer, S. D. and Purdy, G. M. (2002). Microearthquakes on the Endeavour segment of the Juan de Fuca Ridge. *Journal of Geophysical Research-Solid Earth*, **107**, (B12), doi: 10.1029/2001jb000505.

Williams, A. B. and Rona, P. A. (1986). Two new caridean shrimps (bresiliidae) from a hydrothermal field on the Mid-Atlantic Ridge. *Journal of Crustacean Biology*, **6**, (3), 446–462.

Williams, C. M., Tivey, M. A., Schouten, H. and Fornari, D. J. (2008). Central Anomaly Magnetization High documentation of crustal accretion along the East Pacific Rise (9°55′–9°25′ N). *Geochemistry Geophysics Geosystems*, **9**, doi: 10.1029/2007gc001611.

Wilson, D. S., Teagle, D. A. H., Alt, J. C. *et al.* (2006). Drilling to gabbro in intact ocean crust. *Science*, **312**, (5776), 1016–1020.

Wilson, J. T. (1965). A new class of faults and their bearing on continental drift. *Nature*, **207**, 343–347.

Wing, C. G. (1969). MIT vibrating string surface-ship gravimeter. *Journal of Geophysical Research*, **74**, (25), 5882–5894.

Woese, C. R., Kandler, O. and Wheelis, M. L. (1990). Evolution towards a natural system of organisms: proposal for the domains Archaea, Bacteria, and Eucarya. *Proceedings of the National Academy of Science USA*, **87**, 4576–4579.

Wolery, T. J. and Sleep, N. H. (1976). Hydrothermal circulation and geochemical flux at mid-ocean ridges. *Journal of Geology*, **84**, 249–275.

Wolfe, C. J., Purdy, G. M., Toomey, D. R. and Solomon, S. C. (1995). Microearthquake characteristics and crustal velocity structure at 29° N on the Mid-Atlantic Ridge: the architecture of a slow-spreading segment. *Journal of Geophysical Research*, **100**, (B12), 24 449–24 472.

Woods Hole Oceanographic Institution (2011). *Woods Hole Oceanographic Institution: A Brief History*. Woods Hole, Massachusetts: Woods Hole Oceanographic Institution, 2 pp.

Worm, H.-U. (2001). Magnetic stability of oceanic gabbros from ODP Hole 735B. *Earth and Planetary Science Letters*, **193**, 287–302.

Worzel, J. L. (1959). Continuous gravity measurements on a surface ship with the Graf sea gravimeter. *Journal of Geophysical Research*, **64**, (9), 1299–1315.

Worzel, J. L. and Harrison, J. (1963). Gravity at sea. In *The Sea, volume 3*, ed. M. N. Hill. New York: Interscience Publishers, pp. 134–174.

Wright, D. J. (1998). Formation and development of fissures at the East Pacific Rise: implications for faulting and magmatism at mid-ocean ridges. In *Faulting and Magmatism at Mid-Ocean Ridges – Geophysical Monograph 106*, ed. W. R. Buck, P. T. Delaney, J. A. Karson and Y. Lagabrielle. Washington, D.C.: American Geophysical Union, pp. 137–151.

Wylie, J. J., Helfrich, K. R., Dade, B., Lister, J. R. and Salzig, J. F. (1999). Flow localization in fissure eruptions. *Bulletin of Volcanology*, **60**, 432–440.

Xu, M., Canales, J. P., Tucholke, B. E. and DuBois, D. L. (2009). Heterogeneous seismic velocity structure of the upper lithosphere at Kane oceanic core complex, Mid-Atlantic Ridge. *Geochemistry, Geophysics, Geosystems*, **10**, (10), doi:10.1029/2009GC002586.

Yamazaki, T., Seama, N., Okino, K., Kitada, K. and Naka, J. (2003). Spreading process of the northern Mariana Trough: rifting–spreading transition at 22° N. *Geochemistry Geophysics Geosystems*, **4**, doi: 10.1029/2002GC000492.

Yeo, I. (2012). Detailed studies of mid-ocean ridge volcanism – Mid-Atlantic Ridge 45° N and elsewhere. Ph.D. thesis, Durham University, Durham, http://etheses.dur.ac.uk/4944/

Yeo, I. A. and Searle, R. C. (2013). High resolution ROV mapping of a slow-spreading ridge: Mid-Atlantic Ridge 45° N. *Geochemistry Geophysics Geosystems*, **14**, doi: 10.1029/2012GC004436.

Yeo, I., Searle, R. C., Achenbach, K., LeBas, T. and JC24 Shipboard Scientific Party (2012). Eruptive hummocks: building blocks of the upper ocean crust. *Geology*, **40**, (1), 91–94.

Yoerger, D. R., Bradley, A. M., Singh, H. *et al.* (1998). Multisensor mapping of the deep seafloor with the Autonomous Benthic Explorer. In *Proceedings of the 2000 International Symposium on Underwater Technology*. New York: IEEE, pp. 248–253.

Zelt, C. A. and Smith, R. B. (1992). Seismic travel-time inversion for 2-D crustal velocity structure. *Geophysical Journal International*, **108**, 16–34.

# Index

Printed in the United States
by Baker & Taylor Publisher Services